Systems modelling and optimization

Proceedings of the
18th IFIP TC7 Conference

CHAPMAN & HALL/CRC
Research Notes in Mathematics Series

Main Editors

H. Brezis, *Université de Paris*
R.G. Douglas, *Texas A&M University*
A. Jeffrey, *University of Newcastle upon Tyne (Founding Editor)*

Editorial Board

H. Amann, *University of Zürich*
R. Aris, *University of Minnesota*
G.I. Barenblatt, *University of Cambridge*
H. Begehr, *Freie Universität Berlin*
P. Bullen, *University of British Columbia*
R.J. Elliott, *University of Alberta*
R.P. Gilbert, *University of Delaware*
R. Glowinski, *University of Houston*
D. Jerison, *Massachusetts Institute of Technology*
K. Kirchgässner, *Universität Stuttgart*

B. Lawson, *State University of New York at Stony Brook*
B. Moodie, *University of Alberta*
S. Mori, *Kyoto University*
L.E. Payne, *Cornell University*
D.B. Pearson, *University of Hull*
I. Raeburn, *University of Newcastle*
G.F. Roach, *University of Strathclyde*
I. Stakgold, *University of Delaware*
W.A. Strauss, *Brown University*
J. van der Hoek, *University of Adelaide*

Submission of proposals for consideration

Suggestions for publication, in the form of outlines and representative samples, are invited by the Editorial Board for assessment. Intending authors should approach one of the main editors or another member of the Editorial Board, citing the relevant AMS subject classifications. Alternatively, outlines may be sent directly to the publisher's offices. Refereeing is by members of the board and other mathematical authorities in the topic concerned, throughout the world.

Preparation of accepted manuscripts

On acceptance of a proposal, the publisher will supply full instructions for the preparation of manuscripts in a form suitable for direct photo-lithographic reproduction. Specially printed grid sheets can be provided. Word processor output, subject to the publisher's approval, is also acceptable.

Illustrations should be prepared by the authors, ready for direct reproduction without further improvement. The use of hand-drawn symbols should be avoided wherever possible, in order to obtain maximum clarity of the text.

The publisher will be pleased to give guidance necessary during the preparation of a typescript and will be happy to answer any queries.

Important note

In order to avoid later retyping, intending authors are strongly urged not to begin final preparation of a typescript before receiving the publisher's guidelines. In this way we hope to preserve the uniform appearance of the series.

CRC Press UK

Chapman & Hall/CRC Statistics and Mathematics
Pocock House
235 Southwark Bridge Road
London SE1 6LY
Tel: 0171 407 7335

Michael P. Polis
Oakland University

Asen L. Dontchev
Mathematical Reviews

Peter Kall
University of Zurich

Irena Lasiecka
University of Virginia

and

Andrzej W. Olbrot
Wayne State University

(Editors)

Systems modelling and optimization

Proceedings of the 18th IFIP TC7 Conference

CHAPMAN & HALL/CRC

Boca Raton   London   New York   Washington, D.C.

Library of Congress Cataloging-in-Publication Data

Catalog record is available from the Library of Congress.

This book contains information obtained from authentic and highly regarded sources. Reprinted material is quoted with permission, and sources are indicated. A wide variety of references are listed. Reasonable efforts have been made to publish reliable data and information, but the author and the publisher cannot assume responsibility for the validity of all materials or for the consequences of their use.

Apart from any fair dealing for the purposes of research or private study, or criticism or review, as permitted under the UK Copyright Designs and Patents Act, 1988, this publication may not be reproduced, stored or transmitted, in any form or by any means, electronic or mechanical, including photocopying, microfilming, and recording, or by any information storage or retrieval system, without the prior permission in writing of the publishers, or in the case of reprographic reproduction only in accordance with the terms of the licenses issued by the Copyright Licensing Agency in the UK, or in accordance with the terms of the license issued by the appropriate Reproduction Rights Organization outside the UK.

The consent of CRC Press LLC does not extend to copying for general distribution, for promotion, for creating new works, or for resale. Specific permission must be obtained in writing from CRC Press LLC for such copying.

Direct all inquiries to CRC Press LLC, 2000 Corporate Blvd., N.W., Boca Raton, Florida 33431.

Trademark Notice: Product or corporate names may be trademarks or registered trademarks, and are used only for identification and explanation, without intent to infringe.

© 1999 by CRC Press LLC

No claim to original U.S. Government works
International Standard Book Number 0-8493-0607-8
Printed in the United States of America   1 2 3 4 5 6 7 8 9 0
Printed on acid-free paper

Contents

Preface

Organizing Committee

International Program Committee

I. Distributed parameter systems

Control of turbulent flows R Temam, T Bewley and P Moin [temam@iu-math.math.indiana.edu]	3
On superstability of semigroups A V Balakrishnan [bal@ee.ucla.edu]	12
Stability of C_0-semigroups and stabilization of linear control systems V Q Phong [qvu@bing.math.ohiou.edu]	20
A new regularity result in transonic gas dynamics J U Kim [kim@math.vt.edu]	29
On some equations about European option pricing F Colombo, R Monte and V Vespri [vespri@univaq.it]	33
Hidden shape derivative in the wave equation J Cagnol and J-P Zolésio [john.cagnol@sophia.inria.fr]	42
Hidden boundary smoothness in hyperbolic tangential problems on nonsmooth domains M C Delfour and J-P Zolésio [delfour@crm.umontreal.ca]	53
The dam problem: safety factor and shape optimization J Deteix [deteix@crm.umontreal.ca]	62
Stability, observers and compensators for discrete-time delay systems J Karrakchou, F Lahmidi, A Namir and M Rachik [j.karrak@emi.ac.ma]	71
Shape control of elastic plates K Piekarski and J Sokolowski [sokolows@iecn.u-nancy.fr]	80

II. Optimal control and nonsmooth analysis

On Ritz type discretizations for optimal control problems
U Felgenhauer [felgenh@matt.tu-cottbus.de]   91

Nontrivial maximum principle for optimal control problems with state constraints
M M A Ferreira [mmf@fe.up.pt]   100

Abundant subsets of generalized control systems
B Kaskosz [bkaskosz@uriacc.uri.edu]   108

Minimax linear-quadratic controller design
J Z Hu, P D Loewen and S E Salcudean [loew@math.ubc.ca]   117

High-order approximations for abnormal bang-bang extremals
U Ledzewicz and H Schättler [uledzew@siue.edu]   126

Directed sets and differences of convex compact sets
R Baier and E Farkhi [robert.baier@uni-bayreuth.de]   135

Open-loop Nash equilibrium for nonlinear control systems
D A Carlson [dcarlson@math.utoledo.edu]   144

Relaxed minimax optimal control problems. Infinite horizon case
S C Di Marco and R L V Gonzalez [dimarco@unrctu.edu.ar]   153

Conjugate points for optimal control problems with state-dependent control constraints
M d R De Pinho and H Zheng [mrpinho@fe.up.pt]   162

On sufficient conditions for a strong local minimum of broken extremals in optimal control
J Noble and H Schättler [hms@cec.wustl.edu]   171

Dynamical system with state constraints: theory and applications
P Saint-Pierre [saint-pierre@ceremade.dauphine.fr]   180

On variational analysis in infinite dimensions
B S Mordukhovich [boris@math.wayne.edu]   189

On L^1-closed decomposable sets in L^∞
Z Páles and V Zeidan [zeidan@math.msu.edu]   198

Characterizations of weak sharp minima of order one in nonlinear programming
M Studniarski [marstud@imul.uni.lodz.pl]   207

Second-order necessary conditions in nonsmooth programming
D Ward [dw86mthf@miamiu.muoho.edu]   216

Upper Hölder continuity of inverse subdifferentials and the proximal point algorithm
R Zhang [rzhang@nmu.edu]   225

III. Optimization and operations research

Optimization models and algorithms of constructing reliable networks
V P Il'ev and I V Ofenbakh [Iljev@univer.omsk.su] ... 237

Higher order methods for solving sufficient linear complementarity problems
J Stoer, M Wechs and S Mizuno [jstoer@mathematik.uni-wuerzburg.de] ... 245

A linear time approximation algorithm for a storage allocation problem
G Confessore, P Dell'Olmo and S Giordani [confessore@iasi.rm.cnr.it] ... 253

General framework in analyzing mobile terminal tracking protocols
C C F Fong, J C S Lui, M H Wong and E A de Souza e Silva
[edmundo@nce.ufrj.br] ... 261

Risk sensitive optimal control of wear processes: the vector case
M Lefebvre and R Labib [lefebvre@mathappl.polymtl.ca] ... 271

Deadlock in manufacturing systems: application of its necessary and sufficient conditions
R Lipset, P E Deering and R P Judd [lipset@bobcat.ent.ohiou.edu] ... 280

Failure-prone production systems with uncertain demand
J R Perkins and R Srikant [rsrikant@uiuc.edu] ... 289

Transient distributions of cumulative rate and impulse based reward with applications
E de Souza e Silva, H R Gail and J C Guedes [edmundo@nce.ufrj.br] ... 298

IV. Reliability

Reliability based optimization of passive fire protection on offshore topsides
P Thoft-Christensen and F M Jensen [i6ptc@civil.auc.dk] ... 309

Evaluation of the quality of an option compared to its alternatives
D Christozov [dgc@nws.aubg.bg] ... 318

Control charts for monitoring the mean of a multivariate normal distribution
V Dragalin [dragalin@bst.rochester.edu] ... 327

A general and simple approach to economic process control
E von Collani [collani@mathematik.uni-wuerzburg.de] ... 336

Sequential change point detection and estimation for multiple alternative hypothesis
Y Wu [yanhong@fisher.stat.ualberta.ca] ... 345

V. Modelling

Modeling methodologies for modeling environments
S Dresbach [dresbach@aol.com] ... 357

An advanced modeling environment based on a hybrid AI-OR approach
B Funke and H-J Sebastian [sebasti@or.rwth-aachen.de] ... 366

Interactive fuzzy modeling system
Y Nakamori, M Ryoke and K Umayahara [nakamori@edu3.math.konan-u.ac.jp] ... 376

Rule-based design and 2D-analysis of a hybrid system
D Franke [Dieter.franke@UniBw-Hamburg.de] ... 384

A mathematical model for control of flexible robot arms
X Hou and S-K Tsui [tsui@oakland.edu] ... 391

VI. Automotive

Identification of powertrain noise in a low SNR environment using synchronous time averaging
S Amman, M Das, M Blommer and N Otto [amman@e-mail.com] ... 401

Hot wire mass air flow sensor modeling and lag compensation
J R Asik [jasik@ranchero.srl.ford.com] ... 409

Dynamic clustering technique with application to onboard traffic monitoring
Ka C Cheok, S Nishizawa and W J Young [cheok@oakland.edu] ... 418

Modeling of currents in a zirconium oxygen sensor
J Henry, A Viel and J P Yvon [Jacques.Henry@inria.fr] ... 427

Issues in modelling and control of intake flow in variable geometry turbocharged engines
I Kolmanovsky, P Morall, M van Nieuwstadt and A Stefanopoulou
[ikolmano@ford.com] ... 436

Robust controller based on Smith predictor for an electric EGR valve control system
A W Olbrot, J R Asik, and M H Berri [aolbrot@ece.eng.wayne.edu] ... 446

VII Applications

Robust controller design for a VTOL plane
Y-W Choe and G-S Byun [wook@oryukdo.pnut.ac.kr] ... 459

Optimal control of failure prone manufacturing systems with corrective maintenance
J P Kenne and E K Boukas [boukas@meca.polymtl.ca] ... 468

Generation and analysis of a non-linear optimization problem: European Ozone model case study
M Makowski [marek@iiasa.ac.at] ... 477

Adaptive stochastic path planning for robots - real-time optimization by means of neural networks
K Marti and S Qu [kurt.marti@unibw-muenchen.de] 486

Control of molecular dynamics
V Ramakrishna and H Rabitz [vish@utdallas.edu] 495

Preface

The 18th IFIP TC7 Conference on Systems Modelling and Optimization was held at The Westin Hotel in Detroit, Michigan, July 22-25, 1997. It was organized by the School of Engineering and Computer Science at Oakland University, Rochester, Michigan. This proceedings volume represents a selection of papers chosen from among the 250 conference papers presented in 4 plenary and 56 invited and contributed sessions. In making the selection the editors sought high quality papers covering either theoretical aspects of optimization and control or modelling and control applications. The selection process was made very difficult by the tight limit on the number of pages that could economically be printed in this volume. As such, the editors were forced to reject many good papers which simply could not be accommodated due to page limits.

The International Federation of Information Processing (IFIP) was founded in 1960 under the auspices of UNESCO. IFIP is organized around volunteers and operates through a number of technical committees which arrange events and publications. The IFIP TC7 Conference on Systems Modelling and Optimization is held biannually, usually not in the United States, and is generally composed of invited plenary papers and rigorously reviewed contributed papers. The last time the conference was held in the U.S. was in 1981 and to encourage U.S. participation, the 18th IFIP TC7 Conference solicited invited sessions. These sessions were rigorously reviewed and generally were either accepted entirely or rejected. Dr. Asen L. Dontchev of the American Mathematical Society was the invited sessions chair. The success of the conference was due in large part to the quality of these sessions and thus to the following individuals who took time to organize one or more invited sessions: A.V. Balakrishnan, *UCLA* [bal@ee.ucla.edu], J.R. Birge, *University of Michigan* [jrbirge@umich.edu], A.M. Bloch, *University of Michigan* [abloch@math.lsa.umich.edu], E.-K. Boukas, *Ecole Polytechnique of Montreal* [boukas@meca.polymtl.ca], M.C. Delfour, *Universite de Montreal* [delfour@crm.UMontreal.ca], B. Dimitrov, *GMI Egng and Mgmt Institute* [bdimitro@nova.gmi.edu], J.C. Dunn, *North Carolina State University* [dunn@math.ncsu.edu], L. Gawarecki, *GMI Institute of Engineering and Management* [lgawarec@nova.gmi.edu], J. Henry, *INRIA* [jacques.henry@inria.fr], H. Inoue, *Science University of Tokyo* [inoue@ms.kuki.sut.ac.jp], P. Kall, *University of Zurich* [kall@ior.unizh.ch], A. Khapalov, *Washington State University* [khapala@delta.math.wsu.edu], I. Kolmanovsky, *Ford Motor Company* [ikolmano@ford.com], K.O. Kortanek, *University of Iowa* [kort@dollar.biz.uiowa.edu], I. Lasiecka, *University of Virginia* [il2v@amsun41.apma.virginia.edu], U. Ledzewicz, *Southern Illinois University* [uledzew@siue.edu,hms@cec.wustl.edu], F. Lempio, *Universitat Bayreuth* [frank.lempio@uni-bayreuth.de], A.B. Levy, *Bowdoin College* [alevy@bowdoin.edu], K. Marti, *Federal Armed Forces University Munich* [kurt.marti@unibw-muenchen.de], C.

McMillan, *Virginia Polytechnic University* [mcmillan@math.vt.edu], A.S. Nowak, *University of Michigan* [Nowak@engin.umich.edu3], Andrzej W. Olbrot, *Wayne State University* [aolbrot@ece.eng.wayne.edu], H. Schaettler, *Washington University* [hms@cec.wust1.edu], H.-J. Sebastian, *RWTH Aachen* [sebasti@or.rwth-aachen.de], S. Sengupta, *Oakland University* [sengupta@oakland.edu], E. de Souza e Silva, *Federal University of Rio de Janeiro* [edmundo@nce.ufrj.br], J. Stoer, *University of Wurzburg* [jstoer@mathematik.uni-wuerzburg.de], P. Thoft-Christensen, *Aalborg University* [i6ptc@civil.auc.dk], F. Troltzsch, *Technical University Chemnitz-Zwickau* [f.troeltzsch@mathematik.tu-chemnitz.de], R. Van Til, *Oakland University* [vantil@oakland.edu], E. von Collani, *Wuerzburg University* [collani@mathematik.uni-wuerzburg.de], S.W. Wallace, *Norwegian University of Science and Technology* [sww@iot.ntnu.no], D. Ward, *Miami University* [dw86mthf@miamiu.muohio.edu], Keiji Yajima, *Science University of Tokyo* [yajima@ms.kuki.sut.ac.jp,], G. Yin, *Wayne State University* [gyin@math.wayne.edu], V. Zeidan, *Michigan State University* [zeidan@math.msu.edu] and J.-P. Zolesio, *CNRS-INLN, Sophia Antipolis and Ecole des Mines de Paris* [Jean-Paul.Zolesio@sophia.inria.fr]. The 18th IFIP TC7 Conference organizers are indebted to these individuals as well as to all of the authors of the contributed and invited papers which were presented at the conference. Again, the high quality of the papers which were presented made the review process for this proceedings very difficult.

Finally, most exchanges related to the 18th IFIP TC7 Conference were done via email and the email addresses of the corresponding authors and organizers are given in the Table of Contents to facilitate communication.

The Editors

Organizing Committee
General Chairman
Michael P. Polis, *Oakland University* [polis@vela.acs.oakland.edu]

Invited Sessions Chair
Asen L. Dontchev, *American Mathematical Society* [ald@math.ams.org]

Publicity
Ronald Srodawa, *Oakland University* [srowava@oakland.edu]

Local Organization
Naim Kheir, *Oakland University* [kheir@oakland.edu]
Sankar Sengupta, *Oakland University* [sengupta@oakland.edu]
Robert Van Til, *Oakland University* [vantil@oakland.edu]

International Program Committee
P. Kall (Chair), *University of Zurich* [kall@ior.unizh.ch]
A. V. Balakrishnan, *UCLA* [bal@ee.ucla.edu]
Jaroslav Dolezal, *Academy of Sciences of the Czech Republic*
Asen Dontchev, *American Mathematics Society* [ald@math.ams.org]
B. Frankovic, *Slovak Academy of Sciences* [utrrfran@savba.savba.sk]
Jacques Henry, *INRIA* [jacques.henry@inria.fr]
Irena Lasiecka, *University of Virginia* [il2v@amsun41.apma.virginia.edu]
Richard E. Nance, *Virginia Polytechnic University* [nance@vt.edu]
Andrzej W. Olbrot, *Wayne State University* [aolbrot@ece.eng.wayne.edu]
Michael Powell, [mjdp@amtp.cam.ac.uk]
Irwin E. Schochetman, *Oakland University* [schochet@oakland.edu]
Hugo Scolnik, *University of Buenos Aires* [hugo@nobi.uba.ar]
Hans-Juergen Sebastian, *RWTH Aachen* [sebasti@or.rwth-aachen.de]
Edmundo de Souza e Silva, *Federal Univ. of Rio de Janeiro* [edmundo@nce.ufrj.br]
Josef Stoer, *University of Wurzburg* [jstoer@mathematik.uni-wuerzburg.de]
Palle Thoft-Christensen, *Aalborg University* [i6ptc@civil.auc.dk]
Stein W. Wallace, *Norwegian Univ. of Science and Technology* [sww@iot.ntnu.no]
Keiji Yajima, *Science University of Tokyo* [yajima@ms.kuki.sut.ac.jp]
Uwe Zimmermann, *Technical Univ. of Braunschweig* [uz@mo.math.nat.tu-bs.de]

I. Distributed parameter systems

R TEMAM, T BEWLEY AND P MOIN
Control of turbulent flows

It is useful for many industrial applications to be able to control turbulence in fluid flows, either to reduce it or, in some cases, to increase it. Active or passive control procedures are of interest. The problems that we face here are considerable and encompass those related to the control of complex nonlinear systems and those related to the direct numerical simulation of turbulent flows.

Our aim in this lecture is to report on some recent results obtained by the authors in an interactive collaboration between mathematicians and fluid dynamicists, and which represent a small step in the solution of this problem; this includes the mathematical modelling of such control problems, theoretical results (existence of optimal control, necessary conditions of optimality) and the development of effective numerical algorithms.

1. Introduction

It is useful for many industrial applications to be able to control turbulence in fluid flows either to reduce it or, in some cases, to increase it. At this time the most important applications are probably those arising in aeronautics which include the reduction of skin friction drag and the delay of transition to turbulence or separation of the boundary layer. Other important applications may be found in thermohydraulics, magnetohydrodynamics, climate and pollution forecasting. In combustion the objective is to increase turbulence for a better mixing of the fuel and its oxidant.

As in other control problems, such problems need first to be modelled, deciding (choosing) what is costly and what are the objectives; passive and active controls are of interest, and more recently robust control: the objective may be e.g. the design of an airfoil including its surface (shape optimization/passive control) or the active control of small actuators on the surface of the airfoil to properly respond to the coherent structures of the nearwall turbulence.

The difficulties of the problem are considerable and much remains to be done. The description of the "state" of the system amounts to the resolution of the 3D Navier-Stokes equations in a turbulent context; the capacity of the present computers allows the calculation of such flows in simple cases, but this still demands much from the largest available computers in terms of computing power and storage requirements. As solutions to these nonlinear problems must be sought iteratively, we must numerically solve such problems several times, further compounding the computational expense of this procedure.

During the past years, a number of articles have appeared in the engineering and mathematical literatures concerning the control of turbulent flows and treating different aspects of the problem; see e.g. F. Abergel and R. Temam (1990), M. Gunzburger, L. Hou and T.P. Sovobodny (1990), H. Choi, P. Moin and J. Kim (1994), S.S. Sritharan (1991) and the references therein. The presentation which follows is mainly based on the article of

F. Abergel and R. Temam (1990) hereafter referred to as [AT] which sets the problem of controlling turbulence in the framework of control theory in the spirit of J.L. Lions (1968), and on two articles under completion: T.R. Bewley, P. Moin and R. Temam (1998), T.R. Bewley, R. Temam and M. Ziane (1998) hereafter called [BMT] and [BTZ]; see also T.R. Bewley, P. Moin and R. Temam (1996), (1997).

This article is organized as follows. In Section 1 we describe the modelling of the open loop control problem under consideration and give the main theoretical result. In Section 2 we describe the numerical algorithm which has been used without theoretical justification and which has produced a nearly ideal result in some cases in which we obtain an almost complete relaminarization of the flow. In Section 3 we discuss some other issues, namely some conjectures on the theoretical justification of the algorithm which we used, the wall information problem, and some preliminary remarks on the utilization and implementation of robust control.

Although the results obtained in Section 2 are quite significant (nearly optimal), we would like to emphasize that we are still far from practical (industrial) applications with several respects: the geometry is simple, the Reynolds number not too high, full information (and not just wall information) has been used; the practical implementation of the optimal control is not available and extensive calculations have been used which might be difficult to reproduce in real time. Nevertheless there is hope to obtain in the future, for more involved and more realistic problems, a still very useful if not as significant reduction of the cost function.

2. The Channel Flow Problem

We consider the flow of an incompressible fluid in a three dimensional channel as a simplified form of the flow in a wind tunnel. The channel occupies the region $\Omega = (0, \ell_1) \times (0, \ell_2) \times (0, \ell_3)$. The flow is maintained by an unspecified pressure gradient $P = P(\tau)$, in the x_1 (streamwise) direction. The flow will be controlled by the normal velocity of the upper wall Γ_w, $\{x_2 = \ell_2\}$.

Hence, the governing equations are the Navier-Stokes,

(2.1)
$$\frac{\partial u}{\partial t} - \nu \Delta u + (u \cdot \nabla)u + \nabla p = Pe_1 \text{ in } \Omega \times (0, T),$$
$$\nabla \cdot u = 0 \text{ in } \Omega \times (0, T).$$

Here $u = (u_1, u_2, u_3)$, function of x and t is the velocity vector; the pressure is $p(x, t) - x_1 P(t), p, P$ unknown, P accounting for the pressure gradient ($e_1 = (1, 0, 0)$). Periodicity is assumed for u and p in the direction x_1 and x_3 and

(2.2)
$$u = \phi \text{ on } \Gamma_w \times (0, T)$$

whereas $u = 0$ on Γ_ℓ the rest of the lateral boundary of Γ. Finally the flux is fixed and given by

(2.3)
$$\iint_{x_1=0} u_1 \, dx_2 \, dx_3 = F$$

The weak formulation of (2.1)-(2.3), consists in looking for $u = u(x,t)$ which satisfies (2.2), (2.3) and

(2.4) $$\frac{d}{dt}\int_\Omega uv\,dx + \int_\Omega \{\nu\nabla u \cdot \nabla v + [(u \cdot \nabla)u]\}dx = 0,$$

for every (smooth) test function v such that

(2.5) $$\nabla \cdot v = 0, \quad v = 0 \quad \text{on } \Gamma_w \text{ and } \Gamma_\ell \text{ and}$$
$$\iint_{x_1=0} v_1\,dx_2\,dx_3 = 0$$

(see e.g. R. Temam (1984) for many related examples).

Now in the language of control theory, ϕ is the *control*, $u = u(\phi)$ is the *state* of the system, and the *state equation* consists of (2.2)-(2.5).

For the modelling of the control problem, we need to choose/define the cost function J. It consists of two terms $J = J_0 + J_1$. The first term, e.g.

$$J_0(\phi) = \frac{\ell^2}{2}\|\phi\|_X^2 = \frac{\ell^2}{2}\int_0^T\int_{\Gamma_w}|\phi|^2\,dx_1\,dx_3\,dt,$$

accounts for the cost of the control. The second term, e.g.

$$J_{1a}(u) = \frac{1}{2}\int_0^T\int_\Omega |\operatorname{curl} u|^2\,dx\,dt,$$
$$J_{1b}(u) = \int_0^T\int_{\Gamma_w}\frac{\partial u_1}{\partial x_2}\,dx_2\,dx_3,$$
$$J_{1c}(u) = \frac{1}{2}\int_\Omega |u(x,T)|^2\,dx;$$

represents the flow quantity (related to turbulence) which we want to minimize; J_{1a} was used in [AT] for the theoretical study and J_{1b}, J_{1c} are used for the computations in [BMT], J_{1b} representing the terminal value of the turbulent kinetic energy (TKE) and J_{1c} the time-averaged value of the drag.

The corresponding control problems now read (for $i = a, b$ or c):

(2.6) $$\inf_\phi \{J_0(\phi) + J_{1i}(\phi)\}$$

Omitting here certain theoretical questions addressed in [AT], the following results were essentially proved in [AT]:

(2.7) *The control problem (2.6) has a solution, corresponding to the (optimal) control $\bar{\phi}$ and corresponding state $\bar{u} = u_{\bar{\phi}}$.*

(2.8) *The optimal state $\bar{\phi}$ satisfies the necessary condition of optimality*
$$J_0'(\bar{\phi}) + J_{1a}'(\bar{\phi}) = 0,$$
described as usual by an equation for the adjoint state
$$\bar{w} = w_{\bar{\phi}} \text{ (see [AT])}.$$

(2.9) *The gradient algorithm and conjugate gradient algorithm converge to the optimal control $\bar{\phi}$ if the initialization ϕ_0 belongs to a small neighborhood of $\bar{\phi}$ in the space X.*

See [AT] for the details in the case of J_{1a}; the proof easily extends to J_{1b} and J_{1c}, as shown in [BTZ].

3. Numerical Simulations

FIGURE 1. Coherent structures of a turbulent flow at $Re_\tau = 180$.

As motivation to the present work, we show in Figure 1 the coherent structures which appear near the wall in a turbulent channel flow and which we want to annihilate. The figure corresponds to a Reynolds number $Re_\tau = 180$, for which optimally controlled results are still under preparation; the results below correspond to $Re_\tau = 100$.

Taking into account present computational capabilities the algorithms provided by (2.9) are not feasible at this time for large T. Indeed, for the flow to attain some statistical equilibrium, T needs to be sufficiently large and (2.9) implies the resolution of the turbulent Navier-Stokes equations and its adjoint on $(0, T)$, a number of times corresponding to the

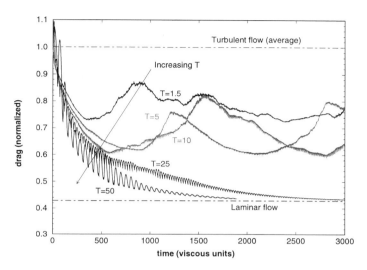

FIGURE 2. Time Evolution of the drag for different values of τ (denoted T in the figure).

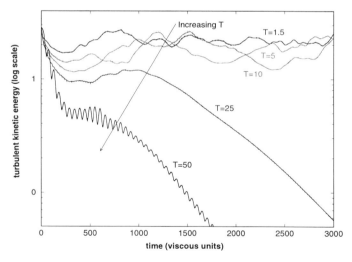

FIGURE 3. Time Evolution of the turbulent kinetic energy for different values of τ (denoted T in the figure).

gradient iterations. Hence, our objective now, will be to look for suboptimal procedures which reduce the cost function, even if not making it minimal.

A very simple feedback law was used as a first step, which produced a drag reduction of approximately 17%. Note that the same procedure was used by H. Choi, R. Temam, P. Moin and J. Kim (1993) for the simpler case of the stochastic Burgers equation driven by a white noise; in this case the reduction of the cost function was approximately 75%.

We looked then for a multi-time-step procedure more adapted to the problem. Indeed there are here two different evolution equations which need to be solved, namely the Navier-Stokes equation itself and the abstract evolution equation of which the gradient algorithms can be seen as a time discretization:

$$\phi^n - \phi^{n-1} + \rho J'(\phi^{n-1}) = 0 \qquad \Longleftrightarrow \qquad \frac{\partial \phi(s)}{\partial s} + J'(\phi(s)) = 0.$$

These two equations produce two different time constants and demand different time steps corresponding to different CFL (Courant-Friedrichs-Lewy) stability conditions.

Briefly described the algorithm implemented in [BMT] consists in dividing the interval $(0,T)$ into intervals of length τ; then on each interval $(m\tau, (m+1)\tau)$, we solve the control problem described in Section 1. At $m\tau$, u is continuous ($u(m\tau + 0) = u(m\tau - 0)$), but however for the gradient algorithms, it appeared best to start the iterations with $\phi^{m,0} = 0$ instead of $\phi^{m,0} = \phi(m\tau - 0)$. On each interval $(m\tau, (m+1)\tau)$ the Navier-Stokes equations are discretized with a time step $\Delta t \ll \tau$.

The evolution of the drag plotted in Figure 2 for different values of τ, at Reynolds number $Re_\tau = 100$, shows a near relaminarization for both $\tau = 25$ and $\tau = 50$ (in viscous time units). For such nearly relaminarized flows, the drag equals about 42 percent of its uncontrolled fully turbulent value (line —·—), which is the best we could achieve using the current approach. The evolution of the turbulent kinetic energy plotted in Figure 3 versus τ shows even more strikingly the relaminarization process initiated for $\tau = 25$ and $\tau = 50$ (in viscous time units). For both these plots, the cost functional used is the terminal value of the turbulent kinetic energy (i.e. $J = J_0 + J_{1c}$). The reason for the lobed behavior of the curves is that the penalty in the cost functional is only on the terminal value of the TKE (i.e., at the end of each optimization interval)—excursions of greater TKE during the middle of each optimization interval are allowed if they lead to reduced values of TKE by the end of the interval. The reader is referred to [BMT] for the details of the calculation and for physical insights which may be drawn from the results.

4. Other issues

(i) Theoretical justification of the algorithm.

Turbulent flows are believed to be statistically stationary and therefore the infinite time horizon, $T \to \infty$, is of physical relevance. In fact stationarity and ergodicity (if proven) imply that the time averages

$$\frac{1}{T} \int_0^T u(x,t) dt$$

converge, as $T \to \infty$, to a measure $\mu = \mu_\phi$ which depends only on ϕ; see C. Foias and R. Temam (1975). Hence we could consider a minimization problem of the form,

$$\inf_\phi \{ \iint_{\Gamma_w} \frac{\partial u_1}{\partial x_2} \, dx \, d\mu_\phi(u) + \frac{\ell^2}{2}[[\phi]]^2 \}$$

(where the norm $[[\cdot]]$ needs to be properly defined), and then compare the gradient algorithm applied to this optimization problem to the procedure described before. For the Burgers equation theoretical issues are addressed in G. DaPrato, A. Debussche and R. Temam (1994) who study the stochastic Burgers equations and in G. Da Prato and A. Debussche (1997) who address the control problem and its relation with the Hamilton-Jacobi equation.

(ii) Wall information.

The previous numerical study was based on the assumption that we have full information on the flow (u known everywhere), which, of course, is not realistic. In practice there will be wall sensors measuring certain quantities at the wall and the control algorithm should be therefore based on wall information only.

Two fundamentally different types of partial-information feedback controllers may be considered for such a purpose. In the first approach, the available flow measurements are fed back through a simple convolution kernel K to compute the control. The problem to be solved here is simply to find the best K which minimizes the flow quantity of interest.

In the second approach, the available flow measurements are fed back through a simple convolution kernel L to compute a forcing term to update the state of an estimator, which is a set of equations which model the evolution of the flow itself. For the sake of analysis, the model equations may be taken simply as the Navier Stokes equation acting on some state estimate $\hat{u}$; however, this approach is particularly attractive from the standpoint that it can maximally utilize simple low-order models, such as POD-based models, for the state estimation of the near-wall turbulence. As an accurate state estimate is developed, the entire state estimate is fed back through a simple convolution kernel K to compute the control. The problem to be solved here is two-fold: i) to find the best L such that the state estimate is an accurate approximation of the state of the flow itself, at least near the wall where the measurements are made and the control is applied, and ii) to find the best K which minimizes the flow quantity of interest.

As discussed in T. Bewley, P. Moin and R. Temam (1996), we may exploit the fact that we know the equations governing the flow to propose a *computationally expensive* adjoint-based technique to optimize the unknown convolution kernels K and L in these systems, which can only be performed on a supercomputer. However, once optimized, the feedback control rules themselves are *much simpler* than the adjoint-based technique used to optimize K and L, and thus may be considered for use in the laboratory.

(iii) Robust control.

Robust control for linear problems is well understood (see, e.g., J.C. Doyle et al., 1989 and K. Zhou, J.C. Doyle, and K. Glover 1996)). Thus, when $(u \cdot \nabla)u$ is dropped or linearized around a stationary laminar flow solution, standard robust (i.e., H_∞) control techniques may be applied. This problem is addressed in a fluid-mechanical context in

T. Bewley and S. Liu (1998), where H_∞ control is used to stabilize a laminar flow to inhibit transition to turbulence. As shown in Figure 4, robust control focuses the control effort on the most unstable mode of the system, not expending control effort on controllable but stable modes of the system. The reduced feedback applied in the H_∞ approach results in reduced opportunity for improper feedback to disrupt the closed-loop system.

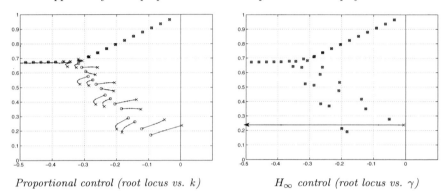

Proportional control (root locus vs. k)   *H_∞ control (root locus vs. γ)*

FIGURE 4. Movement of closed-loop eigenvalues versus control parameters k and γ for proportional and H_∞ (robust) controllers applied to the Orr-Sommerfeld equation for $Re = 10,000$, $k_x = 1$, $k_z = 0$ using blowing/suction as the control variable.

The extension of the robust control concept to nonlinear problems such as turbulence is addressed in T.R. Bewley, P. Moin and R. Temam (1997) from a computational perspective and in [BTZ] from a theoretical perspective. In fact, it boils down to the optimal approach described previously with an additional forcing term ψ added to the RHS of the Navier-Stokes equation (2.1) governing the system. The cost function considered is

$$J = J_1 + \frac{\ell^2}{2} \int_0^T \int_{\Gamma_w} |\phi|^2 \, dx_1 \, dx_3 \, dt - \frac{\gamma^2}{2} \int_0^T \int_\Omega |\psi|^2 \, dx_1 \, dx_2 \, dx_3 \, dt.$$

The cost J is *minimized* with respect to the control ϕ, while simulataneously it is *maximized* with respect to the disturbance ψ, in the spirit of a noncooperative game. Thus, the control ϕ is designed to handle that disturbance ψ which is, in some manner, a *worst case* aggravation to the closed-loop system. By so doing, the control found is effective in the presence of a broad class of disturbances.

5. Conclusions

In this article, we have presented the modeling of a typical control problem in fluid flow, namely, the reduction of the turbulence in a channel. Despite the large size of the system (up to 2.4×10^7 state variables in Ω and 6×10^4 control variables on Γ_w), we have successfully implemented a control procedure which reduces the drag nearly to its absolute minimum, corresponding to laminar flow. Such a relaminarization of this flow by wall-normal blowing and suction has not been possible using *any* other control algorithm.

Further work in this direction will include: a) an attempt for theoretical justification of the algorithm, using probably the stationarity of turbulence, b) the optimization of more practical feedback control algorithms which are computationally inexpensive and depend on wall information only, and c) the design of robust controllers which account for worst-case disturbances in their derivation.

This work was supported in part by the National Science Foundation under Grants NSF-DMS-9400615 and NSF-DMS-9705229, by the Office of Naval Research under grant NAVY-N00014-91-J-1140, by the Air Force Office of Scientific Research under Grant F49620-93-1-0078, and by the Research Fund of Indiana University. Computer time was provided by NASA-Ames Research Center in support of this project.

References

Abergel, F. and Temam, R. (1990), *On some control problems in fluid mechanics*, Theor. and Comp. Fluid Dynamics **1**, 303-325.

Bewley, T.R., Moin, P., and Temam, R. (1996), *A method for optimizing feedback control rules for wall-bounded turbulent flows based on control theory*, ASME FED **237**, p. 279-285..

Bewley, T.R., and Liu, S. (1998), *Optimal and robust control and estimation of linear paths to transition*, accepted for publication in J. Fluid Mech..

Bewley, T.R., Moin, P., and Temam, R. (1997), *Optimal and robust approaches for linear and nonlinear regulation problems in fluid mechanics*, AIAA Paper 97-1872..

Bewley, T.R., Moin, P., and Temam, R. (1998), *Optimal control of turbulence*, to be submitted to J. Fluid Mech..

Bewley, T.R., Temam, R., and Ziane, M. (1998), *A generalized framework for robust control in fluid mechanics*, to be submitted to Physica D..

Choi, H., Moin, P., and Kim, J. (1994), *Active turbulence control for drag reduction in wall-bounded flows*, J. Fluid Mech. **262**, 75-110.

Choi, H., Temam, R., Moin, P. and Kim, J. (1993), *Feedback control for unsteady flow and its application to the stochastic Burgers equation*, J. Fluid Mechanics **253** (1993), 509-543.

DaPrato, G. and Debussche, A. (1997), *in preparation.*

DaPrato, G., Debussche, A., and Temam, R. (1994), *Stochastic Burgers equation,*, Nonlinear Differential Equations and Applications **1**, 389-402.

Doyle, J.C., Glover, K., Khargonekar, P.P., and Francis, B.A. (1989), *State-Space Solutions to Standard H_2 and H_∞ Control Problems*, IEEE Trans. Auto. Control **34**, no. 8, 831-847.

Foias, C. and Temam, R. (1975), *On the stationary statistical solutions of the Navier-Stokes equations*, Publications Mathématiques d'Orsay **120-75-28**.

Gunzburger, M.D., Hou, L., and Svobodny, T.P. (1990), *A numerical method for drag minimization via the suction and injection of mass through the boundary*, In Stabil. of Flexible Structures, Springer.

Lions, J.L. (1968), *Controle Optimal des Systèmes Gouvernés par des Equations aux Dérivées Partielles*, Dunod (English translation, Springer).

Sritharan, S. (1991), *Dynamic programming of the Navier-Stokes equations*, Systems and Control Letters **16**, no. 4, 299-307.

Temam, R. (1984), *Navier-Stokes equations: theory and numerical analysis*, Elsevier Science.

Zhou, K., Doyle, J.C., and Glover, K. (1996), *Robust and Optimal Control*, Prentice-Hall.

A V BALAKRISHNAN
On superstability of semigroups

1. Introduction
This paper presents a brief report on superstable semigroups and applications thereof. The notion of super stability introduced in [Balakrishan, 1996a] is a strengthening of exponential stability and occurs in Timoshenko models of structures with self-straining material using pure (idealized) rate feedback. It is also relevant to the problem of completeness of eigenfunctions of infinitesimal generators under perturbation. We begin with definitions and the associated abstract theory in Section 2. A general technique of constructing superstable semigroups is given in Sections 3. Section 4 deals with applications to "smart structure" theory. The relation to the preservation of the Riesz basis property of eigenfunctions of infinitesimal generators of semigroups under perturbation is outlined in Section 5.

2. Definition and Basic Properties
By a "semigroup," we shall mean a C_0-semigroup over a Hilbert space with compact resolvent, unless otherwise specified. Let $S(t)$, $t \geq 0$ denote the semigroup and $\mathcal{H}$ the Hilbert space. Such a semigroup is said to be exponentially stable if

$$||S(t)|| \leq M \exp{-\sigma t}, \quad t \geq 0$$

for some σ, $M > 0$. The question: "Is there any stronger kind of stability?" has an affirmative answer — at least in infinite-dimensional theory. Thus the semigroup is said to be "superstable" (a notion introduced in [Balakrishan, 1996a]) if for every $\sigma > 0$, $\exists M_\sigma$ such that

$$||S(t)|| \leq M_\sigma \exp{-\sigma t}, \quad t \geq 0. \tag{2.0}$$

The resolvent of a superstable semigroup is defined for every complex number λ by:

$$R(\lambda, A)x = \int_0^\infty e^{-\lambda t} S(t) x \, dt, \quad x \in \mathcal{H}$$

and is thus an entire function — equivalently, the spectrum of A is empty. In particular we see that $\mathcal{H}$ cannot be finite-dimensional — superstability is a truly infinite-dimensional phenomenon — at least for linear systems.

Superstability is equivalent to saying that the stability index ω_0:

$$\omega_0 = \lim_{t \to \infty} \frac{\log ||S(t)||}{t} = -\infty.$$

In particular this is equivalent to saying that $S(t)$ must be quasi-nilpotent for every $t > 0$ — the term superstable is used to emphasize the relation to problems of stability and stabilizability. For the semigroup to be superstable it is necessary that the infinitesimal generator have an empty spectrum or, equivalently, the resolvent is an entire function. However the converse is not true; indeed one has only to construct a C_0 *group* whose generator has an empty spectrum — already described in the 1957 Hille-Phillips treatise [Hille and Phillips, 1957, p. 667]; providing in fact a celebrated example for the failure of the spectral mapping theorem:

$$\sigma[S(t)] \neq e^{t\sigma(A)} \quad \text{plus possibly zero}$$

where A denotes the infinitesimal generator. The point spectrum [Hille and Phillips, 1957, p. 467] satisfies:

$$P\sigma[S(t)] = e^{tP\sigma(A)} \quad \text{plus possibly zero.}$$

It follows that a compact (or eventually compact) semigroup whose generator has an empty spectrum is superstable.

Since we are dealing only with a Hilbert space we should note that a complete characterization is given by the Greiner-Nagel[†] theorem [Nagel, 1980, p. 96]: a semigroup is superstable if and only if

$$\sup_{-\infty < \omega < \infty} ||R(\sigma + i\omega, A)|| < \infty \tag{2.1}$$

for every σ real, where $R(\lambda, A)$ denotes the resolvent of A.

The resolvent of a superstable semigroup being an entire function, an immediate question is whether it is of finite order or not. We have in this connection a crucial result due to A. Sinclair [Sinclair, 1982].

Theorem (Sinclair).[†] Suppose the resolvent $R(\lambda, A)$ is an entire function of exponential type. Then the semigroup is superstable — actually nilpotent. In fact if

$$||R(\lambda, A)|| \leq M e^{|\lambda|T},$$

then

$$S(t) = 0 \quad \text{for } t > T.$$

(and conversely, the converse being trivial).

Proof. The proof relies on a celebrated Paley-Wiener theorem on Fourier transforms, and an outline may be of interest to be included here.

Pick $\sigma > \omega_0$ and define

$$r(z) = R(iz + \sigma, A)$$

[†]I am indebted to J. Martinez for bringing this result to my attention.

so that
$$\|r(z)\| \leq c e^{|z|T}.$$
For x, y in $\mathcal{H}$ and t real:
$$[r(t)x, y] = \int_0^\infty e^{-it\xi} e^{-\sigma\xi} [S(\xi)x, y] \, d\xi$$
and $[r(z)x, y]$ is an entire function such that
$$|[r(z)x, y]| \leq \|x\| \, \|y\| \, c e^{|z|T},$$
$$\int_{-\infty}^\infty |[r(t)x, y]|^2 \, dt < \infty.$$
We can therefore apply the Paley-Wiener theorem (see [Rudin, 1987, p. 375]) and obtain
$$[r(z)x, y] = \int_{-T}^T F(\xi) e^{-i\xi z} \, d\xi$$
$$= \int_0^\infty e^{-iz\xi} e^{-\sigma\xi} [S(\xi)x, y] \, d\xi.$$
Hence
$$[S(\xi)x, y] = 0 \quad \text{for } \xi > T$$
and hence, x, y being arbitrary:
$$S(t) = 0 \quad \text{for } t > T. \quad \text{Q.E.D.}$$

3. Construction of Superstable Semigroups

The simplest example of a superstable (actually nilpotent) semigroup is the shift semigroup on $L_2[0, L]$ for finite L. In [Hille and Phillips, 1957, p. 663, et seq.] Phillips uses Riemann-Liouville fractional integration to construct a class of analytic superstable semigroups. We can generalize this construction considerably. Thus let $S(\cdot)$ denote a *dissipative* superstable semigroup. Define
$$T(z)x = \frac{1}{\Gamma(z)} \int_0^\infty S(\xi) \, x \, \xi^{z-1} \, d\xi, \quad x \in \mathcal{H}, \ \text{Re}\, z > 0. \tag{3.1}$$
This is an analytic semigroup of class $H\left(\frac{-\pi}{2}, \frac{\pi}{2}\right)$ and is superstable along rays,
$$z = r e^{i\theta}, \quad |\theta| < \frac{\pi}{2}$$
since for any $\sigma > 0$
$$\|S(\xi)\| \leq M_\sigma \exp{-\xi\sigma}, \quad \xi \geq 0$$

and consequently:

$$\|T(z)\| \leq M_\sigma \int_0^\infty e^{-\sigma\xi} \left|\frac{\xi^{z-1}}{\Gamma(z)}\right| d\xi \leq M_\sigma \sigma^{-r\cos\theta}.$$

Since
$$T(z) = R(0, A)^z,$$
and $S(\cdot)$ is dissipative, so that $-R(0, A)$ is dissipative, we can invoke the theory of fractional powers of operators [Fattorini, 1983] for alternate expressions for (3.1) for $0 < \mathrm{Re}\, z < 1$. Thus we have:

$$T(z)x = \frac{\sin \pi z}{\pi} \int_0^\infty \lambda^{z-1} (\lambda I + R(0, A))^{-1}(-R(0, A))x\, d\lambda \qquad (3.2)$$

$$= \frac{-\sin \pi z}{\pi} \int_0^\infty \lambda^{-z} R(\lambda, A)x\, d\lambda$$

$$= \frac{1}{\Gamma(-z)} \int_0^\infty \left(e^{-R(0,A)\xi} - I\right) x\, \xi^{-z-1}\, d\xi \qquad (3.3)$$

$$= \frac{1}{\Gamma(-z)} \int_0^\infty \left(e^{-R(0,A)\xi} - \sum_0^{n-1} \frac{(-R(0, A))^k}{k!}\right) \xi^{-z-1}\, d\xi,$$

$$\text{for } n - 1 < \mathrm{Re}\, z < n$$

from which we see in particular that $T(\cdot)$ is compact if $R(0, A)$ is. (The compactness of the resolvent of A is irrelevant for the superstability property.)

Using Kato's formula (see [Yoshida, 1974, p. 260]) we have

$$(\lambda I + T(z))^{-1} = \frac{\sin \pi z}{\pi} \int_0^\infty (rI + R(0, A))^{-1} \left[\frac{r^z}{\lambda^2 - 2\lambda r \cos \pi z + r^{2z}}\right] dr, \quad 0 < \mathrm{Re}\, z < 1 \qquad (3.4)$$

which is of course defined for every $\lambda \neq 0$. Contrast this with the representation using the Mittag-Leffler function in [Hille and Phillips, 1957, p. 668].

Let $\mathcal{A}$ denote the generator of $T(\cdot)$. Then $\mathcal{A}$ has the representation for $x \in \mathcal{D}(\mathcal{A})$:

$$\mathcal{A}x = \gamma x - A \int_0^\infty S(\xi) x \log \xi\, d\xi \qquad (3.5)$$

where γ is Euler's constant; and

$$\mathcal{R}(\lambda, \mathcal{A})x = \int_0^\infty S(\xi)\, x\, E(\lambda, \xi)\, d\xi \qquad (3.6)$$

where

$$E(\lambda, \xi) = \int_0^\infty e^{-\lambda t} \xi^{t-1} \frac{1}{\Gamma(t)}\, dt$$

from which, as already noted in [Hille and Phillips, 1957, p. 667], we see that $\mathcal{R}(\lambda, \mathcal{A})$ is an entire function of infinite order (still satisfying (2.1)) so that in particular $T(z)$ is *not* nilpotent, for $\operatorname{Re} z > 0$.

Remark. Phillips shows, based on some hard analysis results of Kober (see [Hille and Phillips, 1957, p. 665]), that for the special case where $S(\cdot)$ is the left shift on $L_2[0,1]$, the "boundary values"

$$J(\eta)x = \lim_{\sigma \to 0} T(\sigma + i\eta)x, \quad -\infty < \eta < \infty \qquad (3.7)$$
$$= -AT(1+i\eta)x$$

determine a C_0 *group* whose generator $(i\mathcal{A})$ has an empty spectrum. But $J(\cdot)$ is of course not superstable. □

Since
$$R(\lambda, i\mathcal{A}) = -iR(-i\lambda, \mathcal{A})$$
we see that $R(\lambda, i\mathcal{A})$ is also an entire function of λ of infinite order. In particular we see that the resolvent may be an entire function of infinite order and yet the semigroup need *not* be superstable.

Generalization: Hille-Phillips Calculus. We can generalize the construction (3.1) using the Hille-Phillips operational calculus [Hille and Phillips, 1957, Chapter 15]. For example let for each $t > 0$, $a(t, \cdot)$ denote a set function on $[0, \infty)$ and such that (in the notation of [Hille and Phillips, 1957]):

$$\int_0^\infty \|S(\xi)\|\, d|a(t,\xi)| \;<\; \infty$$

for which it is enough if

$$\int_0^\infty e^{-\sigma_0 \xi}\, d|a(t,\xi)| \;<\; \infty \quad \text{for some} \;\; \sigma_0 > 0.$$

Let us assume this. Let

$$\int_0^\infty e^{\lambda \xi}\, da(t,\xi) \;=\; e^{t\phi(\lambda)}, \quad \operatorname{Re}\lambda \leq \sigma_0, \; t > 0.$$

Then under suitable conditions on $a(t, \cdot)$ (equivalently $\phi(\cdot)$):

$$T(t)x \;=\; \int_0^\infty S(\xi)x\, da(t,\xi) \qquad (3.8)$$

defines a C_0-semigroup, whose generator has an empty spectrum. The condition that

$$\int_{-\infty}^\infty \left| e^{t\phi(\sigma + i\tau)} \right|\, d\tau \;<\; \infty, \quad t > 0$$

is sufficient for the semigroup to be superstable. This follows essentially from Lemma 16.3.1 of [Hille and Phillips, 1957] which in turn is a consequence of the representation:

$$T(t)x = \frac{1}{2\pi i} \int_{\gamma-i\infty}^{\gamma+i\infty} e^{t\phi(\lambda)} R(\lambda, A) x \, d\lambda, \quad \operatorname{Re}\lambda \leq -\sigma_0.$$

For example we may consider fractional powers of $(-A)$. Define

$$T(t)x = \int_0^\infty S(\xi) x F(t, \xi) \, d\xi, \quad t > 0 \tag{3.9}$$

where $F(t, \cdot)$ is nonnegative and is defined by the Levy "stable" density:

$$\int_0^\infty e^{\lambda \xi} F(t, \xi) \, d\xi = e^{-t(-\lambda)^\alpha}, \quad \operatorname{Re}\lambda < -\sigma_0, \ 0 < \alpha < 1.$$

The semigroup is superstable with generator

$$-(-A)^\alpha.$$

The superstability can be proved directly from (3.9), since

$$\|T(t)\| \leq M_\sigma e^{-t\sigma^\alpha}.$$

Superstability holds in fact generally for $F(t, \cdot)$ nonnegative.

Let us also note that $(-\mathcal{A}^2)$ generates a C_0-semigroup, when $(i\mathcal{A})$ generates a C_0 group. Moreover using:

$$\mathcal{R}(\lambda, -\mathcal{A}^2) = \frac{R\left(-i\sqrt{\lambda}, \mathcal{A}\right) - R\left(i\sqrt{\lambda}, \mathcal{A}\right)}{2i\sqrt{\lambda}},$$

we can readily verify that (2.1) is satisfied and thence superstability. The semigroup generated has the representation

$$T(t)x = \int_{-\infty}^\infty J(\xi) x \, G(\xi, t) \, d\xi, \quad t > 0$$

where $J(\cdot)$ denotes the group and

$$G(\xi, t) = \frac{1}{2\sqrt{\pi t}} \exp \frac{-\xi^2}{4t}$$

and is actually an analytic semigroup.

4. Applications

An early example in applications of superstability arising in the solution of a boundary value problem for partial dfferential equations — a "disappearing solution," the semigroup being actually nilpotent — is given by [Majda, 1975]. More generally the phenomenon occurs in Timoshenko models[‡] of structures using self-straining material — "smart structures" — for feedback control [Balakrishnan, 1996b; Balakrishnan, 1997]. The simplest such example is the "smart string" described by:

$$\rho\ddot{\theta}(t,s) - c\theta''(t,s) = 0, \quad 0 < s < \ell$$

$$-c\theta'(t,0) + g\dot{\theta}(t,0) = 0; \quad \theta(t,\ell) = 0; \quad c,\rho > 0$$

where the superdots indicate time derivatives and the primes the space derivatives, and $g > 0$ is the variable gain parameter. The semigroup solution (see [Balakrishnan, 1996b]; the semigroup is dissipative) as a function of g becomes superstable at a critical value of g given by

$$g = \sqrt{\rho c}\ .$$

The energy decays to zero in finite time and of course there are no modes. The solution vanishes ("disappears") for $t > 2\ell\sqrt{\rho/c}$ as can be directly verified (see [Balakrishnan, 1996b]). A more realistic version [Balakrishnan, 1997] where we consider both displacement and torsion is given by:

$$m_2\ddot{v}(t,s) - c_4(v''(t,s) - \phi'(t,s)) = 0, \quad 0 < s < \ell; \ 0 < t$$

$$m_6\ddot{\phi}(t,s) - c_2\phi''(t,s) - c_4(v'(t,s) - \phi(t,s)) = 0$$

$$\phi(t,0) = v(t,0) = 0$$

$$c_4(v'(t,\ell) - \phi(t,\ell)) + g\dot{v}(t,\ell) = 0$$

$$c_2\phi'(t,\ell) + g\dot{\phi}(t,\ell) = 0$$

and g is again the variable gain parameter. We have superstability for

$$g = \sqrt{m_6 c_2} = \sqrt{c_4 m_2}\ .$$

The resolvent can be shown to be of exponential type (details will appear elsewhere) and we have thus actually again a "disappearing solution."

5. Relation to Bases

We refer to a subspace as a superstable invariant subspace if it is an invariant subspace for the semigroup which is superstable thereon — in particular such a subspace cannot contain any eigenfunctions of the generator. The latter is the case if we consider the subspace orthogonal to the subspace spanned by the eigenfunctions of the adjoint of the generator. This relationship has already been noted by [Majda, 1975], and can provide a technique for verifying whether Riesz bases are retained under a perturbation [Balakrishan, 1996a].

[‡]It is curious that the phenomenon does not occur in Euler models.

Acknowledgement

I am indebted to J. Martinez and P. Vu for many fruitful discussions and helpful comments.

Research support in part under NASA Grant No. NCC 2-374.

References

A.V. Balakrishnan. "On Superstable Semigroups of Operators," *Dynamic Systems and Applications*, Vol. 5 (1996), pp. 371–384.

A.V. Balakrishnan. "Vibrating Systems with Singualr Mass-Inertia Matrices." In: *First International Conference on: Nonlinear Problems in Aviation & Aerospace.* Ed. S. Sivasundaram. Embry-Riddle Aeronautical University Press, 1996. Pp. 23–32.

A.V. Balakrishnan. "Smart Structures and Super Stability." In: *Computational Science for the 21st Century.* Ed. Jacques Periaux, *et al.* Chichester, England: John Wiley & Sons, Ltd., 1997.

E. Hille and R.S. Phillips. *Functional Analysis and Semigroups.* AMS Colloquium Publications, Vol. 31. 1957.

H.O. Fattorini. *The Cauchy Problem.* Addison-Wesley Publishing Co., 1983.

A. Majda. "Disappearing Solutions for the Dissipative Wave Equation," *Indiana University Math. Journal* (1975), pp. 1119–1133.

R. Nagel, ed. *One-parameter Semigroups of Positive Operators*, Lecture Notes in Mathematics Series. Springer-Verlag, 1980.

W. Rudin. *Real and Complex Analysis*, 3rd Edition. McGraw-Hill, 1987.

A. Sinclair. *Continuous Semigroups in Banach Algebras*, London Mathematical Society Lecture Notes Series, No. 63. Cambridge University Press, 1982.

K. Yosida. *Functional Analysis*, 4th Edition. Springer-Verlag, 1974.

A. V. Balakrishnan
Flight Systems Research Center
UCLA
Los Angeles, CA 90095
U.S.A.

VU QUOC PHONG
Stability of C_0-semigroups and stabilization of linear control systems

1. Introduction

In this paper, we present some results which were obtained in recent years, partly by the author in collaboration with others, on the stability and stabilization problem for linear systems on infinite dimensional spaces. Although some of these results are now known to specialists in the control theory, they are not very well known and we hope that this article will make them more available for applications. Other results are new and did not appear before. The content of this article coincides with the content of a talk given by the author at the 18-th IFIP TC7 Conference on System Modelling and Optimization. The author has decided to preserve the informal style of the talk in the present paper.

Consider the system of linear ordinary differential equations

$$\frac{du}{dt} = Au(t), \quad t \in \mathbf{R}, \tag{1}$$

where A is a matrix (i.e., a linear operator on a finite dimensional space E). The zero solution of Eq.(1) is said to be *asymptotically stable* if $e^{tA}x \to 0$ as $t \to \infty$, for all $x \in E$. It is the classical Lyapunov stability theorem that the zero solution of Eq.(1) is asymptotically stable if and only if all eigenvalues of A have negative real part. In this case, we also have the (uniform) exponential stability, i.e. there exist constants $M \geq 1, \sigma > 0$ such that $\|e^{tA}x\| \leq Me^{-\sigma t}\|x\|, \forall x \in E, t \geq 0$. The Lyapunov theorem gives a very simple and efficient spectral condition for asymptotic stability.

Many problems in physics and engineering can be modelled using equation (1), however, in a infinite dimensional (Hilbert or Banach) space, with a generally unbounded operator A. It is therefore a natural and important problem to generalize the Lyapunov theorem to this infinite dimensional situation. The first question is how to define e^{tA}? If A is a bounded linear operator, then e^{tA} can be defined as in the finite dimensional case, for instance, by using the formula

$$e^{tA} = \sum_{n=0}^{\infty} \frac{t^n A^n}{n!}. \tag{2}$$

If A is unbounded, formula (2) does not have sense, but we can define e^{tA} through the theory of C_0-semigroups. Namely, A must be the generator of a strongly continuous semigroup, i.e. a family $T(t), t \geq 0$, of bounded linear operators on E (which will define $e^{tA} \equiv T(t)$) such that: (i) $T(0) = I$, $T(s+t) = T(s)T(t), \forall s, t \geq$

0; (ii) The function $t \mapsto T(t)x$ is continuous for every $x \in E$. In this case, the domain of A, $D(A)$, is given by $D(A) = \{x \in E : \lim_{t\to 0^+} \frac{1}{t}(T(t)x - x)$ exists$\}$ and $Ax = \lim_{t\to 0^+} \frac{1}{t}(T(t)x - x)$. Recall that a necessary and sufficient condition for an operator A to be generator of a C_0-semigroup, in terms of inequalities of the resolvent of A, is given by the well known Hille-Yosida and Feller-Myadera-Phillips theorems, see e.g. [9].

Therefore, we can assume that the operator A is the generator of a C_0-semigroup $T(t)$. Since there are three types of topology on E - the uniform operator topology, the strong operator topology and the weak operator topology, respectively (which are all distinct if E is infinite dimensional), we can speak about three different notions of asymptotic stability.

First, and the most popular, is the notion of *exponential stability*, which sometimes is called *uniform exponential stability*. A C_0-semigroup $T(t)$ is called *exponentially stable*, if $\lim_{t\to\infty} \|T(t)\| = 0$, or, equivalently, if there exist constants $M \geq 1, \sigma > 0$ such that $\|T(t)\| \leq M e^{-\sigma t}, \forall t \geq 0$. In practical applications, this is the most useful type of asymptotic stability since it is preserved by small perturbation of the generator. However, there are systems which are not exponentially stable and, moreover, can not be exponentially stabilized by any feedback control, see e.g. [3,7,12]. The second notion of asymptotic stability is connected with the strong operator topology. A C_0-semigroup $T(t)$ is called *strongly asymptotically stable* if $\lim_{t\to\infty} \|T(t)x\| = 0, \forall x \in E$. In contrast to the exponential stability, strong asymptotic stability is, in general, vulnerable w.r.t. small perturbations of the generator. This is the reason why there is a widespread opinion among some group of control theorists that the strong asymptotic stability is not a useful property deserving attention. But, on the one hand, the notion of strong asymptotic stability can be interpreted as that "the energy dies away as time goes on", which has a lot of sense physically. On the other hand, as mentioned above, there are systems which can not be exponentially stabilizable, and the best one can hope for is the strong stabilizability, which is in fact possible in a number of situations (see [2,4]).

The third notion of stability is related to the weak operator topology, according to which a semigroup $T(t)$ is called weakly asymptotically stable if $T(t)x$ converges to zero weakly, as $t \to \infty$, for each $x \in E$. However, this notion of stability doesn't make much physical sense, and therefore doesn't attract much attention.

2. Strong asymptotic stability

The main result on strong asymptotic stability, so far, is the following theorem obtained by Lyubich and the author and independently by Arendt and Batty.

Theorem 1. *Suppose that $T(t)$ is a C_0-semigroup on a Banach space E, with the generator A, which satisfies the following conditions:*
  (i) $T(t)$ is uniformly bounded, i.e. $\sup_{t\geq 0} \|T(t)\| < \infty$;
  (ii) $\sigma(A) \cap i\mathbf{R}$ is countable;
  (iii) $\sigma_p(A^*) \cap i\mathbf{R} = \emptyset$.

Then $T(t)$ is strongly asymptotically stable.

We remark that conditions (i) and (iii) are necessary for the strong asymptotic stability. Condition (ii) is not necessary: there are semigroups which are strongly asymptotically stable, yet the imaginary part of the spectrum of the generators is not countable (it can be the whole $i\mathbf{R}$; the typical example is the semigroup of left translations on $L^2(0,\infty)$). Nevertheless, Theorem 1 gives a condition for strong asymptotic stability solely in terms of the spectrum (and the point spectrum) of the generator, which is convenient for applications. (Note that, in view of the Hille-Yosida and Feller-Phillips-Myadera theorems on characterization of bounded semigroup, condition (i) admits a reformulation in terms of the resolvent of A). Two very different proofs of this theorem were given independently in [1] and [12], and later several new proofs were also found. But all the proofs are based on a theorem of Tauberian type. In particular, the proof in [1] is based on some new Tauberian theorems for the Laplace-Stieltjes transforms. The proof in [12] is based on the theorem on non-emptyness of the spectrum of groups of isometric operators on Banach spaces, which is an operator variant of the classical Tauberian theorem, as well as the closely related theorem of Gelfand that an isometric operator whose spectrum is a singleton is a scalar multiple of the identity. The latter is an operator version of the Prime Ideal Theorem, or the theorem that a function $g(t)$ with a single spectrum, $\sigma(g) = \{\lambda_0\}$, must have form $g(t) = e^{\lambda_0 t}$ (see e.g. [17]). To see the Tauberian nature of Theorem 1 let us note that Theorem 1 contains the fact that the spectrum of the generator of an isometric group is non-empty, or even the following stronger statement.

Corollary 2. *Suppose that $T(t)$ is a uniformly bounded C_0-semigroup on a (nonzero) Banach space E, with the generator A, which is not strongly asymptotically stable. Then the spectrum of A is non-empty.*

Let us present basic ideas of the proof of Theorem 1 following the method in [17]. It is useful to locate the following proposition which gives a characterization of strong asymptotic stability.

Proposition 3. *Suppose $T(t)$ is a uniformly bounded C_0-semigroup on E. Then $T(t)$ is strongly asymptotically stable if and only if there does not exist a non-trivial bounded complete trajectory of $T^*(t)$, i.e. a continuous function $g : \mathbf{R} \to E^*$ such that $g(t) = T^*(t-s)g(s), \forall t \geq s, t, s \in \mathbf{R}$.*

The "only if" part can be proved directly. For if $T(t)$ is strongly asymptotically stable and $g : \mathbf{R} \to E^*$ is a bounded nonzero function such that $g(t) = T^*(t-s)g(s)$, then $(g(t))(x) = \lim_{s \to -\infty}(g(s))(T(t-s)x) = 0$ for all $x \in E$ and $t \in \mathbf{R}$, which is an absurd.

To prove the "if" part of Proposition 3, we use the following construction of the *limit isometric semigroup*. Namely, let $l(x)$ be a seminorm in E defined by $l(x) = \limsup_{t \to \infty} \|T(t)x\|, x \in E$, L be the kernel of l, i.e. $L = \{x \in E : \lim_{t \to \infty} \|T(t)x\| = 0\}$, and let $\widehat{E}$ be the quotient space $\widehat{E} = E/L$, equipped with the norm $\hat{l}(\hat{x}) = l(x)$, $\forall \hat{x} \in \widehat{E}$. It is easy to see that the naturally defined operators $\widehat{T}(t)\hat{x} = \widehat{T(t)x}$ are isometric in the norm $\hat{l}$. Thus, we obtain an isometric C_0-semigroup on the normed space $\widehat{E}$ (we can also extend it to an isometric semigroup in the completion of $\widehat{E}$, if needed). To obtain a nontrivial bounded complete trajectory for the dual semigroup $T^*(t)$, we note that if $T(t)$ is not strongly asymptotically stable, then in the construction above the space $\widehat{E}$ is nontrivial ($\widehat{E} \neq \{0\}$). Choose an arbitrary nonzero functional $\psi_0 \in \widehat{E}^*$, and let $\psi(t) = \widehat{T}^*(t)\psi_0, t \geq 0$. It follows by an application of the Hahn-Banach Theorem that $\psi(t)$ can be extended to $t < 0$ such that $\widehat{T}^*(t - s)\psi(s) = \psi(t), \forall s, t \in \mathbf{R}, t \geq s$. Now it remains to "lift" $\psi(t)$ from $\widehat{E}$ to functionals $g(t)$ on E, by putting $g(x) \equiv \psi(\hat{x})$. The identity $\widehat{T}^*(t - s)\psi(s) = \psi(t)$ implies $T(t - s)g(s) = g(t), \forall t, s \in \mathbf{R}, t \geq s$, so that $g(t)$ is a nontrivial bounded complete trajectory of $T^*(t)$.

Now let us outline a proof of Theorem 1 which is based on Proposition 3 and the Tauberian results. Assuming that the conditions in Theorem 1 hold but the semigroup is not strongly asymptotically stable. Then one obtains, by Proposition 3, a nontrivial bounded complete trajectory $g(t)$ of the dual semigroup $T^*(t)$. The spectrum of the function g, $\sigma(g)$, is non-empty, and one can show, using the definition of the spectrum of a bounded function (see [17]) that $\sigma(g) \subset \sigma(A) \cap i\mathbf{R}$, hence $\sigma(g)$ is a nonempty countable subset of $i\mathbf{R}$. One can now use the standard Riesz-Dunford integral in the space spanned by the translations of g, to derive a nontrivial bounded complete trajectory $h(t)$, of $T^*(t)$, with a single spectrum, say $\sigma(h) = \{\lambda_0\}$. Such trajectory must have form $h(t) = e^{\lambda_0 t}h_0$, with $\lambda_0 \in i\mathbf{R}$ and $h_0 \in E^*$. Since $T^*(t)h_0 = e^{\lambda_0 t}h_0$, it follows that $h_0 \in D(A^*)$ and $A^*h_0 = \lambda_0 h_0, h_0 \neq 0$, a contradiction to condition (iii).

Theorem 1 was obtained initially in [16] as a corollary of a more general result about almost periodic semigroups. Since this result will be used for the stabilization problems which we will consider, and since it also makes the stability result more transparent, let us describe it here.

A C_0-semigroup $T(t)$, with the generator A, is called *almost periodic*, if every its trajectory, i.e. every subset $\mathcal{O}(x) = \{T(t)x : t \geq 0\}$, $x \in E$, is relatively compact. The central result in the theory is the De Leeuw-Glicksberg Decomposition Theorem, which states that a C_0-semigroup $T(t)$ is almost periodic if and only if $E = E_0 \dotplus E_1$, where $E_0 = \{x \in E : \lim_{t\to\infty} \|T(t)x\| = 0\}$, and E_1 is spanned by eigenvectors of A corresponding to imaginary eigenvalues. Moreover, the restriction of $T(t)$ to E_1 can be extended to a bounded C_0-group, there are mutually disjoint projections P_λ from E_1 onto the eigensubspaces $M(\lambda) = \{x \in D(A) : Ax = \lambda x\}, \lambda \in i\mathbf{R}$, which form a resolution of the identity on E_1.

From the De Leeuw-Glickssberg Decomposition it follows that a C_0-semigroup $T(t)$ is strongly asymptotically stable if and only if it is almost periodic and its generator A has no purely imaginary eigenvalues. A spectral condition of almost periodicity is given by the following theorem [16].

Theorem 4. Suppose $T(t)$ is a C_0-semigroup, with the generator A, on a Banach space E which satisfies the following conditions:
  (i) $T(t)$ is uniformly bounded;
  (ii) $\sigma(A) \cap i\mathbf{R}$ is countable;
  (iii) $\ker(A - \lambda) + \operatorname{ran}(A - \lambda)$ is dense in E for every $\lambda \in i\mathbf{R}$.
Then $T(t)$ is almost periodic.

As in Theorem 1, condition (i) and (iii) are necessary for the almost periodicity while condition (ii) is not. Let us make a few comments on codition (iii). First, it is automatically fulfiled for all $\lambda \in i\mathbf{R} \setminus \sigma(A)$ so that one need only to check it for $\lambda \in i\mathbf{R} \cap \sigma(A)$. Second, it is equivalent to the statement: If there exists an eigenfunctional $\phi \in E^*, \phi \neq 0$, with an eigenvalue $\lambda \in i\mathbf{R}$, $A^*\phi = \lambda\phi$, then there must exist an eigenvector $x \in E$ such that $\phi(x) \neq 0$ and $Ax = \lambda x$. And, finally, condition (iii) is an ergodicity condition: it means that for all $\lambda \in i\mathbf{R}$, the semigroup $e^{-\lambda t}T(t)$ is mean ergodic. In particular, it is automatically fulfiled for (uniformly bounded) semigroups on refelexive Banach spaces. Thus, we have the following corollary.

Corollary 5. For uniformly bounded C_0-semigroups on reflexive Banach spaces, the countability of the imaginary part of the spectrum of A implies the almost periodicity of $T(t)$.

Theorem 1 follows immediately from Theorem 4 because if condition (iii) in Theorem 1 is fulfiled, then the condition (iii) in Theorem 4 is also fulfiled. Note that, under the conditions of Theorem 1, the generally stronger fact that A has no purely imaginary eigenvalues follows as a consequence of the weaker fact that A^* has no purely imaginary eigenvalues.

Theorem 4 also can be proved similarly to the proof of Theorem 1. Let us outline a proof which is different from both the original method of [16] and the method of the Laplace transform in [5]. This proof is based on the following proposition, which is analogous to Proposition 3 and gives an alternative to almost periodicity.

Proposition 6. Suppose $T(t)$ is a uniformly bounded C_0-semigroup which satisfies condition (iii) of Theorem 4. Then either (i) $T(t)$ is almost periodic, or (ii) there exists a bounded complete trajectory $\phi(t)$ of $T^*(t)$ with uncountable spectrum.

The proof of Proposition 6 is a slight modification of the proof of Proposition 3. The modification is in the definition of the subspace L and the quotient space $\widehat{E} = E/L$. Here, we let $L_0 = \{x \in E : \lim_{t \to \infty} \|T(t)x\| = 0\}$, $L_1 = \overline{\operatorname{span}}\{x \in D(A) : Ax = \lambda x, \lambda \in i\mathbf{R}\}$, and $L = \overline{\operatorname{span}}\{L_0, L_1\}$. Since $T(t)$ is uniformly bounded, it is easy to see that $T(t)|L$ is an almost periodic semigroup so that actually $L = L_0 \dotplus L_1$. Let $\widehat{E} = E/L$, equipped with the norm $\|\|\hat{x}\|\| = \inf_{y \in L_1} \limsup_{t \to \infty} \|T(t)(x - y)\|$.

Assuming that the considered semigroup $T(t)$ is not almost periodic, we have that in the above construction the space $\widehat{E}$ is nonzero, and, arguing as in the proof of Proposition 3, we obtain a nontrivial complete bounded trajectory $\phi(t)$ of the dual semigroup $\widehat{T}^*(t)$. We show that $\sigma(\phi)$ is uncountable. This will imply that the complete bounded trajectory of $T^*(t)$ defined by $g(x) = \phi(\hat{x})$ will also have uncountable spectrum. So assume, on the contrary, that $\sigma(\phi)$ is countable. Then $\sigma(\phi)$ contains an isolated point λ_0. As in the proof of Proposition 3, one can obtain another complete bounded trajectory $\varphi(t)$ of $\widehat{T}^*(t)$ such that $\sigma(\varphi) = \{\lambda_0\}$. Such function φ must have the form $\varphi(t) = e^{\lambda_0 t}\varphi_0, \varphi_0 \in \widehat{E}^*$. This implies that $\widehat{A}^*\varphi_0 = \lambda_0\varphi_0$. Passing to the functional $g_0(x) \equiv \varphi_0(\hat{x})$ we have $A^*g_0 = \lambda_0 g_0$. But, on the other hand, $g_0|L \equiv 0$, and, in particular, $g_0(x) = 0$ for all x such that $Ax = \lambda_0 x$, which is a contradiction to the assumption (condition (iii) in Theorem 4). Theorem 4 now follows immediately from Proposition 6, because under the conditions of Theorem 4, if the semigroup is not almost periodic, then the exists a bounded complete trajectory of $T^*(t)$ with uncountable spectrum, which implies that $\sigma(A^*) \cap i\mathbf{R}$, and hence $\sigma(A) \cap i\mathbf{R}$, is uncountable.

In light of the De Leeuw-Glicksberg Decomposition, Theorem 4 allows to obtain some results on the strong stabilization of linear control systems on Hilbert space, in a quite transparent manner. Recall that a pair (A, B), where A is the generator of a C_0-semigroup on a Hilbert space H and B is a bounded linear operator from another Hilbert space U to H, is called *strongly stabilizable*, if there exists a bounded linear operator $K : H \to U$ such that the semigroup $S_K(t)$ with the generator $A + BK$ is strongly asymptotically stable.

Consider the case when A is dissipative, i.e. $\text{Re}\,(Ax, x) \leq 0, \forall x \in H$, or, equivalently, the semigroup $T(t)$ with the generator A is a contraction semigroup. Then the operator $A - BB^*$ also is dissipative, since $\text{Re}\,((A - BB^*)x, x) \leq \text{Re}\,(Ax, x)$. It can be shown easily that $\sigma(A - BB^*) \cap i\mathbf{R} \subset \sigma(A) \cap i\mathbf{R}$. This implies that the following proposition holds.

Proposition 7. *Suppose $T(t)$ is a contraction C_0-semigroup on a Hilbert space H, with the generator A such that $\sigma(A) \cap i\mathbf{R}$ is countable. Then for every bounded linear operator $B : U \to H$, the semigroup $S(t)$ with the generator $A - BB^*$ is almost periodic.*

Similarly, a simple spectral argument shows that $\sigma_p(A - BB^*) \cap i\mathbf{R} \subset \sigma_p(A) \cap i\mathbf{R}$. Moreover, $(A - BB^*)x = i\lambda x$ for some real λ and $x \in H, x \neq 0$, if and only if $Ax = i\lambda x$ and $B^*x = 0$. In view of the De Leeuw-Glicksberg Decomposition and Proposition 7, $S(t)$ will be strongly asymptotically stable if and only if there does not exist a nonzero $x \in \ker B^*$ and a real λ such that $Ax = i\lambda x$. Thus, the following fact holds (see [5]): *the system (A, B) is strongly stabilizable if and only if $B^*x \neq 0$ for all (nonzero) eigenvectors x of A which correspond to purely imaginary eigenvalues.*

In applications, the operator A usually has a discrete spectrum, $Af_k = \lambda_k f_k, k = 1, 2, ...$, hence the above characterization of strong stabilizability is reduced to: $B^*f_k \neq 0$ for every f_k with $\lambda_k \in i\mathbf{R}$. This is the case, for instance, for the stabilization of a string governed by the wave equation

$$f_{tt} = f_{xx} + R(t)q(x),$$

with a scalar controller $R(t)$ and a distributed control forrce $q \in L^2[0,l]$ (l is the length of the string, see [13]).

However, the previous result on stabilization does not apply to the case when the operator A is not dissipative, e.g., when A has (finitely many) eigenvalues λ such that $\text{Re}\,\lambda > 0$. In this case, one can seek for a stabilizing feedback gain K in the form $K = -B^*P$ where P is a (non-negative definite) solution of the algebraic Riccati equation $PA + A^*P - PBB^*P + R = 0$, where R is also a non-negative definite operator on H. Assuming that the semigroup $S(t)$ with the generator $A - BB^*P$ is uniformly bounded (this may seem to be a strong condition, but it is fulfiled in some instances), then one can prove in a similar way that, *under the countability of $\sigma(A) \cap i\mathbf{R}$, the strong stabilizability of (A,B) is equivalent to the condition:$B^*Px \neq 0$ for all eigenvectors x of A which correspond to purely imaginary eigenvalues* (see [11]).

3. Exponential stability

There are few results available on exponential stability of C_0-semigroups, but the most well known are the following two theorems. The first result is usually known as Datko theorem and gives a characterization of exponential stability for semigroups on arbitrary Banach spaces.

Theorem 8. *Suppose $T(t)$ is a C_0-semigroup on a Banach space E. Then $T(t)$ is exponentially stable if and only if, for some $1 \leq p < \infty$,*

$$\int_0^\infty \|T(t)x\|^p dt < \infty, \; \forall x \in E.$$

The second result gives a characterization of the exponential stability of semigroups on Hilbert space. It was obtained by Gearhart [6] for contraction semigroups and by Herbst [8], Howland [10], Prüss [15] (independently) for general semigroups.

Theorem 9. *Supppose $T(t)$ is a C_0-semigroup on a Hilbert space H, with the generator A. Then: (i) $1 \in \rho(T(1))$ if and only if $2k\pi i \in \rho(A), \forall k \in \mathbf{Z}$ and $\sup_{n \in \mathbf{Z}} \|(2n\pi i - A)^{-1}\| < \infty$;*
(ii) $T(t)$ is exponentially dichotomic if and only if $i\mathbf{R} \subset \rho(A)$ and $\sup_{\lambda \in \mathbf{R}} \|(i\lambda - A)^{-1}\| < \infty$;
(iii) $T(t)$ is exponentially stable if and only if $\{\lambda \in \mathbf{C} : \text{Re}\,\lambda \geq 0\} \subset \rho(A)$ and $\sup_{\text{Re}\,\lambda \geq 0} \|(\lambda - A)^{-1}\| < \infty$.

In this section, we present the following new results which are closely related to Theorems 8 and 9.

Theorem 10. *Suppose H is a Hilbert space, $T(t)$ is a C_0-semigroup on H. If*

$$\sup_{t\geq 0, k\in \mathbf{Z}} \left\| \int_0^t e^{2k\pi i s} T(s)x \right\| < \infty, \quad \forall x \in H, \tag{3}$$

then $1 \in \rho(T(1))$. Conversely, if $T(t)$ is uniformly bounded and $1 \in \rho(T(1))$, then (3) holds.

Theorem 11. *Suppose H is a Hilbert space, $T(t)$, is a C_0-semigroup on H. Then $T(t)$ is exponentially stable if and only if*

$$\sup_{\lambda \in \mathbf{R}, t\geq 0} \left\| \int_0^t e^{i\lambda s} T(s)x\, ds \right\| < \infty, \quad \forall x \in H. \tag{4}$$

Theorem 12. *Suppose E is a Banach space, $T(t)$ is a C_0-semigroup on E, with generator A. If*

$$\sup_{t\geq 0} \left\| \int_0^t e^{i\lambda s} T(s)x\, ds \right\| < \infty \quad \forall x \in E, \quad \lambda \in \mathbf{R}, \tag{5}$$

then
(i) $\|T(t)x\| \to 0$ as $t \to \infty$ for all $x \in D(A^2)$.
(ii) If, in addition to (5), $T(t)$ is uniformly bounded, then $T(t)$ is asymptotically stable.
(iii) If, in addition to (5), $T(t)$ is analytic, then $T(t)$ is exponentially stable.

The proof of Theorems 10-12 is based on the following lemma:

Lemma 13. *Assume that E is a Banach space and $T(t)$, $t \geq 0$, is a C_0-semigroup on E, with generator A. If*

$$\sup_{t\geq 0} \left\| \int_0^t T(s)x\, ds \right\| < \infty, \quad \forall x \in E,$$

then $0 \in \rho(A)$. Moreover, if

$$\left\| \int_0^t T(s)\, ds \right\| \leq M, \quad \forall t \geq 0,$$

then $\|A^{-1}\| \leq M$.

The proof of Lemma 13 is standard, using the Uniform Boundedness Principle, formulas for resolvents, and the integration by parts. From Lemma 13 it follows that if (3) holds, then $2k\pi i \in \rho(A), \forall k \in \mathbf{Z}$ and $\sup_{k\in \mathbf{Z}} \|(2k\pi i - A)^{-1}\| < \infty$, which implies $1 \in \rho(T(1))$, by Theorem 9. Theorems 11 and 12 also follow from Lemma 13 and Theorem 9 in a similar way. Note that it was also shown in [14,Prop. 4.5.3] (implicitely), using a completely different (and more complicated) argument that condition (4) implies $\sup_{\text{Re }\lambda >0} \|(\lambda - A)^{-1}\| < \infty$.

REFERENCES

1. W. Arendt and C.J.K. Batty, *Tauberian theorems and stability of one-parameter semigroups*, Trans. Amer. Math. Soc. **306** (1988), 837–852.
2. G. Avalos and I. Lasiecka, *The strong stability of a semigroup arising from a couple hyperbolic/parabolic system*, Preprint.
3. A.V. Balakrishnan, *Strong stabilizability and the steady state Riccatti equation*, Appl. Math. Optim. **7** (1981), 335–345.
4. A.V. Balakrishnan, *Robust stabilizing compensators for flexible structures with collocated controls*, Appl. Math. Optim. **33** (1996), 35–60.
5. C.J.K. Batty and Vu Quoc Phong, *Stability of individual elements under one-parameter semigroups*, Trans. Amer. Math. Soc. **322** (1990), 805–818.
6. L. Gearhart, *Spectral theory of contraction semigroups on Hilbert space*, Trans. Amer. Math. Soc. **236** (1978), 385–394.
7. J.S. Gibson, *A note on stabilization of infinite dimensional linear oscilators by compact linear feedback*, SIAM J. Control **18:3** (1980), 311–316.
8. I.W. Herbst, *The spectrum of Hilbert space semigroups*, J. Operator Theory **10** (1983), 87–94.
9. E. Hille and R.S. Phillips, *Functional Analysis and Semigroups*, Amer. Math. Soc., Providence, RI, 1957.
10. J.S. Howland, *On a theorem of Gearhart*, Integral Eq. Operator Th. **7** (1984), 138–142.
11. N. Levan, Vu Quoc Phong and S.P. Yung, *Almost periodic semigroups and stabilization via the algebraic Riccati equation*, Preprint.
12. Yu.I. Lyubich and Vu Quoc Phong, *Asymptotic stability of linear differential equations in Banach spaces*, Studia Math. **88** (1988), 37–42.
13. L. Markus, *Introduction to the theory of distributed control systems*, In: Distributed Parameter Control Systems (Minneapolis, MN, 1989), 1–60. Lecture Notes in Pure and Applied Math., vol. 128(1991), Dekker, New York.
14. J.M.A.M. van Neerven, *The asymptotic behaviour of semigroups of linear operators*, Birkhäuser, Basel, 1996.
15. J. Prüss, *On the spectrum of C_0-semigroups*, Trans. Amer. Math. Soc. **284** (1984), 847–857.
16. Vu Quoc Phong and Yu.I. Lyubich, *A spectral criterion for almost periodicity of one-parameter semigroups*, J. Soviet Math. **48** (1990), 644–647. Originally published in Teor. Funktsiĭ, Funktsional. Anal. i Prilozhenia, vol. **47**(1987), 36–41 (in Russian).
17. Vu Quoc Phong, *On the spectrum, complete trajectories and asymptotic stability of linear semi-dynamical systems*, Journal of Differential Equations **105** (1993), 30–45.

Author's address: Department of Mathematics, Ohio University, Athens, OH 45701
E-mail: qvu@bing.math.ohiou.edu

JONG UHN KIM
A new regularity result in transonic gas dynamics

In this paper we present a new result on the regularity of the velocity potential in transonic gas dynamics of two space dimensions. The stationary velocity potential $\phi(x,y)$ satisfies

$$(0.1) \quad (c^2 - \phi_x^2)\phi_{xx} - 2\phi_x\phi_y\phi_{xy} + (c^2 - \phi_y^2)\phi_{yy} = 0, \quad \text{for } (x,y) \in R^2,$$

where $c^2 = 1 - (\gamma-1)(\phi_x^2 + \phi_y^2)/2$, and $\gamma > 1$ is a constant representing the ratio of specific heats. For the derivation of this equation, see Bers [2]. If $c^2 > \phi_x^2 + \phi_y^2$, the equation is elliptic. It becomes hyperbolic if $c^2 < \phi_x^2 + \phi_y^2$. When the set of points for $c^2 = \phi_x^2 + \phi_y^2$ is a curve, it is called a sonic line. For the statement of our main result, we make the following assumptions.

(**i**) Γ is a sonic line, and it is a C^∞-curve in a neighborhood of the point $(x_0, y_0) \in \Gamma$.

(**ii**) $\nabla(c^2 - \phi_x^2 - \phi_y^2) \neq 0$, at (x_0, y_0).

(**iii**) $\nabla\phi$ is not orthogonal to Γ at (x_0, y_0).

A point on the sonic line where either (**ii**) or (**iii**) is not satisfied is called an exceptional point: see Bers [2, p.104].

Under these assumptions, the main result is the following.

THEOREM 0.1. *Suppose that ϕ is a solution of (0.1) which is C^4 in a neighborhood of (x_0, y_0). Then, ϕ is C^∞ in a neighborhood of (x_0, y_0).*

For a quasilinear second order elliptic equation, C^∞-regularity at an interior point is well-known. On the other hand, the singularity of solution of a hyperbolic equation is propagated along bicharacteristic curves. Since the type of the equation (0.1) changes across the sonic line Γ, the above theorem is a new kind of regularity result. We note that a similar result does not hold for an equation of changing type where the sonic line is described by a linear equation of independent variables. For example, we can find a solution of the Tricomi equation which is sufficiently smooth, but is not C^∞, in a neighborhood of a given point on the parabolic line. This can be done by combining the existence theorem of Morawetz [6] and the regularity result in [5], which shows a connection between the regulariy of the boundary value and interior regularity. In the present problem, a salient feature is shown in the equation of Γ, $c^2 = \phi_x^2 + \phi_y^2$, which is crucial in boosting the regularity. We now explain the general strategy of proof of the above theorem. Under the assumption (**i**), let Ω be a neighborhood of (x_0, y_0), and let $\Omega = \Omega^+ \cup (\Gamma \cap \Omega) \cup \Omega^-$, such that the equation is elliptic in Ω^+, and hyperbolic in Ω^-. By means of the the linearized equation on $\Omega^+ \cup (\Gamma \cap \Omega)$, we obtain basic energy estimates with bounds expressed in terms of the tangential derivative of ϕ on Γ. These estimates yield more regularity for the normal derivative of ϕ than for the tangential derivative in Ω^+. By the trace regularity, we find that the regularity of the normal derivative on Γ is stronger than the previous regularity of the tangential derivative. But the equation $c^2 = \phi_x^2 + \phi_y^2$ enforces the equivalent regularity of the normal and tangential derivatives of ϕ on Γ. This enables us to use the bootstrap argument to establish the C^∞-regularity of ϕ on $\Omega^+ \cup (\Gamma \cap \Omega)$ possibly with smaller Ω. The main tool is Hörmander's energy method for the Tricomi operator in [3] with help of the commutator estimates due to Kato and Ponce [4], its generalized version, and the Moser-type estimates proved in Taylor [9]. We then continue C^∞-regularity across Γ into Ω^-. For this, we borrow a result from Berezin [1] who proved the well-posedness of the Cauchy problem for a degenerate hyperbolic equation. A similar result was also obtained by Protter [8]. Finally we note that there are numerous results on the degenerate elliptic equations and the degenerate hyperbolic equations (see the references in [7]). But our result is not covered by any of the known results.

References

[1] Berezin, I.S., On Cauchy's problem for linear equations of the second order with initial conditions on a parabolic line, Mat. Sb. 24(66), 1949, pp. 301-320 ; English transl. Amer. Math. Soc. Transl. (1) 4, 1962, pp. 415-439.

[2] Bers, L., " Mathematical aspects of subsonic and transonic gas dynamics," John Wiley and Sons, Inc., New York, 1958.

[3] Hörmander, L., "The analysis of linear partial differential operators," Vol.3, Springer-Verlag, Berlin-Heidelberg-New York, 1985.

[4] Kato, T. and Ponce, G., Commutator estimates and the Euler and Navier-Stokes equations, Comm. Pure Appl. Math., Vol.41, 1988, pp. 891-907.

[5] Kim, J.U., Interior regularity of solutions of the Tricomi problem, J. Math. Anal. Appl., Vol.192, 1995, pp. 956-968.

[6] Morawetz, C.S., A weak solution for a system of equations of elliptic-hyperbolic type, Comm. Pure Appl. Math., Vol.11, 1958, pp. 315-331.

[7] Oleinik, O.A. and Radkevic, E.V., "Second order equations with nonnegative characteristic form," Amer. Math. Soc., Providence, R.I. and Plenum Press, New York, 1973.

[8] Protter, M.H., The Cauchy problem for a hyperbolic second order equation with data on the parabolic line, Canad. J. Math., Vol.6, 1954, pp. 542-553.

[9] Taylor, M.E., "Pseudodifferential operators and nonlinear PDE," Birkhäuser, Boston, 1991.

by Jong Uhn Kim
Department of Mathematics
Virginia Tech
Blacksburg, VA 24061-0123
e-mail: kim@math.vt.edu

FABRIZIO COLOMBO, ROBERTO MONTE, VINCENZO VESPRI
On some equations about European option pricing*

1 Introduction

The existence and the characterisation of the solution of no-arbitrage pricing problems is a central topic in modern mathematical finance. It is well known that a closed form solution of such a problem is only available under very restrictive assumptions on the stochastic environment and/or on the contingent claim contract. In more general settings we have to resort to numerical techniques, but nevertheless we have first to prove the existence and to characterise the solution.

In a recent paper [3], using the semigroup theory, the authors provide a new set of conditions for the existence of the solution of European contingent claim contracts following the Black & Scholes model. The aim of this note is to show that this new approach can also be used to analyse equations describing European options based on multiple assets with transaction costs in a market model analogous to the Black & Scholes one.

In more details, let us consider a market in which $d+1$ *assets* are traded continuously. One of the assets, called *bond*, is risk-free and its price $X_0(t) \equiv X_0$ evolves according to the deterministic differential equation

$$dX_0(t) = rX_0\, dt, \quad X_0(0) = x_0, \tag{1}$$

where $r > 0$ is the corresponding *interest rate*. The other d assets, called *stocks*, are risky and we suppose (see also [7], [9]) that the evolution of their prices $X_i \equiv X_i(t)$, $i = 1, \ldots, d$, is modelled by the system of linear stochastic differential equations with constant coefficients

$$\begin{cases} dX_i = \mu_i X_i\, dt + \sigma_i X_i\, dW_i, \\ X_i(0) = x_i^0 \end{cases} \quad i = 1, \ldots, d, \tag{2}$$

where for every $i = 1, \ldots, d$ the coefficients μ_i and $\sigma_i > 0$ are the *mean rate of return* and the *volatility* of the i-th asset respectively, and $W_i \equiv W_i(t)$,

---

*This work has been partially supported by CNR, progetto strategico "Modelli e metodi per la matematica e l'ingegneria".

are Wiener processes whose correlation matrix $(\rho_{i,j})_{i,j=1}^d$ has constant entries and is non-singular.

Consider now an investor seeking for a *hedging strategy*, with a consumptionless self-financing portfolio, against a *call option* $C \equiv C(t, X_1, \ldots, X_d)$ based on the above risky assets, whose *payoff* P at the *maturity* T is

$$P \stackrel{def}{=} \max[\sum_{i=1}^d X_i(T) - S, 0], \qquad (3)$$

where S is the *strike price*. Then the investor's *wealth* at time t is

$$\Pi \stackrel{def}{=} \sum_{i=0}^d N_i X_i,$$

where $N_i \equiv N_i(t, X_0, X_1, \ldots, X_d)$ is the *number of shares* of the i-th asset, and $\Pi(0)$ is the *initial endowment*. In *continuous trading* and *absence of arbitrage*, the hedging strategy translates into the condition

$$\Pi(t) = C(t), \qquad (4)$$

at every instant t, and absence of *transaction costs* means

$$dN_0 \, X_0 + \sum_{i=1}^d dN_i \, X_i = 0. \qquad (5)$$

The further assumption that the portfolio has the same *risk profile* as the option yields the *standard hedging strategy conditions*

$$N_i = \frac{\partial C}{\partial X_i}, \quad i = 1, \ldots, d,$$

and it is well known that the above hypotheses imply:

$$\frac{\partial C}{\partial t} + \frac{1}{2} \sum_{i,j=1}^d \frac{\partial^2 C}{\partial X_i \partial X_i} \rho_{i,j} \sigma_i \sigma_j X_i X_j + r \sum_{i=1}^d \frac{\partial C}{\partial X_i} X_i - rC = 0, \qquad (6)$$

which is the celebrated Black & Scholes equation for a European option based on multiple assets. Equation (6) is completed with the terminal condition

$$C(T, X_1, \ldots, X_d) = P. \qquad (7)$$

The assumption of absence of cost in trading is clearly not verified in real markets. For this reason in these years many economists have formulated more realistic models. We consider here a generalisation of the Leland model [4] (see also [7]), assuming that for each risky asset a change dN_i, $i = 1, \ldots, d$ in the number of the asset units involves a cost dK_i, $i = 1, \ldots, d$ given by

$$dK_i = k_i(t) X_i \left(d \langle N_i \rangle \right)^{1/2},$$

where $d \langle N_i \rangle$ is the quadratic variation of the process N_i, and to make compatible continuous rehedging with finite transaction costs in finite time intervals, we suppose further that for every $i = 1, \ldots, d$

$$k_i(t) = \lambda_i \sqrt{dt},$$

where λ_i are suitable constants. In this case we have

$$dN_0 \, X_0 + \sum_{i=1}^{d} dN_i \, X_i = \sum_{i=1}^{d} \lambda_i X_i \left(\sum_{j,k=1}^{d} \sigma_j \sigma_k \rho_{j,k} X_j X_k \frac{\partial N_i}{\partial X_j} \frac{\partial N_i}{\partial X_k} \right)^{1/2} dt, \quad (8)$$

and it easily follows:

$$\frac{\partial C}{\partial t} + \frac{1}{2} \sum_{i,j=1}^{d} \frac{\partial^2 V}{\partial X_i \partial X_j} \rho_{i,j} \sigma_i \sigma_j X_i X_j + r \sum_{i=1}^{d} \frac{\partial C}{\partial X_i} X_i - rC$$

$$- \sum_{i=1}^{d} \lambda_i X_i \left(\sum_{j,k=1}^{d} \sigma_j \sigma_k \rho_{j,k} X_j X_k \frac{\partial N_i}{\partial X_j} \frac{\partial N_i}{\partial X_k} \right)^{1/2} = 0. \quad (9)$$

We study here Equations (6) and (9), applying classical tools of functional analysis and the generation results proved in [3].

2 Notations and Preliminary Material

Given a domain Ω of the Euclidean space $\mathbb{R}^d$, we expect the reader is familiar with the Lebesgue spaces $L^p(\Omega)$, where $1 \leq p \leq \infty$, and the Sobolev spaces $W^{n,p}(\Omega)$, $n \geq 1$. We denote by $C_b(\Omega)$ the subspace of of $L^\infty(\Omega)$ of all continuous functions and we denote by $C_b^n(\Omega)$ the subspace of $W^{n,\infty}(\Omega)$ of all continuously differentiable functions up to order n included. We will write

$\|u\|_{C_b(\Omega)}$ [resp. $\|u\|_{C_b^n(\Omega)}$] instead of $\|u\|_{L^\infty(\Omega)}$ [resp. $\|u\|_{W^{n,\infty}(\Omega)}$] whenever $u \in C_b(\Omega)$ [resp. $u \in C_b^n(\Omega)$]. We also expect the reader is familiar with the Hölder spaces $C^\alpha(\mathbb{R}_+^d)$, where $0 < \alpha < 1$, and $C^{n,\alpha}(\mathbb{R}_+^d)$, where ≥ 1.

Given any locally-integrable function $\xi : \Omega \to \mathbb{R}$ having positive essential infimum, we introduce now, for any $1 \leq p \leq \infty$, the weighted Lebesgue space $L_\xi^p(\Omega)$ as the space of all Borel-measurable functions u such that the product ξu belong to $L^p(\Omega)$, and the weighted Sobolev space $W_\xi^{n,p}(\Omega)$ as the spaces of all the Borel-measurable functions u such that $\xi u \in W^{n,p}(\Omega)$. These spaces will be equipped with the norms

$$\|u\|_{L_\xi^p(\Omega)} \stackrel{def}{=} \|\xi u\|_{L^p(\Omega)}, \text{ and } \|u\|_{W_\xi^{n,p}(\Omega)} \stackrel{def}{=} \|\xi u\|_{W^{n,p}(\Omega)},$$

respectively. Similarly, given any locally-integrable functions $\psi_i : \Omega \to \mathbb{R}$, $i = 1, \ldots, d$, and $\gamma : \Omega \to \mathbb{R}$ such that $\operatorname{ess\,inf} \gamma^2 > 0$, we introduce the weighted Sobolev space $W_{\psi,\gamma^2}^{1,p}(\Omega)$, as the space of all functions $u \in L_{\gamma^2}^p(\Omega)$ such that the products $\gamma \psi_i D_i u$ belong to $L^p(\Omega)$ for all $i = 1, \ldots, d$, and the weighted Sobolev space $W_{\psi,\gamma^2}^{2,p}(\Omega)$, as the space of all functions $u \in W_{\psi,\gamma^2}^{1,p}(\Omega)$ such that $\psi_i \psi_j D_{i,j} u \in L^p(\Omega)$ for all $i,j = 1, \ldots, d,$. For these spaces we provide the respective norms

$$\|u\|_{W_{\psi,\gamma^2}^{1,p}(\Omega)} \stackrel{def}{=} \|\gamma^2 u\|_{L^p(\Omega)} + \sum_{i=1}^d \|\gamma \psi_i D_i u\|_{L^p(\Omega)}$$

and

$$\|u\|_{W_{\psi,\gamma^2}^{2,p}(\Omega)} \stackrel{def}{=} \|u\|_{W_{\psi,\gamma^2}^{1,p}(\Omega)} + \sum_{i,j=1}^d \|\psi_j \psi_i D_{i,j} u\|_{L^p(\Omega)}.$$

Finally we introduce the weighted Hölder space $C_\psi^\alpha(\Omega)$ of all functions $u \in C_b(\Omega)$ satisfying

$$[u]_{\Omega,\alpha,\psi} \equiv \sum_{i=1}^d \sup_{x \in \Omega, \lambda > 0} \frac{|\psi_i(x+\lambda e_i) u(x+\lambda e_i) - \psi(x) u(x)|}{\lambda^\alpha} < \infty,$$

and the weigthted Hölder space $C_\psi^{n,\alpha}(\Omega)$ of all $u \in C_b^n(\Omega)$ such that, for every multiindex $\kappa \equiv (\kappa_1, \ldots, \kappa_d)$ of height n,

$$[D^\kappa u]_{\Omega,\alpha,\psi} \equiv \sum_{|\iota|=n} \sup_{x \in \Omega, \lambda > 0} \frac{|\Psi^{\iota/n}(x+\lambda e_i) D^\kappa u(x+\lambda e_i) - \Psi^{\iota/n}(x) D^\kappa u(x)|}{\lambda^\alpha} < \infty,$$

where $\iota \equiv (\iota_1, \ldots, \iota_d)$ is another multiindex and the symbol $\Psi^{\iota/n}$ is shorthand for $\psi_1^{\iota_1/n} \cdots \psi_d^{\iota_d/n}$. The latter spaces will be endowed with the respective norms

$$\|u\|_{C^\alpha_\psi(\Omega)} \stackrel{def}{=} \|u\|_{C_b(\Omega)} + [u]_{\Omega,\alpha,\psi}, \quad \text{and} \quad \|u\|_{C^{n,\alpha}_\psi(\Omega)} \stackrel{def}{=} \|u\|_{C^n_b(\Omega)} + \sum_{|\kappa|=n} [D^\kappa u]_{\Omega,\alpha,\psi}.$$

It is easy to see that all above-introduced spaces are Banach.

In the sequel we shall need some generation results which have been proven in [3] for a wider class of operators than we are concerning here.

Let A be the operator given by

$$Au \stackrel{def}{=} \frac{1}{2} \sum_{i,j=1}^d \rho_{i,j} \sigma_i \sigma_j x_i x_j D_{i,j} u + r \sum_{i=1}^d x_i D_i u - ru,$$

and, frow now on, let $\xi_\alpha = \left(1 + \sum_{i=1}^d x_i^2\right)^{-\alpha}$ for $\alpha \geq 0$ and $\psi = (x_1, \ldots, x_d)$. We have:

Theorem 1 *Assume $f \in L^p_{\xi_\alpha}(\mathbb{R}^n_+)$ where $2 \leq p < \infty$. Then there exists $\omega > 0$, depending only on p and α, such that the resolvent equation*

$$\lambda u - Au = f, \tag{10}$$

has a unique solution u for $\mathrm{Re}\,\lambda \geq \omega$. Moreover the following estimate holds:

$$|\lambda| \|u\|_{L^p_{\xi_\alpha}(\mathbb{R}^d_+)} + |\lambda|^{1/2} \|\xi_\alpha u\|_{W^{1,p}_{1,\psi}(\mathbb{R}^d_+)} + \|\xi_\alpha u\|_{W^{2,p}_{1,\psi}(\mathbb{R}^d_+)} \leq C \|f\|_{L^p_{\xi_\alpha}(\mathbb{R}^d_+)}, \tag{11}$$

where C is a positive constant independent of λ.

Assume $f \in L^\infty_{\xi_\alpha}(\mathbb{R}^d_+)$. Then there exists $\omega > 0$, depending only on α, such that Equation (10) has a unique solution u for $\mathrm{Re}\,\lambda \geq \omega$. Moreover the following estimate holds:

$$|\lambda| \|u\|_{L^\infty_{\xi_\alpha}(\mathbb{R}^d_+)} + |\lambda|^{1/2} \|\xi_\alpha u\|_{W^{1,\infty}_{1,\psi}(\mathbb{R}^d_+)} \tag{12}$$

$$+ |\lambda|^{d/(2p)} \sup_{x_0 \in \mathbb{R}^d_+} \sum_{i,j=1}^d \|\xi_\alpha \psi_i \psi_j D_{i,j} u\|_{L^p(B(x_0, |\lambda|^{-1/2}))} \leq C \|f\|_{L^\infty_{\xi_\alpha}(\mathbb{R}^d_+)},$$

where C is a positive constant independent of λ and $B(x_0, |\lambda|^{-1/2})$ denotes a ball of $\mathbb{R}^d_+$, centred in x_0 and having radius $|\lambda|^{-1/2}$.

Theorem 1 gives the desired generation results in $L^p_{\xi_\alpha}(\mathbb{R}^d_+)$ for $2 \leq p \leq \infty$.

3 Interpolation spaces

The characterisation of the interpolation spaces is an essential tool in applying the semigroup theory in concrete cases. Here we just recall the essential definitions, referring the reader to [5] for further details.

Let X be a Banach space endowed with the norm $\|\cdot\|$ and let $A : D(A) \to X$ be an operator which generates an analytic semigroup on X. For each $0 < \theta < 1$ we consider the interpolation space $V(\infty, \theta, D(A), X)$ as the space of all Borel-measurable functions $v : \mathbb{R}_+ \to X$ such that the maps

$$t \in \mathbb{R}_+ \mapsto v_\theta(t) \stackrel{def}{=} t^\theta v(t) \quad \text{and} \quad t \in \mathbb{R}_+ \mapsto w_\theta(t) \stackrel{def}{=} t^\theta v'(t),$$

belong to $L^\infty(\mathbb{R}_+; D(A))$ and $L^\infty(\mathbb{R}_+; X)$ respectively, where $D(A)$ is a Banach space under the graph norm. The space $V(\infty, \theta, D(A), X)$ is also Banach if endowed with the norm

$$\|v\|_{V(\infty,\theta,D(A),X)} \stackrel{def}{=} \|v_\theta\|_{L^\infty(\mathbb{R}_+;D(A))} + \|w_\theta\|_{L^\infty(\mathbb{R}_+;X)}.$$

We write e^{tA} for the analytic semigroup generated by $A : D(A) \to X$, and we denote $D_A(\theta, \infty)$ the further interpolation space between D_A and X, defined in terms of the semigroup e^{tA} as the manifold of all $x \in X$ such that the function

$$t \in \mathbb{R}_+ \mapsto v(t) \stackrel{def}{=} t^{1-\theta} \|A e^{tA} x\| \tag{13}$$

is bounded. The space $D_A(\theta, \infty)$ is Banach under the norm

$$\|u\|_{D_A(\theta,\infty)} \stackrel{def}{=} \|x\| + \sup_{t \in \mathbb{R}_+} v(t),$$

We focus now our attention on the operator $A_{\infty,\xi_\alpha} : D(A_{\infty,\xi_\alpha}) \to L^\infty_{\xi_\alpha}(\mathbb{R}^d_+)$ given by

$$A_{\infty,\xi_\alpha} u \stackrel{def}{=} Au$$

for every $u \in D(A_{\infty,\xi_\alpha})$, which is the linear submanifold of $L^\infty_{\xi_\alpha}(\mathbb{R}^d_+)$ whose elements u, owing to (13), satisfy the following conditions:

$$\xi_\alpha \psi_i \psi_j D_{i,j} u \in L^p_{loc}(\mathbb{R}^d_+), \quad \text{for all } i,j = 1, \ldots, d$$
$$\psi_i D_i u \in L^\infty_{\xi_\alpha}(\mathbb{R}^d_+), \quad \text{for every } i = 1, \ldots, d,$$
$$Au \in L^\infty_{\xi_\alpha}(\mathbb{R}^d_+).$$

We are finally in a position to establish the main result of this section

Theorem 2 *We have:*

$$D_{A_{\infty,\xi_\alpha}}(\theta,\infty) = \begin{cases} L^\infty_{\xi_\alpha}(\mathbb{R}^d_+) \cap C^{2\theta}_{\psi^{2\theta}}(\mathbb{R}^d_+) & \text{if } 0 < \theta < 1/2 \\ L^\infty_{\xi_\alpha}(\mathbb{R}^d_+) \cap C^{1,2\theta-1}_{\psi^{2\theta}}(\mathbb{R}^d_+) & \text{if } 1/2 < \theta < 1; \end{cases}$$

where $\psi^\theta \equiv (x_1^\theta, \ldots, x_n^\theta)$.

Proof. For each $x_0 \equiv (x_1^0, \ldots, x_d^0) \in \mathbb{R}^d_+$ having non null entries we introduce the change of variables $T_{x_0} : \mathbb{R}^d_+ \to \mathbb{R}^d_+$ given by

$$T_{x_0}(x) \stackrel{def}{=} ((x_1 - x_1^0)/x_1^0, \ldots, (x_d - x_d^0)/x_d^0).$$

Then, applying a standard localisation procedure, we reduce ourselves to the non-degenerate elliptic operator case. This allow us to apply locally a Lunardi result (see [6, Theor. 2.10, p.306]) and to obtain a local characterisation of the interpolation spaces. Finally, changing the variable back and following the pattern scheme in [8, sec. 5], our theorem follows. □

4 Application to Financial Equations

In this section we employ our interpolation results to obtain some regularity properties of the solutions of Equations (6) and (9).

Our first result follows from a straightforward application of well known theorems on parabolic equations (see for instance [5]).

Theorem 3 *Let v be the solution of Equation (6) with the terminal condition*

$$v(T,x) = \max[\sum_{i=1}^n x_i - S, 0], \tag{14}$$

Then for each $0 < \beta < 1$ the function v belongs to $C([0,T]; C^\beta_{\psi^\beta,\xi_{1/2}}(\mathbb{R}^d_+)) \cap C^1([0,T]; C^\beta_{\psi^\beta,\xi_{1/2}}(\mathbb{R}^d_+)) \cap C([0,T]; C^{2,\beta}_{\psi^{2+\beta},\xi_{1/2}}(\mathbb{R}^d_+))$.

Next, we want establish an useful estimate for the solution v of Equation (9) with the same terminal condition (14). To this end, we need a preliminary result which follows from the celebrated Dore-Venni theorem (see [2]).

Lemma 1 Let $p > 2$, $\alpha \geq \frac{p+n+1}{p}$ and let $v : [0,T] \times \mathbb{R}_+^d \to \mathbb{C}$ be the solution of the backward problem

$$\begin{cases} v_t = \frac{1}{2} \sum_{i,j=1}^d \rho_{i,j} \sigma_i \sigma_j x_i x_j D_{i,j} v + r \sum_{i=1}^d x_i D_i v - rv + g \\ v(0,x) = max[\sum_{i=1}^d x_i - E, 0] \end{cases}$$

where $g \equiv g(t)$ satisfies $\|\sqrt{t} g \xi_\alpha\|_{L^p([0,T] \times \mathbb{R}_+^d)} < \infty$. Then we have

$$\|v\|_{W_{\xi_\alpha}^{2,p,\sqrt{t}}([0,T] \times \mathbb{R}_+^d)} \leq C(\|\sqrt{t} g \xi_\alpha\|_{L^p([0,T] \times \mathbb{R}_+^d)} + \|v(x,0) \xi_\alpha\|_{W^{1,p}(\mathbb{R}_+^d)}), \quad (15)$$

where

$$\|v\|_{W_{\xi_\alpha}^{2,p,\sqrt{t}}([0,T] \times \mathbb{R}_+^d)} \stackrel{def}{=} \sum_{i,j=1}^d \|\sqrt{t} x_i x_j D_{i,j} v \xi_\alpha\|_{L^p([0,T] \times \mathbb{R}_+^d)}$$
$$+ \sum_{i=1}^d \|x_i D_i v \xi_\alpha\|_{L^p([0,T] \times \mathbb{R}_+^d)} + \|v \xi_\alpha\|_{L^p([0,T] \times \mathbb{R}_+^d)}$$

and $C > 0$ is a constant depending only on p and α.

We have finally:

Theorem 4 Let $p > 2$ and $\alpha \geq \frac{p+n+1}{p}$. Then there exists a constant $\Lambda > 0$, depending only on p and α, such that if $\sum_{i=1}^n \lambda_i^2 \leq \Lambda$, then (9) has a unique solution $v : [0,T] \times \mathbb{R}_+^d \to \mathbb{C}$. Moreover we have

$$\|u\|_{W_{\xi_\alpha}^{2,p,\sqrt{t}}([0,T] \times \mathbb{R}_+^d)} \leq C \|u(0,x) \xi_\alpha\|_{W^{1,p}(\mathbb{R}_+^d)}, \quad (16)$$

where C is a constant depending only on p and α.

Proof. Indeed, let us introduce the map T from $W_{\xi_\alpha}^{2,p,\sqrt{t}}([0,T] \times \mathbb{R}_+^d)$ into itself given by

$$Tv \stackrel{def}{=} w,$$

where w is a solution of the backward problem

$$\begin{cases} w_t = \frac{1}{2} \sum_{i,j=1}^d \rho_{i,j} \sigma_i \sigma_j x_i x_j D_{i,j} w + r \sum_{i=1}^d x_i D_i w - rw \\ - \sum_{i=1}^d \lambda_i x_i \left(\sum_{j,k=1}^n \sigma_j \sigma_k \rho_{j,k} x_j x_k D_{i,j} v D_{i,k} \right)^{1/2} = 0 \\ v(0,x) = max[\sum_{i=1}^d x_i - E, 0] \end{cases}.$$

Then, choosing Λ small enough, by Estimate (15), T turns out to be contractive, and it is easy to verify that the unique fixed point is the solution of (9) and satisfies Estimate (16). □

References

[1] F. Black, M. Scholes: The pricing of options and corporate liabilities. *Journal of Political Economy* 81, 637–654 (1973).

[2] G. Dore, A. Venni: On the closedness of the sum of two closed operators. *Math. Z.* 196, 189-201 (1987).

[3] F. Gozzi, R. Monte, V. Vespri: *Generation of analytic semigroups for degenerate elliptic operators arising in financial mathematics* Preprint Universitá di Pisa.

[4] H.E. Leland: Option pricing and replication with transaction costs. *Journal of Finance*, 5, 1283–1301 (1985).

[5] A. Lunardi *Analytic Semigroups and Optimality Regularity in Parabolic Problems*. Birkhäuser Verlag, Berlin (1995).

[6] A. Lunardi Interpolation Spaces Between Domains of Elliptic Operators and Spaces of Continuous Functions with Applications to Nonlinear Parabolic Equations *Math. Nachr.*, 121, 295-318 (1985)

[7] G. Pacelli, M.C. Recchioni, F. Zirilli: *A hybrid method for pricing European options based on multiple assets with transaction costs*. Preprint Universitá di Roma.

[8] V. Vespri: Analytic semigroups, Degenerate Elliptic Operators and Applications to Nonlinear Cauchy Problems. *Ann. Mat. Pura e Appl.*, 145, 353-388 (1989).

[9] P. Wilmott, J. Dewynne, S. Howison: *Option Pricing: Mathematical Models and Computation*. Oxford Financial Press, Oxford, UK (1995).

JOHN CAGNOL AND JEAN-PAUL ZOLÉSIO
Hidden shape derivative in the wave equation

1 Introduction

The shape differentiability is well known for elliptic and parabolic problems where the proof is obtained from the implicit function theorem. However, in the hyperbolic situation, we do not have the "nice" isomorphism between the data and the solution anymore, hence that method fails. Nevertheless we use a more technical approach to prove a comparable result.

Let $D \subset \mathbb{R}^N$ be a domain with ∂D piecewise C^2 and let Ω be an open bounded domain in D of class C^2. We note $I = [0,T]$ with $T < \infty$ and $\Gamma = \partial \Omega$, $\Sigma = I \times \Gamma$, $Q = I \times \Omega$. Let K be a coercive symmetric matrix with $K_{i,j} \in L^\infty(I, W^{2,\infty}(D)) \cap W^{1,\infty}(I, L^\infty(D))$. We note λ the coercivity constant and $\kappa = \max_{i,j} \|\partial_t K_{i,j}\|_{L^\infty(I \times D)}$. We request that $\kappa < \frac{\lambda}{2T^2}$.

We consider the problem

$$\begin{cases} \partial_t^2 y - \operatorname{div}(K\nabla y) = f & \text{on } Q \\ y = 0 & \text{on } \Sigma \\ y(0) = \Phi_0 & \text{on } \Omega \\ \partial_t y(0) = \Phi_1 & \text{on } \Omega \end{cases} \qquad (1)$$

where $\operatorname{supp} \Phi_0 \subset\subset \Omega$ and $\operatorname{supp} \Phi_1 \subset\subset \Omega$ and either $(H1)$ or $(H2)$.

$(H1) \quad f \in L^1(I, H^1(D)) \cap W^{1,1}(I, L^2(D)), \quad \Phi_0 \in H^2(D), \quad \Phi_1 \in H^1(D)$

$(H2) \quad f \in L^1(I, L^2(D)), \quad \Phi_0 \in H^1(D), \quad \Phi_1 \in L^2(D)$

We refer to [2, theorem 2.3] or to [1, theorem 2.2] and [1, section 4] for the existence and regularity of the solution to (1). Let n is the unit outward normal derivative to Γ, then

$(H1) \Rightarrow y \in C(I, H^2(\Omega)) \cap C^1(I, H^1(\Omega)) \cap C^2(I, L^2(\Omega))$ and $\dfrac{\partial y}{\partial n} \in H^1(\Sigma)$

$(H2) \Rightarrow y \in C(I, H^1(\Omega)) \cap C^1(I, L^2(\Omega))$ and $\dfrac{\partial y}{\partial n} \in L^2(\Sigma)$

We will denote by y_s the solution to the problem in the perturbed domain Q_s and by $\tilde{\phi}$ the extension of ϕ by 0 to D. We shall prove the

Theorem 1 *Under (H1) there exists $Y \in L^2(I \times D)$ such that*

$$\frac{\widetilde{y_s} - \widetilde{y}}{s} \to Y \text{ in } L^2(I \times D) \text{ as } s \to 0$$

Moreover $y' = Y|_Q$ belongs to $\in C(I, H_0^1(\Omega)) \cap C^1(I, L^2(\Omega))$ and is solution to the problem

$$\begin{cases} \partial_t^2 y' - \mathrm{div}\,(K \nabla y') = 0 & \text{on } Q \\ y' = -\frac{\partial y}{\partial n} V(0).n & \text{on } \Sigma \\ y'(0) = 0 & \text{on } \Omega \\ \partial_t y'(0) = 0 & \text{on } \Omega \end{cases} \qquad (2)$$

When f is less regular we will prove the shape derivative continues to exist in $C(I, L^2(\Omega))$, using the hidden regularity provided by [1].

Theorem 2 *Under (H2) and $K = I$ there exists $Y \in L^2(I \times D)$ such that*

$$\frac{\widetilde{y_s} - \widetilde{y}}{s} \rightharpoonup Y \text{ weakly in } \sigma\, L^2(I \times D) \text{ as } s \to 0$$

moreover $y' = Y|_Q$ belongs to $C(I, L^2(\Omega)) \cap C(I, H^{-1}(\Omega))$ and is solution to (2).

Throughout this paper, we will denote by $\phi(t)$ the function $\phi(t, \cdot)$ when ϕ is a function of both x and t, and $P = \frac{\partial^2}{\partial t^2} - \mathrm{div}\,(K \nabla)$.

2 Transformation of Domains

2.1 Definitions and Transported Problem

Consider a non-negative real S. For each $s \in [0, S[$, let T_s be a one-to-one mapping from $\bar{D}$ onto $\bar{D}$ such that $T_0 = I$ and the functions $(s, x) \mapsto T_s(x)$ and $(s, x) \mapsto T_s^{-1}(x)$ belong to $C^1([0; S[, C^2(\bar{D}; \bar{D}))$. We will also denote by T_s the function $(t, x) \mapsto (t, T_s(x))$. This defines a family of domains $\{\Omega_s\}$ given by $\Omega_s = T_s(\Omega)$. We will respectively denote by Γ_s, Q_s and Σ_s the boundary of Ω_s, the set $I \times \Omega_s$ and its boundary. The speed vector field $V(s)$ at the point x has the form $V(s)(x) = (\frac{\partial}{\partial s} T_s) \circ T_s^{-1}(x)$. The vector field defined as $V(s, x) = V(s)(x)$ satisfies $V \in C([0; S[; C^2(\bar{D}, \mathbb{R}^N))$. For every $s \in [0; S[$ we consider the problem

$$P(y_s) = f \text{ on } Q_s, \quad y_s = 0 \text{ on } \Sigma_s, \quad y_s(0) = \Phi_0 \text{ on } \Omega_s, \quad \partial_t y_s(0) = \Phi_1 \text{ on } \Omega_s \qquad (3)$$

If S is small enough we have $\mathrm{supp}\,\Phi_0 \subset\subset \Omega_s$ and $\mathrm{supp}\,\Phi_1 \subset\subset \Omega_s$. From now S will be chosen to satisfy this condition. This is not restrictive since we are interested in the situation as s tends to 0.

We will call (3) the *transported problem*, and y_s the *transported solution*. The existence and the regularity of the transported solution is obtained, as well, in [1]. Let n_s is the unit outward normal derivative to Γ_s, under (H1) we have

$$y_s \in C(I, H^2(\Omega_s)) \cap C^1(I, H^1(\Omega_s)) \cap C^2(I, L^2(\Omega_s)) \quad \text{and} \quad \frac{\partial y_s}{\partial n_s} \in H^1(\Sigma_s)$$

We note $y^s = y_s \circ T_s^{-1}$, $w_s = \frac{1}{s}(y_s - y \circ T_s^{-1})$ and $w^s = w_s \circ T_s$. We obtain

$$w_s \in C(I, H^2(\Omega_s)) \cap C^1(I, H_0^1(\Omega_s)) \cap C^2(I, L^2(\Omega_s))$$

$$w^s \in C(I, H^2(\Omega)) \cap C^1(I, H_0^1(\Omega)) \cap C^2(I, L^2(\Omega))$$

Corresponding results hold with $(H2)$.

2.2 Properties of the Transformation and Limits as $s \to 0$

Let $\gamma_s = |\det(DT_s)|$ and $K_s = \gamma_s(DT_s)^{-1}(K \circ T_s)(^*DT_s^{-1})$. We note $P_s = \frac{\partial^2}{\partial t^2} - \text{div}(K_s \nabla)$.

Proposition 1 *Let Ψ be in $\mathcal{D}'(\Omega)$ then $(P\Psi) \circ T_s = P_s(\Psi \circ T_s)$*

Proposition 2 *Let $\varepsilon_K = \frac{1}{2}(DV(0).K + {}^*(DV(0).K))$ and DK be the third-order tensor $(\partial_{x_k} K_{i,j}(t,x))_{i,j,k}$ we have, as $s \to 0$*

$$K_s \to K \quad \text{in} \quad L^\infty(I, W^{1,\infty}(\Omega, \mathcal{M}_N(\mathbb{R})))$$

$$\frac{K_s - K}{s} \to (\text{div } V(0)).K - 2\varepsilon_K + DK.V(0) \quad \text{in} \quad L^\infty(I, W^{1,\infty}(\Omega, \mathcal{M}_N(\mathbb{R})))$$

The next proposition can be proved using arguments similar to [3, prop. 2.32] and [3, prop. 2.33].

Proposition 3 *Assume $(s,x) \mapsto G_s \in C([0;S], H^1(\mathbb{R}^N)) \cap C^1([0;S], L^2(\mathbb{R}^N))$ then $s \mapsto G_s \circ T_s$ is differentiable in $L^2(\mathbb{R}^N)$ and the derivative is given by*

$$\left(\frac{\partial}{\partial s}(G_s \circ T_s)\right)_{s=0} = \left.\frac{\partial G_s}{\partial s}\right|_{s=0} + \nabla G.V(0)$$

Corollary 1 *Assume $G \in H^1(\mathbb{R}^N)^N$ then $s \mapsto G \circ T_s$ is differentiable in $L^2(\mathbb{R}^N)^N$ and the derivative is given by $(\frac{\partial}{\partial s}(G \circ T_s))_{s=0} = DG.V(0)$.*

Remark 1 *When the function T_s is replaced by $(t,x) \mapsto (t, T_s(x))$ the results of this section hold with $I \times \bar{D}$ instead of $\bar{D}$ and $\mathbb{R}^{N+1}$ instead of $\mathbb{R}^N$.*

3 Material Derivative for smooth data ($H1$)

3.1 Preliminaries

Lemma 1 Let $\Psi \in C(I, H^2(\Omega)) \cap C^1(I, H_0^1(\Omega)) \cap C^2(I, L^2(\Omega))$. We note
$$a = \|P\Psi\|_{L^1(I,L^2(\Omega))}$$
$$b = \kappa\|\nabla\Psi\|_{L^1(I,L^2(\Omega))}^2 + \int_\Omega (K(0,x)\nabla\Psi(0,x).\nabla\Psi(0,x) + (\partial_t\Psi)^2(0,x))dx$$
then
$$\|\partial_t\Psi\|_{L^\infty(I,L^2(\Omega))} \leq a + \sqrt{a^2 + 2b} \tag{4}$$

$$\|\Psi\|_{L^\infty(I,H_0^1(\Omega))} \leq \frac{\sqrt{2}}{\sqrt{\lambda}}\sqrt{a^2 + b + a\sqrt{a^2+2b}} \tag{5}$$

Proof – Let $E(\Psi,t) = \frac{1}{2}\int_\Omega (K(t,x)\nabla\Psi(t,x).\nabla\Psi(t,x) + (\partial_t\Psi)^2(t,x))dx$. We compute the derivative of E with respect to t. Thru a technical process and an integration, we obtain

$$E(\Psi,\tau) \leq \|\partial_t\Psi\|_{L^\infty(I,L^2(\Omega))}\|P\Psi\|_{L^1(I,L^2(\Omega))} + \kappa\|\nabla\Psi\|_{L^1(I,L^2(\Omega))}^2 + E(\Psi,0) \tag{6}$$

From $\frac{1}{2}\|\partial_t\Psi(\tau)\|_{L^2(\Omega)}^2 \leq E(\Psi,\tau)$ and (6) we get (4). From $\frac{\lambda}{2}\|\Psi(\tau)\|_{H_0^1(\Omega)}^2 \leq E(\Psi,\tau)$, (6) and (4), we obtain (5). ∎

Let $e_s = \int_\Omega K_s(0,x)\nabla w^s(0,x).\nabla w^s(0,x)\,dx$ and $\delta_s = \max_{i,j}\|(DT_s^{-1}{}^*DT_s^{-1})_{i,j}\|_{L^\infty(\Omega)}$. We note $a_s = \|\sqrt{\gamma_s}P_s w^s\|_{L^1(I,L^2(\Omega))}$ and $b_s = \kappa\delta_s\|\sqrt{\gamma_s}w^s\|_{L^1(I,H_0^1(\Omega))}^2 + e_s$.

Proposition 4 $\forall \eta \in]0;1[, \exists \varepsilon^* > 0, \forall \varepsilon \in]0;\varepsilon^*[, \exists \alpha \in]1-\eta, 1+\eta[, \forall s \in [0;\varepsilon[$,

$$\alpha\|w^s\|_{L^\infty(I,H_0^1(\Omega))} \leq \frac{\sqrt{2}}{\sqrt{\lambda}}\sqrt{a_s^2 + b_s + \sqrt{a_s^2 + 2b_s}}$$

Proof – Lemma 1 applies to w_s and gives (5) where a, b and e are replaced by $\bar{a}_s = \|Pw_s\|_{L^1(I,L^2(\Omega))}$, $\bar{e}_s = \int_{\Omega_s} K(0,x)\nabla w_s(0,x).\nabla w_s(0,x)\,dx$ and $\bar{b}_s = \kappa\|w_s\|_{L^1(I,H_0^1(\Omega))} + \bar{e}_s$. The change of variables $x = T_s(X)$ in $\bar{a}_s$ yields $\bar{a}_s = a_s$. Similarly we obtain $\bar{e}_s = e_s$ and $\bar{b}_s \leq b_s$. On the other hand we prove $\forall \eta \in]0;1[, \exists \varepsilon^* > 0, \forall \varepsilon \in]0;\varepsilon^*[, \exists \alpha \in]1-\eta, 1+\eta[, \|w_s\|_{L^\infty(I,H_0^1(\Omega))} \geq \alpha\|w^s\|_{L^\infty(I,H_0^1(\Omega_s))}$. ∎

Similarly, we have the

Proposition 5 $\forall \eta \in]0;1[, \exists \varepsilon^* > 0, \forall \varepsilon \in]0;\varepsilon^*[, \exists \alpha \in]1-\eta, 1+\eta[, \forall s \in [0;\varepsilon[$,

$$\alpha\|\partial_t w^s\|_{L^\infty(I,L^2(\Omega))} \leq a_s + \sqrt{a_s^2 + 2b_s}$$

Lemma 2 $P_s(w^s) = \frac{f\circ T_s - f}{s} + \gamma_s^{-1}\text{div}\left(\frac{K-K_s}{s}\nabla y\right) + \frac{1-\gamma_s^{-1}}{s}\text{div}(K\nabla y)$

Proof – We apply proposition 1 to $P_s y^s$. ∎

3.2 Existence of w

Let U be an appropriate neighborhood of 0.

Lemma 3 *When $s \in U$, we have $a_s \leq M_a$ where M_a is independent of s*

Proof – $\frac{a_s}{s} \leq \|\gamma_s\|_{L^\infty(Q)}^{\frac{1}{2}} \|P_s w^s\|_{L^1(I, L^2(\Omega))}$. Since $\|\gamma_s\|_{L^\infty(Q)} \to 1$ as $s \to 0$ we just need to prove that $P_s w^s$ converges in $L^1(I, L^2(\Omega))$ as $s \to 0$. We compute the limit of each term of the expression given in lemma 2. ∎

Lemma 4 *When $s \in U$, we have $e_s \leq M_e$ where M_e is independent of s*

Proof – The limit of $\nabla w^s(0) = \frac{1}{s}(\,^*DT_s \nabla \Phi_0 \circ T_s - \nabla \Phi_0)$ as $s \to 0$ is $DV(0).\nabla \Phi_0 - D\nabla \Phi_0.V(0)$ in $L^2(\Omega)$ as $s \to 0$. ∎

Proposition 6 $\|w^s\|_{L^\infty(I, H_0^1(\Omega))}$ *is bounded when $s \in U$.*

Proof – We have $b_s \leq \kappa \delta_s \|\gamma_s\|_{L^\infty(Q)} T^2 \|w^s\|_{L^\infty(I, H_0^1(\Omega))}^2 + M_e$. Let us note $X = \|w^s\|_{L^\infty(I, H_0^1(\Omega))}$. Since $\lim_{s=0} \delta_s \|\gamma_s\|_{L^\infty(Q)} = 1$ there exist a non negative real C independent of s such that $b_s \leq \kappa C T^2 X^2 + M_e$. Proposition 4 yields

$$\alpha X \leq \frac{\sqrt{2}}{\sqrt{\lambda}} \left(M_a^2 + \kappa C T^2 X^2 + M_e + M_a \sqrt{M_a^2 + 2\kappa C T^2 X^2 + 2M_e} \right)^{\frac{1}{2}} \quad (7)$$

The condition $\kappa < \frac{\lambda}{2T^2}$ and (7) yield X is bounded. ∎

Corollary 2 *When $s \in U$ we have $b_s \leq M_b$ where M_b is independent of s*

Proposition 7 $\|w^s\|_{H^1(Q)}$ *is bounded and $\exists w \in L^\infty(I, H_0^1(\Omega)) \cap W^{1,\infty}(I, L^2(\Omega))$ s.t.*

   i) $w^{n_k} \rightharpoonup w$ *weakly σ $H^1(Q)$.*

   ii) $w^{n_k} \to w$ *strongly $L^2(Q)$.*

Proof – $\|w^s\|_{H^1(Q)}$ is bounded because of lemma 3 and corollary 2. The existence of w in $L^\infty(I, H_0^1(\Omega)) \cap W^{1,\infty}(I, L^2(\Omega))$ and (i) derives. So does (ii). ∎

3.3 Uniqueness of w

For each sequence (s_n) of the corollary 7 there is a function w. The aim of this section is to prove the

Proposition 8 $w \in L^\infty(I, H_0^1(\Omega)) \cap C(I, H^1(\Omega)) \cap C^1(I, L^2(\Omega))$ and is unique.

The three next lemmae will show that w satisfies (8), where $F \in L^2(Q)$.

$$P(w) = F \text{ on } Q, \quad w = 0 \text{ on } \Sigma, \quad w(0) = \Phi_0.V(0) \text{ on } \Omega, \quad \partial_t w(0) = \Phi_1.V(0) \text{ on } \Omega \quad (8)$$

Remark 2 $w = 0$ on Σ is a consequence of corollary 7.

Lemma 5 For any sequence (s_n), $w(0) = \Phi_0.V(0)$.

Proof – $\frac{y^s - y}{s} = \frac{\Phi_0 \circ T_s^{-1} - \Phi_0}{s}$ on Ω. Proposition 3 applies on the weak formulation of that equality, it follows $\forall \phi \in C_0^\infty(\Omega), \int_Q w\phi = \int_Q \Phi_0.V(0)\phi$ therefore $w = \Phi_0.V(0)\phi$ on Ω. ∎

The same method can be used to prove the

Lemma 6 For any sequence (s_n), $\partial_t w(0) = \Phi_1.V(0)$.

Lemma 7 There exists $F \in L^2(Q)$ such that for any sequence (s_n), $Pw = F$

Proof – Let $\psi \in C_0^\infty(Q_s)$, we perform the change of variable $X = T_s(x)$ in $\int_{Q_s} P(y_s)\psi - f\psi = 0$. It follows

$$\forall \phi \in C_0^\infty(Q), \int_Q -\gamma_s \partial_t y^s \partial_t \phi + K_s \nabla y^s \nabla \phi - \gamma_s f \circ T_s \phi = 0 \quad (9)$$

On the other hand

$$\forall \phi \in C_0^\infty(Q), \int_Q -\partial_t y \partial_t \phi + K \nabla y \nabla \phi - f\phi = 0 \quad (10)$$

Subtracting (10) from (9) yields $\forall \phi \in C_0^\infty(Q)$

$$-\int_Q \gamma_s \partial_t \left(\frac{y^s - y}{s}\right) \partial_t \phi - \int_Q \frac{\gamma_s - 1}{s} \partial_t y \partial_t \phi + \int_Q K_s \nabla \left(\frac{y^s - y}{s}\right) \nabla \phi \\ + \int_Q \frac{K_s - K}{s} \nabla y \nabla \phi - \int_Q \gamma_s \left(\frac{f \circ T_s - f}{s}\right) \phi - \int_Q \frac{\gamma_s - 1}{s} f\phi = 0 \quad (11)$$

Let $F = -\text{div}(V(0))\partial_t^2 y + \text{div}((\text{div}(V(0))K - 2\varepsilon_K + DK.V(0))\nabla y) + \nabla f.V(0) + \text{div}(V(0))f$ then (11) yields $\forall \phi \in C_0^\infty(Q), \int_Q (Pw)\phi - F\phi = 0$ thus $Pw = F$ on Q and F belongs to $L^2(Q)$ as each term of F is in $L^2(Q)$. ∎

w is a solution to (8). We refer to [1, theorem 2.1] and [1, section 4] to prove that w is unique and belongs to $C(I, H^1(\Omega)) \cap C^1(I, L^2(\Omega))$. This proves proposition 8.

3.4 Definition of the Material Derivative

Proposition 9 *There exists $\dot{y} \in \cap C(I, H^1(\Omega)) \cap C^1(I, L^2(\Omega))$ such that*

$$\left\| \frac{y^s - y}{s} - \dot{y} \right\|_{L^2(I \times D)} \to 0 \ as \ s \to 0$$

$\dot{y}$ *is solution to the problem*

$$P(\dot{y}) = F \ on \ Q, \ \dot{y} = 0 \ on \ \Sigma, \ \dot{y}(0) = \Phi_0.V(0) \ on \ \Omega, \ \partial_t \dot{y}(0) = \Phi_1.V(0) \ on \ \Omega \quad (12)$$

Definition 1 $\dot{y}$ *is the* material derivative *of y.*

4 Shape Derivative for smooth data ($H1$)

4.1 Definition of the Shape Derivative

Let $\tilde{y}$, $\tilde{y}_s$ and $\widetilde{y^s}$ be the extensions by 0 of y, y_s and y^s to D.

Proposition 10 $\frac{\tilde{y}_s - \tilde{y}}{s} \to Y$ *in $L^2(I \times D)$ as s tends to 0.*

Proof – One can write $\frac{\tilde{y}_s - \tilde{y}}{s} = \frac{\tilde{y}_s - \widetilde{y^s}}{s} + \frac{\widetilde{y^s} - \tilde{y}}{s}$. Proposition 3 yields $\frac{\tilde{y}_s - \widetilde{y^s}}{s} \to \nabla \tilde{y}.V(0)$ in $L^2(D)$ as $s \to 0$. Moreover from the definition of $\dot{y}$ we have $\frac{\widetilde{y^s} - \tilde{y}}{s} \to \tilde{\dot{y}}$ in $L^2(Q)$ as s tends to 0. ∎

Definition 2 $y' = Y|_Q$ *is the shape derivative of y.*

Corollary 3 $y' = \dot{y} - \nabla y.V(0)$ *on Q*

Notation 1 *We denote by y'_σ the shape derivative in σ. ($y' = y'_0$.)*

4.2 Characterization of the shape derivative

Lemma 8 *For each $s \in [0; S[$ let F_s be $\in L^1(Q_s)$. We request that $s \mapsto F_s(t, x)$ is differentiable for all $(t, x) \in D$. Let $F = F_0$ then*

$$\frac{\partial}{\partial s} \left(\int_{Q_s} F_s(x, t) \, dx \, dt \right)_{s=0} = \int_Q F'_s(x, t)|_{s=0} \, dx \, dt + \int_\Sigma F(x, t) \langle V(0), n \rangle \, d\Gamma \, dt$$

Proposition 11 $P(y') = 0$ *on Q*

Proof – $P(y_s) = f$ on Q_s therefore $\forall \phi \in C_0^\infty(D)$, $\frac{\partial}{\partial s}\int_{Q_s} \partial_t^2 y_s \phi - \text{div}\,(K\nabla y_s)\phi - f\phi = 0$. Lemma 8 yields for all $\phi \in C_0^\infty(D)$

$$\int_Q \left(\partial_t^2 y_s \phi - \text{div}\,(K\nabla y_s)\phi - f\phi\right)'_{s=0} + \int_\Sigma \left(\partial_t^2 y \phi - \text{div}\,(K\nabla y)\phi - f\phi\right)\langle V(0), n\rangle = 0$$

but $Py = f$ gives $\partial_t^2 y\phi - \text{div}\,(K\nabla y)\phi - f\phi = 0$ hence $\forall \phi \in C_0^\infty(D)$, $\int_Q \partial_t^2 y'\phi - \text{div}\,(K\nabla y')\phi = 0$ this leads to $P(y') = 0$ on Q. ∎

Corollary 4 *The shape derivative y' is solution to*

$$P(y') = 0 \text{ on } Q,\ y' = -\frac{\partial y}{\partial n}V(0).n \text{ on } \Sigma,\ y'(0) = 0 \text{ on } \Omega,\ \partial_t y'(0) = 0 \text{ on } \Omega \quad (13)$$

Proof – We have $y = 0$ on Σ hence $y' = -\nabla y.V(0)$ on Σ. Moreover $\dot{y}(0) = \Phi_0.V(0)$ and $\Phi_0.V(0) - \nabla y(0).V(0) = 0$ on Ω. This leads to $y'(0) = 0$. The same method applies to prove that $\partial_t y'(0) = 0$. ∎

Remark 3 *With the very same method, one can prove that y'_σ is solution to*

$$P(y'_\sigma) = 0 \text{ on } Q_\sigma,\ y'_\sigma = -\frac{\partial y_\sigma}{\partial n_\sigma}V(\sigma).n_\sigma \text{ on}\Sigma_\sigma,\ y'_\sigma(0) = 0 \text{ on } \Omega_\sigma,\ \partial_t y'_\sigma(0) = 0 \text{ on } \Omega_\sigma$$

Proposition 10 and corollary 4 prove theorem 1.

5 Shape Derivative for non smooth data ($H2$)

We suppose in this section that $f \in L^1(I, L^2(\Omega))$ and $K = I$. According to [1, theorem 2.1] and [1, section 4] we have $\frac{\partial y}{\partial n} \in L^2(\Sigma)$ thus using [1, theorem 2.3] and [1, section 4] on (13) we obtain that y' exists, is unique, and belongs to $C(I, L^2(\Omega)) \cap C^1(I, H^{-1}(\Omega))$. The aim of this section is to prove that y' is the shape derivative of the solution to (1).

[1, theorem 2.1] and [1, section 4] say the solution to (1) exists, is unique and belongs to $C(I, H^1(\Omega)) \cap C^1(I, L^2(\Omega))$. Let y be this solution.

We denote by ϕ a function in $L^2(I, L^2(D))$. Let us introduce the functions g and $\bar{g}$

$$g(s) = \int_{Q_s} y_s \phi\, dx\, dt \quad \text{and} \quad \bar{g}(s) = \int_{Q_s} y'_s \phi\, dx\, dt$$

Proposition 12 $\bar{g}(0) = g'(0)$

Proof – We first prove in section 5.1 that g is absolutely continuous, then we prove in section 5.2 that $\bar{g}$ is continuous in 0. ∎

Theorem 2 is a consequence of proposition 12 and corollary 4.

5.1 Absolute Continuity

Proposition 13 *The function g is absolutely continuous. More precisely*

$$g(s) = g(0) + \int_0^s \bar{g}(\sigma)\, d\sigma$$

Let $(\rho_n)_n$ be a mollifier sequence. We denote by y^n the solution to

$$P(y^n) = f * \rho_n \text{ on } Q, \quad y^n = 0 \text{ on } \Sigma,$$
$$y^n(0) = \Phi_0 * \rho_n \text{ on } \Omega, \quad \partial_t y^n(0) = \Phi_1 * \rho_n \text{ on } \Omega \tag{14}$$

According to theorem 1, there is a shape derivative $y^{n\prime}$ for each n. We consider the functions

$$g_n(s) = \int_{Q_s} y_s^n \phi\, dx\, dt \quad \text{and} \quad \bar{g}_n(s) = \int_{Q_s} y_s^{n\prime} \phi\, dx\, dt$$

We proved y^n was differentiable in $L^2(I \times D)$ therefore g_n is differentiable. Thus g_n is absolutely continuous and $g_n' = \bar{g}_n$. That is

$$g_n(s) = g_n(0) + \int_0^s \bar{g}_n(\sigma)\, d\sigma \tag{15}$$

From [1, remark 2.2], we get $y^n \to y$ in $C(I, L^2(\Omega))$ as $n \to +\infty$. Hence $g_n(s) \to g(s)$ and $g_n(0) \to g(0)$ as n tends to $+\infty$. Therefore, to prove $g(s) = g(0) + \int_0^s \bar{g}(\sigma)\, d\sigma$ it is sufficient to prove the following lemma.

Lemma 9 *We have*

$$\int_0^s \bar{g}_n(\sigma)\, d\sigma \to \int_0^s \bar{g}(\sigma)\, d\sigma \text{ as } n \to +\infty$$

Proof – According to [1, remark 2.2], this function is continuous

$$L^1(I, L^2(\Omega)) \times H^1(\Omega) \times L^2(\Omega) \to L^2(\Sigma) \; : \; (f, \phi_0, \phi_1) \mapsto \frac{\partial y}{\partial n}$$

This same remark applies to (13) and yields this function is continuous

$$L^2(\Sigma) \to C(I, L^2(\Omega)) \; : \; \frac{\partial y}{\partial n} \mapsto y'$$

thus $\exists C, \|y^{n\prime}\|_{C(I,L^2(\Omega))} \leq C \left(\|f * \rho_n\|_{L^1(I,L^2(\Omega))} + \|\phi_0 * \rho_n\|_{H^1(\Omega)} + \|\phi_1 * \rho_n\|_{L^2(\Omega)} \right)$, it follows $\|y^{n\prime}\|_{L^2(Q)} \leq C \left(\|f\|_{L^1(I,L^2(D))} + \|\phi_0\|_{H^1(D)} + \|\phi_1\|_{L^2(D)} \right)$.

Similarly, and using the continuity of C with respect to the domain, we have $\|y_\sigma^{n'}\|_{L^2(Q_s)} \leq M$ this leads to $\int_{Q_\sigma} y_\sigma^{n'} \phi \, dx \, dt \leq M \|\phi\|_{L^2(Q)}$ hence

$$\forall n \in \mathbb{N}, \forall s \in [0; S], \int_0^s \bar{g}_n(\sigma) d\sigma \leq SM\|\phi\|_{L^2(Q)} \tag{16}$$

On the other hand from the continuity of $(f, \phi_0, \phi_1) \mapsto y'$ in $C(I, L^2(\Omega))$ and from

$$f * \rho_n \to f \text{ in } L^1(I, L^2(\Omega)), \qquad \phi_0 * \rho_n \to \phi_0 \text{ in } H^1(\Omega), \qquad \phi_1 * \rho_n \to \phi_1 \text{ in } L^2(\Omega)$$

as n tends to $+\infty$, we obtain $y^{n'} \to y'$ in $C(I, L^2(\Omega))$ as $n \to +\infty$. Therefore we have $\int_Q y^{n'} \phi \, dx \, dt \to \int_Q y' \phi \, dx \, dt$ as $n \to +\infty$. The same method holds to prove

$$\forall \sigma \in [0; s], \int_0^s \bar{g}_n(\sigma) d\sigma \to \int_0^s \bar{g}(\sigma) d\sigma \text{ as } n \to +\infty \tag{17}$$

With (16) and (17) we can use the theorem of Lebesgue to prove the lemma. ■

5.2 Continuity of $\bar{g}$

Lemma 10 *Let* $W(s) = \frac{1}{2}\gamma_s \|^*DT_s^{-1}n\|^{-1}DT_s^{-1}V(s) \circ T_s$ *and let* Λ_s *be the solution to*

$$P(\Lambda_s) = \phi \text{ on } Q_s, \quad \Lambda_s = 0 \text{ on } \Sigma_s, \quad \Lambda_s(T) = 0 \text{ on } \Omega_s, \quad \Lambda_s(T) = 0 \text{ on } \Omega_s \tag{18}$$

then

$$\bar{g}(s) = \int_\Sigma \left(\left(\frac{\partial}{\partial n}(y^s + \Lambda^s)\right)^2 - \left(\frac{\partial}{\partial n} y^s\right)^2 - \left(\frac{\partial}{\partial n} \Lambda^s\right)^2 \right) \langle W(s), n \rangle \, d\Gamma \, dt$$

Proof – For the sake of simplicity we will suppose div $(W(0)) = 0$. From [1] we have $\Lambda_s \in C(I, L^2(\Omega_s)) \cap C^1(I, H^{-1}(\Omega))$, using this adjoint variable we obtain

$$\bar{g}(s) = \int_{\Sigma_s} \frac{\partial y_s}{\partial n_s} \frac{\partial \Lambda_s}{\partial n_s} \langle V(s), n_s \rangle \, d\Gamma_s \, dt$$

we use the change of variable $X = T_s(x)$ and $\frac{\partial y^s}{\partial n} \frac{\partial \Lambda^s}{\partial n} = \frac{1}{2}\left(\frac{\partial}{\partial n}(y^s + \Lambda^s)\right)^2 - \frac{1}{2}\left(\frac{\partial y^s}{\partial n}\right)^2 - \frac{1}{2}\left(\frac{\partial \Lambda^s}{\partial n}\right)^2$ to prove the lemma. ■

Lemma 11 *The function* $s \mapsto \int_\Sigma (\frac{\partial}{\partial n}\Psi^s)^2 \langle W(s), n \rangle \, d\Gamma \, dt$ *is continuous in 0 for* $\Psi = y$, $\Psi = \Lambda$ *and* $\Psi = y + \Lambda$.

Proof – Let $W(T) = 0$, this request is not restrictive since T can be taken as large as we want. We use the following identity

$$\int_\Sigma \left(\frac{\partial \Psi^s}{\partial n}\right)^2 \langle W(s), n\rangle \, d\Gamma \, dt =$$

$$-\int_Q P_s(\Psi^s) \langle \nabla \Psi^s, W(0)\rangle \, dx \, dt - \frac{1}{2}\int_Q (\partial_t \Psi^s)^2 \mathrm{div}\,(\|{}^*DT_s^{-1}n\|^{-2}W(0)) \, dx \, dt$$

$$+ \int_Q \partial_t \Psi^s \langle \nabla \Psi^s, \partial_t W(0)\rangle \, dx \, dt + \int_\Omega \partial_t \Psi^s(0) \langle \nabla \Psi^s, W(0)\rangle \, dx$$

$$+ \int_Q \langle {}^*DT_s^{-1}\,{}^*D(\|{}^*DT_s^{-1}n\|^{-2}W(0))\nabla \Psi^s, {}^*DT_s^{-1}W(0)\rangle \, dx \, dt$$

$$- \frac{1}{2}\int_Q \|{}^*DT_s^{-1}W(0)\|^2 \mathrm{div}\,(\|{}^*DT_s^{-1}n\|^{-2}W(0)) \, dx \, dt$$

$$- \frac{1}{2}\int_Q \langle (\nabla(DT_s^{-1}\,{}^*DT_s^{-1})W(0).\nabla \Psi^s, \nabla \Psi^s\rangle \, dx \, dt$$

By density, $\partial_t \Psi^s$ and $\nabla \Psi^s$ are continuous as $s \to 0$. Moreover ${}^*DT_s^{-1}W(0)$ and $\nabla(DT_s^{-1}\,{}^*DT_s^{-1})W(0)$ are bounded. ■

From the two previous lemmae follows the

Proposition 14 $\bar{g}$ *is continuous in 0.*

References

[1] I Lasiecka, J.-L. Lions, and R. Triggiani. Non homogeneous boundary value problems for second order hyperbolic operatores. *Journal de Mathématiques pures et Appliquées*, 65, 1986.

[2] I. Lasiecka and R. Triggiani. Recent advances in regularity of second-order hyperbolic mixed problems, and applications. *Dynamics reported. Expositions in dynamical systems. New series.*, 3:104,162, 1994.

[3] Jan Sokolowski and Jean-Paul Zolésio. *Introduction to Shape Optimization.* Springer-Verlag, 1991.

John Cagnol and Jean-Paul Zolésio
CMA, Ecole des Mines de Paris
INRIA – 2004 route des Lucioles – B.P. 93
06902 Sophia Antipolis Cedex, France

MICHEL C DELFOUR[*] AND JEAN-PAUL ZOLÉSIO[†]
Hidden boundary smoothness in hyperbolic tangential problems on nonsmooth domains

1. Wave equation in cylindrical domains

The boundary smoothness has been first proved for a disk in [7] and then for smooth domains in the *Euclidean space* $\mathbf{R}^N$ in [6]. For instance it is assumed that the boundary is C^2. We extend these results to some hyperbolic equations on bounded open domains in a $C^{1,1}$ *submanifolds* of $\mathbf{R}^N$ with a Lipschitzian type of boundary. We first revisit the Euclidean case and recall the *extractor approach* introduced in [3] and [4]. Given a fixed open *hold-all* or *universe* $D \subset \mathbf{R}^N$, consider open sets Ω in D and smooth one-parameter families of transformations $T_t = T_t(V)$, $t \geq 0$, of D, arising from the flow associated with a sufficiently smooth vector field V defined over D with zero normal component on the boundary of D. For $V \in W^{1,\infty}(D)$, the *extractor* is defined for all $\phi \in H_0^1(\Omega)$ as

$$E(V).\phi \stackrel{\text{def}}{=} \frac{\partial}{\partial t}(\|\phi \circ T_t(V)^{-1}\|^2_{H_0^1(\Omega_t)})|_{t=0}, \tag{1}$$

where $\Omega_t = T_t(\Omega)$. Recall the following properties

Proposition 1. *For all $\phi \in H^1(\Omega)$, we have*

$$E(V).\phi = \int_\Omega < [\text{div}(V)I_d - 2\varepsilon(V)].\nabla\phi, \nabla\phi > dx. \tag{2}$$

Moreover, if $\phi \in H^2(\Omega)$, then

$$E(V).\phi = \int_{\partial\Omega} \|\nabla\phi\|^2 < V(0), n > -2 < V(0), \nabla\phi > \frac{\partial\phi}{\partial n} d\Gamma + 2\int_\Omega \Delta\phi < \nabla\phi, V > dx.$$

Corollary 1. *For all $\phi \in H^2(\Omega) \cap H_0^1(\Omega)$, we get*

$$E(V).\phi = -\int_{\partial\Omega} \left|\frac{\partial\phi}{\partial n}\right|^2 < V(0), n > d\Gamma + 2\int_\Omega \Delta\phi < \nabla\phi, V > dx. \tag{3}$$

For all $\phi \in H_0^1(\Omega) \cap H^2(\Omega)$ we have

$$\begin{aligned}&\int_{\partial\Omega} \left|\frac{\partial}{\partial n}\phi\right|^2 < V(0), n > d\Gamma \\ &= 2\int_\Omega \Delta\phi < \nabla\phi, V > dx - \int_\Omega < [\text{div}(V)I_d - 2\varepsilon(V)].\nabla\phi, \nabla\phi > dx\end{aligned} \tag{4}$$

We make use of the following division result

---

[*]Centre de recherches mathématiques et Département de mathématiques et de statistique, Université de Montréal, CP 6128, Succ Centre-ville, Montréal (Qc), Canada H3C 3J7, E-mail: delfour@crm.UMontreal.ca. The research of the first author has been supported by National Sciences and Engineering Research Council of Canada research grant A-8730 and by a FCAR grant from the Ministère de l'Éducation du Québec.

[†]CNRS Institut Non Linéaire de Nice, 1361 route des Lucioles, 06904 Sophia Antipolis Cedex, France and Centre de Mathématiques Appliquées, Ecole des Mines (CMA/MEIJE, INRIA), 2004 route des Lucioles, BP 93, 06902 Sophia Antipolis Cedex, France, E-mail: Jean-Paul.Zolesio@sophia.inria.fr

Proposition 2. Let Ω be a domain in $\mathbf{R}^N$ such that its complement Ω^c is of positive reach, that is there exists $h > 0$ such that
$$d^2_{\Omega^c} \in C^{1,1}(U_h(\Gamma)), \quad \Gamma \stackrel{\text{def}}{=} \partial\Omega.$$
Then there exists a constant $c > 0$ such that for any $\phi \in H^1_0(\Omega)$
$$\int_{U_h(\Gamma) \cap \Omega} \left|\frac{\phi}{d_{\Omega^c}}\right|^2 dx \leq c \int_{U_h(\Gamma) \cap \Omega} |<\nabla\phi, \nabla d_{\Omega^c}>|^2 dx$$

Proof. We prove the result for $\phi \in \mathcal{D}(\Omega)$ and extend to $H^1_0(\Omega)$ by density. For $0 \leq r \leq 1$, consider the change of variable
$$T_r(x) \stackrel{\text{def}}{=} x + (r-1)\frac{1}{2}\nabla d^2(x) : U_h \to U_h, \quad d \stackrel{\text{def}}{=} d_{\Omega^c}, \ U_h \stackrel{\text{def}}{=} U_h(\Gamma) \cap \Omega.$$
For $0 < h < 1$, T_r is invertible. It is $C^{1,1}(U_h)^N$ by assumption and
$$DT_r(x) = I + \frac{r-1}{2}D^2 d^2(x)$$
belongs to $L^\infty(U_h)$. Since $\nabla d(x)$ exists in almost every x, in any such point define the function $f(r) = (\phi \circ T_r)(x)$. Then
$$f(1) = \phi(x), \quad f(0) = \phi(x - \frac{1}{2}\nabla d^2(x)) = \phi(p(x)) = 0, \quad p = p_{\Omega^c}$$
$$\frac{df}{dr}(r) = <\nabla\phi \circ T_r(x), \frac{1}{2}\nabla d^2(x)> = <\nabla\phi \circ T_r(x), d(x)\nabla d(x)>.$$
Therefore
$$\frac{\phi(x)}{d(x)} = \int_0^1 <\nabla\phi \circ T_r, \nabla d(x)> dr = \int_0^1 (<\nabla\phi, \nabla d>) \circ T_r \, dr$$
since $\nabla d \circ T_r = \nabla d$ and
$$\left|\frac{\phi(x)}{d(x)}\right| \leq \int_0^1 |(<\nabla\phi, \nabla d>) \circ T_r| \leq \{\int_0^1 |(<\nabla\phi, \nabla d>) \circ T_r|^2 \, dr\}^{1/2}.$$
Finally
$$\int_{U_h} \left|\frac{\phi(x)}{d(x)}\right|^2 dx \leq \int_{U_h} \int_0^1 |(<\nabla\phi, \nabla d>) \circ T_r|^2 \, dr \, dx$$
$$= \int_0^1 \int |(<\nabla\phi, \nabla d>)|^2 |\det(DT_r)| \, dx \, dr \leq c \int_{U_h} |<\nabla\phi, \nabla d>|^2 \, dx$$
The final result follows by density. □

We have a similar result for the solution ϕ to the evolution problem.

Proposition 3. For all $\phi \in L^2(0, \tau, H^2(\Omega) \cap H^1_0(\Omega)) \cap H^2(0, \tau, L^2(\Omega))$, the normal derivative $\partial\phi/\partial n$ on the lateral boundary of the evolution domain verifies

$$\int_0^\tau \int_{\partial\Omega} \left|\frac{\partial\phi}{\partial n}\right|^2 <V(0), n> dt \, d\Gamma = -2 \int_\Omega [\phi_t(0) <\nabla\phi(0), V(0)> - \phi_t(\tau) <\nabla\phi(\tau), V(\tau)>] dx$$

$$- \int_0^\tau \int_\Omega <[\text{div}(V)I_d - 2\varepsilon(V)].\nabla\phi, \nabla\phi> dt \, dx \quad (5)$$

$$- 2 \int_0^\tau \int_\Omega (\phi_{tt} - \Delta\phi) <\nabla\phi, V> dt \, dx$$

$$+ \int_0^\tau \int_\Omega (\phi_t^2 \text{div}(V) - 2\phi_t <\nabla\phi, V_t>) dt \, dx.$$

The proof is straightforward with the following identity

Lemma 1.
$$-2\int_0^\tau\int_\Omega \phi_{tt} <\nabla\phi,V> dx\,dt = 2\int_\Omega [\phi_t(0)<\nabla\phi(0),V(0)> - \phi_t(\tau)<\nabla\phi(\tau),V(\tau)>]\,dx$$
$$+ \int_0^\tau\int_\Omega 2\phi_t <\nabla\phi,V_t> - (\phi_t)^2 \,\mathrm{div}(V)\,dx\,dt \qquad (6)$$
$$+ \int_0^\tau\int_\Gamma (\phi_t)^2 <V,n>\,d\Gamma\,dt.$$

When ϕ is a "non smooth" element, say $\phi \in L^2(0,\tau,H_0^1(\Omega))\cap H^1(0,\tau,L^2(\Omega))$ by the use of the Green's theorem the normal derivative $\partial\phi/\partial n$ is weakly defined on the lateral boundary for ϕ such that $(\phi_{tt}-\Delta\phi) \in L^2(0,\tau,L^2(\Omega))$. Let
$$\mathcal{H} \stackrel{\mathrm{def}}{=} \{\phi \in L^2(0,\tau,H_0^1(\Omega))\cap H^1(0,\tau,L^2(\Omega))\ \text{with}\ (\phi_{tt}-\Delta\phi)\in L^2(0,\tau,L^2(\Omega))\}$$
The normal derivative is then weakly defined as follows: for all $\psi \in L^2(0,\tau,H^1(\Omega))\cap H^1(0,\tau,L^2(\Omega))$,
$$\int_0^\tau <\frac{\partial\phi(t)}{\partial n},\psi>_{H^{-\frac{1}{2}}(\Gamma)\times H^{\frac{1}{2}}(\Gamma)}\,dt + <\phi_t(0),\psi(0)>_{H^{-\frac{1}{2}}(\Omega)\times H^{\frac{1}{2}}(\Omega)}$$
$$- <\phi_t(\tau),\psi(\tau)>_{H^{-\frac{1}{2}}(\Omega)\times H^{\frac{1}{2}}(\Omega)} = \int_0^\tau\int_\Omega (<\nabla\phi,\nabla\psi> - \phi_t\psi_t) + (\Delta\phi-\phi_{tt})\psi\,dx\,dt. \qquad (7)$$

We shall extend this identity to an estimate of the normal derivative in the L^2-norm of the lateral boundary for any element in $\mathcal{H}$. For that purpose we shall use a density argument that we prove to be true under only weak smoothness of the domain boundary which need not be C^1. A domain Ω is *locally starshaped* if at each point $z \in \bar{\Omega}$ there exists a ball $B(x,r)$, $x \in \Omega$, such that $z \in B(x,r)$ and $\bar{\Omega}\cap B(x,r)$ is starshaped with respect to x, that is that for all $\xi \in B(x,r)\cap \bar{\Omega}$ and λ, $0 \leq \lambda \leq 1$, $x+\lambda(\xi-x) \in B(x,r)\cap\bar{\Omega}$.

Proposition 4. *Let Γ be a Lipschitzian boundary and Ω be locally starshaped. Assume that there exists a tubular neighborhood $\mathcal{U}$ of $\partial\Omega$ such that $\Delta b_\Omega \in L^{3+\varepsilon}(\mathcal{U})$, $\varepsilon > 0$. Then*
$$L^2(0,\tau,H^2(\Omega)\cap H_0^1(\Omega))\cap H^2(0,\tau,L^2(\Omega))\ \text{is dense in}\ \mathcal{H}.$$

Proof. The result is very easy for an element $\phi \in \mathcal{H}$ with compact support in Q. In that case consider the sequence $\{\phi_n = \rho_n *_{t,x} \phi\}$ for an appropriate mollifier ρ_n. This sequence solves the density question. Now consider the case where the support of the function ϕ is contained in a tubular neighborhood $[0,\tau]\times\mathcal{U}$ of the lateral boundary Σ_h. First assume the domain to be starshaped with respect to the origin (then we also assume that $0 \in \Omega$). As $\phi(t,.)$ belongs to $H_0^1(\Omega)$, we know that
$$\phi_b(t,.) \stackrel{\mathrm{def}}{=} b_\Omega^{-1}(.)\phi(t,.) \in H^1(\Omega)$$
where b_Ω is the oriented distance function to Ω. It is equal to the negative of the distance to the boundary in Ω. Associate with a positive sequence $\{\alpha_n\}$ converging to zero, the sets
$$\Omega_n = \{x \in \mathbf{R}^N\ :\ (1+\alpha_n)x \in \Omega\}$$
and the extensions $\phi_b^n(t,x)$ of $\phi_b(t,x)$ to the larger domain $Q_n =]0,(1+\alpha_n)\tau[\times\Omega_n$. The time extension, for $\tau \leq t \leq (1+\alpha_n)\tau$ is the element $\phi_b^n(t,.) = \phi_b((1+\alpha_n)^{-1}t,.)$ while the x-extension is the classical one for a starshaped domain with respect to the origin
$$\phi_b^n(t,x) = \phi_b((1+\alpha_n)^{-1}t,(1+\alpha_n)^{-1}x)$$
and the extension by 0 to D. That element belongs to $L^2(0,2\tau,H_0^1(D))\cap L^2(0,2\tau,L^2(D))$.

Given the usual mollifier $\rho_n(t,x)$, set
$$\phi_n(t,x) = b_\Omega(x)\, \rho_n *_{t,x} \phi_b^n(t,x).$$
$\phi_n(t,.)$ is an element of $H_0^1(\Omega)$ and
$$((\phi_n)_{tt} - \Delta\,\phi_n)(t,x) = (1+\alpha_n)^{-2}\, b_\Omega\, \rho_n\, *_{t,x}\, ((\phi_{tt})_b)^n(t,x) - \Delta b_\Omega\; \rho_n * (\phi_b)^n$$
$$- 2\nabla b_\Omega\; \rho_n * \nabla((\phi_b)^n) - b_\Omega\; \rho_n * \Delta((\phi_b)^n).$$
and we have the following expressions
$$\nabla(\phi_b^n) = (1+\alpha_n)^{-1}\,(-\frac{1}{b^2}\nabla b\, \phi + \frac{1}{b}\nabla\phi)^n$$
$$\Delta(\phi_b^n) = (1+\alpha_n)^{-2}\,(\frac{2}{b^3}\phi - \frac{2}{b^2}\nabla b\nabla\phi - \frac{1}{b^2}\phi\Delta b + \frac{1}{b}\Delta\phi)^n,$$
where $(\ )^n$ indicates the x-extension with respect to α_n. So that finally
$$\left[\frac{\partial^2}{\partial t^2} - \Delta\right](\phi_n - \phi) = (1+\alpha_n)^{-2}\, b_\Omega\, \rho_n *_{t,x}\, (b_\Omega^{-1}(\frac{\partial^2}{\partial t^2}\phi - \Delta\phi))^n\; + R_n$$
where the remainder R_n is given by
$$R_n = -\Delta b\; \rho_n * (\phi_b)^n - 2\nabla b\; \rho_n * \nabla((\phi_b)^n)$$
$$- b\, \rho_n * ((1+\alpha_n)^{-2}\,(\frac{2}{b^3}\phi - \frac{2}{b^2}\nabla b\nabla\phi - \frac{1}{b^2}\phi\Delta b))$$
that is,
$$R_n = -\Delta b\; \rho_n * (\phi_b)^n - 2\nabla b\; \rho_n * [(1+\alpha_n)^{-1}\,(-\frac{1}{b^2}\nabla b\, \phi + \frac{1}{b}\nabla\phi)^n]$$
$$- b\, \rho_n * ((1+\alpha_n)^{-2}\,(\frac{2}{b^3}\phi - \frac{2}{b^2}\nabla b\nabla\phi - \frac{1}{b^2}\phi\Delta b))^n.$$
It can be verified that $R_n \to 0$ in $L^2(Q)$ as $n \to \infty$. The "worst term" in R_n is
$$-\Delta b_\Omega\, \rho_n * (\phi_b)^n\; + \; (1+\alpha_n)^{-2} b_\Omega\, \rho_n * (b_\Omega^{-2}\Delta b_\Omega\, \phi)^n.$$
Since $b_\Omega^{-1}\phi \in H^1(\Omega) \subset L^6(\Omega)$ (assuming $N \leq 3$), that term converges to zero in $L^2(Q)$ for $\Delta b_\Omega \in L^{3+\varepsilon}(\Omega)$. Indeed we have the following boundedness, as we consider the convergence of the restrictions to Ω, by the choice of α_n such that supp $\rho_n \subset B(0,\alpha_n/2)$: for $x \in \Omega$,
$$\exists c > 0 \text{ such that } \forall x \in \bar\Omega,\, \forall y \in \text{supp } \rho_n(x-.),\quad |b_\Omega(x)|/|(b_\Omega)^n(y)| \geq c.$$
In the general case, for a bounded domain Ω, its closure can be covered by a finite family $O_1,...,O_p$ of starshaped open sets, each open set $O_i = B(x_i, r_i) \cap \Omega$ is starshaped with respect to the point x_i lying in the open domain Ω. Let r_i be a smooth function with compact support in $B(x_i, r_i)$, $0 \leq r_i \leq 1$ and $r_1 + ... + r_p = 1$ on Ω. Set
$$\phi_n = b_\Omega\, \rho_n * (r_i(x)\, b_\Omega^{-1}\phi)^{i,n})$$
where $(\psi)^{i,n}$ stands for the $1+\alpha_n$ dilation in the ball $B_i^n = B(x_i, r_i(1+\alpha_n))$ with respect to x_i which is in $\mathcal{H}_i$. Then from the first part of the proof, there exists a sequence $\{\phi_{i,m}\} \subset L^2(0,\tau, H^2(O_i) \cap H_0^1(O_i)) \cap H^1(0,\tau, L^2(O_i))$ converging to ϕ_i in the $\mathcal{H}_i$-norm. Then set $\phi_m(t,x) = \sum_{i=1,...,p} \phi_{i,m}(t,x)$. That sequence converges as $m \to \infty$ to ϕ in the $\mathcal{H}$-norm. □

Using that density argument, we get the following theorem

Theorem 1. Let $\phi \in L^2(0,\tau, H_0^1(\Omega)) \cap H^1(0,\tau, L^2(\Omega))$ such that
$$\phi_{tt} - \Delta\phi \in L^2(0,\tau, L^2(\Omega)).$$

Then for any field $V \in C^1([0,\tau], \mathrm{Lip}(D,D))$ with $<V(t,.), n(.)> \geq 0$ on ∂D and $V(t,.).n(.) \geq \alpha(t)$ on the lateral boundary Σ_h with $\alpha \in C^0([0,\tau])$, $\alpha \geq 0$, $\alpha(0) = \alpha(\tau) = 0$, we have

$$\int_0^\tau \alpha(t) \int_\Gamma \left|\frac{\partial \phi}{\partial n}\right|^2 d\Gamma\, dt \leq -\int_0^\tau \int_\Omega <[\mathrm{div}(V)I_d - 2\varepsilon(V)].\nabla\phi, \nabla\phi> dt\, dx$$

$$-2\int_0^\tau \int_\Omega (\phi_{tt} - \Delta\phi) <\nabla\phi, V> dt\, dx$$

$$+2\int_0^\tau \int_\Omega \left(\frac{1}{2}\phi_t^2 \mathrm{div}(V) - \phi_t <\nabla\phi, V_t>\right) dt\, dx.$$

2. Wave equation in non cylindrical domains

Let V and W in $C^n([0,\tau], W^{1,\infty}(D, \mathbf{R}^N))$ be two non-autonomous vector fields. Assume in this section that the non-cylindrical evolution domain Q is in the following form:

$$Q = \bigcup_{0<t<\tau} \{t\} \times \Omega_t(W), \quad \Omega_t(W) = T_t(W)(\Omega). \tag{8}$$

Then from the first part of the corollary we get

Corollary 2. *For any t and $\phi \in H^2(\Omega_t(W)) \cap H^1_0(\Omega_t(W))$ with $\Delta\phi \in L^2(\Omega_t(W))$, we have*

$$\int_{\partial\Omega_t(W)} \left|\frac{\partial\phi}{\partial n_t}\right|^2 V(t).n_t\, d\Gamma_t$$
$$= -\int_{\Omega_t(W)} 2\Delta\phi <\nabla\phi, V(t)> - <[\mathrm{div}(V(t))I_d - 2\varepsilon V(t)].\nabla\phi, \nabla\phi> dx$$

For the time derivative we need the Hilbert space

$$\mathcal{E}(Q) \stackrel{\mathrm{def}}{=} \{\phi \in L^2(Q) : \phi_t, \phi_{tt} \in L^2(Q), \nabla\phi \in L^2(Q)^N, D^2\phi \in L^2(Q)^{N^2}$$
$$\phi(t,.) \in H^2(\Omega_t(W)) \cap H^1_0(\Omega_t(W))\}.$$

Lemma 2. *Assuming that $\phi \in \mathcal{E}(Q)$ and $V(\tau) = 0$, $(V\phi_t \nabla\phi)(0) = 0$, we have*

$$\int_0^\tau \int_{\Omega_t(W)} \phi_{tt} <\nabla\phi, V(t)> dt\, dx = \int_0^\tau \int_{\Omega_t(W)} \left(\frac{1}{2}|\phi_t|^2 \mathrm{div}(V(t)) - \phi_t <\nabla\phi, V_t(t)>\right) dt\, dx$$

$$+ \frac{1}{2}\int_0^\tau \int_{\Gamma_t(W)} \left|\frac{\partial\phi}{\partial n_t}\right|^2 (<W(t), n_t>)^2 <V(t), n_t> dt\, d\Gamma_t.$$

Proof.

$$\int_0^\tau \int_{\Omega_t(W)} \phi_{tt} <\nabla\phi, V(t)> dt\, dx = \int_0^\tau \partial_t(\int_{\Omega_t(W)} \phi_t <\nabla\phi, V> dx)\, dt$$

$$-\int_0^\tau \int_{\Omega_t(W)} (\phi_t <\nabla\phi_t, V> + \phi_t <\nabla\phi, V_t>)\, dx\, dt$$

$$-\int_0^\tau \int_{\Gamma_t(W)} \phi_t <\nabla\phi, V><W, n_t> d\Gamma_t\, dt.$$

But

$$\int_{\Omega_t(W)} (\phi_t <\nabla\phi_t, V> + \phi_t <\nabla\phi, V_t>)\, dx = \int_{\Omega_t(W)} -\frac{1}{2}(\phi_t)^2 \operatorname{div}(V)\, dx + \int_\Gamma \frac{1}{2}(\phi_t)^2 <V, n_t> d\Gamma_t.$$

On the other hand as ϕ is zero on the lateral boundary we get

$$\phi_t = -\frac{\partial}{\partial n_t}\phi <W, n_t> \text{ on } \Sigma_h = \bigcup_{0<t<\tau} \{t\} \times \Gamma_t(W).$$

□

Proposition 5. *For any "smooth function ϕ in $L^2(0,\tau, H^1_0(\Omega_t(W)))$", that is, $\phi \in \mathcal{E}(Q)$, assuming $V(\tau) = 0$ and $[V(t)\phi_t \nabla\phi]_{|t=0} = 0$, we have*

$$\int_0^\tau \int_{\partial\Omega_t(W)} |\frac{\partial\phi}{\partial n_t}|^2 (1 + (<W(t), n_t>)^2 <V(t), n_t> dt\, d\Gamma_t$$

$$= \int_0^\tau \int_{\Omega_t(W)} 2(\phi_{tt} - \Delta\phi) <\nabla\phi, V(t)> dt\, dx - \int_0^\tau \int_{\Omega_t(W)} <[\operatorname{div}(V(t))I_d - 2\varepsilon(V(t))].\nabla\phi, \nabla\phi> dt\, dx$$

$$- \int_0^\tau \int_{\Omega_t(W)} ((\phi_t)^2 \operatorname{div}(V(t)) - 2\phi_t <\nabla\phi, V_t(t)>)\, dt\, dx.$$

3. Wave equation for the Laplace-Beltrami operator

We now turn to the similar results for the tangential problem associated with the Laplace-Beltrami wave operator on the global cylindrical evolution domain $Q_0 =]0,\tau[\times\partial\Omega$ where Ω is a bounded open domain in $\mathbf{R}^N$ of class $C^{1,1}$. Assume now that ω is a smooth open set in $\partial\Omega$ and denote by γ its relative boundary while $\Sigma_h =]0,\tau[\times\gamma$ will denote the lateral evolution boundary. In this case the evolution domain Q is restricted to $Q =]0,\tau[\times\omega \subset Q_0$. In this section the field V is assumed to be tangent to the surface $\Gamma = \partial\Omega$ so that the associated flow transformation $T_t(V)$ maps the surface Γ onto itself and then in the definition of the tangential E_Γ the transported element $u \circ T_t(V)^{-1}$ will be defined in the open set $\omega_t(V) \subset \Gamma$, e.g. the moving domains will remain in the same surface Γ.

Proposition 6. *Let $u \in L^2(0,\tau, H^1_0(\omega)) \cap H^2(0,\tau, L^2(\omega))$ with $[V(t)u_t]_{|t=0} = 0$, $V(\tau) = 0$ and verify the smoothness assumption*

$$u_{tt} - \Delta_\Gamma u \in L^2(0,\tau, L^2(\omega)).$$

Then we have

$$\int_{\Sigma_h} u_{tt} <V, \nabla_\Gamma u> d\Sigma_h = \int_{\Sigma_h} (-u_t < \frac{\partial V}{\partial t}, \nabla u> + \frac{1}{2}(u_t)^2 \text{div}_\Gamma(V_\Gamma)) d\Sigma_h. \tag{9}$$

The main idea is now to follow the same technique as in the Euclidean case without curvature. Consider the tangential extractor defined as follows. For any element $u \in H_0^1(\omega)$ we know that the transported element $u \circ T_t(V)^{-1}$ belongs to $H_0^1(\omega_t(V))$ and we consider the element

$$E_\Gamma(V).u = \frac{\partial}{\partial t} \|u \circ T_t(V)^{-1}\|_{H_0^1(\omega_t(V))}\big|_{t=0} \tag{10}$$

that is:

$$E_\Gamma(V).u = \frac{\partial}{\partial t} \int_{\omega_t(V)} \|\nabla_{\Gamma_t}(u \circ T_t(V)^{-1})\|^2 d\Gamma\big|_{t=0}. \tag{11}$$

We now turn to the computation of this tangential extractor. Consider the unitary tangential vector field ν to $\Gamma = \partial\Omega$ which is normal to $\gamma = \partial_\Gamma \omega$ and exterior to ω. Then we have the following result.

Proposition 7.

$$E_\Gamma(V).u = -\int_\gamma (<\nabla_\Gamma u, \nu>)^2 <V_\Gamma, \nu> d\Gamma + 2\int_\Gamma \Delta_\Gamma u <V, \nabla_\Gamma u> d\Gamma \tag{12}$$

Proof. By boundary change of variable we get

$$E_\Gamma(V).u = \frac{\partial}{\partial t} \int_\omega <w(t) DT_t^{-1}.DT_t^{-*}.\nabla_\Gamma u, \nabla_\Gamma u> d\Gamma, \tag{13}$$

where

$$w(t) = \frac{\det(DT_t).DT_t^{-*}}{\|\det(DT_t).DT_t^{-*}\|}.$$

We make use of the integration by parts formula on the open set ω whose (relative) boundary in Γ is γ:

$$\int_\omega \nabla_\Gamma <f, E_\Gamma> d\Gamma = -\int_\omega \text{div}_\Gamma(E_\Gamma) f\, d\Gamma + \int_\gamma f <E_\Gamma, \nu> d\Gamma \tag{14}$$

to get

$$E(V).u = \int_\omega < (\text{div}_\Gamma(V) I_d - 2\varepsilon_\Gamma(V).\nabla_\Gamma u, \nabla_\Gamma u> d\Gamma \tag{15}$$

and we have

$$\int_\omega <\varepsilon_\Gamma(V).\nabla_\Gamma u, \nabla_\Gamma u> d\Gamma$$
$$= -\int_\omega <V, \nabla_\Gamma u> \Delta_\Gamma u\, d\Gamma - \int_\omega <D_\Gamma(\nabla_\Gamma u).\nabla_\Gamma u, V> d\Gamma$$
$$+ \int_\gamma (\nabla_\Gamma u.\nu)^2 <V_\Gamma, \nu> dl.$$

But as u belongs to $H_0^1(\omega)$, the trace on the (relative) boundary γ of the tangential gradient ∇_Γ is given via the normal tangential derivative $\frac{\partial u}{\partial \nu} = <\nabla_\Gamma u, \nu>$ and then we get

$$\int_\omega <\varepsilon_\Gamma(V).\nabla_\Gamma u, \nabla_\Gamma u> d\Gamma$$
$$= -\int_\omega <V, \nabla_\Gamma u> \Delta_\Gamma u - <D_\Gamma(\nabla_\Gamma u).\nabla_\Gamma u, V> d\Gamma + \int_\gamma \left|\frac{\partial u}{\partial \nu}\right|^2 <V_\Gamma, \nu> dl.$$

We have the same behavior for the divergence term

$$\int_\omega \text{div}_\Gamma \|\nabla_\Gamma u\|^2 \, d\Gamma$$
$$= \int_\omega - <V, \nabla_\Gamma(\|\nabla_\Gamma u\|^2)> + H \|\nabla_\Gamma u\|^2 <V, n> \, d\Gamma + \int_\gamma \left|\frac{\partial u}{\partial \nu}\right|^2 <V_\Gamma, \nu> \, dl.$$

The field V is chosen to be tangent on Γ, $<V, n> = 0$, where n is the usual unitary normal field to the surface Γ, exterior to the domain Ω. Then the second integral on the right-hand side is zero, where H is the *mean curvature* of Γ. On the other hand using the identity

$$\nabla_\Gamma(<\nabla_\Gamma u, \nabla_\Gamma u>) = 2D_\Gamma(\nabla_\Gamma u).\nabla_\Gamma u \qquad (16)$$

we get the following

$$\int_\Gamma (<\text{div}_\Gamma(V)I_d - 2\varepsilon_\Gamma(V).\nabla_\Gamma u, \nabla_\Gamma u> \, d\Gamma$$
$$= \int_\Gamma - <v, \nabla_\Gamma(<\nabla_\Gamma u, \nabla_\Gamma u>)> + 2 <D_\Gamma(\nabla_\Gamma u).\nabla_\Gamma u, V> + 2 <V, \nabla_\Gamma> \Delta_\Gamma u \, d\Gamma$$
$$- \int_\gamma \left|\frac{\partial}{\partial \nu}u\right|^2 <V, \nu> \, dl.$$

$\square$

Corollary 3.

$$\int_\gamma \left|\frac{\partial u}{\partial \nu}\right|^2 <V, \nu> \, dl = \int_\Gamma 2 <V, \nabla_\Gamma u> \Delta_\Gamma u \, d\Gamma$$
$$- \int_\Gamma <(\text{div}_\Gamma(V)I_d - 2\varepsilon_\Gamma(V)).\nabla_\Gamma u, \nabla_\Gamma u> \, d\Gamma.$$

For the tangential wave operator we get

Proposition 8. Let u in $L^2(0, \tau, H^2 \cap H^1_0(\omega)) \cap H^2(0, \tau, L^2(\omega))$ be such that

$$u_{tt} - \Delta_\Gamma u \in L^2(0, \tau, L^2(\omega))$$

and assume that $[V(t)u_t]_{t=0}$ and $V(\tau) = 0$. Then

$$- \int_0^\tau \int_\gamma (\frac{\partial u}{\partial \nu})^2 <V, \nu> \, dt\, dl = 2\int_0^\tau \int_\Gamma <V, \nabla_\Gamma u> (u_{tt} - \Delta_\Gamma u) \, dt\, d\Gamma$$
$$- \int_0^\tau \int_\Gamma <(\text{div}_\Gamma(V)I_d - 2\varepsilon_\Gamma(V)).\nabla_\Gamma u, \nabla_\Gamma u> \, dt\, d\Gamma$$
$$+ \int_0^\tau \int_{\Sigma_h} 2u_t <\frac{\partial V}{\partial t}, \nabla u> - (u_t)^2 \text{div}_\Gamma(V_\Gamma) \, dt\, d\Sigma_h.$$

Using that identity and the corresponding density result we shall derive the tangential estimate for the normal derivative. Consider the space

$$\mathcal{H}(\omega) = \{ u \in H^1(0, \tau, L^2(\omega)) \cap L^2(0, \tau, H^1_0(\omega)) : \frac{\partial^2}{\partial t^2} - \Delta_\Gamma u \in L^2(Q_\Gamma)\}$$

Denote by Σ_h the lateral boundary

$$\Sigma_h = \{x \in \mathcal{U}_h : p(x) \in \gamma\}$$

and by p_{Σ_h} the *projection mapping* onto Σ_h. Introduce the distance function b_{S_h} (see [1] and [2]) defined over $\mathcal{U}$ and not to be confused with b_Ω, where
$$S_h = \{x \in \mathcal{U}_h : p(x) \in \omega, \text{ and } |b_\Omega(x)| \leq h\}$$
is the *normal cylinder* to ω. For $u \in \mathcal{H}(\omega)$ set $\phi = u \circ p$. Then ϕ is an element of
$$\mathcal{H}(S_h) = \{\phi \in H^1(0, \tau, L^2(S_h)) \cap L^2(0, \tau, H^1(S_h)),$$
$$\phi = 0 \text{ on } \Sigma_h, \frac{\partial^2 \phi}{\partial t^2} - A.\phi \in L^2(0, \tau, L^2(S_h))\}$$
where the operator A is defined in divergence form in S_h as follows
$$A.\phi = \operatorname{div}\left([I + b_\Omega D^2 b_\Omega \circ p] . \nabla(\phi \circ p) \circ p\right) + a_0 \phi \circ p,$$
where a_0 is a first order operator. The Hilbert space $\mathcal{H}(S_h)$ is endowed with the graph norm. The element
$$\phi_{b_{S_h}} = b_{\Sigma_h}^{-1} \, u \circ p \in H^1(S_h).$$
Assume that the cylinder S_h is locally starshaped in R^N so that
$$\phi_n = b_{S_h} \rho_n *_{t,x} (r_i \phi_{b_{\Sigma_h}})^{n_i}$$
converges to ϕ in $\mathcal{H}(S_h)$.

Remark 1. *In all the constructions of the paper the fact that $\phi/b \in H^1(\Omega)$ or $H^1(\omega)$ is not essential. In fact for $N \leq 3$ it is sufficient that given a small $\varepsilon > 0$ there exists some $\alpha > 0$ such that $\phi/(-b)^\alpha \in L^{6-\varepsilon}(\Omega)$ or $L^{6-\varepsilon}(\omega)$.*

1. M.C. Delfour and J.-P. Zolésio, *Shape analysis via distance functions*, J. Funct. Anal. **123** (1994), 129–201.
2. ———— *Shape analysis via distance functions: local theory*, CRM Report 2299, March 1996, Université de Montréal.
3. ———— *On the design and control of systems governed by differential equations on submanifolds*, Control and Cybernetics 25 (1996), 497-514.
4. ———— *Hidden boundary smoothness for some classes of differential equations on submanifolds*, in "Proc. 1996 Joint Summer Research Conference on Optimization Methods in PDE's", S. Cox and I. Lasiecka, eds, Contemporary Mathematics, AMS Publications, Providence, R.I., 1997.
5. ———— *Intrinsic differential geometry and theory of thin shells*, Quaderni, Scuola Normale Superiore, Pisa (Italy), to appear; also as lecture notes, version 1.0, August 1996.
6. I. Lasiecka, J.L. Lions, and R. Triggiani, *Non Homogeneous Boundary Value Problems for Second Order Hyperbolic Operators*, J. Math. pures et appl. 65 (1986), 149–192.
7. D.L. Russell, *Controllability and stabilizability theory for linear partial differential equations: recent progress and open questions*, SIAM Review 20 (1978), 639–739.

JEAN DETEIX
The dam problem:
safety factor and shape optimization

1. Introduction.
Many free boundary and shape optimization problems can be found in soil mechanics: coastal erosion, landslides, pollutants seepage, etc. The problem that we propose originate from an article of De Mello. In his article, [4], De Mello compare 3 dams showing that certain shapes and positions of a filter (a filter is a part of the dam composed of a material with different properties from the rest of the dam) should be preferred in relation with downstream failures. From this simple fact we can create a shape optimization problem: how can we distribute two materials (one of them taken as the filter) or what is the shape of the filter so that downstream ruptures are practically impossible. We will present a model and formulation of this shape problem and briefly present some theoretical and numerical results.

Free boundary and shape optimization problems are generally solved using ad hoc approach (e.g. [2] for the dam) or using an a priori knowledge of the geometry of the optimal shape (e.g. boundary perturbations in [11]). Here we propose to use an approach which seems to solve a great variety of shape and free boundary problems.

In this approach instead of a shape we look for a relaxed characteristic function (meaning that $\chi \in \{0,1\}$ a.e. become $\chi \in [0,1]$ a.e.). As in any optimization problem we are mainly confronted with a problem of topology and compactness. In our approach the definition of the set of admissible functions and the topology on this set is based on the compactness for the Caccioppoli sets [7]. The peculiarity of our approach is the systematic use of relaxed characteristic functions which link this method to the work of Kohn [8]. This relaxation makes the conception of numerical schemes easier, [5], and may be used for the existence of a solution.

Among the first use of this approach we can mention [1] for the dam problem and [6] for the Céa-Malanowski problem. In both cases no hypothesis on the shape or the boundary are made so the solution with this approach is global (compare the hypothesis made by [2] and [1] for the dam). Finally not only does this technique gives us a sound mathematical basis and theoretical results but it allows us to obtain, in a very natural way, numerical schemes using only one grid ([8], [10], [5]).

We need a tool to compare the dams, a cost function, associated with downstream failures. Following [4] we will use a measure of the capacity of resistance to ruptures called a safety factor. Here we will define this factor as a ratio of two moments (limit equilibrium method). This definition will take into account the nature of the material and the stress in the dam. Following [12] we will have to define the interstitial pressure in the dam. For the pressure head (and the interstitial pressure) we will follow the works of Alt [1] (the pressure is solution of a free boundary problem).

2. The physical quantities and their constraints.

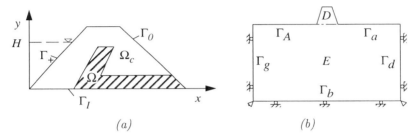

FIGURE 1.

Let D be a two-dimensional model of the dam (Figure 1(a)), $\Omega \subset D$ be the filter and $\Omega_c = D \setminus \Omega$. The base of D, Γ_I, is impervious, H is the level of the reservoir and the two materials are isotropics and homogeneous. We neglect evaporation but we take into account capillary effects. Let p_c be the capillary pressure, p_A the atmospheric pressure ($p_c \leq p_A$, neglecting the capillary effects correspond to $p_c = p_A = 0$.) Let χ_Ω be the characteristic function of Ω, K_1 and K_2 be constants associated to the permeability of the materials composing Ω and Ω_c respectively. Introducing $K_\Omega(x) = K_1 \chi_\Omega(x) + K_2(1 - \chi_\Omega(x))$, p_+ the pressure along Γ_+ (see [5] for details) and

$$M = \left\{ v \in H^1(D) | \ v = p^+ \text{ on } \Gamma_+ \ \ v \leq p_A \text{ on } \Gamma_0 \right\}, \ \vec{e} = \rho_l \vec{g}, \ \vec{g} = g \begin{pmatrix} 0 \\ 1 \end{pmatrix}$$

the free boundary problem for the dam become [1]

$$\begin{cases} \text{find } p \in M \text{ and } S \in L^\infty(D) \text{ such that} \\ \int_D \nabla(v - p) \cdot K_\Omega (\nabla p + S\vec{e}) \ dx \geq 0 \qquad \forall v \in M, \\ p \geq p_c \text{ and } \chi(p > p_c) \leq S \leq 1 \qquad \text{in } D. \end{cases} \quad (1)$$

For the effective stress we consider $D_+ = D \cup E$ (Figure 1(b)) where E is made of an impervious, homogeneous and isotropic material (see [5] or [12]). On $\Gamma_g \cup \Gamma_d \cup \Gamma_b$ we assume a zero horizontal or vertical displacement and on Γ_A a pressure g_A caused by the water body. Let E^1_{ijkh}, E^2_{ijkh}, E^E_{ijkh} and ρ_1, ρ_2, ρ_E be the elastic tensor and the densities of the materials (for Ω, Ω_c and E resp.). Posing $E^\Omega_{ijkh}(x) = \chi_\Omega(x) E^1_{ijkh} + (1 - \chi_\Omega(x)) E^2_{ijkh}$, $\rho_\Omega(x) = \rho_1 \chi_\Omega(x) + \rho_2(1 - \chi_\Omega(x)) + S(x)\rho_l$ and

$$V = \left\{ (v_1, v_2) \in [H(D_+)]^2 \mid v_1 = 0 \text{ on } \Gamma_g \cup \Gamma_d, \ v_2 = 0 \text{ on } \Gamma_b \right\},$$

the effective stress tensor σ_{ij} come from the solution of the problem (we used the repeated indices convention)

$$\begin{cases} \text{find } u_i(x,y) \ i=1,\ 2 \text{ in V such that } \forall v \in V \\ \int_{D_+} \epsilon_{ij}(v)\sigma_{ij}(u) \ dx = -\int_D v \cdot (\nabla p + \rho_\Omega \vec{g}) dx - \int_E v \cdot \rho_E \vec{g} dx - \int_{\Gamma_A} v \cdot g_A \vec{n} \ d\Gamma, \\ \sigma_{ij}(u) = (E^E_{ijkh}\chi_E + E^\Omega_{ijkh}\chi_D)\epsilon_{kh}(u) \quad \epsilon_{ij}(u) = \frac{1}{2}\left(\frac{\partial u_i}{\partial x_j} + \frac{\partial u_j}{\partial x_i}\right) \quad i,j = 1,\ 2. \end{cases} \quad (2)$$

Finally, if we have a filter made of air, then the maximal safety factor is clearly attain for $\Omega = D$. To insure us of a non-trivial ($\Omega \neq D$) solution we will add a discharge constraint or more simply a volume constraint

$$0 \leq \int_D \chi_\Omega \ dx \leq \alpha < \int_D dx. \tag{3}$$

3. Safety factor and formulation of the problem.

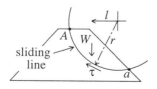

FIGURE 2.

Suppose that all failures are landslides along circular curves, caused by the water pressure. Let M_M be the moment associated with the shear force and M_R the shear strength moment along the curve $A - a$ (Figure 2).

M_R is associated with the maximal resistance to motion and from the Coulomb equation [3] we have

$$M_M = r \int_a^A \tau \, ds, \qquad M_R = r \int_a^A \tau_r \, ds = r \int_a^A c - \sigma_n \, \mathrm{tg}\, \phi \, ds$$

where r is the radius of the circle, τ the shear stress, σ_n the normal stress. The cohesion c and ϕ the angle of internal friction are parameters depending uniquely on the material (size of grain, etc). Let c_i ϕ_i $i = 1, 2$ be the parameters for Ω, Ω_c respectively, $c_\Omega = \chi_\Omega c_1 + (1 - \chi_\Omega) c_2$, $\phi_\Omega = \chi_\Omega \phi_1 + (1 - \chi_\Omega) \phi_2$ and

$$F = \frac{M_R}{M_M} = \frac{\int_a^A (c_\Omega - \sigma_n \, \mathrm{tg}\, \phi_\Omega) \, ds}{\int_a^A \tau \, ds}.$$

$F \leq 1$ gives us a sliding line. Since we don't know if there is a sliding line and where it is; we have to consider all the curves that are possible sliding line. Let C_{ad} be the set of circular curves associated with downstream landslide (with a bound on the radius C_{ad} correspond to a compact of $\mathbb{R}^3$). The dam is defined as stable when F is greater then 1 for all curves in C_{ad}. This suggest a definition for the safety factor F_s and since the factor is a function of Ω (by (1)-(2), p, σ and F are functions of Ω) we denote

$$F_s(\Omega) = \min_{\mathcal{C} \in C_{ad}} F(\mathcal{C}) = \min_{\mathcal{C} \in C_{ad}} \frac{\int_\mathcal{C} (c_\Omega - \sigma_n \, \mathrm{tg}\, \phi_\Omega) \, ds}{\int_\mathcal{C} \tau \, ds}.$$

Now the shape optimization problem can be formulate as

$$\max_{\Omega \subset D} F_s(\Omega)$$

under the condition that (1), (2) and (3) are satisfied.

4. Relaxation of the shape problem.

We formulated our problem in conditions such that only Ω can modify F_s. Moreover the relations to Ω are completely expressed by the use of χ_Ω. To complete our formulation we have to describe the admissibles sets (or equivalently the set of characteristic functions noted $X(D)$) and then we have to relaxe this set (which will give us the set noted $X_r(D)$). The major point in the construction of those sets is that we don't want to impose severe constraint on admissible shape (such as connexity or a limit on the number of holes) but we want to be able to have a topology on those sets insuring an existence result (at least for $X_r(D)$). For a discussion concerning this construction see [5]. Those requirements lead us to the use of finite perimeter or Cacciopoli sets. Let

$$BV(D) = \{\phi \in L^1_{loc}(D) \mid \mathcal{P}(\phi) < \infty\} \qquad \mathcal{P}(\phi) = \sup_{\substack{\zeta \in C^1_0(D) \\ \|\zeta\|_\infty \leq 1}} \int_D \phi \operatorname{div} \zeta \, dx.$$

and

$$X(D) = \{\chi \in BV(D) \mid \chi(x) \in \{0,1\} \text{ a.e. in D}, \mathcal{P}(\chi) \leq r\}.$$

For $\chi \in X(D)$ we have $\Omega = \{x \in D \mid \chi(x) = 1\}$ and the shape problem is to maximize $F_s(\chi)$ for $\chi \in X(D)$ under the condition that (1)-(3) are satisfied. [7] gives us the strong compactness of $X(D)$ in $L^1(D)$ (for any $r > 0$). Applying the relaxation to this set we have

$$X_r(D) = \{\chi \in BV(D) \mid \chi(x) \in [0,1] \text{ a.e. in D}, \mathcal{P}(\chi) \leq r\}.$$

$X_r(D)$ is a bounded, convex subset, compact in $L^1(D)$-strong and by [6] this set is compact in $L^p(D)$-strong for $p \in [1, \infty[$. We want a topology on $X_r(D)$ such that F_s is semi-continuous (at least) and $X_r(D)$ compact. Weak semi-continuity of F_s implies weak convergence of the solution of (1)-(2). According to [9] it seems unreasonable to hope for such convergence (we can have existence in spite of [9], see [6] for example). From this we choose to use the $L^2(D)$-strong topology for $X_r(D)$.

The choice of $X(D)$ and $X_r(D)$ brings us a new difficulty: F_s is not well defined. A priori for $\chi \in X(D)$ or $\chi \in X_r(D)$ the integral along $\mathcal{C} \in \mathcal{C}_{ad}$ of $\sigma_n \operatorname{tg} \phi_\Omega$ have no meaning. To solve this new problem we consider

$$\mathcal{C}_\delta = \{x \in D \mid d(\mathcal{C}, x) < \delta\} \qquad \overline{n} = \frac{x - x_\varrho}{\varrho} \in C^\infty(\overline{D})$$

where $d(\mathcal{C}, x)$ is the distance from $\mathcal{C}$ to x and x_ϱ is the center of the circle associated with $\mathcal{C}$. We replace the integrals on $\mathcal{C}$ by the integrals over $\mathcal{C}_\delta$:

$$\int_\mathcal{C} f(s) ds \rightsquigarrow \frac{\int_\mathcal{C} ds}{\int_{\mathcal{C}_\delta} dx} \int_{\mathcal{C}_\delta} f(x) \, dx = \ell(\mathcal{C}) \fint_{\mathcal{C}_\delta} f(x) \, dx.$$

Introduicing

$$\sigma_n = \sigma n \cdot n \rightsquigarrow \sigma \overline{n} \cdot \overline{n} = \overline{\sigma}_n, \qquad \tau \rightsquigarrow \left(|\sigma \overline{n}|^2 - (\overline{\sigma}_n)^2\right)^{1/2} = \overline{\tau}$$

to F we can substitute

$$F^\delta(\mathcal{C}, \chi) = \frac{\ell(\mathcal{C})\displaystyle\oint_{\mathcal{C}_\delta} c_\Omega - \overline{\sigma}_n \operatorname{tg} \phi_\Omega \, dx}{\ell(\mathcal{C})\displaystyle\oint_{\mathcal{C}_\delta} \overline{\tau} \, dx} = \frac{\displaystyle\int_D (c_\Omega - \overline{\sigma}_n \operatorname{tg} \phi_\Omega)\chi_{\mathcal{C}_\delta} \, dx}{\displaystyle\int_D \overline{\tau}\chi_{\mathcal{C}_\delta} \, dx}.$$

Introducing $J(\chi) = \min_{\mathcal{C} \in \mathcal{C}_{ad}} F^\delta(\mathcal{C}, \chi)$ the shape problem is replaced by the problem (we call this problem a relaxed shape problem):

$$\max_{\chi \in X_r(D)} J(\chi)$$

under the constraint that (1)-(3) are satisfied.

5. Theoretical considerations.

In this formulation the shape we are seeking is contained in D which is finite. D is called a control volume or hold-all since the "maximal" shape is $\Omega = D$. The presence of D allow us to use some simple functional analysis tools. In general D is not present and we will have to impose a control volume D if we want to use those tools.

For an existence result the plan is very simple. For $\chi \in X_r(D)$ we want existence and uniqueness for (1) and (2), continuity for (1)-(3), semi-continuity of $F^\delta(\mathcal{C}, \chi)$ in $\mathcal{C}$ and χ and with those results we will have the semi-continuity of J.

Existence of a solution for (1) is given in [1] and is a classical result for (2). Uniqueness is more difficult. It is possible to show that (1) has a unique solution when $\operatorname{div}(K_\Omega(x)e) \geq 0$ and if the base of D is a monotone graph ([5]). For the general case the problem of uniqueness is still open. Since in our case the base of D, Γ_I, is flat and since uniqueness seems to be based on the geometry of D we will make the conjecture that we have uniqueness. It should be mentioned that numerically this assumption seems to be verified. Continuity of the solution of (1)-(3), continuity of $F^\delta(\mathcal{C}, \chi)$ in χ and in $\mathcal{C}$ and semi-continuity of J are easily obtained (see [5]). Finally the compactness of $X_r(D)$ gives us

THEOREM 1. *Let $\delta > 0$ and*

$$J(\chi) = \min_{\mathcal{C} \in \mathcal{C}_{ad}} F^\delta(\mathcal{C}, \chi) = \min_{\mathcal{C} \in \mathcal{C}_{ad}} \frac{\displaystyle\int_D c_\Omega \chi_{\mathcal{C}_\delta} - \overline{\sigma}_n \operatorname{tg} \phi_\Omega \chi_{\mathcal{C}_\delta} \, dx}{\displaystyle\int_D \overline{\tau}\chi_{\mathcal{C}_\delta} \, dx}$$

then there exists $\hat{\chi}_\delta \in X_r(D)$ such that $J(\hat{\chi}_\delta) \geq J(\chi) \quad \forall \chi \in X_r(D)$.

The relaxation of characteristic functions is not necessary for the existence result which means that we have existence of a solution in $X(D) \subset X_r(D)$ ([5]). In our case we don't have a proof that the solution is in fact a characteristic function (i.e. $\hat{\chi}_\delta \in \{0,1\}$ a.e.). Thus, a priori, D is divided in three parts: $\{\chi = 1\}$, $\{\chi = 0\}$ and a "fuzzy" region where $\{\chi \in]0,1[\}$. For this zone the elastic constants are linear functions of χ so we can't, as in [8], associate this region with a composite material with periodic structure. However it is possible to consider (at a low cost theoretically and numerically) a penalization of J that gives suboptimal solution in $X(D)$.

The continuity result with respect to $X_r(D)$ allow us to obtain existence results for a great number of problems. We can consider problems using more then two materials (non-binary problems). Those problems correspond (see [5]) to the replacement of a characteristic function by a vector of characteristic functions, for example the optimization of the shape of the dam (D) and of the filter (Ω) simultaneously. The continuity results for (1)-(3) allow us to have existence of a solution for this non-binary problem as long as the new cost function is semi-continuous.

6. Numerical results.

This approach of the shape problem allow us to generate very easily numerical scheme that use only one grid. We can produce a scheme of finite element type, [5], that gives us a solution to the problem while ignoring the conditions imposed for mathematical (theoretical) reasons (modifications of F). In this case the relaxation is used only to avoid the use of operational research technique to obtain numerical results (for the motivation of this point see [5]).

The numerical method proposed in [5] is based on the approximation of S and χ by constant on barycentric cells, p and u by piecewise linear functions. For (1) the variational inequality need a special treatment (see [10]). For a complete analysis of the discretization and resulting scheme see [5]. Space being limited we present here only two cases where C_{ad} is composed of a unique curve.

$Test\ 1:\ E_2 = 1.5E_1,\ \nu_2 = 2\nu_1, K_1 = 10K_2,\ c_1 = 1.5c_2,\ \phi_1 = \phi_2,\ \alpha = 0.25\,\text{mes}(D)$.

$Test\ 2:\ E_1 = 1.5E_2,\ \nu_1 = 2\nu_2, K_1 = 10K_2,\ c_2 = 1.5c_1,\ \phi_1 = \phi_2,\ \alpha = 0.25\,\text{mes}(D)$.

where $E_i\ i = 1,2$ is the Young module, $\nu_i\ i = 1,2$ is the Poisson coefficient for the filter and the rest of the dam resp. and we use the plane strain hypothesis. In [5] can be found many other tests and a detailed analysis of the results.

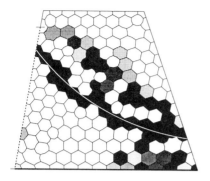 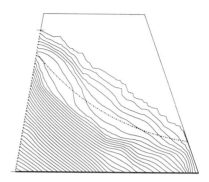

FIGURE 3.

In the first test, Figure 3, the filter is placed on the supposed sliding line (i.e. $\chi = 1$ for all the cells that touch the curve). We can see that the rest of the material is placed along the free boundary of the liquid (see the pressure head, Figure 3 right). Finally the upper bound for (3) is attained, so we use the material 1 as much as possible.

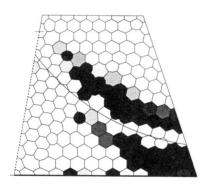 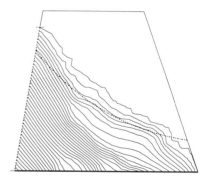

FIGURE 4.

For the second test we take materials with reverse mechanical properties (with respect with the first test). In Figure 4, the solution seems to be to reinforce the base of the dam in the downstream zone while avoiding to put the filter on the sliding line (i.e. $\chi = 0$ on cells that touch the curve). From the pressure head (Figure 4 right) again we have part of the filter along the free boundary of the liquid. As for the other test we use all the material allowed i.e. the upper bound in (3) is attained.

JEAN DETEIX (deteix@crm.umontreal.ca)
Université de Montréal, Dept. Mathématiques et Statistique
CP 6128, Succ Centre-ville, Montréal, Canada, H3C 3J7

REFERENCES

[1] W. H. ALT, *Strömungen durch inhomogene poröse medien mit freiem rand.*, J. Reine Angew. Math., **305** (1979), 89-115.

[2] C. BAIOCCHI, V. COMINCIOLI, E. MAGENES AND G. A. POZZI, *Free boundary problems in the theory of fluid flow through porous media: existence and uniqueness theorems.*, Ann. Mat. Pura Appl., **96** 4 (1973), 1-82.

[3] C. B. BROWN AND L. E. GOODMAN, *Dead load stresses and instability of slopes.*, J. Soil Mech. Found. Div., Am. Soc. Civil Engrs., **83** (1963), SM3,

[4] V. F. B. DE MELLO, *Reflections on design decisions of practical significance to embankment dams*, Géotechnique, **27** (1977), 3, 279-355.

[5] J. DETEIX, *Conception optimale d'une digue à deux matériaux*, Thèse de Doctorat, Université de Montréal, 1997.

[6] M. C. DELFOUR AND J. P. ZOLESIO, *The optimal design problem of Céa and Malanowski revisited.*, Progr. Systems Control Theory, 19, Birkhäuser Boston, 1995, 133-150.

[7] GIUSTI, *Minimal surfaces and functions of bounded variation*, Birkhäuser, Boston, Basel, Stuttgart, 1984.

[8] R. KOHN, *Numerical structural optimization via a relaxed formulation.*, Shape optimization and free boundaries (Montreal, 1990), 173-210, NATO Adv. Sci. Inst. Ser. C Math. Phys. Sci., 380, Kluwer, Dordrecht, 1992.

[9] F. MURAT, *Un contre-exemple pour le problème du contrôle dans les coefficients.* C. R. Acad. Sci. Paris Sér.A-B **273** 1971 A708–A711.

[10] P. PIETRA, *An up-wind finite element method for a filtration problem.*, R.A.I.R.O. Analyse numérique, **16** (1982), 4, 463-481.

[11] O. PIRONNEAU, *Optimal shape design with applications to aerodynamics.*, Shape optimization and free boundaries (Montreal, 1990), 211-252, NATO Adv. Sci. Inst. Ser. C Math. Phys. Sci., 380, Kluwer, Dordrecht, 1992.

[12] O. C. ZIENKIEWICZ, *Stress analysis of hydraulic structures including pore pressure effects*, Water Power, vol. 15 (March 1963), 104-108.

J KARRAKCHOU, F LAHMIDI, A NAMIR AND M RACHIK
Stability, observers and compensators for discrete-time delay systems[1]

1. Introduction

The theory of linear discrete-time systems is one of the most important notions in systems analysis. A number of contributions can be found in the engineering and mathematical literature (see [2,3,6 ...]). However, in most of the studies available, several parameters, especially the delays, are neglected. In this work, we consider a class of linear discrete-time distributed systems described by the following delay equation

$$\begin{cases} \xi_{i+1} = \sum_{j=0}^{m} A_j \xi_{i-j} + \sum_{j=0}^{q} B_j u_{i-j}, \quad i \in \mathbb{N}, \\ u_k = \mu_k \text{ for } -q \leq k \leq -1 \text{ and } \xi_k \text{ is given for } -m \leq k \leq 0, \end{cases} \quad (1)$$

with the output equation

$$y_i = \sum_{j=0}^{p} C_j \xi_{i-j}, \ i \in \mathbb{N}, \text{ with } p \leq m, \quad (2)$$

where $\xi_i \in \mathcal{X}$, $A_j \in L(\mathcal{X}, \mathcal{X})$, $B_j \in L(U, \mathcal{X})$ and $C_j \in L(\mathcal{X}, Y)$; $(\mu_{-q}, \mu_{-q+1}, \ldots, \mu_{-1})$ is a given sequence in U and $y_i \in Y$ ($\mathcal{X}, U, Y$ are Hilbert spaces).

Our objective is to study and to analyse the concepts of stability, stabilizability, observers and compensators of such systems. The work contains five sections. In section 2, we show that the delay equation (1) (with $(u_i)_{i \geq -q} \equiv 0$) can be transformed to a system of difference equation on a product space. Sufficient and necessary conditions for the stability of the new system are then established. Stabilizability is treated in section 3. As in section 2, we rewrite the delay equation (1) in an extended space so that the delays disappear in the new state equation. To examine the stabilizability of the new system, we decompose its state space X to $X = X_u \oplus X_s$, where X_u and X_s are subspaces of X such that the projection onto X_s is stable. The idea therefore is to stabilize the projection onto X_u without upsetting the stability properties on X_s. We will present a precise formulation of this result. Section 4 is devoted to the study of observers. Luenberger (see [4]) introduced observers for linear distributed systems, and many authors had since extended that theory. Our contribution consists

---
[1] Supported in part by the CNRPST, Morocco.

of investigating the existence of observers for the systems whose states evolutions are described by equations (1),(2). Sufficient conditions for the existence of an observer are presented. The case of an identity observer is also considered and we establish a method to construct such observer. In section 5 we treat compensators. A compensator is a system that takes the observation and the control of the initial system as, respectively, its input and output. We will focus our study on the compensator of finite order for which we present a method of its construction.

2. Stability

Through this section, we assume that the system (1) is not controlled $((u_i)_{i\geq -q} \equiv 0)$.

Definition 1. *The system (1) is said to be stable if* $\sum_{i=-m}^{\infty} \| \xi_i \|^2 < \infty$.

To the delay equation (1), we associate the following difference system

$$\begin{cases} x_{i+1} = Ax_i, & i \in \mathbb{N}, \\ x_0 \in X, \end{cases} \quad (3)$$

with $X = \mathcal{X}^{m+1}$ and $A = \begin{pmatrix} A_0 & A_1 & \cdots & \cdots & A_m \\ I & 0 & \cdots & \cdots & 0 \\ 0 & I & \ddots & & \vdots \\ \vdots & \ddots & \ddots & \ddots & \vdots \\ 0 & \cdots & 0 & I & 0 \end{pmatrix}$.

Proposition 2. *Let $(\xi_i)_{i\geq -m}$ be the solution of the system (1) corresponding to the initial data $(\xi_k)_{-m\leq k\leq 0}$, then $x_i = (\xi_i, \ldots, \xi_{i-m})^T$, $i \geq 0$, is the solution of (3) for the initial data $x_0 = (\xi_0, \ldots, \xi_{-m})^T$.*

The following proposition gives a characterization of the stability of system (1).

Proposition 3. *The following statements are equivalent*
(a) the system (1) is stable,
(b) the system (3) is stable,
(c) $\sigma(A) \subset \{\lambda \in \mathbb{C} / |\lambda| < 1\}$, where $\sigma(A)$ denotes the spectrum of A,
(d) $r_\sigma(A) < 1$, $r_\sigma(A)$ is the spectral radius of A defined by $r_\sigma(A) := \lim_{i\to\infty} \| A^i \|^{1/i}$,
(e) there exist two real constants $M \geq 1$, $k \in]0,1[$ such that $\| A^i \| \leq Mk^i$ for all $i \geq 0$
(f) $\lim_{i\to\infty} \| A^i \| = 0$.

<u>Proof</u> (a)⇔(b) by the proposition 2., the proof of (b)⇒(c) is similar to the one of Th4.1.1 [1]. By direct computation we can prove the implications (c)⇒ (e) ⇒(f) ⇒(c), (e)⇒ (f) and (c)⇔ (d).

3. Stabilizability

Definition 4. *The system (1) is stabilizable if for any initial data $(\xi_j)_{-m \leq j \leq 0}$, there exists a control $(u_i)_{i \geq 0}$ such that the corresponding solution satisfies $\sum_{i=-m}^{\infty} \| \xi_i \|^2 < \infty$.*

To (1) we associate the related system

$$\begin{cases} x_{i+1} = Ax_i + Bu_i, & i \in \mathbb{N}, \\ x_0 \in X, \end{cases} \quad (4)$$

with $X = \mathcal{X}^{m+1} \times U^q$, $A \in \mathcal{L}(X)$ and $B \in \mathcal{L}(U, X)$ given by

$$A = \begin{pmatrix} A_0 & A_1 & \cdots & \cdots & A_m & B_1 & \cdots & \cdots & \cdots & B_q \\ I & 0 & \cdots & & 0 & 0 & \cdots & \cdots & \cdots & 0 \\ 0 & \ddots & \ddots & & \vdots & \vdots & & & & \vdots \\ \vdots & \ddots & \ddots & \ddots & \vdots & \vdots & & & & \vdots \\ 0 & \cdots & 0 & I & 0 & 0 & \cdots & \cdots & \cdots & 0 \\ 0 & \cdots & \cdots & \cdots & 0 & 0 & \cdots & \cdots & \cdots & 0 \\ \vdots & & & & \vdots & I & \ddots & & & \vdots \\ \vdots & & & & \vdots & 0 & \ddots & \ddots & & \vdots \\ \vdots & & & & \vdots & \vdots & \ddots & \ddots & \ddots & \vdots \\ 0 & \cdots & \cdots & \cdots & 0 & 0 & \cdots & 0 & I & 0 \end{pmatrix} \text{ and } B = \begin{pmatrix} B_0 \\ 0 \\ \vdots \\ \vdots \\ 0 \\ I_U \\ 0 \\ \vdots \\ \vdots \\ 0 \end{pmatrix}.$$

Proposition 5. *Let $(\xi_i)_{i \geq -m}$ be the solution of the system (1) with the initial data $(\xi_k)_{-m \leq k \leq 0}$ and $(\mu_k)_{-q \leq k \leq -1}$, then the state $x_i = (\xi_i, \ldots, \xi_{i-m}, u_{i-1}, \ldots, u_{i-q})^T$, $i \geq 0$, is the solution of (4) for the initial state $x_0 = (\xi_0, \ldots, \xi_{-m}, \mu_{-1}, \ldots, \mu_{-q})^T$.*

Proposition 6. *If the system (4) is stabilizable then the system (1) is stabilizable.*

<u>Proof</u> Let $x_0 = (\xi_0, \ldots, \xi_{-m}, \mu_{-1}, \ldots, \mu_{-q})^T$. Since the system (4) is stabilizable, there exists a control sequence $(u_i)_{i \geq 0}$ such that the solution $(x_i)_{i \geq 0}$ of the system (4) satisfies $\sum_{i=0}^{\infty} \| x_i \|^2 < \infty$. By Proposition 5., we have $x_i = (\xi_i, \ldots, \xi_{i-m}, u_{i-1}, \ldots, u_{i-q})^T$. Hence $\sum_{i=0}^{\infty} \| x_i \|^2 = \sum_{i=0}^{\infty} (\sum_{j=0}^{m} \| \xi_{i-j} \|^2 + \sum_{j=1}^{q} \| u_{i-j} \|^2) < \infty$. Thus $\sum_{i=-m}^{\infty} \| \xi_i \|^2 < \infty$. ∎

Because of Proposition 6., we will focus our study on stabilizing the system (4). The method that we use is via the decomposition of the state space X.

Definition 7. *The operator A is said to satisfy the spectrum decomposition assumption if $\sigma_u(A) := \{\lambda \in \sigma(A) / |\lambda| \geq 1\}$ is a bounded, closed and open subset in the relative topology of $\sigma(A)$.*

In this case (see [5], prop 4.11):
(1) the space X can be decomposed to $X = X_u \oplus X_s$, where X_u and X_s are subspaces of X such that $AX_u \subset X_u$, $AX_s \subset X_s$, $X_u = PX$, $X_s = (I - P)X$, where P is the projection on X_u and P is bounded.
(2) If A_u and A_s are the restrictions of A to X_u and X_s respectively, then $\sigma(A_u) = \sigma_u$ and $\sigma(A_s) = \sigma(A) - \sigma_u$.

In the following we will assume that the operator A satisfies the spectrum decomposition assumption. The operators A and B reduce to

$$A = \begin{pmatrix} A_u & 0 \\ 0 & A_s \end{pmatrix} \text{ and } B = \begin{pmatrix} B_u \\ B_s \end{pmatrix}.$$

Proposition 8. *The system (4) is stabilizable if the following hold*
(a) the operator A satisfies the spectrum decomposition assumption,
(b) there exists an operator $D_u \in \mathcal{L}(X_u, U)$ such that $A_u + B_u D_u$ is stable.

<u>Proof.</u> Let $D := (D_u \ 0) : x \in X \mapsto D_u x_u \in U$. Then

$$A + BD = \begin{pmatrix} A_u + B_u D_u & 0 \\ B_s D_u & A_s \end{pmatrix}.$$

The operator $A + BD$ is stable and hence $u_i = Dx_i$ stabilize the system (4). ∎

Example 1 Consider the system

$$\begin{cases} \xi_{i+1} = A_1 \xi_{i-1} + B_0 u_i, & i \in \mathbb{N}, \\ \xi_{-1} \text{ and } \xi_0 \text{ are given}, \end{cases} \quad (5)$$

where $\xi_i \in \mathcal{X} = L^2(0,1); U = \mathbb{R}$;

$$A_1 : x \in \mathcal{X} \mapsto \sum_{n=1}^{\infty} e^{-(n-1)^2 \pi^2 \delta} <x, e_n> e_n, \quad e_n = \sqrt{2}\sin(n\pi.), \quad \delta > 0,$$

$$B_0 : \alpha \in \mathbb{R} \mapsto \alpha e_1 \in L^2(0,1).$$

The related system is given by

$$\begin{cases} x_{i+1} = Ax_i + Bu_i, & i \geq 0, \\ x_0 \in X, \end{cases} \quad (6)$$

where $X = \mathcal{X}^2$, $A = \begin{pmatrix} 0 & A_1 \\ I & 0 \end{pmatrix}$ and $B = \begin{pmatrix} B_0 \\ 0 \end{pmatrix}$. The spectrum of A is given by

$$\sigma(A) = \{0\} \cup \{-e^{-\frac{1}{2}(n-1)^2\pi^2\delta}, e^{-\frac{1}{2}(n-1)^2\pi^2\delta}; \ n \geq 1\}.$$

Hence $\sigma_u(A) = \{-1, 1\}$ and A satisfies the spectrum decomposition assumption. The space X can be decomposed to $X = X_u \oplus X_s$ where

$$X_u = \left\{ \begin{pmatrix} \alpha e_1 \\ \beta e_1 \end{pmatrix} / \alpha, \beta \in \mathbb{R} \right\}, \quad X_s = \left\{ \begin{pmatrix} x \\ y \end{pmatrix} \in X / <x, e_1> = <y, e_1> = 0 \right\}.$$

One can easily check that the operators A_u and B_u are given by

$$A_u = \begin{pmatrix} 0 & I \\ I & 0 \end{pmatrix} \text{ and } B_u : \alpha \in \mathbb{R} \mapsto \begin{pmatrix} \alpha e_1 \\ 0 \end{pmatrix}.$$

Let $D_u : \begin{pmatrix} \alpha e_1 \\ \beta e_1 \end{pmatrix} \in X_u \mapsto -k\beta \in \mathbb{R}$, where $k \in]0, 1[$. It is easy to see that $r_\sigma(A_u + B_u D_u) = (1-k)^{1/2} < 1$, and hence the operator $A_u + B_u D_u$ is stable. Let $D := (D_u \ 0)$. By Proposition 8, the control $u_i = Dx_i, i \geq 0$ stabilize the system (6), and by Proposition 6., it also stabilize the system (5).
As a numerical example, pose $\xi_{-1} = e_1$, $\xi_0 = 2e_1$ and $\delta = 0.1$. For $k = 0.5$ and $k = 0.7$ respectively, we obtain the following numerical results

	N	4	10	13	18	21
$k = 0.5$	$\sum_{i=-1}^{N} \|\xi_i\|^2$	6.562500	6.665039	6.666504	6.666660	6.666666
$k = 0.7$	$\sum_{i=-1}^{N} \|\xi_i\|^2$	5.490500	5.494502	5.494505	5.494505	5.494505

4. Observers

Through this section we assume that $p = m$. Consider the system

$$\begin{cases} z_{i+1} = \sum_{j=0}^{m} F_j z_{i-j} + G y_i + \sum_{j=0}^{q} H_j u_{i-j}, \quad i \in \mathbb{N}, \\ z_k \in Z, \text{ for } k = -m, \ldots, 0, \end{cases} \quad (7)$$

where Z is a Hilbert space, $F_j \in L(Z, Z)$, $G \in L(Y, Z)$ and $H_j \in L(U, Z)$.

Definition 9. *For a given $T \in \mathcal{L}(X, Z)$, the system (7) is said to be an observer of (1),(2) if $\lim_{i \to \infty} (z_i - T\xi_i) = 0$.*

Denote by $\tilde{F}$ the operator on $\mathcal{X}^{m+1}$ given by

$$\tilde{F} = \begin{pmatrix} F_0 & F_1 & \cdots & \cdots & F_m \\ I & 0 & \cdots & \cdots & 0 \\ 0 & I & \ddots & & \vdots \\ \vdots & \ddots & \ddots & \ddots & \vdots \\ 0 & \cdots & 0 & I & 0 \end{pmatrix}.$$

Proposition 10. *The system (7) is an observer of (1),(2) if the following hold*
(a) $H_j = TB_j$; $j = 0, \ldots, q$,
(b) $TA_j - F_jT = GC_j$; $j = 0, \ldots, m$,
(c) the operator $\tilde{F}$ is table.

<u>Proof.</u> Let $e_i = z_i - T\xi_i$. By (a) and (b) $e_{i+1} = \sum_{j=0}^{m} F_j e_{i-j}$. Hence $\lim_{i \to \infty} e_i = 0$. ∎

4.1. Identity observer

We assume now that $Z = \mathcal{X}$ and $T = Id_\mathcal{X}$. The relations (a) and (b) reduce respectively to $H_j = B_j$, $0 \le j \le q$ and $F_j = A_j - GC_j$, $0 \le j \le m$. After resolving these equations, one can define an identity observer for the system (1). In the following, we develop another approach. It consists of constructing an identity observer for the related system (4) and then to deduce an asymptotic estimator of ξ_i.

Consider the output equation (2). Pose $C = (C_0 \ C_1 \ \ldots \ C_m \ \underbrace{0 \ \ldots \ 0}_{q \text{ times}})$, where 0 denotes the zero mapping of $\mathcal{L}(U, Y)$. Then (2) can be written as

$$y_i = Cx_i, \quad i \ge 0. \tag{8}$$

Proposition 11. *Assume that*
(a) the operator A satisfies the spectrum decomposition assumption,
(b) the spaces X_u and Y are finite-dimensional,
(c) there exists $D_u \in \mathcal{L}(X_u, Y)$ such that $A_u^T + C_u^T D_u$ is stable.
Then an identity observer exists for the system (4).

<u>Proof</u> Since X_u and Y are finite-dimensional, the operators A_u, C_u and D_u can be considered as matrices. One can easily see that an identity observer can be defined by

$$\begin{cases} \bar{z}_{i+1} = (A - GC)\bar{z}_i + Gy_i + Bu_i, \\ \bar{z}_0 \in X, \end{cases}$$

where G is given by $G : y \in Y \mapsto -D_u^T y \in X$.

Remark Denote by $\bar{z}_i^0 \in \mathcal{X}$ the first component of $\bar{z}_i$. Since $\lim_{i \to \infty}(\bar{z}_i - x_i) = 0$, we deduce that $\lim_{i \to \infty}(\bar{z}_i^0 - \xi_i) = 0$, and hence $\bar{z}_i^0$ is an asymptotic estimator of ξ_i.

Example 2 Consider again the system (5) described in example 1, with the output equation $y_i = C_0\xi_i$; $i \in \mathbb{N}$, where $y_i \in Y = \mathbb{R}$ and $C_0 : \xi \in L^2(0, 1) \mapsto <\xi, e_1> \in \mathbb{R}$. The related system is then given by

$$\begin{cases} x_{i+1} = Ax_i + Bu_i, \ x_0 \in X, \\ y_i = Cx_i, \end{cases}$$

where

$$A = \begin{pmatrix} 0 & A_1 \\ I & 0 \end{pmatrix}, \ B = \begin{pmatrix} B_0 \\ 0 \end{pmatrix} \text{ and } C = (C_0 \ 0).$$

The space $X = L^2(0,1) \times L^2(0,1)$ can be decomposed to $X = X_u \oplus X_s$, where X_u and X_s are specified in example 3. The subspace X_u is of dimension 2 and can be identified with $\mathbb{R}^2$. The operators A_u and C_u can then be written as

$$A_u = \begin{pmatrix} 0 & 1 \\ 1 & 0 \end{pmatrix} \text{ and } C_u = (1 \ 0).$$

Let $k \in]0,1[$ and define $D_u \in \mathcal{L}(\mathbb{R}^2, \mathbb{R})$ by $D_u = (0 \ -k)$. The operator $A_u^T + C_u^T D_u$ is stable with $\sigma(A_u^T + C_u^T D_u) = \{-(1-k)^{1/2}, (1-k)^{1/2}\}$. Define $G: \alpha \in \mathbb{R} \mapsto -D_u^T \alpha$. The identity observer is then given by

$$\begin{cases} \bar{z}_{i+1} = (A - GC)\bar{z}_i + Gy_i + Bu_i, \\ \bar{z}_0 \in X. \end{cases}$$

Let, for example, $x_0 = \begin{pmatrix} \sum_{n=1}^{\infty} \frac{1}{3^n} e_n \\ \sum_{n=1}^{\infty} \frac{1}{n^3} e_n \end{pmatrix}$, $\bar{z}_0 = \begin{pmatrix} \sum_{n=1}^{\infty} \frac{1}{n^2} e_n \\ \sum_{n=1}^{\infty} \frac{1}{2^n} e_n \end{pmatrix}$, $(u_i)_{i \geq 0} \equiv 0$ and $\delta = 0.1$.

For $k = 0.5$ and $k = 0.7$ respectively, we obtain the following numerical results.

N	4	10	13	18	21
$k = 0.5$ $\|\|\bar{z}_i^0 - \xi_i\|\|^2$	0.028150	0.000435	0.000061	0.000002	0.000000
$k = 0.7$ $\|\|\bar{z}_i^0 - \xi_i\|\|^2$	0.003972	0.000004	0.000000	0.000000	0.000000

5. Stabilizing compensator

Through this section, the spaces U and Y will be assumed to be finite-dimensional. A compensator will be a new system of the form

$$\begin{cases} w_{i+1} = Nw_i + My_i, \\ u_i = Lw_i, \end{cases} \tag{9}$$

with $w_i \in W$ (a Hilbert space), $N \in \mathcal{L}(W,W)$, $M \in \mathcal{L}(Y,W)$ and $L \in \mathcal{L}(W,U)$. The systems (4), (8) and (9) together lead to the following equation

$$\begin{pmatrix} x_{i+1} \\ w_{i+1} \end{pmatrix} = A_e \begin{pmatrix} x_i \\ w_i \end{pmatrix}, \text{ where } A_e = \begin{pmatrix} A & BL \\ MC & N \end{pmatrix}.$$

Definition 12. *The system (9) is said to be a stabilizing compensator for the system (1),(2) if the operator A_e is stable on $X \oplus W$. The order of the compensator is the dimension of its state space W.*

The proposition below gives sufficient conditions for the existence of a compensator.

Proposition 13. *Assume there exist linear mappings $F \in \mathcal{L}(X,U)$ and $G \in \mathcal{L}(Y,X)$, together with a finite-dimensional subspace $V \subset X$ such that the following hold*
(a) $A + BF$ *is stable,*
(b) $A + GC$ *is stable,*
(c) $(A + BF)x \in V$ *for all $x \in V$,*
(d) $Im\, G \subset V$.
Then there exists a stabilizing compensator of order k, where $k = dim\, V$.

<u>Proof</u> Let W be a Hilbert space isomorphic to V and $R : V \to W$ the mapping that provides the isomorphism. Pose $L = FR^{-1}$, $M = -RG$ and $N = R(A+BF+GC)R^{-1}$ and introduce the following subspace and isomorphisms :

$$\mathcal{M} = \{\begin{pmatrix} x \\ Rx \end{pmatrix} / x \in V\} = \{\begin{pmatrix} R^{-1}w \\ w \end{pmatrix} / w \in W\},$$

$$T : w \in W \mapsto \begin{pmatrix} R^{-1}w \\ w \end{pmatrix} \in \mathcal{M} \text{ and } H : \begin{pmatrix} x \\ w \end{pmatrix} \in X \oplus W \mapsto \begin{pmatrix} x - R^{-1}w \\ Tw \end{pmatrix} \in X \oplus \mathcal{M}.$$

We have $HA_e H^{-1} = \begin{pmatrix} A + GC & 0 \\ -TRGC & TR(A+BF)R^{-1}T^{-1} \end{pmatrix}$. Hence A_e is stable. ∎

A method to construct the subspace V is given by

Proposition 14. *Let $F \in \mathcal{L}(X,U)$ and $G \in \mathcal{L}(Y,X)$ such that*
(a) $A + BF$ *is stable,*
(b) *the eigenvectors of $A + BF$ are complete,*
(c) $A + GC$ *is stable.*
Then there exists a stabilizing compensator of finite order for the system (1),(2).

<u>Proof</u> By (c) and Proposition 3., there exist $b \geq 1$ and $\gamma \in]0,1[$ such that $\|(A+GC)^i\| \leq b\gamma^i$ for all $i \geq 0$. Let $k \in]\gamma, 1[$, $\delta = k - \gamma$ and $\{y_i : 1 \leq i \leq p\}$ an orthonormal basis of Y and pose $g_i = Gy_i$, $i = 1,\ldots,p$ and $\varepsilon = b^{-1}p^{-1/2}\|C\|^{-1}\delta$. By the completeness assumption on the eigenvectors of $A + BF$, there exists for every $i = 1,\ldots,p$ a finite set $\{x_{i1},\ldots,x_{iN_i}\}$ of generalized eigenvectors of $A + BF$ such that $\|g_i - \sum_{j=1}^{N_i}\alpha_{ij}x_{ij}\| < \varepsilon$, for suitable numbers $\alpha_{ij}(i = 1,\ldots,p; j = 1,\ldots,N_i)$. To every pair of indices (i,j), there exists a $\lambda_{ij} \in \mathbb{C}$ and an $n_{ij} \in \mathbb{N}$ such that $(\lambda_{ij} - (A+BF))^{n_{ij}}x_{ij} = 0$. Now define the subspace V as follows

$$V := \text{span}\{(\lambda_{ij} - (A+BF))^l x_{ij}/i = 1,\ldots,p; j = 1,\ldots,N_i; l = 0,\ldots,n_{ij}-1\}.$$

V is a finite-dimensional subspace of X and it is invariant under $A + BF$. Write $\hat{g}_i := \sum_{j=1}^{N_i}\alpha_{ij}x_{ij}$ and define $\hat{G} : y_i \in Y \mapsto \hat{g}_i \in X$. Then $Im\,\hat{G} \subset V$ and $\|G - \hat{G}\| \leq \varepsilon p^{1/2} = \delta b^{-1}\|C\|^{-1}$. On the other hand,

$$\|(A+\hat{G}C)^i\| \leq \|(A+GC)^i\| + \cdots + \|\hat{G}-G\|^i\|C\|^i \leq bk^i, \quad i \geq 0.$$

Hence, by Proposition 3., $A + \hat{G}C$ is stable. Using the subspace V and the mappings F and $\hat{G}$, we conclude, by Proposition 13., that there exists a stabilizing compensator of finite order. ∎

Now, we give the main result of this section which proof is based on the following Lemma

Lemma 15. *Consider the system (4) under the assumptions*
(a) $\sigma(A)$ consists only of eigenvalues with finite multiplicities,
(b) the eigenvectors of A are complete,
(c) $\sigma_u(A)$ consists of a finite number of isolated eigenvalues,
(d) there exists $F_u \in \mathcal{L}(X_u, U)$ such that $\sigma(A_u + B_u F_u) \subset \{\lambda \in \mathbb{C}/r < |\lambda| < 1\}$, where $r = \sup\{|\lambda|, \lambda \in \sigma_s(A)\}$.
Then there exists a linear bounded mapping $F : X \to U$ such that $A + BF$ is stable, each element of $\sigma(A+BF)$ is an eigenvalue with finite multiplicity and the eigenvectors of $A + BF$ are complete.

Proof(see [5] Lemma 5.4)

Proposition 16. *Consider the system (4),(8) under the assumptions*
(a) the spectrum of A consists only of eigenvalues with finite multiplicities,
(b) the eigenvectors of A are complete,
(c) $\sigma_u(A)$ consists of a finite number of isolated eigenvalues,
(d) there exists $G_u \in \mathcal{L}(Y, X_u)$ and $F_u \in \mathcal{L}(X_u, U)$ such that $A_u + G_u C_u$ is stable and
$$\sigma(A_u + F_u B_u) \subset \{\lambda \in \mathbb{C}/r < |\lambda| < 1\},$$
where $r = \sup\{|\lambda|, \lambda \in \sigma_s(A)\}$. Then a compensator of finite order exists for the system (1).

Proof Let $G : y \in Y \mapsto G_u y \in X$. Then
$$A + GC = \begin{pmatrix} A_u + G_u C_u & G_u C_s \\ 0 & A_s \end{pmatrix}.$$

Hence $A + GC$ is stable. Using the precedent results, the conclusion is immediate. ∎

References

[1] A.V.BALAKRISHNAN, *Applied Functional Analysis*, Spring Verlag, 1981.
[2] J.KARRAKCHOU, M.RACHIK, *Optimal control of discrete distributed systems with delays in the control: the finite horizon case*. Arch.Control.Sc, Vol 4(XL),1-2, 1995.
[3] K.Y.LEE, S.CHOW AND R.O.BARR, *On the control of discrete-time distributed parameter system*. SIAM J. Control, 10 (1972), No.2.
[4] D.G.LUENBERGER, *Observing the state of linear system*. IEEE vol, MIL-8, pp, 74-80,1964.
[5] J.M.SCHUMACHER, *Dynamic feedback in finite- and infinite-dimensional linear systems*, M. C. Tracts No. 143, Mathematish Centrum, Amsterdam, 1982.
[6] J.ZABCZYK, *Remarks on the control of discrete-time distributed parameter systems*. SIAM J. Control, 12(1974), No. 4.

K PIEKARSKI AND J SOKOŁOWSKI
Shape control of elastic plates

1 Introduction

In the paper a class of shape optimization problems associated with the optimal location of controls is considered. Two-level parametric optimization problem with control constraints for the Kirchhoff plate model is formulated. In general, such a problem may not have any solution. A regularization technique is applied to assure the existence of an optimal geometrical domain ω for the problem under considerations. Shape sensitivity analysis with respect to the location of controls is performed. First order optimality conditions for parametric problems are derived. A relaxation of the shape optimization problem is introduced and the necessary optimality conditions for the relaxed problem are derived.

We refer the reader to (Sokolowski, 1988) for the related results on the boundary control problems for parabolic equations. Control problems in solid mechanics are analysed in the monograph (Khludnev, Sokolowski, 1997).

In section 2 the parametric optimization problem is formulated. The first order necessary optimality conditions are derived in section 4. Finally, in section 5 a relaxation of shape optimization problem under considerations is proposed and the second order necessary optimality conditions are established. We refer the reader to (Bednarczuk, Pierre, Rouy, Sokolowski) for the related results on tangent sets in L^∞.

2 Domain Optimization Problem

Let D be a bounded open subset of $\mathbb{R}^2$, with a smooth boundary ∂D. We denote by $\Omega(D)$ the family of all open subsets of D such that $\omega \in \Omega(D)$ is mesurable (Lebesgue), $\omega = \bigcup_{i=1}^{N} \omega_i$, $\overline{\omega_i} \cap \overline{\omega_j} = \emptyset$ for $i \neq j$, and the volume of ω is prescribed, $|\omega| = \alpha << |D|$; α is given. Let $\partial D = \partial D_1 \cup \partial D_2$, $\partial D_1 \cap \partial D_2 = \emptyset$.

Consider a Kirchhoff plate equation defined in D along with the following boundary and initial conditions

$$\frac{\partial^2 y}{\partial t^2} + Ay = \sum_{i=1}^{N} \left(u_i \chi_\omega^i \right)(x,t) \quad \text{in } Q = (0,T) \times D , \tag{1}$$

$$y = 0 \text{ in } \Sigma = \partial D \times (0,T), \tag{2}$$

$$\frac{\partial y}{\partial n} = 0 \quad \text{in } \Sigma_1 = \partial D_1 \times (0,T), \tag{3}$$

$$M_n^y = 0 \quad \text{in } \Sigma_2 = \partial D_2 \times (0,T), \tag{4}$$

$$y(0,x) = y_0(x), \frac{\partial y}{\partial t}(0,x) = y_1(x) \quad \text{in } D, \tag{5}$$

with ω given in $\Omega(D)$, where y_0 and y_1 are given functions defined in D. We denote by $\chi_\omega^i, i = 1,...,N$, the characteristic functions of $\omega_i \subset D$, $\frac{\partial y}{\partial n}$ is the normal derivative, M_n^y is the bending moment for the normal displacement y of the plate and $(0,T)$ is the time interval. We assume that $A \in \mathcal{L}(U;U')$ is an elliptic operator of the fourth order, where $U = \{v | v \in H^2(D), v_{|\partial D} = 0, \frac{\partial v}{\partial n}_{|\partial D_1} = 0\}$.

Let us denote by $y(u,\omega)$, the solution of the equations (1)–(5), where $u \in L^2(Q) = L^2(0,T;L^2(D))$, $u(x,t) = \Sigma_{i=1}^N (u_i \chi_\omega^i)(x,t)$ a.e. in Q, $u_i \in L^2(0,T;L^2(\omega_i))$, $i = 1,...,N$. Define the cost functional

$$J(u,\omega) = \|u\|_{L^2(0,T;L^2(\omega))}^2 + \|y(u,\omega) - y_d\|_{L^2(Q)}^2 \tag{6}$$

with y_d given in $L^2(Q)$.

The following minimization problem is considered

$$\text{Min}\{J(u,\omega) | u \in U_{\text{ad}}, \ \omega \in \mathcal{O}_{\text{ad}}\}, \tag{7}$$

where the set of admissible controls U_{ad} is a convex and closed subset of the space $L^2(0,T;L^2(\omega))$. For a given set ω we shall use the same symbol $u(x,t)$ to denote an admissible control $u \in L^2(0,T;L^2(\omega))$ and the function $u\chi_\omega = \Sigma_{i=1}^N u_i \chi_\omega^i \in L^2(0,T;L^2(D))$. In the particular case, the controls u_i supported on ω_i can be selected as the functions $u_i = u_i(t)$ of the variable $t \in (0,T)$, and independent of the space variable x.

In order to assure the existence of a solution to the problem (7), the family of admissible domains $\mathcal{O}_{\text{ad}} \subset \Omega(D)$ should satisfy the following compactness assumption.

For any sequence $\omega_k \in \mathcal{O}_{\text{ad}}, k = 1,2,...$ there exists a subset $\hat{\omega} \in \mathcal{O}_{\text{ad}}$ and a subsequence of the sequence $k = 1,2,...$, still denoted by $k = 1,2,...$, such that the sequence of characteristic functions χ_{ω_k} is convergent to a characteristic function,

$$\chi_{\omega_k} \to \chi_{\hat{\omega}} \tag{8}$$

weak–$(*)$ in $L^\infty(D)$ with $k \to \infty$.

The above assumption is satisfied, in particular, if the perimeter of the set ω in D, defined by

$$P_D(\omega) = \sup \left\{ \int_\omega \text{div } \phi dx | \phi \in \mathcal{D}^1(D, \mathbb{R}^N); \max_{x \in D} \|\phi(x)\|_{\mathbb{R}^N} = 1 \right\}, \tag{9}$$

is bounded ie., there exists a constant M such that $P_D(\omega) \leq M$. Let

$$J_\epsilon(u,\omega) = J(u,\omega) + \epsilon P_D(\omega) \tag{10}$$

where ϵ is a constant. For $\epsilon > 0$ we consider the regularized minimization problem,

$$\text{Min}\{J_\epsilon(u,\omega)|u \in U_{\text{ad}},\ \omega \in \Omega(D),\ P_D(\omega) < \infty\} \tag{11}$$

The optimization problem is formulated as a two-level minimization problem. First, for a given set $\omega \subset D$ the unique optimal control $\chi_\omega u^* = \Sigma_{i=1}^N u_i^* \chi_\omega^i$ is determined for the following optimal control problem

$$J(u^*,\omega) = \|u^*\|^2_{L^2(0,T;L^2(\omega))} + \|y(u^*,\omega) - y_d\|^2_{L^2(Q)} \leq J(u,\omega) \tag{12}$$

for all $u \in U_{\text{ad}}$. The optimal control is denoted by $u^* = u(\omega)$, and the optimal value of the cost functional by $I(\omega) = J(u^*,\omega) = J(u(\omega),\omega)$.

Then, the minimization of the resulting shape functional is performed with respect to ω,

$$\text{Min}\{I(\omega) + \epsilon P_D(\omega)|\omega \in \Omega(D),\ P_D(\omega) < \infty\} \tag{13}$$

3 Optimality System

Let us consider the control problem (12) and denote by $\mathcal{P}$ the metric projection in $L^2(\omega \times (0,T))$ onto the set U_{ad}. The unique optimal control $u^* \in U_{\text{ad}}$ is given by the following nonlinear relation (Lions, 1968),

$$u^* = \mathcal{P}(-p\chi_\omega), \tag{14}$$

where p is the adjoint state given by the unique solution to the following equation along with the boundary and initial conditions,

$$\frac{\partial^2 p}{\partial t^2} + Ap = 2(y - y_d) \quad \text{in } Q, \tag{15}$$

$$p = 0 \quad \text{in } \Sigma, \quad \frac{\partial p}{\partial n} = 0 \quad \text{in } \Sigma_1, \quad M_n^p = 0 \quad \text{in } \Sigma_2, \tag{16}$$

$$p(T,x) = \frac{\partial p}{\partial t}(T,x) = 0 \quad \text{in } D, \tag{17}$$

where M_n^p is the bending moment for p. The optimality system for the optimal control problem (12) includes (15)–(17) and the state equation in the following form

$$\frac{\partial^2 y}{\partial t^2} + Ay = \mathcal{P}(-p\chi_\omega) \quad \text{in } Q \tag{18}$$

along with the boundary conditions (2)–(4) and the initial conditions (5). The optimal value of the cost functional takes the following form,

$$I(\omega) = \|\mathcal{P}(-p\chi_\omega)\|^2_{L^2(0,T;L^2(\omega))} + \|y - y_d\|^2_{L^2(Q)}.$$

4. Shape Sensitivity Analysis

The sensitivity analysis of optimal value of the cost functional with respect to the perturbations of the set ω is performed.

The material derivative method is used for the shape sensitivity analysis. Let $T_s, s \in [0,\delta), \delta > 0$, be a family of smooth transformations $T_s : \mathbb{R}^2 \to \mathbb{R}^2$, $T_s(X) = x(s)$. We assume that T_0 is the identity transformation, T_s is bijective, continuously differentiable for each $s \in [0,\delta)$ and T_s^{-1} enjoys the same properties. The family of transformations T_s is defined in the form of the flow of the vector field $V(s, x(s)) = \left(\frac{\partial T_s}{\partial s} \circ T_s^{-1}\right)(x(s))$. We assume that the vector field V is sufficiently smooth, that is $V(\cdot,\cdot) \in C^1([0,\delta); C^2(\mathbb{R}^2, \mathbb{R}^2))$. Let us denote $\omega_s = T_s(\omega)$ and $\partial\omega_s = T_s(\partial\omega)$. The functional $I(\omega)$ can be rewritten as follows

$$I(\omega) = \int_0^T \int_D (y(u^*,\omega) - y_d)^2 dx dt + \sum_{i=1}^N \int_0^T \int_D (u_i^*)^2 \chi_\omega^i dx dt, \qquad (19)$$

For $s > 0$ we have

$$I(\omega_s) = \int_0^T \int_D (y_s(u_s^*,\omega_s) - y_d)^2 dx dt + \sum_{i=1}^N \int_0^T \int_D (u_{s,i}^*)^2 \chi_{\omega_s}^i dx dt \qquad (20)$$

where $u_s^*(t,x) = \Sigma_{i=1}^N \left(u_{s,i}^* \chi_{\omega_s}^i\right)(x,t)$ is the optimal solution to the optimal control problem $\text{Min}\{J(u_s, \omega_s) | u_s \in U_{ad}^s\}$ for given $s \geq 0$.

The form of directional derivative $dI(\omega; V)$ can be obtained assuming the Hadamard differentiability of the metric projection $\mathcal{P}$ onto the set of admissible controls and using the material derivatives of solutions to the state and adjoint state equations. We refer the reader to (Sokolowski, 1988) for the detailed description of the method in the case of boundary control of parabolic equations.

Our aim is to obtain the expression for $dI(\omega; V)$.

Proposition 1 *Assume that the metric projection $\mathcal{P}$ is directionally differentiable in the sense of Hadamard, let us denote its directional derivative by $\mathcal{P}'$, then the directional derivative $dI(\omega; V)$ is given by*

$$dI(\omega; V) = 2\int_0^T \int_D \dot{y}(y^* - y_d) dx dt + 2\int_0^T \int_D [\chi_\omega \mathcal{P}(-p\chi_\omega) \mathcal{P}'(-\dot{p}\chi_\omega)] dx dt,$$

where the material derivatives $\dot{y}$ and $\dot{p}$ are solutions to the following systems

$$\frac{\partial^2 \dot{y}}{\partial t^2} + A\dot{y} + \text{div}V\frac{\partial^2 y}{\partial t^2} + A'y = \chi_\omega \mathcal{P}(-p\chi_\omega)\text{div}V + \chi_\omega \mathcal{P}'(-\dot{p}\chi_\omega) , \tag{21}$$

$$\dot{y} = 0 \quad in\ \Sigma\ , \quad \frac{\partial \dot{y}}{\partial n} = 0 \quad in\ \Sigma_1\ , \quad M_n^{\dot{y}} = 0 \quad in\ \Sigma_2\ , \tag{22}$$

$$\dot{y}(0,x) = \nabla y_0 \cdot V(0), \quad \frac{\partial \dot{y}}{\partial t}(0,x) = \nabla y_1 \cdot V(0) \quad in\ D\ , \tag{23}$$

and

$$\frac{\partial^2 \dot{p}}{\partial t^2} + A\dot{p} + \frac{\partial^2 p}{\partial t^2}\text{div}V + A'p = 2\dot{y} + 2(y - y_d)\text{div}V\ , \tag{24}$$

$$\dot{p} = 0 \quad in\ \Sigma\ , \quad \frac{\partial \dot{p}}{\partial n} = 0\ in\ \Sigma_1\ , \quad M_n^{\dot{p}} = 0 \quad in\ \Sigma_2\ , \tag{25}$$

$$\dot{p}(T,x) = \frac{\partial \dot{p}}{\partial t}(T,x) = 0 \quad in\ D \tag{26}$$

A' denotes the derivative of the elliptic operator A which is obtained in the standard way (Sokolowski, Zolesio, 1992).

Proposition 1 follows by the following results on the shape sensitivity analysis of the optimal control problem for the plate equation and the standard argument (Sokolowski, 1988).

Lemma 1 *Assume that the metric projection $\mathcal{P}$ is directionally differentiable in the sense of Hadamard and denote by $\mathcal{P}'$ its directional derivative. Then,*

$$\frac{y_s \circ T_s(u_s^*, \omega_s) - y(u^*, \omega)}{s} \rightharpoonup \dot{y} \quad weakly\ in\ L^2(Q)\ , \tag{27}$$

$$\frac{p_s \circ T_s(u_s^*, \omega_s) - p(u^*, \omega)}{s} \to \dot{p} \quad strongly\ in\ L^2(Q)\ , \tag{28}$$

$$\chi_\omega \frac{u_s^* \circ T_s - u^*}{s} \to \chi_\omega \dot{u} = \chi_\omega \mathcal{P}'(-\dot{p}\chi_\omega) \quad strongly\ in\ L^2(Q)\ . \tag{29}$$

In general, we can only expect the so-called conical differentiability of $\mathcal{P}$. We refer reader eg. to (Sokołowski, 1985) for the related results on the directional differentiability of the metric projection.

Theorem 1. *There exists an optimal domain ω which minimizes the shape functional $\mathcal{J}_\epsilon(\cdot)$,*

$$\mathcal{J}_\epsilon(\omega) = I(\omega) + \epsilon P_D(\omega), \tag{30}$$

subject to $|\omega| = \alpha$.

Furthermore, if the optimal domain ω is sufficiently regular, then for any vector field V with the compact support in D and such that $\text{div}\ V = 0$ it follows that

$$d\mathcal{J}_\epsilon(\omega; V) = dI(\omega; V) + \epsilon \int_{\partial \omega} \kappa(x) V(0,x) \cdot n(x) d\Gamma(x) = 0, \tag{31}$$

here κ denotes the tangential divergence of the normal vector field n, $\kappa = -2\mathcal{H}$, where $\mathcal{H}$ is the mean curvature of $\partial\omega$.

5. Parametric Optimization - Relaxation of Shape Optimization Problems

Consider a parametric optimization problem which is similar to the shape optimization problem formulated in section 2 but with the following differences: we assume that characteristic function χ_ω is replaced by function $\eta \in L^\infty(D)$ and control $u = u(t,x)\eta(x)$, $u(t,x) \in L^2(Q)$. The only constraints are those for function η:

$$\mathcal{U}_{ad} = \{\eta \in L^\infty(D) | 0 \leq \eta(x) \leq 1, \int_D \eta(x)dx = \alpha\}. \tag{32}$$

The set $\mathcal{U}_{ad}$ is convex in $L^\infty(D)$ dislike the set of characteristic functions. Plate equilibrium equation is now as follows

$$\frac{\partial^2 y}{\partial t^2} + Ay = u(t,x)\eta(x) \quad \text{in } Q, \tag{33}$$

$$y = 0 \quad \text{in } \Sigma, \quad \frac{\partial y}{\partial n} = 0 \quad \text{in } \Sigma_1, \quad M_n^y = 0 \quad \text{in } \Sigma_2, \tag{34}$$

$$y(0,x) = y_0(x), \frac{\partial y}{\partial t}(0,x) = y_1(x) \quad \text{in } D. \tag{35}$$

Functional $I(\omega)$ for the unconstrained optimal control problem can be written as

$$I(\omega) = \|p\chi_\omega\|^2_{L^2(Q)} + \|y - y_d\|^2_{L^2(Q)} \tag{36}$$

Introduce the functional

$$h(\eta) = \|p^*\eta^{\frac{1}{2}}\|^2_{L^2(Q)} + \|y^* - y_d\|^2_{L^2(Q)} \tag{37}$$

where y^* and p^* are solutions of the systems

$$\frac{\partial^2 y^*}{\partial t^2} + Ay^* = -p^*\eta \quad \text{in } Q, \tag{38}$$

$$y^* = 0 \quad \text{in } \Sigma, \quad \frac{\partial y^*}{\partial n} = 0 \quad \text{in } \Sigma_1, \quad M_n^{y^*} = 0 \quad \text{in } \Sigma_2, \tag{39}$$

$$y^*(0,x) = y_0(x), \frac{\partial y^*}{\partial t}(0,x) = y_1(x) \quad \text{in } D, \tag{40}$$

$$\frac{\partial^2 p^*}{\partial t^2} + Ap^* = 2(y^* - y_d) \quad \text{in } Q, \tag{41}$$

$$p^* = 0 \quad \text{in } \Sigma, \quad \frac{\partial p^*}{\partial n} = 0 \quad \text{in } \Sigma_1, \quad M_n^{p^*} = 0 \quad \text{in } \Sigma_2, \tag{42}$$

$$p^*(T,x) = \frac{\partial p^*}{\partial t}(T,x) \quad \text{in } D. \tag{43}$$

Equations (38)–(43) constitute the optimality system for the optimal control problem, and the resulting parametric optimization problem takes the following form

$$\min\{h(\eta)|\eta \in \mathcal{U}_{ad}\} . \qquad (44)$$

The above minimization problem is a relaxation of the shape optimization problem defined in section 2 as for $\eta = \chi_\omega$ and $u = \sum_{i=1}^{N} u_i(t)\chi_\omega^i(x)$ we have $I(\omega) = h(\chi_\omega)$.

Remark 1 *Let us observe, that the optimal value $h(\chi_\omega) = I(\omega)$ of the cost functional $\tilde{J}(u,\chi_\omega) = J(u,\omega)$ with respect to u can be defined for any function $\eta \in L^\infty(D)$, $\eta \geq 0$, $\int_D \eta > 0$. It means that for such function η,*

$$h(\eta) = \inf\{\tilde{J}(u,\eta)|u \in L^2(Q)\} \geq 0 .$$

On the other hand, for any $\eta \in L^\infty(D), \eta < 0$ on a set of positive Lebesque measure, we can expect in general that $\inf\{\tilde{J}(u,\eta)|u \in L^2(Q)\} = -\infty$. Therefore, the functional $h(\chi_\omega) = I(\omega)$ can be extended to the set of functions $\eta \in L^\infty(D)$, $\eta \geq 0$, $\int_D \eta > 0$. This observation is important for the derivation of the necessary optimality conditions for the relaxed optimization problem. To this end the following scalar function is introduced

$$\Psi(s) = h(\eta + sv + \frac{s^2}{2}w_n) ,$$

where v is an element of the first order tangent cone to the set $\mathcal{U}_{ad}$ at $\eta \in \mathcal{U}_{ad}$ and the sequence $\{w_n\}$ satisfies the condition

$$\eta + sv + \frac{s^2}{2}w_n \in \mathcal{U}_{ad}$$

for $s > 0$, s small enough, ie., the limit w in L^∞ of the sequence $\{w_n\}$ is an element of the second tangent set.

In this setting

$$\Psi'(0^+) = dh(\eta;v) ,$$
$$\Psi''(0^+) = d^2h(\eta;v,v) + dh(\eta;w_n) ,$$

therefore, denoting by w the limit of the sequence $\{w_n\}$, it follows that for a local minimum at η,

$$dh(\eta;v) \geq 0$$
$$d^2h(\eta;v,v) + dh(\eta;w) \geq 0 \quad \text{for } dh(\eta;v) = 0 .$$

We refer to (Bednarczuk, Pierre, Rouy, Sokolowski) for the related results on tangent sets in L^∞ and to (Penot, preprint) for the second order necessary optimality conditions in general setting.

Proposition 2 *There exists a solution to the parametric optimization problem* $\min\{h(\eta)|\eta \in \mathcal{U}_{ad}\}$.

Let us observe that in the case of unconstrained control problem, an optimal solution to the parametric optimization problem takes the following form

$$\eta^*(x) = \frac{\alpha}{|D|} .$$

Therefore, control constraints are required to obtain nontrivial solutions to the parametric optimization problem.

We present the first order necessary optimality conditions for the parametric optimization problem. Similar optimality conditions can be derived for the control constrained parametric optimization problem, however in such a case only the directional derivative $dh(\eta; v)$ exists, in general.

Proposition 3 *A solution $\eta^* \in \mathcal{U}_{ad}$ of a parametric optimization problem (44) satisfies the following necessary optimality conditions:*

$$dh(\eta^*; v - \eta^*) \geq 0 \qquad \forall v \in \mathcal{U}_{ad} \qquad (45)$$

where

$$dh(\eta; v) = 2\int_0^T \int_D (\eta p'(v) + vp^*)p^* dx dt + 2\int_0^T \int_D y'(v)(y^* - y_d) dx dt , \qquad (46)$$

and y' and p' are solutions of the systems

$$\frac{\partial^2 y'(v)}{\partial t^2} + Ay'(v) = -(\eta p'(v) + vp^*) \quad in \ Q , \qquad (47)$$

$$y'(v) = 0 \quad in \ \Sigma , \quad \frac{\partial y'(v)}{\partial n} = 0 \quad in \ \Sigma_1 , \ M_n^{y'(v)} = 0 \quad in \ \Sigma_2 , \qquad (48)$$

$$y'(v)(0, x) = \frac{\partial y'(v)}{\partial t}(0, x) = 0 \quad in \ D , \qquad (49)$$

$$\frac{\partial^2 p'(v)}{\partial t^2} + Ap'(v) = 2y'(v) \quad in \ Q , \qquad (50)$$

$$p'(v) = 0 \quad in \ \Sigma , \quad \frac{\partial p'(v)}{\partial n} = 0 \quad in \ \Sigma_1 , \ M_n^{p'(v)} = 0 \quad in \ \Sigma_2 , \qquad (51)$$

$$p'(v)(T, x) = \frac{\partial p'(v)}{\partial t}(T, x) = 0 \quad in \ D . \qquad (52)$$

References

[1] Bednarczuk E., Pierre M., Rouy E., and Sokołowski J., Calculating tangent sets to certain sets in functional spaces, *INRIA-Lorraine, Rapport de Recherche* **3190** (1997).

[2] Khludnev A.M. and Sokołowski J. (1997): *Modelling and Control in Solid Mechanics*, Birkhauser Verlag, Basel, 1997, ISNM Vol. 122 (1997).

[3] Penot J.-P., *Optimality conditions in mathematical programming* preprint

[4] Sokołowski, J., Sensitivity analysis of control constrained optimal control problems for distributed parameter systems, *SIAM Journal Control and Optimization*, **25**, 1542–1556 (1987).

[5] Sokołowski, J., Shape sensitivity analysis of boundary optimal control problems for parabolic systems, *SIAM Journal on Control and Optimization* **26**, 763-787 (1988).

[6] Sokołowski, J., Displacement derivatives in shape optimization of thin shells, *Optimization Methods in Partial Differential Equations*, S. Cox, I. Lasiecka (Eds.), Contemporary Mathematics **209**, American Mathematical Society, 247–266 (1997).

[7] Sokołowski, J. and Zolesio, J.-P., *Introduction to Shape Optimization. Shape sensitivity analysis.* Springer Verlag, New York, (1992).

[8] Zolesio, J.-P.(1984): Optimal location of a control in a parabolic equation, *Preprints of the 9th World Congress of IFAC*, **5**, 64–67 (1994).

K. PIEKARSKI
INSTITUTE OF MATHEMATICS AND PHISICS
TECHNICAL UNIVERSITY OF BIAŁYSTOK
UL. WIEJSKA 45 A, 15-351 BIAŁYSTOK, POLAND

J. SOKOŁOWSKI
INSTITUT ELIE CARTAN, LABORATOIRE DE MATHÉMATIQUES,
UNIVERSITÉ HENRI POINCARÉ NANCY I, B.P. 239,
54506 VANDOEUVRE LÈS NANCY CEDEX, FRANCE
AND SYSTEMS RESEARCH INSTITUTE OF THE POLISH ACADEMY OF SCIENCES,
UL. NEWELSKA 6, 01-447 WARSZAWA, POLAND
E-MAIL: SOKOLOWS@IECN.U-NANCY.FR

II. Optimal control and nonsmooth analysis

URSULA FELGENHAUER
On Ritz type discretizations for optimal control problems

1. Problem and discretization method

Consider the nonlinear constrained optimal control problem

(**P**) $$J(x,u) = \int_0^1 r(t,x(t),u(t))\,dt \longrightarrow \min,\qquad(1)$$

subject to:
$$\dot{x}(t) = f(t,x(t),u(t)) \quad \text{for } a.e.\ t \text{ in } [0,1];\qquad(2)$$

$$x(0) = a,\quad x(1) = b;\qquad(3)$$

$$g(t,x(t),u(t)) \leq 0 \quad \text{for } a.e.\ t \text{ in } [0,1];\qquad(4)$$

where $x \in W^1_\infty([0,1];\mathbb{R}^n)$, $u \in L_\infty([0,1];\mathbb{R}^m)$, and r, f, g are C^{2+1}-functions of all arguments. Denote $W = \{(t,\xi,v) : g(t,\xi,v) \leq 0\}$. Further, let the HAMILTON function be given as

$$H(t,x,u,\eta) = r(t,x,u) + \eta^T f(t,x,u).$$

Characterizations of optimal solutions for problem (**P**) may be derived from a duality approach proposed by R. KLÖTZLER, which has been successfully used for the investigation of several control problems in a series of papers by S. PICKENHAIN et al., see e.g. [13]. For problem (**P**) the dual problem may be derived as follows: Given $S: \mathbb{R}_+ \times \mathbb{R}^n \to \mathbb{R}$, consider

(**D**) $$\mathcal{L}(S) = \int_0^1 \frac{d}{dt} S(t,x(t))\,dt \longrightarrow \max,\qquad(5)$$
subject to:
$$\phi = H(t,x,u,\nabla_x S(t,x)) + S_t(t,x) \geq 0 \quad a.e.\ \text{on } W.\qquad(6)$$

Here we assume that $S(\cdot,x)$ is LIPSCHITZ continuous on $[0,1]$, $S(t,\cdot)$ and $S_t(t,\cdot)$ are supposed to be C^2 functions on $\mathbb{R}^n$.

If $z = (x,u)$ is feasible for (**P**) and S for (**D**) resp., then $J(x,u) \geq \mathcal{L}(S)$ follows by direct calculation. The duality gap becomes zero iff

$$\int_0^1 (H^* + S_t^*)\,dt = 0\qquad(7)$$

holds for x^*, u^* and S^*. This condition represents at the same time a sufficient optimality condition for the solution (x^*, u^*) of (**P**), see [12], Theorem 3.1.
This fact suggests to investigate the auxiliary problem

(**OP**) $\quad \Phi(x,u) = \int_0^1 [H(t,x(t),u(t),\nabla_x S(t,x(t))) + S_t(t,x(t))]\, dt \longrightarrow \min$, $\quad$ (8)

$\qquad$ subject to: $\qquad G(x,u) \in -K$,

where $G(x,u)(\cdot) = g(\cdot, x(\cdot), u(\cdot))$, and K consists of all L_∞ functions with nonnegative essential infimum.

The approach has been also used in [13] but with a pointwise formulation of the HAMILTON-JACOBI condition (6), (7) given here in integrated form. In the forthcoming, problem (**OP**) will serve as the starting point for our RITZ type discretization.

Remark that (**OP**) depends on the dual parameter S which plays the role of a *verification function*: If S is such that (**OP**) has a locally unique solution (x_0, u_0) and Φ satisfies certain local quadratic growth estimation then the pair (x_0, u_0) is a strict local minimizer of problem (**P**).

With the LAGRANGE functional $L(x,u,\mu) = \Phi(x,u) + <\mu, G(x,u)>$, the KARUSH-KUHN-TUCKER system may be derived for (**OP**) and we remark, that under certain regularity assumptions and with the notation $\nabla_x S(t, x_0(t)) = p(t)$, we arrive at the well-known canonical system for (**P**):

$$L_x = \dot{p} + \hat{H}_x = 0, \; L_u = \hat{H}_u = 0, \; L_p = f - \dot{x} = 0 \qquad (9)$$

where $\hat{H}$ stands for the *augmented* HAMILTONian related to the constrained formulation (**P**), and p coincides with the adjoint variable. The equations have to be completed by the complementarity relations

$$L_\mu = g \in -K, \; \mu \in K, \; <\mu, g> = 0. \qquad (10)$$

Following [11] e.g., this system allows an interpretation as a *generalized equation*

(**GE**) $\qquad\qquad\qquad 0 \in \Theta(w) + \partial \psi_K(w)$, $\qquad\qquad\qquad$ (11)

where $\partial \psi_K$ is the *normal cone operator* related to K, cf. [14], and Θ consists of the partial FRÉCHET derivatives of L resp.

The regularity properties of (**GE**) have been analyzed in [11] in detailed form. Remark that in particular for the FRÉCHET differentiability of Θ it is important to consider the problem (**P**) in a L_∞-setting rather than in L_2-based function spaces (cf. [10] for the *two-norm-discrepancy* phenomenon). The assumptions suitable in our discretization context will be in essence analogous to those given for the EULER discretization in [11], see also [2] and [3]:

(**A1**) $\qquad$ The pair $(x_0, u_0) \in Z = W^1_\infty \times L_\infty$ is a locally unique solution of (**P**) with continuously differentiable state function x_0 and continuous on $[0,1]$ control u_0. There exist $(p_0, \mu_0) \in C^1 \times C^0 \subset Z$ which together with z_0 yield a solution of the first order system (**GE**).

The second condition concerns the constraint qualification: let $g^\sigma(t)$ denote the restriction of $g(t, x_0(t), u_0(t))$ to those components with $g^i(t) \geq -\sigma$, i.e. the σ- or *nearly active* constraints. Denote now $G_z^\sigma = g_z^\sigma(\cdot, x_0(\cdot), u_0(\cdot))$, and $A_z = f_z(\cdot, x_0(\cdot), u_0(\cdot))$. We require:

(A2)   The operator C given by

$$C(y, v) = \begin{pmatrix} \dot{y} - A_x y - A_u v \\ G_x^\sigma y + G_u^\sigma v \end{pmatrix}$$

is a surjective mapping from $Z = W_\infty^1 \times L_\infty$ onto $V^\sigma = L_\infty \times Y^\sigma$ (where Y^σ is the product of L_∞-spaces with supports
$\Omega_i^\sigma = \{t : g^i(t, x_0(t), u_0(t)) \geq -\sigma\}$.)
There exist a positive constant κ and a linear map $C^+ : V^\sigma \to Z$ such that $(C\, C^+)$ is the identical mapping on V^σ, and $\|C^+\| \leq \kappa$.

In [3] one can find a description of **(A2)** in terms of the so-called *invertibility* and the *controllability* conditions. Their stable formulations are enclosed in [11] too.

The last assumption concerns a second order sufficient optimality condition. To this end define the following tangent space to "active constraints" related to **(OP)** :

$$T_\delta(t) = \{\zeta : \nabla_z g^i \zeta = 0 \quad \forall i \ \text{ with } \mu_0^i(t) \geq \delta\}\,.$$

Denote $\nabla_x^2 S(t, x_0(t)) = Q(t)$ and $M = \nabla_z^2 L(x_0, u_0, \mu_0)$. The Second Order Condition may be interpreted as a condition on Q (resp. on S) now:

(A3)   There exist a bounded, continuously differentiable matrix function Q and a constant $\alpha > 0$ such that
$$\zeta^T M(t) \zeta \geq \alpha \|\zeta\|^2 \quad \text{for all } \zeta \in T_\delta(t)\setminus\{0\}\,, \ \forall t \in [0,1]\,.$$

The above relation in terms of the "unknown" Q then has the form of a matrix RICCATI differential inequality, for details see [13] and [12].

Now we come to the discretization: Let the interval $[0, 1]$ be divided into N subintervals $\{\omega_j\}_{j=1,\ldots,N} =: \omega$ of length $h = 1/N$ respectively, denote $t_1 = 0$; $t_{\sigma+1} = t_\sigma + h$; $t_{N+1} = 1$, T – the vector of all nodes t_σ, $\sigma = 1, \ldots, (N+1)$. In addition, we use the notation $\bar{T}$ for the vector of $\bar{t}_j = 0.5(t_j + t_{j+1}), j = 1, \ldots, N$.
In correspondence to the smoothness assumptions on (x_0, u_0) we define finite-dimensional subspaces $X_h \subset X = W_\infty^1$, $V_h \subset V = L_\infty$ as the sets of all functions which are linear resp. constant on every ω_j, $j = 1, \ldots, N$. Obviously, the spaces X_h, V_h are homomorphic to finite-dimensional EUCLIDean vector spaces.

Let the projection operators $P_h^1 : X \to X_h$ and $P_h^0 : V \to V_h$ be given by $(P_h^1 x)(T) = x(T)$ and $(P_h^0 u)(\bar{T}) = \bar{u}$ with $\bar{u}_j = h^{-1} \int_{\omega_j} u(t)\,dt$. These projections describe the best approximation operators from $\hat{Z} = W_2^1 \times L_2$ to $Z_h = X_h \times V_h$ if

W_2^1 is supplied with the norm $\|x\|^2 = |x(0)|^2 + \|\dot{x}\|_2^2$. For functions $(y, v) \in C^1 \times C^0$ we have (with $\rho(\eta, h) = \max\{|\eta(t_1) - \eta(t_2)| : |t_1 - t_2| \le h\}$):

$$\| P_h^1 y - y \|_{1,\infty} \le (1+h)\rho(\dot{y}, h) ; \quad \| P_h^0 v - v \|_\infty \le \rho(v, h) .$$

For $z = (x, u) \in Z$ the projection $P_h z \in Z_h$ is defined respectively componentwise. Besides of the projection operator we fix a summation formula for evaluating the integral in the auxiliary problem. With the rectangular formula the RITZ method leads to the following approximation of **(OP)**

$$\Phi_\omega(x, u) = \int_0^1 R_h \phi(t, P_h^1 x, P_h^0 u) \, dt = \sum_j |\omega_j| \phi(\bar{t}_j, \bar{x}_j, \bar{u}_j) ;$$

$$G_\omega(x, u)(\cdot) = R_h g(\cdot, P_h^1 x(\cdot), P_h^0 u(\cdot)) .$$

Thus we arrive at the discrete version of **(OP)**,

(OP$_\omega$) $\quad \Phi_\omega(x, u) \longrightarrow \min \quad$ s.t. $\quad x_1 = a, \quad x_{N+1} = b, \quad G_\omega(x, u) \in -K_\omega \quad$ (12)

where the cone K_ω stands for $K \cap V_h$. The consistency analysis for the discrete approximation (12) to **(OP)** has been carried out in [4].

The LAGRANGian related to **(OP$_\omega$)** takes the form

$$L_\omega = \Phi_\omega + <\mathcal{M}, G_\omega>,$$

and the KKT system yields the discrete analogy to the canonical system, i.e.

(GE$_\omega$) $\qquad 0 \in \Theta_\omega(w) + \partial \psi_{K_\omega}(w) ,$ (13)

Remark: The setting of the discrete generalized equation differs from the formulation in [11] mainly in the inclusion of terms depending on S (resp. Q or Q_h).

2. Convergence result

Our aim is to prove that near the solution point $w_0 = (x_0, u_0, p_0, \mu_0) \in Z \times Z = \mathcal{W}$ and for sufficiently small h the discrete problems **(GE$_\omega$)** resp. **(OP$_\omega$)** have unique solutions which converge in $\mathcal{W}$ and in the limit provide a solution of **(P)**. The idea of the proof is to apply S. ROBINSON's theory of *strongly regular* generalized equations [14] to the sequence of discrete problems. Recently K. MALANOWSKI [11] this way has obtained an abstract convergence result for discretized generalized equations. He showed that besides of certain differentiability conditions on Θ_ω and approximation error bounds the crucial question consists in the verification of the *uniform strong regularity* of the problems **(GE$_\omega$)** for $h \to 0$ (see [11], conditions I.1 - I.3). This property is described by:

For $w_h = P_h w_0$ let s_h be such that $\quad 0 \in \Theta_\omega(w_h) + s_h + \partial \psi_{\mathcal{K}_\omega}(w_h)$.
Consider the linearized problem

$$\delta \in \Theta_\omega(w_h) + D_w \Theta_\omega(w_h)(w - w_h) + s_h + \partial \psi_{\mathcal{K}_\omega}(w).$$

$\exists\ h_1, r_0, r_1 > 0$ such that for arbitrary positive $h < h_1$ and δ : $\|\delta\|_{\mathcal{W}^*} \le r_0$ there exists an unique in $B_1 = \{w \in \mathcal{W}_h : \|w - w_h\| \le r_1\}$ solution $w = w_h(\delta)$, and

$$\| w_h(\delta') - w_h(\delta'') \|_{\mathcal{W}} \le l \, \| \delta' - \delta'' \|_{\mathcal{W}^*} \tag{14}$$

for all δ', δ'' such that $\|\delta\|_{\mathcal{W}^*} \le r_0$, with a constant l independent of h.

In [10] the theoretical convergence theorem was applied then to EULER's method. For the proof of (14) the underlying discrete control problems have been used.
To prove the uniform strong regularity in the context given by the RITZ method we analyze the finite-dimensional mathematical programs $(\mathbf{OP}_\omega)$.
The main steps will be explained shortly in the following:

First remember that regularity criteria for mathematical programs have been already derived in [14]. So, the proofs of Theorem 3.1 and Theorem 4.1 in [14] show that the regularity, and particularly the LIPSCHITZ property (14) depend on the following two conditions:

- *linear independence* of the binding constraints:
  If the gradients of the active constraints are assembled into the matrix $\bar{C}$, then require $\|\bar{C}^+\| \le \bar\kappa$.

- *strong* second order sufficient condition:
  Let T_+ denote the tangent space to equality and to those active inequality constraints corresponding to positive LAGRANGE multipliers.
  Then: $\quad \zeta^T \nabla^2 L \, \zeta \ge \bar\alpha \, \|\zeta\|^2$ for all $\quad \zeta \in T_+$.

Repeating the conclusions in [14] and adding step by step estimates for the appearing matrices, it is easy to see, that the LIPSCHITZ constant l depends on $\bar\kappa$ and $\bar\alpha$ together with general bounds for the derivatives of Θ_ω only (cf. Cor. 4, 5 [7]). Thus it is sufficient to find *uniform* w.r.t h bounds for these constants in the problems $(\mathbf{OP}_\omega)$, $h \to 0$.

Using again the consistency properties of the RITZ approximation we obtain (see [7], Lemma 6):

LEMMA 1 *Let w_0 be a solution of* (**GE**) *and suppose the assumptions* (**A1**) *and* (**A2**) *hold. Then for arbitrary $q \in (0, 1/\kappa)$ there exists a constant $\bar h > 0$ such that $\forall\, h < \bar h\ \ C_\omega\,:\,Z_h \to V_h'$ is surjective, and $\quad \| C_\omega^+ \| \le \kappa/(1 - \kappa q)$.*

(*Proof:* The proof first shows that in elements where the discrete constraints are active the original constraints are nearly active. Comparing then the operator C_ω with an appropriate restriction C' of C the consistency of the projection mappings ensure that $\|C' - C_\omega\|$ tends to zero if $h \to 0$. Then the estimate follows directly from a perturbation lemma for surjective linear operators. *q.e.d.*)

Starting from the coercivity assumption (**A3**) one can derive the desired uniform coercivity property for the matrices $M_\omega = \nabla_z^2 L_\omega$ then ([7], Lemma 9):

LEMMA 2 *There exist $\alpha' > 0$ and $h'_0 > 0$ such that for all $h < h'_0$ from $\zeta \in T_+ \subset Z_h$ it follows that*

$$< \zeta, M_\omega \zeta > \geq \alpha' \|\zeta\|^2 \quad \text{and} \quad \zeta(t)^T M_\omega(\bar{t}_j) \zeta(t) \geq \alpha' |\zeta(t)|^2 \qquad (15)$$

whenever $t \in \omega_j$.

(*Proof:* The formulation of assumption (**A3**) in particular ensures that for small h the elements of T_+ are "nearly-tangential" w.r.t. T_δ. By continuity arguments, M is positive definite then on T_+ too. Using the projection consistency, for sufficiently small h the definiteness of M_ω follows. *q.e.d.*)

The last two lemmas imply the uniform strong regularity condition to hold (see also condition I.3 and propos. 5.6 in [11]), so that we are able to apply MALANOWSKI's convergence theorem (Theor. 2.2). Thus we come to the final estimate ([7], Theor. 3):

THEOREM 1 *Let the assumptions (**A1**)-(**A3**) hold for problem (**P**) and the solution point (x_0, u_0). Further, let w_0 be the corresponding solution of (**GE**). There exist positive constants r_0, c_0 and $\bar{h}_0$ such that for every $h < \bar{h}_0$ there exists an unique in $V_0 = \{ w \in W_h : \|w - w_h\|_W \leq r_0 \}$ solution w_h^* of (**GE**$_\omega$), and the solutions suffice the error estimate*

$$\| w_h^* - w_o \| \leq c_0 \, (h + \rho(w_0, h)) \, \|w_0\|$$

(where $\rho(w_0, h)$ is the maximal of the variations $\rho(\phi, h)$, $\phi \in \{\dot{x}_0, u_0, \dot{p}_0, \mu_0\}$).

The convergence of the first order RITZ type method (as could be desired) therefore turns out to be comparable with the behavior of the EULER method. The advantage of the given approach may be seen in its direct applicability for problems with general boundary conditions including also boundary constraints of inequality type (cf. [7]). Another important motivation was given by the fact that projection type discretizations may be easily generalized and adapted to problems with multiple integrals in the objective functional (see [13] e.g. for the theory). The described scheme of the convergence proof will work in this multi-dimensional situation analogously.

3. Solution regularity discussion

In the preceding section the local convergence of the discretizations was analyzed in L_∞-based function spaces under the assumption that the optimal control in the nominal solution is a continuous function. This condition may seem to be rather restrictive because the practical experience knows many applications with discontinuous optimal control regimes. Furthermore, in recent publications A. DONTCHEV obtained approximation estimates for EULER's method under the central assumption of RIEMANN integrable controls only ([2], [1]).

For a better understanding of the problem let us consider first an abstract approximation problem for L_∞ elements: Suppose that a sequence $\{V_h\}$ of subspaces of piecewise constant functions in L_∞ is given corresponding to the discretizations $\{\omega(h)\}$, $h \to 0$. We ask for the properties of functions which may be approximated by a sequence $\{v_h\}$, $v_h \in V_h$. Using as a theoretical tool the so-called *essential limit sets* introduced in [6] resp. [8] for analyzing the local behavior of L_∞-functions, we get:

THEOREM 2 *Assume that in a given point* $t \in [0,1]$ *the function* $v \in L_\infty$ *is* essentially discontinuos, *i.e. there exist at least two values* $w_1 \neq w_2$ *and sets* M_1, M_2 *such that for* $i = 1, 2$ a) $\operatorname{meas}(M_i \cap U_\delta(t)) > 0$ *for all* $\delta > 0$,   *and* b) $v(\tau)|_{M_i} \to w_i$ *as* $\tau \to t$.
If $v_h \in V_h$ *is continuous at* t *then* $\|v - v_h\|_\infty \geq 0.5\,|w_1 - w_2|$ *holds independently of the size of* h.

(*Proof:* Consider $\Delta = \operatorname{ess\,sup} |v - v_h| \geq \max\{\sup_{M_1} |v - v_h|, \sup_{M_2} |v - v_h|\}$. Restricting the latter sets to appropriately small neighborhoods $U_\delta(t)$, we see that for arbitrary $\epsilon > 0$
$\Delta > \max\{|v_h(t) - w_1|, |v_h(t) - w_2|\} - \epsilon$ which yields the desired estimate. q.e.d.)
This Theorem shows a natural limitation for the approximation accuracy within the given function classes and for an a-priori fixed grid refinement strategy.

Consider now the structure of optimal control functions for the following modification of (**P**): Replace the boundary conditions by an initial condition $x(0) = a$, and the restriction (4) by the control constraint

$$u(t) \in U \qquad \text{a.e. in } [0,1] ,$$

where U is a convex closed set in $\mathbb{R}^m$.

Under appropriate coercivity assumptions it could be proved in [6] that the optimal control function represents a selection of LIPSCHITZ continuous curves ([6], Theor. 1)

THEOREM 3 *Let* $\bar{u}$ *be an optimal control of the problem with convex control constraints satisfying the strong second order coercivity condition w.r.t.* $(U - U)$. *Then*

there exists a positive constant h_0 such that on every interval $[t_1, t_2] \subseteq [0,T]$ of length $h < h_0$ the structure of the function $\bar{u}$ can be described as follows:
There exists a finite number ν of uniformly bounded, LIPSCHITZ continuous arcs V^j on $[t_1, t_2]$ such that $\bar{u}(t) \in \{V^j(t)\}_{1 \leq j \leq \nu}$ a.e. in $[t_1, t_2]$.

Further, if $\bar{t}$ is a point where $\bar{u}$ is essentially continuous then there exists a constant $\delta > 0$ with $\bar{u}(t) = V^j(t)$ a.e. on $U_\delta(\bar{t})$ for some fixed index j. If in particular the control function is equivalent to a continuous function then the continuous representative is LIPSCHITZ continuous on $[0,T]$.

The above problem has been also considered by A. DONTCHEV in [1] but under the additional assumption that the data functions r and f do not explicitly depend on t (autonomous setting). We will refer to this problem as (**P'**).
For the EULER method it was shown that the L_∞-distance of the approximations to the solution set of the original problem tends to zero as $h \to 0$ (Theor. 1 in [1]), and this property holds for any regular discretization. As a particular conclusion [1] contains the result that the uniqueness of the optimal control together with the coercivity condition imply the RIEMANN integrability of the optimal u.

By means of the above theorems it is now possible to strengthen this statement: Indeed, varying the discretization we see that the approximation accuracy in terms of the L_∞-norm by Theorem 2 can be achieved only in the case of a continuous control function. Therefore, with Theorem 3 we obtain the assertion

Corollary 3 *If the optimal control problem* (**P'**) *has an unique solution in $W^1_\infty \times L_\infty$ and assumptions of the strong coercivity type hold then the optimal control is* LIPSCHITZ *continuous on* $[0, 1]$.

(For details see [1] and [6], Cor. 4).

Remark that this conclusion is a generalization of the related statements in W. HAGER's paper [9] obtained there for particular convex optimal control problems.

Returning to the problem formulation (**P**) it is easy to verify that for this general, non-autonomous case examples with discontinuous (e.g. piecewise continuous) control functions exist which are locally unique and strict minimizers. Theorem 2 says that it is impossible to find a L_∞-convergent discretization without the knowledge of jumps location. This should be a motivation for a convergence analysis in L_2-related norms although one has to expect the well-known difficulties due to the *two-norm discrepancy* then. An encouraging argument may be given by the fact that discrete versions of Second Order sufficient conditions guarantee the optimality of an at least piecewise continuous limit solution already under L_2-convergence assumptions, cf. [4], [5].

References

[1] Dontchev A.L. (1996) *An a priori estimate for discrete approximations in nonlinear optimal control*, SIAM J.Contr. Optim. 34:1315-1328.

[2] Dontchev A.L., Hager W.W. (1993) *Lipschitzian stability in nonlinear control and optimization*, SIAM J.Contr. Optim. 31:569-603.

[3] Dontchev A.L., Hager W.W., Poore A.B., Yang B. (1995) *Optimality, Stability, and convergence in Nonlinear Control*, Appl. Math. and Optim., 31:297-326.

[4] Felgenhauer U. (1996) *Numerical optimality test for control problems*, in: Proc. IV. Conference on "Parametric Optimization and Related Topics", Enschede 1995; eds.: J.Guddat, H.Th.Jongen, G.Still, F.Twilt; 85-101.

[5] Felgenhauer U. (1996) *Discretization based optimality test for certain parametric problems*, preprint, BTU Cottbus, Reihe Mathematik, M-01/1996.

[6] Felgenhauer U. (1997) *Regularity properties of optimal controls obtained via discrete approximation*, preprint, BTU Cottbus, Reihe Mathematik, M-03/1997.

[7] Felgenhauer U. (1997) *On optimality criteria for control problems. Part II: Convergence of the Ritz method*, preprint, BTU Cottbus, Reihe Mathematik, M-11/1997.

[8] Felgenhauer U. Wagner M. (1998) *Essential properties of L^∞-functions*, Z. Anal. Anw., 17: 229-242.

[9] Hager W.W. (1979) *Lipschitz continuity for constrained processes*, SIAM J. Control Optim., 17:321-338.

[10] Malanowski K. (1995) *Stability and sensitivity analysis of solutions to nonlinear optimal control problems*, Appl. Math. and Optim., 32:111-141.

[11] Malanowski K., Büskens C., Maurer H. (1997) *Convergence of approximations to nonlinear control problems*, in: Mathematical Programming with Data Perturbation, ed.: A.V.Fiacco, Lect. Notes in Pure Appl. Math., 195, Marcel Dekker Inc., New York, 253-284.

[12] Maurer H., Pickenhain S. (1995) *Second Order Sufficient Conditions for Optimal control problems with Mixed Control-State Constraints*, J.Optim. Theor. Appl. 86:649-667.

[13] Pickenhain S. (1992) *Sufficiency Conditions for Weak Local Minima in Multidimensional Optimal Control Problems with Mixed Control-State Restrictions*, Z. Anal. Anw. 11:559-568.

[14] Robinson S.M. (1980) *Strongly regular generalized equations*, Math. Oper. Res. 5:43-62.

Dr. U. Felgenhauer
Institut für Mathematik
Brandenburgische Technische Universität Cottbus
PF 101344, 03013 Cottbus, Germany
e-mail: felgenh@math.tu-cottbus.de

M M A FERREIRA[1]
Nontrivial maximum principle for optimal control problems with state constraints

1 Introduction.

Consider the following optimal control problem with state constraints:

$$(Q) \quad \begin{aligned} &\text{Minimize } g(x(b)) \\ &\text{subject to} \\ &\dot{x}(t) = f(t, x(t), u(t)) \quad \text{a.e. } t \in [a, b] \qquad (1.1)\\ &x(a) = x_0 \\ &u(t) \in \Omega \quad \text{a.e. } t \in [a, b] \\ &h(t, x(t)) \leq 0 \quad \text{all } t \in [a, b] \\ &e(x(b)) = 0 \qquad (1.2) \end{aligned}$$

Here $f : [a, b] \times \mathbb{R}^n \times \mathbb{R}^m \longmapsto \mathbb{R}^n$ and $h : [a, b] \times \mathbb{R}^n \longmapsto \mathbb{R}$, $e : \mathbb{R}^n \longmapsto \mathbb{R}^k$ are given functions, $\Omega \subset \mathbb{R}^m$ is a given set and $x_0 \in \mathbb{R}^n$ is a fixed point.

The aim of this paper is to get nondegenerate necessary conditions for the optimal control problem above when the initial state lies in the boundary of the state constraint set region. In such situation there's always a set of trivial nonzero multipliers satisfying the maximum principle conditions and from which no information about the optimal solution can be obtained. Our goal is to give hypotheses on the data of the problem, the so-called constraint qualifications, in order to ensure the existence of a set of nontrivial multipliers in addition to the trivial ones.

Throughout this paper an admissible process is a pair of functions comprising an absolutely continuous function $x \in AC^1([a,b]; \mathbb{R}^n)$ and $u \in U$, where
$U := \{ \text{ measurable functions } u : [a, b] \longmapsto \mathbb{R}^m, \ u(t) \in \Omega \text{ a. e. } \}$,

for which the differential equation (1.1) and all the constraints of problem (Q) are satisfied. A minimizer is an admissible process for which the minimum of $g(x(b))$ over other admissible processes (x, u) is achieved.

Under suitable hypotheses the maximum principle asserts that if $(\bar{x}, \bar{u})$ is a minimizer for problem (Q), then there exist $(p, \mu, \sigma, \lambda)$, with $p \in AC^1([a,b]; \mathbb{R}^n)$, μ is a measure representing a point in $C^*([a,b])(= C^*([a,b]; \mathbb{R}))$, $\sigma \in \mathbb{R}^k$, $\lambda \in \mathbb{R}$ such that

$$\mu \geq 0, \ \lambda \geq 0,$$

---
[1] This author was supported by J.N.I.C.T. Project PICT/CEG/2438/95, Portugal.

$$-\dot{p}(t) = \left(p(t) + \int_{[a,t)} h_x(s, \bar{x}(s))\mu(ds)\right) \cdot f_x(t, \bar{x}(t), \bar{u}(t)), \quad \text{a.e. } t \in [a,b],$$

for almost every $t \in [a, b]$, the function

$$u \longmapsto \left(p(t) + \int_{[a,t)} h_x(s, \bar{x}(s))\mu(ds)\right) \cdot f(t, \bar{x}(t), u)$$

is maximized over Ω at $u = \bar{u}(t)$,

$$supp\{\mu\} \subset \{t \in [a, b] : h(t, \bar{x}(t)) = 0\}$$

$$-\left(p(b) + \int_{[a,b]} h_x(s, \bar{x}(s))\mu(ds)\right) = \sigma^T e_x(\bar{x}(b)) + \lambda g_x(\bar{x}(b)). \tag{1.3}$$

and

$$\|\mu\|_{C^*} + |\sigma| + \lambda > 0. \tag{1.4}$$

When the initial state lies on the boundary of the state constraint set region, in the sense that:

$$h(a, x_0) = 0.$$

it can easily be checked that the above necessary conditions are always satisfied with the trivial nonzero multipliers:

$$p(t) \equiv -h_x(a, x_0), \quad \mu = \delta_{\{a\}}, \quad \text{and } \lambda = 0, \tag{1.5}$$

in which $\delta_{\{a\}}$ denotes the unit measure concentrated on the point $\{a\}$.

In this case the maximum principle may degenerate and it is possible that no information about the optimal solution may be obtained from those conditions.

In [2] Ferreira and Vinter consider this problem of degeneracy when the constraint (1.2) takes the form of an inequality, $d(x) \leq 0$, where $d : \mathbb{R}^n \longrightarrow \mathbb{R}^k$. A constraint qualifications is imposed on the data of the problem ensuring the existence of nontrivial multipliers besides the trivial ones associated to the optimal solution.

Results concerning the nondegeneracy of the state constraint maximum principle were also derived by Arutyunov et al in [4], [5], [6], [8] and [7]. Their approach is essencially based on penalization techniques, quite different from ours. Their constraint qualifications are of interest since they do not depend on the optimal solution of the problem. A relevant difference between the results of these authors and ours ([2] and the present paper) lies on the regularity conditions of the dynamic of the problem for time dependence, considerably weaker on our work.

The approach used in [2] is not, as a whole, applicable when right endpoint equality constraints are present as in problem (Q). In particular, the definition of some sets that enable us to apply a convenient separation theorem must be reformulated. In this paper such reformulation will be done. The constraint qualifications imposed in [2] are again exactly the same we need to get nondegenerate necessary conditions for (Q). An outline of the proof of this result is given. We refer to [2] for a detailed analysis of the comum steps.

Finally, we give a brief description of the last developments done jointly by R. B. Vinter, F. A. Fontes and myself. Details will be reported on a later paper.

2 Main Results.

It is assumed henceforth that the following hypotheses are in force:

(H1): $x \longmapsto f(t,x,u)$ is continuously differentiable for fixed (t,u), $(t,u) \longmapsto f(t,x,u)$ is $\mathcal{L} \times \mathcal{B}$-measurable for fixed x, and f and f_x are bounded on bounded sets.
(H2): g and e are continuously differentiable.
(H3): Ω is a bounded, Borel set.
(H4): $x \longmapsto h(t,x)$ is differentiable for fixed t, and h and h_x are continuous.
(H5): $f(t,x,\Omega)$ is convex for each t,x.

The main results derived in [2] are now validated to (Q).
Theorem 2.1 *Suppose that*
$(CQ)_1$: $\quad h(a,x_0) = 0$ *and there exists* $u' \in U$ *such that,*

$$\lim_{\epsilon \to 0^+} \operatorname*{ess\,sup}_{a \leq s < a + \epsilon} h_x(a,x_0) \cdot [f(s,x_0,u'(s)) - f(s,x_0,\bar{u}(s))] < 0 .$$

Then $(\bar{x},\bar{u})$ *satisfies the conditions of the Maximum Principle with multipliers* (p,μ,σ,λ) *for which*

$$\int_{(a,b]} \mu(ds) + |\sigma| + \lambda \neq 0. \tag{2.1}$$

Theorem 2.2 *The assertions of Theorem 2.1 remain valid when the constraint qualification* $(CQ)_1$ *is replaced by*

$(CQ)_2$: *there exists* $\bar{t} \in (a,b]$ *such that* $h(t,\bar{x}(t)) < 0 \quad \forall t \in (a,\bar{t}\,]$.

The important feature of condition 2.1 is that the measure μ is integrated over the *half-open* interval $(a,b]$; it therefore rules out multipliers for which $\lambda = 0$, $p(t) \equiv -h_x(a,x_0)$ and μ is concentrated at $\{a\}$.

$(CQ)_1$ imposes the existence of a control function which pulls the state trajectory away from the state constraint boundary more rapidly than does $\bar{u}$. On the other hand, when $h(a,x_0) = 0$, $(CQ)_2$ requires that the minimizing state trajectory $\bar{x}$ itself leaves the boundary immediately.

3 Present and Future Work.

Vinter, Fontes and Ferreira have recently extended the results stated above so as to cover more general problems. In this section, we give a brief review of their developments (see [9]).

Consider the problem (Q) with the endpoint constraint (1.2) replaced by the more general constraint $x(b) \in C$, where C is a closed set. Assume that the continuous differentiability on the data of the problem is replaced by Lipschitz continuity and that the hypothesis on the convexity of the velocity set, $f(t, x, \Omega)$, is dropped. Lipschitz continuity on h as a function of x enables us to consider constraints on the state variable in the implicit form $x(t) \in X(t)$, where $X(\cdot)$ is a given multifunction.

The treatment of such problem must be done in a framework of nonsmooth analysis and the following necessary conditions can be established:

There exist an arc $p : [a, b] \mapsto \mathbb{R}^n$, a measurable function γ, a nonnegative Radon measure $\mu \in C^*([a, b], \mathbb{R})$, and, a scalar $\lambda \geq 0$ such that

$$\| \mu \|_{C^*} + \| p \| + \lambda > 0, \tag{3.1}$$

$$-\dot{p}(t) \in \bar{\partial}_x H(t, \bar{x}(t), p(t) + \int_{[a,t]} \gamma(s)\mu(ds), \bar{u}(t)) \quad a.e., \tag{3.2}$$

$$-\left(p(b) + \int_{[a,b]} \gamma(s)\mu(ds)\right) \in N_C(\bar{x}(b)) + \lambda \partial g(\bar{x}(b)), \tag{3.3}$$

$$\gamma(t) \in \partial_x^> h(t, \bar{x}(t)) \quad \mu - a.e., \tag{3.4}$$

$$\mathrm{supp}\{\mu\} \subset \{t \in [a, b] : h(t, \bar{x}(t)) = 0\}, \tag{3.5}$$

and, for almost every $t \in [a, b]$, $\bar{u}(t)$ maximizes, over $\Omega(t)$

$$u \mapsto H(t, \bar{x}(t), p(t) + \int_{[a,t]} \gamma(s)\mu(ds), \bar{u}(t)). \tag{3.6}$$

In the above $H(t, x, p, u)$ denotes the Hamiltonian which is defined as

$$H(t, x, p, u) = p \cdot f(t, x, u).$$

$N_C(x)$ represents the limiting normal cone:

$$N_C(x) := \{\lim y_i : \text{there exist } x_i \xrightarrow{C} x, \{M_i\} \subset \mathbb{R}^+ \text{ s.t.}$$
$$y_i(z - x_i) \leq M_i |z - x_i|^2 \text{ for all } z \in C\},$$

The set ∂f is the limiting subdifferential of a Lipschitz function f,

$$\partial f(x) := \{y : (y, -1) \in N_{\mathrm{epi} f}(x, f(x))\},$$

the set $\bar{\partial} f$ is the Clarke's subdifferential, which can be given by the convex hull of the limiting subdifferential, $\bar{\partial} f := \mathrm{co}\partial f$, and $\partial_x^> h(t,x)$ is the hybrid partial subdifferential of h in the x-variable

$$\partial_x^> h(t,x) := \mathrm{co}\{\xi: \text{there exist } (t_i, x_i) \to (t,x) \text{ s.t.}$$
$$h(t_i, x_i) > 0, \ h(t_i, x_i) \to h(t,x), \text{ and } \nabla_x h(t_i, x_i) \to \xi\}$$

If $h(a, x_0) = 0$ and $\partial_x^> h(a, x_0) \neq \emptyset$ we have again the trivial multipliers:

$$(\lambda, \mu, p) = (0, \delta_{\{a\}}, -\zeta), \text{ with } \zeta \in \partial_x^> h(a, x_0).$$

For this problem, assume that:

CQ If $h(a, x_0) = 0$ then there exist positive constants K_u, ϵ, ϵ_1, δ, and a control $\tilde{u} \in U$ such that for a. e. $t \in [a, a+\epsilon)$

$$\|f(t, x_0, \bar{u}(t))\| \leq K_u, \quad \|f(t, x_0, \tilde{u}(t))\| \leq K_u,$$

and

$$\zeta \cdot [f(t, x_0, \tilde{u}(t)) - f(t, x_0, \bar{u}(t))] < -\delta$$

for all $\zeta \in \partial_x^> h(s,x)$, $s \in [a, a+\epsilon)$, $x \in \{x_0\} + \epsilon_1 B$.

Under this additional condition (CQ), a nondegenerate maximum principle, which rules out the trivial nonzero multipliers described, was recently established. Further developments on the subject are currently under way.

4 Proof of the Main Results.

The proof of Theorem 2.2 follows exactly the same arguments of the corresponding result on [2]. Small changes due to the fact that the multiplier σ may take positive or non positive values are minor and do not affect the main ideas.

Next, a sketch of the proof of Theorem 2.1 is given. For details see [10].

Proof of Theorem 2.1. Define the convex set

$$\mathcal{T}_k = \{\theta \in \mathbb{R}^k : \theta_j \geq 0, \ j = 1, ..., k, \ \sum_{j=1}^k \theta_j \leq 1\},$$

and y_{u_j} as the unique solution on [a,b] of the linearized system,

$$\dot{y}(t) = f_x(t, \bar{x}(t), \bar{u}(t)) \cdot y(t) + [f(t, \bar{x}(t), u_j(t)) - f(t, \bar{x}(t), \bar{u}(t))] \quad (4.1)$$
$$y(a) = 0$$

For $v \in \Omega$ denote

$$\Delta f_v(t) := f(t, \bar{x}(t), v) - f(t, \bar{x}(t), \bar{u}(t)). \quad (4.2)$$

We deduce that the following proposition holds.

Proposition 4.1 *Let $(\bar{x}(\cdot), \bar{u}(\cdot))$ be an admissible process for problem (Q). Take arbitrary control functions $\{u_1(.), \ldots, u_k(.)\} \subset U$, $k \in \mathcal{N}$, finite. Then, there exists $\mathcal{T}' \subset \mathcal{T}_k$, neighborhood of $\theta = 0$ relative to $\mathcal{T}_k$, and a function $w : \mathcal{T}' \to [0, \infty)$ satisfying $\lim_{\theta \to 0} w(\theta) = 0$ with the following property:*
Corresponding to any $\theta \in \mathcal{T}'$ a process $(x_\theta(.), u_\theta(.))$ may be found for which

$$\| h(t, x_\theta(t)) - h(t, \bar{x}(t)) - \sum_{j=1}^{k} \theta_j \, h_x(t, \bar{x}(t)) y_{u_j}(t) \| \leq (t - a) \, \|\theta\| \, w(\theta).$$

We begin with a restriction of the set of control functions:

$$U^* := \{ \, u \in U : t = a \text{ is a right Lebesgue point of } t \to h_x(a, x_0). \triangle f_u(t) \, \}$$

Take any $u \in U^*$. It is a simple matter to check that

$$(t - a)^{-1} h_x(t, \bar{x}(t)) \cdot y_u(t) \longmapsto h_x(a, x_0) \cdot \triangle f_u(a) \quad \text{as } t \downarrow a \,.$$

Write $\eta_u : [a, b] \longmapsto \mathbb{R}$ for the continuous extension of $(t - a)^{-1} h_x \cdot y_u(t)$ on $(a, b]$ to $[a, b]$, namely

$$\eta_u(t) := \begin{cases} (t - a)^{-1} h_x(t, \bar{x}(t)) \cdot y_u(t) & t \in (a, b] \\ h_x(a, x_0) \cdot \triangle f_u(a) & t = a \end{cases}$$

Define

$$G_0 = \{t \in [a, b] : h(t, \bar{x}(t)) = 0\} \,.$$

Notice that, in view of $(CQ)_1$, G_0 is non-empty. Because of the properties of h this set is also compact.

Considering the restriction of $\eta_u(t)$ to the set G_0, define

$$\mathcal{W} = \{ \, (\eta_u(t), \, e_x(\bar{x}(b)) y_u(b), \, g_x(\bar{x}(b)) \cdot y_u(b) \,) \, : \, u \in U^* \} \subset C([G_0]) \times \mathbb{R}^k \times \mathbb{R}$$

where $C([G_0])$ represents the set of continuous functions defined on G_0 with values on $\mathbb{R}$.

It is easy to see that $\mathcal{W}$ is a convex set. Set $C = \{c(t) \in C([G_0]) : c(t) < 0 \, \forall t \, \}$. Then, C is open and convex. Moreover, $0 \in \bar{C}$ and $0 \in \mathcal{W}$.

An application of Theorem V.2.1, [3], allow us to deduce that either

(a) there exists $\varphi = (\varphi_0, \sigma, \lambda) \in (C([G_0]))^* \times \mathbb{R}^k \times [0, \infty)$ such that $\varphi \neq 0$,

$$\varphi(z) = \varphi_0(z_0) + \sigma^T \cdot z_1 + \lambda z_2 \geq 0 \qquad \forall \, z = (z_0, z_1, z_2) \in \mathcal{W}$$

and

$$\varphi_0(c) \leq 0 \qquad \forall c \in \bar{C} \,, \tag{4.3}$$

or

(b) there exist points $\xi^i = (\xi_0^i, \xi_1^i, \xi_2^i) \in \mathcal{W}$ and numbers $\beta^i > 0$ $(i = 0, ...k)$ such that
$\sum_{i=0}^k \beta^i = 1$, $\xi_0^i \in C$, the set $\{(\xi_1^0, \xi_2^0), ..., (\xi_1^k, \xi_2^k)\}$ is linearly independent, $\xi_2^i < 0$ and
$$\sum_{i=0}^k \beta^i \xi_1^i = 0$$

Proposition 4.1 and minor alterations to the proof of Theorem V.2.2 in [3] allow us to conclude that for our problem the following result holds.

Proposition 4.2 *Suppose that there exist points $\xi^i = (\xi_0^i, \xi_1^i, \xi_2^i) \in \mathcal{W}$ and numbers $\beta^i > 0$
$(i = 0, ...k)$ such that $\sum_{i=0}^k \beta^i = 1$, $\xi_0^i \in C$, the set $\{(\xi_1^0, \xi_2^0), ..., (\xi_1^k, \xi_2^k)\}$ is independent, $\xi_2^i < 0$ and $\sum_{i=0}^k \beta^i \xi_1^i = 0$.
Then there exist $\tilde{\gamma} > 0$ and a function $\gamma \to \tilde{\theta}(\gamma) : (0, \tilde{\gamma}] \to \mathcal{T}' \subset \mathcal{T}_{k+1}$ such that for each $\gamma \in (0, \tilde{\gamma}]$ an admissible process $(x_{\tilde{\theta}(\gamma)}, u_{\tilde{\theta}(\gamma)})$ can be found with*

$$g(x_{\tilde{\theta}(\gamma)}(b)) < g(\bar{x}(b)).$$

From (b) we can choose an admissible process with lower cost than the optimal. This contradiction shows that (b) is never satisfied. The first alternative, (a), occurs. This allows us to establish the existence of multipliers $(\tilde{\nu}, \sigma, \lambda) \in C^*([a, b]) \times \mathbb{R}^k \times \mathbb{R}$ such that

$$\tilde{\nu} \geq 0, \ \lambda \geq 0$$
$$supp\{\tilde{\nu}\} \subset \{t : h(t, \bar{x}(t)) = 0\}$$
$$\tilde{\nu}\{(a, b]\} + |\sigma| + \lambda \neq 0,$$

and

$$\int_{G_0} \eta_u(t) \tilde{\nu}(dt) + \sigma^T e_x(\bar{x}(b)) y_u(b) + \lambda g_x(\bar{x}(b)) \cdot y_u(b) \geq 0 \tag{4.4}$$

for all $u \in U^*$.

Extend $\tilde{\nu}$ as a regular Borel measure on the Borel subsets of $[a, b]$ as $\nu(B) = \tilde{\nu}(B \cap G_0)$. We get a nonegative measure with support in $G_0 = \{t \in [a, b] : h(t, \bar{x}(t)) = 0\}$. As proved in [2], $(CQ)_1$ implies the existence of a control $\tilde{u} \in U$ such that

$$\lim_{t \to a^+} h_x(a, x_0).y_{\tilde{u}}(t) < 0,$$

where $y_{\tilde{u}}$ is the solution of 4.1 when $u_j = \tilde{u}$. This solution and eq. (4.4) allow us to conclude that $(\nu, \sigma, \lambda) \neq (\rho \delta_{\{a\}}, 0, 0)$, for any $\rho > 0$.

The function $s \to (s - a)^{-1}$ is integrable with respect to the regular measure ν. Defining $\mu(A) = \int_{(a,b] \cap A} (s - a)^{-1} \nu(ds)$ and the function $p(t)$ in an appropriate way (see [2] for details), we get the conclusions of Theorem 2.1. ∎

References

[1] Clarke, F H (1983) *Optimization and Nonsmooth Analysis*, John Wiley, New York

[2] Ferreira, M M A and Vinter, R B *When is the Maximum Principle for State Constrained Problems Nondegenerate?*, J. of Mathematical Analysis and Applications, **187**, no. 2, 438-467, 1994

[3] Warga, J (1972), *Optimal Control of Differential and Functional Equations*, Academic Press, New York

[4] Arutyunov, A V and Tynyanskiy, N T *The Maximum Principle in a Problem with Phase Constraints*, Izv. Akad. Nauk SSSR Tekhn. Kibernet., **4**, 60-80 (1984); English transl: Soviet J. Comput. Systems Sci., **23**, no. 1, 28-35 (1985)

[5] Arutyunov, A V *On the Theory of the Maximum Principle in Optimal Control Problems with Phase Constraints*, Soviet Math. Dokl., **39**, no. 1, 1-4 (1989)

[6] Arutyunov, A V, Aseev, S M and Blagodat-Skikh, V I *First Order Necessary Conditions in the Problem of optimal Control of a Differential Inclusion with Phase Constraints*, Russian Acad. Sci. Sb. Math., **79**, no. 1, 117-139 (1994)

[7] Arutyunov, A V and Aseev, S M *Investigation of the Degeneracy Phenomenon of the Maximum Principle for Optimal Control Problems with State Constraints*, SIAM J. Control and Opt., **35**, no. 3, 930-952 (1997)

[8] Arutyunov, A V and Blagodat-Skikh, V I *The Maximum Principle for Differential Inclusions with Phase Constraints*, Proc. of the Steklov Institute of Math., no. 2, 3-25 (1993)

[9] Ferreira, M M A, Fontes, F A and Vinter R B *Nondegenerate Necessary Conditions for Nonconvex Optimal Control Problems with State Constraints* (submitted to publication)

[10] Ferreira, M M A *Novo Princípio do Máximo para Problemas de Controlo Óptimo com Restrições de Estado* PhD Thesis, 1994.

BARBARA KASKOSZ
Abundant subsets of generalized control systems

1. Introduction

A "generalized control system" which we consider in this paper is a collection of vector fields which are measurable in the time variable and Lipschtzian in the state variable. The collection is assumed to satisfy certain boundedness and decomposability conditions, but need not have any additional structure. For such systems we define a concept of an "abundant subset". Our definition follows the definition of an abundant set of control functions introduced by Warga [11].

In the next section we define a generalized control system and an abundant subset. We then present an extremality-controllability theorem which says, in essence, that either a given trajectory of the system satisfies a type of the maximum principle, or a neighborhood of its endpoint can be covered using only trajectories corresponding to vector fields of any given abundant subset.

In Sec. 3 we apply the theorem to control systems in the classical formulation and obtain a controllability-extremality result which is stronger, in some respects, than the classical maximum principle and all related versions of the maximum principle as in [12], [14], [5]. We give an example which demonstrates that.

In Sec. 4 we apply our main theorem to a differential inclusion problem. We show how the Pontryagin-type maximum principle for non-convex inclusions, like the one proved by Tuan [10], follows as an easy corollary. In our choice of an abundant subset, in this context, we use an idea inspired by the work of Zhu [15].

Control systems of a similar nature, which are simply collections of vector fields, were considered previously by the author and Lojasiewicz [4], and recently by Sussmann [9], who studies them without the assumption of Lipschitz continuity.

2. The Main Theorem

In this and in all subsequent sections, V is a given open subset of R^n, $C \subseteq V$ is a closed subset, $g : V \to R^m$ is a given Lipschitzian mapping. Let **S** be a collection of functions $f(t,x)$, $f : [0,1] \times V \to R^n$ which satisfies the following conditions:

(A1) $f(\cdot, x)$ is measurable for every $f \in \mathbf{S}$ and every $x \in V$.

(A2) There exists a function $k(\cdot) \in L^1([0,1]; R)$ such that for every $f \in \mathbf{S}$ and almost every $t \in [0,1]$, $f(t, \cdot)$ is $k(t)$-Lipschitzian in V.

(A3) There exists a function $l(\cdot) \in L^1([0,1]; R)$ such that for every $f \in \mathbf{S}$ and almost all $t \in [0,1]$, $|f(t,x)| \leq l(t)$ for all $x \in V$.

(A4) For any measurable subset $A \subseteq [0,1]$ and any $f_1, f_2 \in \mathbf{S}$ the function defined for $t \in [0,1]$, $x \in V$ by $\chi_A(t) f_1(t,x) + \chi_{[0,1]-A}(t) f_2(t,x)$ belongs again to $\mathbf{S}$, where χ_K denotes the characteristic function of any given set K.

Denote by $B(x,r)$ the closed ball with center x and radius r in R^n, $B = B(0,1)$. To simplify the presentation of this paper we shall make an additional assumption (A5). All results of this and the next section can be derived without this assumption but formulation becomes more complicated. We treat the general case in [6].

(A5) For every $x_0 \in C$, $B(x_0, \int_0^1 l(t) dt) \subset V$.

Denote $\widetilde{\mathbf{S}} = \mathbf{S} \times C$. For every pair $(f, x_0) \in \widetilde{\mathbf{S}}$ consider the initial value problem:

$$x'(t) = f(t, x(t)) \quad \text{a.e. in} \quad [0,1], \quad x(0) = x_0. \tag{2.1}$$

Under the assumptions (A1)-(A5) there exists a unique solution of (2.1) in [0,1]. We denote it by $x_{f,x_0}(\cdot)$. We shall refer to $\widetilde{\mathbf{S}}$ as a generalized control system or, simply, a control system, and to $x_{f,x_0}(\cdot)$ as a trajectory of $\widetilde{\mathbf{S}}$. We define the *reachable set* of $\widetilde{\mathbf{S}}$ at time 1 to be $\mathcal{R}_{\widetilde{\mathbf{S}}}(1) = \{x_{f,x_0}(1) : (f, x_0) \in \widetilde{\mathbf{S}}\}$.

Let $x(\cdot)$ be a trajectory of $\widetilde{\mathbf{S}}$. We say that the system $\widetilde{\mathbf{S}}$ is *locally g-controllable* around $x(1)$, if $g(x(1)) \in \text{int} g(\mathcal{R}_{\widetilde{\mathbf{S}}}(1))$. In this paper we are concerned with sufficient conditions for $\widetilde{\mathbf{S}}$ to be locally g-controllable around a given trajectory. (Or, equivalently, with necessary conditions for a trajectory to be a boundary trajectory relative to g, that is, such that $g(x(1))$ belongs to the boundary of $g(\mathcal{R}_{\widetilde{\mathbf{S}}}(1)))$. This is an important problem because of its connection to many optimization problems whose solutions, after some reformulation, must be boundary trajectories.

In our extremality-controllability theorem we use a variation on the notion of an "abundant set of controls" introduced by Warga [11], [12]. Similarly as in the original definition of Warga, our definition of an "abundant subset" of the control system $\widetilde{\mathbf{S}}$ is technical and complicated. It proves, however, to be useful in a variety of applications as it captures, in a distilled form, a property of a subset of a control system crucial in proving variants of the maximum principle.

For a given $N \in \{1, 2, ...\}$ denote

$\Delta_N = \{(\lambda_0, \lambda_1, ..., \lambda_N) \in R^{N+1} : \sum_{j=0}^{N} \lambda_j = 1, \quad \lambda_j \geq 0 \quad \text{for} \quad j = 0, 1, ...N\}$.

Let $f_0, f_1, ..., f_N \in \mathbf{S}$ and a mapping $c(\cdot) : \Delta_N \to C$ be given. For any

$\lambda = (\lambda_0, ..., \lambda_N) \in \Delta_N$ consider the initial value problem

$$x'(t) = \sum_{j=0}^{N} \lambda_j f_j(t, x(t)) \quad \text{a.e. in} \quad [0,1], \quad x(0) = c(\lambda). \tag{2.2}$$

It follows from (A1)-(A5) that (2.2) has a unique solution in [0,1], (although $\sum_{j=0}^{N} \lambda_j f_j$ may not be in **S**).

DEFINITION 2.1. Let $\tilde{\mathbf{L}} \subseteq \tilde{\mathbf{S}}$, $\eta : V \to R^m$ be a continuous mapping. We say that $\tilde{\mathbf{L}}$ is an η-abundant subset of $\tilde{\mathbf{S}}$ if for every $N \in \{1, 2, ...\}$, every choice of $f_0, f_1, ..., f_N \in \mathbf{S}$, every continuous function $c(\cdot) : \Delta_N \to C$ and every $\epsilon > 0$ there exists a mapping $\theta : \Delta_N \to \tilde{\mathbf{L}}$ such that

(i) $|\eta(x_{\theta(\lambda)}(1)) - \eta(x_\lambda(1))| < \epsilon$ for every $\lambda \in \Delta_N$,

where $x_\lambda(\cdot)$ is the solution of (2.2), $x_{\theta(\lambda)}(\cdot)$ is the trajectory corresponding to the pair $\theta(\lambda) \in \tilde{\mathbf{L}}$,

(ii) the mapping $\lambda \to \eta(x_{\theta(\lambda)}(1))$ from Δ_N to R^m is continuous.

We say that $\tilde{\mathbf{L}}$ is an abundant subset of $\tilde{\mathbf{S}}$ if it is η-abundant with η equal to the identity mapping on V.

It follows easily from [2, Lemma 4.2] that $\tilde{\mathbf{S}}$ is an abundant subset of itself. It is worth noticing that an abundant subset is assumed to be only a subset, and not a subsystem of $\tilde{\mathbf{S}}$. In particular, even if $\tilde{\mathbf{L}}$ is of the form $\mathbf{L} \times C$, for some $\mathbf{L} \subseteq \mathbf{S}$, $\mathbf{L}$ does not have to satisfy the decomposability condition (A4).

As most versions of the maximum principle, our extremality-controllability theorem involves a tangent cone to the set of initial conditions C. Recall the definition of a regular tangent cone (see [8],[10]).

DEFINITION 2.2. Let C be a closed subset of R^n, $x_0 \in C$. A closed convex cone $T_C(x_0) \subseteq R^n$ is said to be a regular tangent cone to C at x_0 if for every $h > 0$ there exists a continuous mapping $O_h(\cdot) : T_C(x_0) \cap B \to \mathbb{R}^n$ such that

$$\lim_{h \to 0^+} \max_{v \in T_C(x_0) \cap B} \frac{|O_h(v)|}{h} = 0, \quad x_0 + hv + O_h(v) \in C \quad \text{for} \quad v \in T_C(x_0) \cap B.$$

Clearly, a regular tangent cone is not unique. If C is convex, the intermediate cone, which coincides in this case with the Clarke and the contingent cones, is regular, (see [8]).

For a cone $T \subseteq R^n$ denote by T^* its negative polar; that is, $T^* = \{v \in R^n :< v, w > \leq 0 \text{ for all } w \in T\}$, where $< \cdot, \cdot >$ denotes the scalar product in R^n. For any collection of matrices A we denote by A^T the collection of transpose matrices. Elements of an Euclidean space are regarded as column vectors for purposes of matrix multiplication. By "∂" below we denote the Clarke generalized gradient or Jacobian with respect to the x variable, (see [1] for definitions and properties).

For any given trajectory $z(\cdot)$ of we say that an element $\bar{f} \in \mathbf{S}$ *generates* $z(\cdot)$ if $z(\cdot) = x_{\bar{f},z(0)}(\cdot)$, that is, $z'(t) = \bar{f}(t, z(t))$ a.e. in [0,1].

THEOREM 2.1. *Assume that $\tilde{\mathbf{L}}$ is a g-abundant subset of $\tilde{\mathbf{S}}$. Let $z(\cdot)$ be a trajectory of $\tilde{\mathbf{S}}$, let $T_C(z(0))$ be a regular tangent cone to C at $z(0)$. Then either*

(a) *for every element $\bar{f} \in \mathbf{S}$ which generates $z(\cdot)$ there exist an absolutely continuous function $p(\cdot) : [0,1] \to R^n$ and a vector $v \in R^m$, $|v| = 1$, such that*

$$p(0) \in T_C^*(z(0)), \quad p(1) \in \partial g(z(1))^T v, \tag{2.3}$$

$$-p'(t) \in \partial \bar{f}(t, z(t))^T p(t) \quad \text{a.e. in } [0,1], \tag{2.4}$$

and for every $f \in \mathbf{S}$:

$$< p(t), z'(t) > \geq < p(t), f(t, z(t)) > \quad \text{a.e. in } [0,1], \tag{2.5}$$

or

(b)

$$g(z(1)) \in \text{int}\{g(x_{f,x_0}(1)) : (f, x_0) \in \tilde{\mathbf{L}}\}. \tag{2.6}$$

Note that Theorem 2.1 implies, in particular, that if the maximum principle (a) is not satisfied, then $g(z(1)) \in \text{int} g(\mathcal{R}_{\tilde{\mathbf{S}}}(1))$.

The proof of Theorem 2.1 is given in [6]. It relies on the open mapping principle by Warga [13]. Theorem 2.2 below, whose proof is also contained in [6], gives a sufficient condition for a subset $\mathbf{L} \subseteq \mathbf{S}$ to be such that $\tilde{\mathbf{L}} = \mathbf{L} \times C$ is abundant in $\tilde{\mathbf{S}}$. Denote by $\|\cdot\|$ the norm in $C([0,1]; R^n)$.

THEOREM 2.2 *Assume that a subset $\mathbf{L}$ of $\mathbf{S}$ is such that*
(i) *for every $f \in \mathbf{S}$, $x_0 \in C$ and every $\epsilon > 0$ there exists $h \in \mathbf{L}$ such that $\| x_{f,x_0}(\cdot) - x_{h,x_0}(\cdot) \| < \epsilon$,*
(ii) *if $h_1, h_2 \in \mathbf{L}$, I_1, I_2 are disjoint subintervals such that $I_1 \cup I_2 = [0,1]$, then $\chi_{I_1}(t) h_1(t, x) + \chi_{I_2}(t) h_2(t, x) \in \mathbf{L}$.*
Then $\tilde{\mathbf{L}} = \mathbf{L} \times C$ is an abundant subset of $\tilde{\mathbf{S}}$.

3. Application to Control Systems in the Classical Formulation

Let V, C, g be as in the previous section. Let $U \subset R^q$ be a given compact set, $\phi : [0,1] \times V \times U \to R^n$. Consider the control system

$$x'(t) = \phi(t, x(t), u(t)) \quad \text{a.e. in } [0,1], \quad x(0) = x_0, \tag{3.1}$$

where $x_0 \in C$, control functions $u(\cdot)$ are required to be measurable and to satisfy $u(t) \in U$ for $t \in [0,1]$. Denote by $\mathcal{U}$ the set of all such control functions. We shall make the following assumptions about the system:

(H1) $\phi(\cdot, x, u)$ is measurable for all $(x, u) \in V \times U$.

(H2) $\phi(t, \cdot, \cdot)$ is continuous for each $t \in [0, 1]$.

(H3) There exists a function $r(\cdot) \in L^1([0,1]; R)$ such that for almost all $t \in [0,1]$ and all $u \in U$, $\phi(t, \cdot, u)$ is $r(t)$-Lipschitzian in V.

(H4) There exists a function $l(\cdot) \in L^1([0,1]; R)$ such that for almost all $t \in [0,1]$ we have $|\phi(t, x, u)| \leq l(t)$ for all $x \in V$, $u \in U$.

Again, for simplicity of presentation we shall make an additional assumption

(H5) For every $x_0 \in C$, $B(x_0, \int_0^1 l(t)dt) \subset V$.

The last assumption together with (H1)-(H4) ensures that for each control function $u(\cdot) \in \mathcal{U}$ and each $x_0 \in C$ there exists a unique trajectory of the system (3.1) in [0,1]. We shall denote it by $x_{u(\cdot), x_0}(\cdot)$. Consider the reachable set of the system (3.1) at time 1, that is, $\mathcal{R}_\mathcal{U}(1) = \{x_{u(\cdot), x_0}(1) : u(\cdot) \in \mathcal{U}, x_0 \in C\}$. Let $z(\cdot)$ be a given trajectory of (3.1). We say that a control $\bar{u}(\cdot)$ *generates* the trajectory $z(\cdot)$ if $z(\cdot) = x_{\bar{u}(\cdot), z(0)}(\cdot)$, that is, $z'(t) = \phi(t, z(t), \bar{u}(t))$ a.e. in $[0, 1]$.

Define a generalized control system $\tilde{\mathbf{S}}$ associated with the system (3.1) and its subset $\tilde{\mathbf{L}}$ as follows.

Let a function $k(\cdot) \in L^1([0,1]; R)$ be given, $k(t) \geq r(t)$ for $t \in [0,1]$. Let $\mathbf{S}$ be the collection of all functions $f(\cdot, \cdot) : [0, 1] \times V \to R^n$ such that:

(S1) $f(t, x) \in \overline{co}\phi(t, x, U)$ for $t \in [0, 1]$, $x \in V$,

(S2) $f(\cdot, x)$ is measurable for all $x \in V$,

(S3) for almost all $t \in [0, 1]$, $f(t, \cdot)$ is $k(t)$-Lipschitzian in V.

It follows from (H1)-(H5), (S1)-(S3) that the collection satisfies (A1)-(A5). Set $\tilde{\mathbf{S}} = \mathbf{S} \times C$.

We say that a control $u(\cdot)$ is piecewise constant if there exist mutually disjoint intervals $I_1, I_2, ..., I_s$ whose union is [0,1] such that $u(\cdot)$ is constant on each I_i. Denote by $\mathcal{P}$ the set of all piecewise constant controls. Let $\mathcal{R}_\mathcal{P}(1) = \{x_{u(\cdot), x_0}(1) : u(\cdot) \in \mathcal{P}, x_0 \in C\}$ be the reachable set at time 1 by piecewise constant controls. Define $\mathbf{L}$ to be the collection of all functions $f(\cdot, \cdot)$ for which there exists a control $u(\cdot) \in \mathcal{P}$ such that $f(t, x) = \phi(t, x, u(t))$ for $t \in [0,1]$, $x \in V$. Clearly, $\mathbf{L} \subseteq \mathbf{S}$. Let $\tilde{\mathbf{L}} = \mathbf{L} \times C$. The next proposition follows easily from Theorem 2.2.

PROPOSITION 3.1 $\tilde{\mathbf{L}}$ *is an abundant subset of* $\tilde{\mathbf{S}}$.

$\tilde{\mathbf{L}}$ is an abundant and, therefore, a g-abundant subset of $\tilde{\mathbf{S}}$. As $\mathbf{S}$ contains all elements of the form $\phi(t, x, u(t))$, $u(\cdot) \in \mathcal{U}$, any trajectory of (3.1) is a trajectory of $\tilde{\mathbf{S}}$. We apply Theorem 2.1 and obtain the following

THEOREM 3.1 *Let $z(\cdot)$ be a trajectory of (3.1), $T_C(z(0))$ be a regular tangent cone to C at $z(0)$. Then either*

(a) *for every function $\bar{f}(\cdot,\cdot)$ satisfying* (S1)-(S3) *and such that*

$$z'(t) = \bar{f}(t, z(t)) \quad a.e. \ in \quad [0,1] \tag{3.2}$$

there exist an absolutely continuous function $p(\cdot) : [0,1] \to R^n$ *and a vector* $v \in R^m$, $|v| = 1$, *such that*

$$p(0) \in T^*_C(z(0)), \quad p(1) \in \partial g(z(1))^T v, \tag{3.3}$$

$$-p'(t) \in \partial \bar{f}(t, z(t))^T p(t) \quad a.e. \ in \quad [0,1], \tag{3.4}$$

$$< p(t), z'(t) > = \max_{u \in U} < p(t), \phi(t, z(t), u) > \quad a.e. \ in \quad [0,1], \tag{3.5}$$

or

(b)
$$g(z(1)) \in \mathrm{int} g(\mathcal{R}_\mathcal{P}(1)). \tag{3.6}$$

Note that **S** contains all elements of the form $f(t,x) = \int_U \phi(t,x,u)\sigma(t)(du)$, where $\sigma(\cdot)$ is a relaxed control (see Warga [11] for the definition and properties). Hence, Theorem 3.1 remains true for any trajectory $z(\cdot)$ of the relaxed system.

Possible elements $\bar{f}(\cdot,\cdot)$ in part (a) include all functions $\bar{f}(t,x) = \phi(t,x,\bar{u}(t))$, where $\bar{u}(\cdot)$ is any control generating $z(\cdot)$, as well as all functions of the form

$$\bar{f}(t,x) = \int_U \phi(t,x,u)\bar{\sigma}(t)(du), \tag{3.7}$$

for any relaxed control $\bar{\sigma}(t)$ which generates $z(\cdot)$, that is, for which $\bar{f}(t,x)$ given by (3.7) satisfies (3.2). Possible functions $\bar{f}(\cdot,\cdot)$ in part (a) include more, however. There may be selections which satisfy (S1)-(S3) and (3.2) which are not of the form (3.7). This strenghtens part (a) (see Example 3.1 below) and makes the maximum principle of Theorem 3.1 stronger than results which employ only ordinary controls or relaxed controls of the form (3.7), as in Warga [12]. (For systems without phase constrains, the results of [12] include phase constraints which we do not treat). Theorem 3.1 strenghtens also the results of the papers [14], [5], which do involve selections of the convexified right-hand side $\mathrm{co}\phi(t,x,U)$, because part (b) of our theorem provides a stronger condition than $g(z(1)) \in \mathrm{int} g(\mathcal{R}_\mathcal{U}(1))$ obtained in [14], [5].

EXAMPLE 3.1 Consider the following system on the plane

$$x'_1(t) = u_1(t)(1 - (u_2(t) - t)^2), \quad x'_2(t) = |x_1(t)| + u_3(t), \quad x_1(0) = x_2(0) = 0, \tag{3.8}$$

where $t \in [0,1]$, $u_1 \in [-1,1]$, $u_2 \in [0,1]$, $u_3 \in \{-1,0\}$. We have here $V = R^2$, $C = \{(0,0)\}$, $x = (x_1, x_2)$, $u = (u_1, u_2, u_3)$, $U = [-1,1] \times [0,1] \times \{-1,0\}$, $\phi(t,x,u) = (u_1(1 - (u_2 - t)^2), |x_1| + u_3)$. Let g be the identity on R^2. Consider the trajectory $z(t) \equiv (0,0)$. We have $T_C(z(0)) = \{(0,0)\}$. It is easy to see that for any ordinary control $\bar{u}(\cdot)$ or any relaxed control $\bar{\sigma}(\cdot)$ which generate $z(\cdot)$ the corresponding right-hand side $\phi(t,x,\bar{u}(t))$ or $\bar{f}(t,x)$ given by (3.7) is $(0,|x_1|)$. The maximum principle (a) is satisfied with the adjoint function $p(t) \equiv (0,1)$. It is easy to check that the Hamiltonian-type maximum principle of Clarke also holds for the trajectory $z(\cdot)$ with the same adjoint function $p(\cdot)$. Thus, neither the Hamiltonian nor the Pontryagin-type maximum principle using only ordinary or open-loop relaxed controls allow us to draw any conclusion as to $z(1)$ being a boundary or an interior point of the reachable set.

Take now $\bar{f}(t,x)$ to be a selection of $\operatorname{co}\phi(t,x,U)$ defined as $\bar{f}(t,x) = \bar{f}(x) = (0,x_1)$ for x in a neighborhood of the origin and all t, extended in an arbitrary way to a Lipschitzian selection of $\operatorname{co}\phi(t,x,U)$ for all $x \in R^2$, $t \in [0,1]$. The maximum principle (a) does not hold for this selection. Indeed, if $p(t) = (p_1(t), p_2(t)))$ satisfies (3.5), then $p_1(t) \equiv 0$. The adjoint equation implies then $p_2(t) = 0$, thus (3.3) cannot hold. We conclude therefore, that the system (3.8) is locally controllable around $z(1)$, moreover a neighborhood of $z(1)$ can be covered using only piecewise constant controls. Observe, that for the system (3.8), $\mathcal{R}_\mathcal{P}(1) \neq \mathcal{R}_\mathcal{U}(1)$. For example, the point $(1,1/2)$ can be reached only when $u_1(t) = 1$, $u_2(t) = t$, $u_3(t) = 0$ a.e. in $[0,1]$.

4. Application to Differential Inclusions

As another example of how our extremality-controllability theorem can be applied, we derive from it the Pontryagin-type maximum principle for nonconvex inclusions proved previously by Tuan [10]. Consider the differential inclusion

$$x'(t) \in F(t,x(t)), \quad \text{a.e.} \quad t \in [0,1], \quad x(0) \in C, \tag{4.1}$$

where $F(\cdot,\cdot) : [0,1] \times V \to 2^{R^n}$. Assume the following:

(B1) $F(t,x)$ is a nonempty, compact subset of R^n for all $t \in [0,1]$, $x \in V$.

(B2) $F(\cdot,x)$ is a measurable multifunction for each $x \in V$.

(B3) There exists a function $r(\cdot) \in L^1([0,1]; R)$ such that for almost all $t \in [0,1]$, $F(t,\cdot)$ is $r(t)$-Lipschitzian in V in the Hausdorff metric.

(B4) There exists $l(\cdot) \in L^1([0,1]; R)$ such that for almost all $t \in [0,1]$ and all $x \in V$, $\max\{|v| : v \in F(t,x)\} \leq l(t)$.

(B5) For every $x_0 \in C$, $B(x_0, \int_0^1 l(t)dt) \subset V$.

Denote by $\mathcal{T}_F$ the set of all trajectories of the inclusion (4.1); that is, all solutions to (4.1) and by $\mathcal{R}_F(1)$ the corresponding reachable set $\mathcal{R}_F(1) = \{x(1) : x(\cdot) \in \mathcal{T}_F\}$.

Consider the convexified problem:

$$x'(t) \in \operatorname{co}F(t,x(t)) \quad \text{a.e.} \quad t \in [0,1], \quad x(0) \in C. \tag{4.2}$$

Denote the set of all trajectories of (4.2) by $\mathcal{T}_{\mathrm{co}F}$.

Let $k(\cdot) \in L^1([0,1]; R)$ be such that $k(t) \geq 4nr(t)$ for $t \in [0,1]$. Define $\mathbf{S}$ to be the set of all functions $f(\cdot,\cdot) : [0,1] \times V \to R^n$ such that
(C1) $f(t,x) \in \mathrm{co}F(t,x)$ for $t \in [0,1]$, $x \in V$,
(C2) $f(\cdot, x)$ is measurable for each fixed $x \in V$,
(C3) for almost all $t \in [0,1]$, $f(t,\cdot)$ is $k(t)$-Lipschitzian in V.

The collection $\mathbf{S}$ satisfies (A1)-(A5). Take again $\widetilde{\mathbf{S}} = \mathbf{S} \times C$. Denote by $\mathcal{T}_{\widetilde{\mathbf{S}}}$ the set of all trajectories $x_{f,x_0}(\cdot)$ of $\widetilde{\mathbf{S}}$. The next proposition follows from a selection theorem due to Lojasiewicz (an easy proof can be found in [2]).

PROPOSITION 4.1 Let $x(\cdot) \in \mathcal{T}_{\mathrm{co}F}$, $\psi(\cdot)$ be a measurable function such that $\psi(t) \in \mathrm{co}F(t,x(t))$ a.e. in $[0,1]$. Then there exists a selection $f(\cdot,\cdot)$ satisfying (C1)-(C3) with $k(t) = 4nr(t)$ such that $\psi(t) = f(t,x(t))$ a.e. in $[0,1]$.

The theorem implies that $\mathcal{T}_{\mathrm{co}F} = \mathcal{T}_{\widetilde{\mathbf{S}}}$. Define $\widetilde{\mathbf{L}}$ as follows

$$\widetilde{\mathbf{L}} = \{(f,x_0) \in \widetilde{\mathbf{S}} : x'_{f,x_0}(t) \in F(t, x_{f,x_0}(t)) \text{ a.e. in } [0,1]\}.$$

In other words, $(f,x_0) \in \widetilde{\mathbf{L}}$ if $x_{f,x_0}(\cdot) \in \mathcal{T}_F$. Proposition 4.1 implies that $\mathcal{T}_{\widetilde{\mathbf{L}}} = \mathcal{T}_F$, where $\mathcal{T}_{\widetilde{\mathbf{L}}} = \{x_{f,x_0}(\cdot) : (f,x_0) \in \widetilde{\mathbf{L}}\}$.

PROPOSITION 4.2 $\widetilde{\mathbf{L}}$ is an abundant subset of $\widetilde{\mathbf{S}}$.

Indeed, let $f_0, f_1, ..., f_N \in \mathbf{S}$, $c(\cdot) : \Delta_N \to C$ as in Definition 2.1 be given. Let, for $\lambda \in \Delta_N$, $x_\lambda(\cdot)$ be the solution of (2.2). Then $x_\lambda(\cdot) \in \mathcal{T}_{\mathrm{co}F}$, and the mapping $\lambda \to x_\lambda(\cdot)$ from Δ_N to $C([0,1]; R^n)$ is continuous. Let an $\epsilon > 0$ be given. From the result of Fryszkowski and Rzeuchowski [3, Theorem 2], there exists a mapping $\lambda \to y_\lambda(\cdot)$ from Δ_N to $\mathcal{T}_F$ continuous in $C([0,1]; R^n)$ topology and such that $\| y_\lambda(\cdot) - x_\lambda(\cdot) \| < \epsilon$, $y_\lambda(0) = c(\lambda)$. As $\mathcal{T}_F = \mathcal{T}_{\widetilde{\mathbf{L}}}$, for every $\lambda \in \Delta_N$ there exists $f^\lambda \in \mathbf{S}$ such that $(f^\lambda, c(\lambda)) \in \widetilde{\mathbf{L}}$, $y_\lambda(\cdot) = x_{f^\lambda, c(\lambda)}(\cdot)$. We define $\theta(\lambda) = (f^\lambda, c(\lambda))$. Both conditions of Definition 2.1 are satisfied and the proposition is proved.

By Proposition 4.2, $\widetilde{\mathbf{L}}$ is g-abundant in $\widetilde{\mathbf{S}}$. We can apply Theorem 2.1 and obtain the following theorem:

THEOREM 4.1. Let $z(\cdot)$ be a trajectory of the inclusion (4.2), $T_C(z(0))$ a regular tangent cone to C at $z(0)$. Then either

(a) for every single-valued selection $\bar{f}(t,x)$ of $\mathrm{co}F(t,x)$ which satisfies (C1)-(C3) and such that $z'(t) = \bar{f}(t, z(t))$ a.e. in $[0,1]$ there exist an arc $p(\cdot) : [0,1] \to R^n$ and a vector $v \in R^m$, $|v| = 1$, such that

$$p(0) \in T_C^\star(z(0)), \quad p(1) \in \partial g(z(1))^T v, \quad -p'(t) \in \partial \bar{f}(t,z(t))^T p(t) \quad a.e. \text{ in } [0,1],$$

$$< p(t), z'(t) > = \max_{w \in F(t,z(t))} < p(t), w > \quad a.e. \text{ in } [0,1], \qquad (4.3)$$

or

(b) $g(z(1)) \in \text{int} g(\mathcal{R}_F(1))$.

The maximum condition (4.3) follows easily from (2.5) via the Castaing representation of $\text{co} F(t, z(t))$.

References

1. Clarke, F.H., *"Optimization and Nonsmooth Analysis"*, Wiley, New York 1983.
2. Frankowska, H. and Kaskosz, B., *Linearization and boundary trajectories of nonsmooth control systems,* Can. J. Math, 40 (1988), 589-609.
3. Fryszkowski, A. and Rzezuchowski, T., *Continuous version of Filipov-Wazewski relaxation theorem,* J. Diff. Eq., 94 (1991), 254-265.
4. Kaskosz, B. and Lojasiewicz, S.,Jr., *A maximum principle for generalized control systems,* Nonlinear Analysis, 9 (1985), 109-130.
5. Kaskosz, B., *A maximum principle in relaxed controls,* Nonlinear Analysis, 14 (1990), 357-367.
6. Kaskosz, B., *Extremality, controllability and abundant subsets of generalized control systems,* to appear.
7. Mordukhovich, B.S., *Necessary optimality and controllability conditions for nonsmooth control systems,* in Proceedings of the 33rd CDC, Lake Buena Vista, FL, 1994, 3992-3997.
8. Polovkin, E.S. and Smirnov, G.V., *An approach to the differentiation of many-valued mappings, and necessary conditions for optimization of solutions of differential inclusions,* Differentsial'nye Uravneniya, 22 (1986), 944-954.
9. Sussmann, H.J., *A strong version of the maximum principle under weak hypotheses,* in Proceedings of the 33rd CDC, Lake Buena Vista, FL, 1994, 1950-1956.
10. Tuan, H.D., *Controllability and extremality in nonconvex differential inclusions,* J. Optim. Theory and Applic., 85 (1995), 435-472.
11. Warga, J., *"Optimal Control of Differential and Functional Equations"*, Academic Press, New York, 1972.
12. Warga, J., *Controllability, extremality and abnormality in nonsmooth optimal control,* J. Optim. Theory and Applic., 41 (1983), 239-260.
13. Warga, J., *Optimization and controllability without differentiability assumptions* SIAM J. Control. Optim., 21 (1983), 837-855.
14. Warga, J., *An extension of the Kaskosz maximum principle,* Appl. Math. Optim., 22 (1990), 61-74.
15. Zhu, Q.J., *Necessary optimality conditions for nonconvex differential inclusions with endpoints constraints,* J. Diff. Eq., 124 (1996), 186-204.

Author's address: Department of Mathematics, University of Rhode Island, Kingston, RI 02881.

JOHN Z HU, PHILIP D LOEWEN, S E SALCUDEAN
Minimax linear-quadratic controller design

1. Problem Statement. In this paper we propose reasonable feedback control laws for the following family of minimax problems:

$$v(\xi) := \min_u \left\{ \max_i J_i[u;\xi] \,:\, u \in L^2([0,\infty);\mathbb{R}^m),\ i = 1,\ldots,N \right\}. \qquad P(\xi)$$

The objective functionals J_i whose maximum provides our problem's performance index are defined with the aid of the linear time-invariant system

$$\dot{x}(t) = Ax(t) + Bu(t) \text{ a.e. } t > 0, \quad x(0) = \xi. \tag{1}$$

Given a control function u in $L^2([0,\infty);\mathbb{R}^m)$ and a starting point ξ in $\mathbb{R}^n$, the differential equation (1) has a unique solution x, which then defines

$$J_i[u;\xi] = \int_0^\infty \left(x(t)' Q_i x(t) + u(t)' R_i u(t) \right) dt, \qquad i = 1,\ldots,N. \tag{2}$$

Here the symmetric cost coefficient matrices Q_i and R_i are given, and we assume throughout that

(H1) $Q_i > 0$ and $R_i > 0$ for all i,
(H2) (A, B) is controllable.

Note that (H1) makes each pair (Q_i, A) observable, hence detectable, so all the standard results of the LQ theory hold in their strongest forms: see [**12**, Chap. 12].

We will use the following notation:

$$\begin{aligned} J[u;\xi] &= \max_i J_i[u;\xi], \\ \vec{J}[u;\xi] &= (J_1[u;\xi],\ldots,J_N[u;\xi]), \\ \Sigma_N &= \left\{ \lambda = (\lambda_1,\ldots,\lambda_N) \,:\, \lambda_i \geq 0\ \forall i,\ \sum \lambda_i = 1 \right\}. \end{aligned} \tag{3}$$

Case $N = 1$ of the minimax problem is the famous and well-understood linear-quadratic regulator, whose solution and basic properties are well known (see, for example, Wonham [**12**]), and play an important role in our discussion below. Cases $N \geq 2$ of related problems have been studied by several authors, particularly Li [**7**], Boyd and Barratt [**1**], and Boyd et al. [**2**].

2. Scalarization and Duality. Given any vector λ in $\mathbb{R}^N$, write

$$J_\lambda[u;\xi] = \lambda \cdot \vec{J}[u;\xi] = \sum_{i=1}^N \lambda_i J_i[u;\xi]. \tag{4}$$

(J_i is short for $J_{\widehat{e}_i}$.) Clearly, with Σ_N from (3), $\max_{i=1,\ldots,N} J_i[u;\xi] = \max_{\lambda \in \Sigma_N} J_\lambda[u;\xi]$, so problem $P(\xi)$ can be written as
$$v(\xi) = \min_{u \in L^2} \max_{\lambda \in \Sigma_N} J_\lambda[u;\xi]. \tag{5}$$
Interchanging the order of the min and max operations here makes everything work. Upon forming the related problem
$$w(\xi) = \max_{\lambda \in \Sigma_N} \min_{u \in L^2} J_\lambda[u;\xi], \tag{6}$$
we define the dual objective $\phi(\lambda;\xi) := \min_u J_\lambda[u;\xi]$ and proceed in two steps:

1. Maximize $\phi(\lambda;\xi)$ over λ in Σ_N. That is, find $\widehat{\lambda}$ satisfying
$$\widehat{\lambda} \in \Sigma_N, \quad \phi(\widehat{\lambda};\xi) = \min_u J_{\widehat{\lambda}}[u;\xi] \geq \min_u J_\lambda[u;\xi] = \phi(\lambda;\xi) \; \forall \lambda \in \Sigma_N. \tag{7}$$

2. Given any vector $\widehat{\lambda}$ as in (7), find $\widehat{u}$ in L^2 for which $\phi(\widehat{\lambda};\xi) = J_{\widehat{\lambda}}[\widehat{u};\xi]$.

This procedure is computationally feasible, and the control $\widehat{u}$ it generates is certain to solve $P(\xi)$.

With regard to computational feasibility, notice first that ϕ is concave, being the minimum of a family of functions linear in λ. Furthermore, for fixed $\lambda \in \Sigma_N$, evaluating $\phi(\lambda;\xi)$ amounts to solving a standard single-objective linear-quadratic problem with dynamics (1) and quadratic cost J_λ of the same form as J_i, except that the coefficient matrices are now $R_\lambda = \sum \lambda_i R_i$ and $Q_\lambda = \sum \lambda_i Q_i$. Thus
$$\phi(\lambda;\xi) = \xi' S(\lambda) \xi \quad \forall \lambda \in \Sigma_N,$$
where $S(\lambda)$ is the unique positive-definite solution of the Algebraic Riccati Equation
$$SA + A'S - SBR_\lambda^{-1}B'S + Q_\lambda = 0. \tag{8}$$
Efficient evaluation of $S(\lambda)$ is possible (see, e.g., [10]). Furthermore, a result of Delchamps [3] (see also [5]) implies that the map $\lambda \mapsto S(\lambda)$ is analytic at every point in Σ_N, so the function ϕ is not only convex in λ, but also smooth. (Assumption (H1) guarantees that $R_\lambda > 0$ and $Q_\lambda > 0$ for every λ in Σ_N, so Delchamps's result applies even at boundary points.) This makes the minimization of ϕ numerically straightforward. In the results presented below, we used the optimization routines in Matlab's Optimization Toolbox; for a discussion of other sample problems and larger-scale methods, including the constraint-dropping cutting plane method of Elzinga and Moore [4], see Hu [6].

Step 2 is also straightforward: given $\widehat{\lambda}$ as in (7), let
$$K(\widehat{\lambda}) = R_{\widehat{\lambda}}^{-1} B' S(\widehat{\lambda}). \tag{9}$$
Then the feedback law $u = -K(\widehat{\lambda})x$ produces evolution optimal with respect to $J_{\widehat{\lambda}}$ for every initial point; in particular, the minimum defining $\phi(\widehat{\lambda};\xi)$ is attained by the control $\widehat{u}(t) = -K(\widehat{\lambda})\widehat{x}(t)$, where $\widehat{x}(t)$ is the solution of the (stable) linear system
$$\dot{x}(t) = (A - BK(\widehat{\lambda}))x(t), \; x(0) = \xi. \tag{10}$$

To show that this method works in principle, we prove the following.

Lemma. *The control law $\widehat{u}$ produced in Step 2 gives the minimum in problem $P(\xi)$; the minimum value is $v(\xi) = \phi(\widehat{\lambda}; \xi) = J_{\widehat{\lambda}}[\widehat{u}; \xi]$.*

Proof. Clearly $\phi(\lambda; \xi) \leq J_\lambda[\widehat{u}; \xi]$ for any λ in $\mathbb{R}^N$, while $\phi(\widehat{\lambda}; \xi) = J_{\widehat{\lambda}}[\widehat{u}; \xi]$ by construction. Thus

$$\phi(\lambda; \xi) - \phi(\widehat{\lambda}; \xi) \leq J_\lambda[\widehat{u}; \xi] - J_{\widehat{\lambda}}[\widehat{u}; \xi] = \vec{J}[\widehat{u}; \xi] \cdot \left(\lambda - \widehat{\lambda}\right) \quad \forall \lambda \in \mathbb{R}^N.$$

Recalling that ϕ is differentiable at $\widehat{\lambda}$, we deduce that $\nabla \phi(\widehat{\lambda}; \xi) = \vec{J}[\widehat{u}; \xi]$. But since the point $\widehat{\lambda}$ maximizes ϕ over the polyhedral convex set Σ_N, it must be the case that

$$0 \geq \nabla \phi(\widehat{\lambda}; \xi) \cdot (\lambda - \widehat{\lambda}) = \vec{J}[\widehat{u}; \xi] \cdot (\lambda - \widehat{\lambda}) \quad \forall \lambda \in \Sigma_N.$$

It follows readily that a component $\widehat{\lambda}_i$ of $\widehat{\lambda}$ can be positive only if the corresponding index is active in defining the max-function J, i.e., only if $J_i[\widehat{u}; \xi] = J[\widehat{u}; \xi]$. Consequently $J_{\widehat{\lambda}}[\widehat{u}; \xi] = J[\widehat{u}; \xi]$, and we deduce

$$J[\widehat{u}; \xi] = J_{\widehat{\lambda}}[\widehat{u}; \xi] = \min_u J_{\widehat{\lambda}}[u; \xi] = \max_{\lambda \in \Sigma_N} \min_u J_\lambda[u; \xi]$$
$$\leq \min_u \max_{\lambda \in \Sigma_N} J_\lambda[u; \xi] = \min_u J[u; \xi].$$

This completes the proof. ////

An alternative approach to the solution of $P(\xi)$ can be given in terms of Linear Matrix Inequalities (LMI's). The relationship between a single LMI and the algebraic Riccati equation (8) was noted by Willems [11] well before fast algorithms for LMI-based problems created such widespread interest in the topic. The value $v(\xi)$ can be computed by solving the following maximization problem, where the choice variables are a symmetric matrix S and a vector λ, and the constraints are readily interpreted as LMI's. (Here $\lambda_N = 1 - \lambda_1 - \ldots - \lambda_{N-1}$ is used in defining Q_λ, R_λ as above.)

$$v(\xi) = \max_{S, \lambda}\Big\{\xi'S\xi \ : \ \begin{bmatrix} A'S + SA + Q_\lambda & SB \\ B'S & R_\lambda \end{bmatrix} \geq 0,$$
$$\lambda_1 \geq 0, \ldots, \lambda_{N-1} \geq 0, 1 - \lambda_1 - \ldots - \lambda_{N-1} \geq 0\Big\}.$$

3. Symmetry and Scaling. Given an initial point ξ and control u with corresponding trajectory x, consider any scalar α. The uniqueness of solutions to (1) guarantees that using the control αu with starting point $\alpha \xi$ in the system dynamics will generate the evolution αx, for x as above. Since each cost functional J_i is purely quadratic, we will have $J_i[\alpha u; \alpha \xi] = \alpha^2 J_i[u; \xi]$ for each i; componentwise maximization then gives $J[\alpha u; \alpha \xi] = \alpha^2 J[u; \xi]$. It follows that the function v defined by $P(\xi)$ is homogeneous of degree two:

$$v(\alpha \xi) = \alpha^2 v(\xi) \quad \forall \xi \in \mathbb{R}^n, \ \forall \alpha \in \mathbb{R}. \tag{11}$$

In particular, choosing $\alpha = |\xi|^{-1}$ gives

$$v(\xi) = |\xi|^2 v\left(\frac{\pm\xi}{|\xi|}\right), \forall \xi \neq 0,$$

so the function v is completely characterized by its values on any half of the spherical shell $|\xi| = 1$ in $\mathbb{R}^n$.

Similar analysis applies in the dual problem (7): if we write $\Lambda(\xi)$ for the set of $\widehat{\lambda}$-values obeying (7), then this set is nonempty, compact, and convex for every ξ, and $\Lambda(\alpha\xi) = \Lambda(\xi)$ for all real α.

4. Sample Problems. We will illustrate our results using a simple SISO system whose state evolves in the plane, taking

$$A = \begin{bmatrix} 0 & 1 \\ 0 & 0 \end{bmatrix}, \quad B = \begin{bmatrix} 0 \\ 1 \end{bmatrix}.$$

Three objectives are involved, with cost coefficients given by

$$Q_1 = \begin{bmatrix} 0.156073 & 0.936435 \\ 0.936435 & 5.774685 \end{bmatrix}, \quad R_1 = 0.156073;$$

$$Q_2 = \begin{bmatrix} 1.241033 & 1.241033 \\ 1.241033 & 1.266361 \end{bmatrix}, \quad R_2 = 0.400000;$$

$$Q_3 = \begin{bmatrix} 2.167149 & 0.021671 \\ 0.021671 & 0.000433 \end{bmatrix}, \quad R_3 = 0.021671.$$

Figure 1 shows polar plots of $r = \min_u J_i[u;\xi(\theta)]$, for $i = 1,2,3$, where the initial point $\xi(\theta)$ is $(\cos\theta, \sin\theta)$.

Figure 2 provides a graphic depiction of the set of maximizers $\Lambda(\xi(\theta))$ defined in (7), in the form of a graph of each component function $\lambda_i(\theta)$, $i = 1,2,3$, versus $\theta \in [0, 2\pi]$.

Together, Figures 1 and 2 show that the three objectives described above conflict for a wide range of initial states: only for a small sector near $\theta = 1$ rad (and, by symmetry, $\theta = 1 + \pi$) does the comparatively large magnitude of J_2 swamp the influences of J_1 and J_3.

Having tabulated the values of $\Lambda(\xi)$ for points on the unit circle, it is a simple matter to evaluate the control feedback matrix and minimum objective value, as described above. The outermost curve in Figure 1 shows the minimax value computed in this way. This is the best possible performance one can expect against the given minimax criterion: the remaining question is, how can we design a reasonable controller to approximate or achieve it in practice?

5. Receding Horizon Control. In the single-objective problem ($N = 1$), where $\Sigma_1 = \{1\}$, the set of dual maximizers $\Lambda(\xi)$ defined as in (7) must be constant, so the

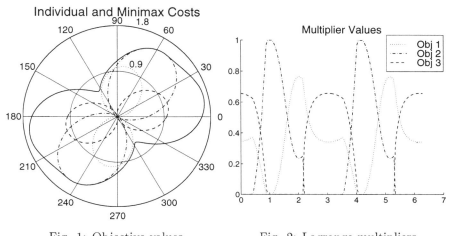

Fig. 1: Objective values vs initial orientation

Fig. 2: Lagrange multipliers

matrix K defined in (9) is independent of the initial point ξ. This makes the feedback law $u = -Kx$ globally optimal. When $N > 1$, however, Λ need not be constant, and hence the matrix $K(\widehat{\lambda})$ in (9) will change as ξ varies. Thus the specification $u(t) = -K(\widehat{\lambda})x(t)$ used in (10) must be regarded as an implicit definition of an open-loop control, whose validity is limited to the single trajectory starting from the given initial point.

This raises the prospect of defining a feedback control law for (1) by taking a continuous selection $\lambda(x)$ of $\Lambda(x)$ and setting

$$U(x) = -K(\lambda(x))x, \qquad x \in \mathbb{R}^n. \qquad (12)$$

This corresponds to treating every point x like a starting point in the minimax problem, and applying the control value that would be optimal at time 0 for x, computed as above. The term "receding-horizon controller" (see Mayne and Michalska [8]) neatly summarizes the philosophy behind this approach, which produces a feedback that is sure to stabilize the system (1). A direct proof of this assertion can be fashioned following [8, 9]: the idea is to use the function $v(\xi)$ defined in the statement of $P(\xi)$ as a Lyapunov function, noting that for any evolution of the system under (1) and (12), every $\xi \neq 0$ obeys

$$\nabla v(\xi) \left[A - BR_\lambda^{-1}B'S_\lambda\right]\xi = 2\xi'S(\lambda)\left[A - BR_\lambda^{-1}B'S_\lambda\right]\xi$$
$$= \xi'\left[-Q - S_\lambda BR_\lambda^{-1}B'S_\lambda\right]\xi \quad \text{(by (8))}$$
$$< 0.$$

Consequently $t \mapsto v(x(t))$ is strictly decreasing along trajectories of the system under receding-horizon control; since the function v is obviously positive-definite, it follows that all trajectories converge to the origin as t approaches infinity.

To make the preceding argument rigorous, we need to know that the set of multipliers $\Lambda(\xi)$ admits a continuous selection. A sufficient condition for this is the hypothesis we now introduce:

(H3) For each fixed $\xi \neq 0$ in $\mathbb{R}^n$, the function $\phi(\lambda;\xi)$ defined above is strictly concave on Σ_N.

This condition implies that the multifunction $\xi \mapsto \Lambda(\xi)$ is single-valued. Since it is straightforward to see that this multifunction takes nonempty closed convex subsets of Σ_N as values, and that the graph of Λ is a closed subset of $\mathbb{R}^n \times \Sigma_N$, it follows that Λ can be considered as a continuous function on $\mathbb{R}^n \setminus \{0\}$.

Assumption (H3) also guarantees the differentiability of the function v at nonzero points in $\mathbb{R}^n$: indeed, the representation $v(\xi) = \max\{\xi'S(\lambda)\xi : \lambda \in \Sigma_N\}$ provided by the Lemma above displays v as the maximum of a family of functions convex in ξ. Standard results on the subgradients of max-functions, combined with the uniqueness of the λ-value giving the maximum defining v, produce the gradient formula used in the calculation above:

$$\nabla v(\xi) = 2\xi'S(\Lambda(\xi)), \text{ for all } \xi \neq 0.$$

Implementing the feedback controller U of (12) requires efficient identification of a vector $\lambda(x)$ in $\Lambda(x)$ for any given initial point x. Since $\Lambda(x) = \Lambda(\alpha x)$ for all α, it is enough to know $\lambda(x)$ when x is a unit vector in a certain half-space of $\mathbb{R}^n$ preset by convention: in problems with very few states, a collection of such λ-values can be tabulated off-line for later reference by interpolation; in larger problems, the comparatively simple dual minimization problem (7) can be solved iteratively at each position.

The receding-horizon control law U cannot be expected to produce optimal evolutions with respect to J for every initial point. Indeed, it is not difficult to find points on the unit circle in our sample problem for which the optimal state paths cross non-tangentially on their way to the origin. Since both trajectories are unique minimizers and their velocities at the intersection point disagree, it follows that no continuous feedback law can produce optimal evolution—in particular, the receding-horizon control law cannot be optimal. A second informal explanation runs like this: for any initial point ξ on the unit circle, the control law u optimal for $P(\xi)$ maintains a linear relationship with the state trajectory x, namely $u(t) = -Kx(t)$, and the matrix $K = K(\lambda(\xi))$ is constant along the whole trajectory. By contrast, the matrix K in (12) is constant along lines through the origin in $\mathbb{R}^n$, but typically varies from one line to another. So it is usually not constant along the trajectories generated by using control law U in (1), and tends to generate suboptimal prescriptions for $\dot{x}$ at all times except $t = 0$.

Running the receding-horizon controller of (12) on the sample system described above produces the total costs outlined in Figure 3. Here for each initial vector $\xi(\theta) = (\cos\theta, \sin\theta)$, markers indicate the observed values of J_i for $i = 1, 2, 3$, and the largest of these is circled. A solid curve traces the optimum minimax values, shown earlier in Figure 1. The receding-horizon objective values are, of course, never less than the corresponding optima, but the overestimate is at most 30% in this example, and often significantly less.

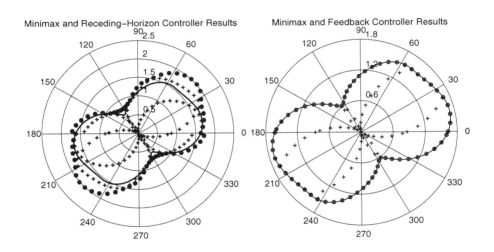

Fig. 3: Objective Values from Receding-Horizon Controller

Fig. 4: Objective Values from Optimal Feedback Controller

6. Optimal Feedback. As noted above, no simple function of the system state will provide a control law that realizes the infimum in $P(\xi)$ regardless of ξ. This problem originates with the non-separability of the objective in $P(\xi)$: at some instant $\tau > 0$ and position ξ, the portion of the minimax problem in the remaining interval $t > \tau$

is not identical in form to the problem $P(x(\tau))$, because the accumulated integral contributions to the objectives J_i may differ from one component to another. The easiest way to see this and account for it is to embed problem $P(\xi)$ as $Q(\xi, 0)$ in the following infinite-horizon Mayer problem with $n + N$ states—the original state vector x, plus a new state z that tracks the integral-so-far for each objective:

$$Q(\xi, \zeta) \qquad v(\xi, \zeta) := \min\Big\{\max\{z_1(\infty), \ldots, z_N(\infty)\} :$$
$$\dot{x}(t) = Ax(t) + Bu(t) \text{ a.e. } t > 0, \ x(0) = \xi,$$
$$\dot{z}_i(t) = x(t)' Q_i x(t) + u(t)' R_i u(t) \text{ a.e. } t > 0, \ z_i(0) = \zeta_i\Big\}.$$

Keeping track of the cost-so-far for each objective need not be difficult in practice, so we propose a feedback law depending on the current values of both x and z. Notice that $Q(\xi, \zeta)$ is decomposable, so a feedback law can be expected to exist. Indeed, under the dynamic constraint (1), a scalarization and dualization procedure just as above reveals

$$v(\xi, \zeta) = \min_u \max_i \{\zeta_i + J_i[u; \xi]\}$$
$$= \max_{\lambda \in \Sigma_N} \min_u \left\{\lambda \cdot (\zeta + \vec{J}[u; \xi])\right\} \qquad (20)$$
$$= \max_{\lambda \in \Sigma_N} (\zeta \cdot \lambda + \phi(\lambda; \xi)),$$

for the dual objective ϕ introduced just before (7). It follows that the optimal control at a state position ξ and cost-so-far vector ζ can be determined in two steps:

1. Solve the dual problem on the right in (20), i.e., find $\widehat{\lambda}$ in $\mathbb{R}^N$ such that

$$\widehat{\lambda} \in \Sigma_N, \quad \zeta \cdot \widehat{\lambda} + \min_u J_{\widehat{\lambda}}[u; \xi] \geq \zeta \cdot \lambda + \min_u J_\lambda[u; \xi] \ \forall \lambda \in \Sigma_N. \qquad (21)$$

2. Given any vector $\widehat{\lambda}$ as in (21), find $\widehat{u}$ in L^2 for which $\phi(\widehat{\lambda}; \xi) = J_{\widehat{\lambda}}[\widehat{u}; \xi]$.

A succinct formulation of the resulting feedback law, analogous to (12), is

$$U(x, z) = -K(\lambda(x, z))x, \qquad x \in \mathbb{R}^n, \ z \in \mathbb{R}^N, \qquad (22)$$

where K is defined in (8)–(9) and $\lambda(x, z)$ is the unique (by (H3)) solution of (21) for $x = \xi$, $z = \zeta$. The results of this control strategy in the context of the sample problem introduced in Section 4 are shown in Figure 4. Like Figure 3, this diagram shows the values of all three integral objectives accumulated from a variety of inital points, with the largest one circled. In spite of comparatively loose tolerances on the approximate minimization in step 1, the numerical integration of the six-dimensional system of dynamic equations, and the truncation of the interval of integration from $[0, \infty)$ to $[0, 10]$, the minimax objective value produced by this procedure agrees very well with that predicted by the calculations above: among 32 initial points equally spaced around the upper half of the unit circle the largest difference between prediction and observation was below 0.30%, and the average was 0.10%.

References.

[1] Boyd, S. P., and C. H. Barratt, Linear Controller Design: Limits of Performance. Englewood Cliffs: Prentice-Hall, 1991.

[2] Boyd, S. P. et. al., *Linear Matrix Inequalities in System and Control Theory*, Philadelphia: SIAM, 1994.

[3] Delchamps, D. F., A note on the analyticity of the Riccati metric, in C. I. Byrnes and C. F. Martin, eds., *Algebraic and Geometric Methods in Linear Systems Theory*, AMS Lectures in Applied Mathematics, vol. 18, 1980, 37–41.

[4] Elzinga, J., and T. Moore, A central cutting plane algorithm for the convex programming problem, Math. Programming Studies **8**(1975), 134–145.

[5] Gahinet, P., and A. Laub, Computable bounds for the sensitivity of the algebraic Riccati equation, SIAM J. Control Optim. **28**(1990), 1461–1480.

[6] Hu, Z., *Multiple Objective Control with Applications to Teleoperation*, Dissertation, Electrical Engineering, University of British Columbia, 1996.

[7] Li, D., On general multiple linear-quadratic control problems, IEEE Trans. Auto. Control **AC-38**(1993), 1722–1727.

[8] Mayne, D. Q., and H. Michalska, Receding horizon control of nonlinear systems, IEEE Trans. Auto. Control **AC-35**(1990), 814–824.

[9] Michalska, H., and R. B. Vinter, Receding horizon control for nonlinear time-varying systems, Proc. 30th Conference on Decision and Control–December 1991, 75–76.

[10] Patel, R. V., A. J. Laub, and P. M. Van Dooren, eds., *Numerical linear algebra techniques for systems and control*, IEEE, 1994.

[11] Willems, J. C., Least squares stationary optimal control and the algebraic Riccati equation, IEEE Trans. Auto. Control **AC-16**(1971), 621–634.

[12] Wonham, W. M., *Linear Multivariable Control: A Geometric Approach*, 3/e. New York: Springer-Verlag, 1985.

John (Zhongzhi) Hu, Systems Design and Analysis, GE Aircraft Engines, 1000 Western Avenue, Lynn, MA 01910: `john.hu@ae.ge.com`

Philip D. Loewen, Department of Mathematics, University of British Columbia: `loew@math.ubc.ca`

S. E. Salcudean, Department of Electrical and Computer Engineering, University of British Columbia: `tims@ee.ubc.ca`

The support of Canada's Natural Science and Engineering Research Council is gratefully acknowledged.

URSZULA LEDZEWICZ[1] AND HEINZ SCHÄTTLER[2]
High-order approximations for abnormal bang-bang extremals

1. Introduction

If one formulates the optimal control problem for ordinary differential equations with terminal constraints in the framework of optimization theory, then the dynamical equations together with their respective boundary conditions define an operator F which describes an equality constraint Q. Necessary conditions for optimality are typically derived using tangent approximations to this set Q. If the Fréchet derivative $F'(x_0)$ at a reference point x_0 is onto, the so-called surjectivity condition, then the classical Lyusternik theorem describes the tangent space to Q at x_0 as the kernel of $F'(x_0)$. However, this does not apply if $F'(x_0)$ is not onto. Yet the associated necessary conditions for optimality can be satisfied trivially by choosing a multiplier which annihilates the image of $F'(x_0)$. This generates so-called *abnormal extremals*. While the nomenclature, which has its origin in Calculus of Variations, suggests that abnormal processes are an aberration, for optimal control problems this is not the case. The phenomenon is quite general, can be observed in a multitude of problems in optimal control and abnormal extremals cannot be excluded from optimality apriori. For instance, there exist optimal abnormal trajectories for the problem of stabilizing the harmonic oscillator time-optimally in minimum time, a simple time-invariant linear system.

Since the conventional necessary conditions for optimality provide conditions which are trivially satisfied for abnormal processes, it becomes necessary to develop theories which are tailored to abnormal processes. This has been a quite active area of research recently, e.g. [1, 2, 4, 12]. In this paper we follow an approach initiated by Avakov [3] which relies on an extension of the Lyusternik theorem in which the operator F is analyzed further using higher order derivatives. Replacing linear approximations by polynomial approximations of degree p we obtained a generalization of the classical Lyusternik Theorem which describes the structure of p-order tangent directions to the equality constraint $Q = \{x \in X : F(x) = 0\}$ in a Banach space also for non-regular operators [7]. This result has then been applied to derive generalized first- and second-order necessary conditions for optimality for general optimization problems in Banach spaces [8]. Based on these results both a local and global version of an extended p-order

---
[1] supported in part by NSF Grant DMS-9622967, SIUE Research Scholar Award and 1997 Summer Fellowship
[2] supported in part by NSF Grant DMS-9503356

Maximum Principle have been given [9, 10]. Our conditions differ from other high-order necessary conditions for optimality like for instance classical results of Hoffmann and Kornstaedt who consider general polynomial approximations for smooth problems [5] or also more recent results like those of Páles and Zeidan who consider nonsmooth systems [13] in the sense that they are non-trivial for both normal and abnormal cases.

In this paper we apply this extended version of the Maximum Principle corresponding to $p = 2$ [6] to describe second-order approximations for the case of abnormal bang-bang extremals. Specifically, we will evaluate the subspace of critical directions and formulate conditions which allow to exclude the optimality of abnormal bang-bang extremals based on our results in [6]. This will be done in Section 3. The results and constructions are for a general dimension, but for simplicity and lack of space we only present the three-dimensional case. In Section 2 we briefly recall the second-order extended Maximum Principle.

2. An Extended Global Maximum Principle

We formulate a strong version of the extended maximum principle [6] for the following time-invariant problem (P) : minimize the functional

$$I(x, u) = \int_0^T L(x(t), u(t)) dt \qquad (1)$$

under the constraints

$$\dot{x}(t) = f(x(t), u(t)), \quad x(0) = 0, \quad q(x(T)) = 0 \qquad (2)$$
$$u(\cdot) \in \mathcal{U} = \{u(\cdot) \in L_\infty^r(0, T) : u(t) \in U \text{ for } t \in [0, T]\}. \qquad (3)$$

where $x(\cdot) \in \overline{W}_{11}^n(0, T) = W_{11}^n(0, T) \cap \{x(\cdot) : x(0) = 0\}$, $u(\cdot) \in L_\infty^r(0, T)$, $L(x, u) \in \mathbb{R}$, $f(x, u) \in \mathbb{R}^n$, and $q(x) \in \mathbb{R}^k$. The terminal time T is free and the control set U is an arbitrary subset of $\mathbb{R}^m$. We make the following *smoothness assumptions*: for every $u \in M$ the function $L(\cdot, u)$ is twice and $f(\cdot, u)$ is thrice continuously differentiable in x; for every $x \in \mathbb{R}^n$ the functions $L(x, \cdot)$ and $f(x, \cdot)$ are continuous; all partial derivatives are bounded on compact subsets of (x, u)-space; q is thrice continuously differentiable in x.

The result below formulates an extended version of the Maximum Principle which also gives non-trivial conditions for abnormal processes. It is obtained from a local version by means of singular time transformations. As a result, it is parametrized by critical directions $(h, z) \in \overline{W}_{11}^n(0, T) \times L_\infty^1(0, T)$ along which an operator F which describes the dynamics of an auxiliary reparametrized problem used in the proof is non-regular. This operator is given by

$$F = (F_1, F_2) \; : \; \overline{W}_{11}^n(0, T) \times L_\infty^1(0, T) \to \overline{W}_{11}^n(0, T) \times \mathbb{R}^k \qquad (4)$$

$$(x, w) \mapsto F(x, w) = \left(x(\cdot) - \int_0^{(\cdot)} w(r) f(x(r), u_*(r)) dr, q(x(T)) \right)$$

where $(x_*(\cdot), u_*(\cdot))$ denotes the reference trajectory and $w \geq 0$ becomes a new control variable allowing for reparametrizations in time. The class $\mathcal{C} = \mathcal{C}(x_*, u_*)$ of *critical directions* consists of all pairs of functions $(h, z) \in \overline{W}_{11}^n(0, T) \times L_\infty^1(0, T)$ which satisfy the following conditions:

$$(i) \quad \dot{h} = f_x(x_*, u_*)h + zf(x_*, u_*), \quad q_x(x_*(T))h(T) = 0,$$

$$(ii) \quad \left(-\int_0^{(\cdot)} (f_{xx}(x_*, u_*)(h, h) + 2zf_x(x_*, u_*)h) \, dt, \right.$$

$$q_{xx}(x_*(T))(h(T), h(T))) \in Im \ B$$

$$(iii) \quad \int_0^T (L_x(x_*, u_*)h + zL(x_*, u_*)) \, dt = 0,$$

where B is the linear operator given by

$$B : \overline{W}_{11}^n(0, T) \times L_\infty^1(0, T) \to \overline{W}_{11}^n(0, T) \times \mathbb{R}^k \tag{5}$$

$$(h, z) \mapsto B(h, z) = \left(h(\cdot) - \int_0^{(\cdot)} f_x(x_*, u_*)h + zf(x_*, u_*) dr, q_x(x_*(T))h(T) \right).$$

Conditions (i) and (ii) are related to the dynamics, i.e. the equality constraints of the problem. Condition (i) is equivalent to $(h, z) \in Ker \ B$, while condition (ii) is a necessary condition for the existence of a second-order approximation. Condition (iii) states that (h, z) is also a critical direction for the objective.

In the theorem below we write covectors like ψ as row vectors and we denote the space of row vectors in $\mathbb{R}^n$ by $(\mathbb{R}^n)^*$. This is consistent with the multiplier interpretation of the adjoint variable.

Theorem 1 *(Extended Global Maximum Principle,[6])* Suppose the admissible process (x_*, u_*) is optimal for problem (P). Then for every $(h, z) \in \mathcal{C}(x_*, u_*)$ there exist multipliers $\nu_0 = \nu_0(h, z) \geq 0$, $a = a(h, z) \in (\mathbb{R}^k)^*$, $b = b(h, z) \in (\mathbb{R}^k)^*$, functions $\rho(\cdot) = \rho(h, z)(\cdot)$, $\psi(\cdot) = \psi(h, z)(\cdot)$ from $[0, T]$ into $(\mathbb{R}^n)^*$, not vanishing simultaneously, such that the following conditions are satisfied a.e. on $[0, T]$ (all partial derivatives are evaluated along $(x_*(t), u_*(t))$):

$$\dot{\psi}(t) = -\nu_0 L_x(x_*, u_*) - \psi(t) f_x(x_*, u_*) - \rho(t) \left(f_{xx}(x_*, u_*) h(t) + z(t) f_x(x_*, u_*) \right) \tag{6}$$

with terminal condition

$$\psi(T) = -aq_x(x_*(T)) - bq_{xx}(x_*(T))h(T). \tag{7}$$

The function $\rho(\cdot)$ and the vector b satisfy

$$\dot{\rho}(t) = -\rho(t)f_x(x_*, u_*), \qquad \rho(T) = -bq_x(x_*(T)), \tag{8}$$

$$\rho(t)f(x_*, u_*) = 0. \tag{9}$$

and the following minimum condition is satisfied for every $u \in M$

$$0 \equiv \nu_0 L(x_*, u_*) + \psi(t)f(x_*, u_*) + \rho(t)f_x(x_*, u_*)h, \tag{10}$$
$$0 \leq \nu_0 L(x_*, u) + \psi(t)f(x_*, u) + \rho(t)f_x(x_*, u)h, \tag{11}$$

3. Application to abnormal bang-bang extremals

We now consider the problem (T) of time optimal control to a terminal point q for a system of the form

$$\dot{x} = f(x) + ug(x), \qquad x \in \mathbb{R}^3, \qquad |u| \leq 1. \tag{12}$$

Here f and g are C^3 vector fields. The conditions of the Maximum Principle for this problem state that, if $\gamma = (x_*, u_*)$ defined over $[0, T]$ is time optimal, then there exist a constant $\lambda_0 \geq 0$ and an absolutely continuous function $\lambda : [0, T] \to (\mathbb{R}^n)^*$, called adjoint state, which do not vanish simultaneously, such that $\dot\lambda = -\lambda(Df(x_*) + u_* Dg(x_*))$, $u_*(t) = -sgn \langle \lambda(t), g(x_*(t)) \rangle$ and $\lambda_0 + \langle \lambda(t), f(x_*(t)) + u_*(t)g(x_*(t)) \rangle \equiv 0$. The function $H(\lambda_0, \lambda, x_*, u_*) = \lambda_0 + \langle \lambda, f(x) + ug(x) \rangle$ is called the Hamiltonian function for the time-optimal control problem and the expression $\varphi(t) = \langle \lambda(t), g(x_*(t)) \rangle$ which determines the optimal control is called the switching function. Also denote the vector fields which correspond to the constant controls $u \equiv -1$ and $u \equiv 1$ by $X = f - g$ and $Y = f + g$ respectively.

We now consider a bang-bang extremal with 2 switchings at times $0 < t_1 < t_2 < T$ and switching points $p_1 = x_*(t_1)$ and $p_2 = x_*(t_2)$. Without loss of generality suppose $u \equiv -1$ on $[0, t_1]$ and denote the times along the trajectories by $s_1 = t_1$, $s_2 = t_2 - t_1$ and $s_3 = T - t_2$. We write XYX for trajectories of this type. It follows from the maximum condition that we have $\langle \lambda(t_i), g(x_*(t_i)) \rangle = 0$ at every switching time t_i, $i = 1, 2$. We now combine these conditions by moving the vector $g(p_1)$ forward along the flow to the last switching p_2. This is done by integrating the variational equation of the system along the reference trajectory with initial condition $g(p_1)$ at time t_1. Using exponential notation we denote the result by

$$\exp(-s_2 adY)(g(p_1)) = (\exp(-s_2 adY)g)(p_2)). \tag{13}$$

The covector λ is moved forward along the flow of Y by integrating the covariational equation along the reference trajectory. But this is precisely the adjoint equation on

λ and thus we simply get $\lambda(t_2)$. Furthermore, it follows for the same reason that the function
$$t \mapsto \langle \lambda(t), \exp(-(t-t_1)adY)g(x_*(t))\rangle \tag{14}$$
is constant on the interval $[t_1, t_2]$. Hence we have
$$\begin{aligned}0 &= <\lambda(t_1), g(p_1)> = <\lambda(t_2), \exp(-s_2 adY)g(p_1)> \\ &= <\lambda(t_2), (\exp(-s_2 adY)g)(p_2)>\end{aligned} \tag{15}$$
and thus $\lambda(t_2)$ vanishes against both $g(p_2)$ and $(\exp(-s_2 adY)g)(p_2)$. Let
$$V = lin\ span\{g(p_2), (\exp(-s_2 adY)g)(p_2)\} \tag{16}$$
and henceforth we assume that V is two-dimensional.

This XYX extremal is abnormal if $\lambda_0 = 0$. In this case the Maximum principle implies that we have $\langle \lambda(t_i), f(x_*(t_i))\rangle = 0$ at the switching times t_1 and t_2. Hence, if $f(p_2) \notin V$, then $\lambda(t_2)$ must vanish. But this contradicts the nontriviality of the multiplier (λ_0, λ) and therefore necessarily $f(p_2) \in V$. Noticing that $f(p_1) = Y(p_1) - g(p_1)$, it also follows that
$$\begin{aligned}(\exp(-s_2 adY)f)(p_2) &= (\exp(-s_2 adY)Y)(p_2) - (\exp(-s_2 adY)g)(p_2) \\ &= Y(p_2) - (\exp(-s_2 adY)g)(p_2) \in V,\end{aligned} \tag{17}$$
i.e. moving the vector $f(p_1)$ with the flow of Y to the point p_2 gives a redundant condition. Conversely, if $f(p_2) \in V$, choosing a solution λ of the adjoint equation such that $\lambda(t_2) \perp V$ generates an abnormal extremal. Thus *the XYX extremal is abnormal if and only if* $f(p_2) \in V$.

The constructions of section 2 allow to set-up non-trivial high-order approximations along abnormal trajectories. We need to analyze the cone of *critical directions* for this problem, i.e. the set of vectors $(h, z) \in \mathcal{C}(x_*, u_*)$ such that conditions (i)-(iii) for our problem (T) are satisfied. Condition (i) reads
$$\dot{h} = A(t)h + b(t)z, \quad h(0) = 0, \quad h(T) = 0, \tag{18}$$
where
$$A(t) = Df(x_*(t)) + u_*(t)Dg(x_*(t)) \tag{19}$$
is the dynamics of the variational equation along the abnormal XYX extremal and
$$b(t) = f(x_*(t)) + u_*(t)g(x_*(t)) \tag{20}$$
is the dynamics of the reference trajectory. By a well-known result about linear systems, the corresponding single-input time-varying linear system (18) is *not completely controllable* if and only if there exists a nonzero multiplier ρ such that
$$\dot{\rho} = -\rho A(t) \quad \text{and} \quad \rho b(t) \equiv 0 \quad \text{on}\ [0,T]. \tag{21}$$

This condition of non-controllability for the variational equation (18) is equivalent to the condition that the operator $B : \overline{W}_{11}^n(0,T) \times L_\infty^1(0,T) \to \overline{W}_{11}^n(0,T) \times \mathbb{R}^k$, which now reads

$$(h, z) \longmapsto \left(h(\cdot) - \int_0^{(\cdot)} (A(t)h(t) + b(t)z(t))dt, \; h(T)\right), \tag{22}$$

is not onto. Notice also that the condition (18) can be rewritten in the form $(h, z) \in \ker B$.

A multiplier ρ which satisfies (21) is given by the adjoint vector from the maximum principle. Indeed, since $\dim V = 2$ there exists a nonzero vector $\overline{\rho}$ such that $\overline{\rho} \perp V$, i.e

$$0 = \langle \overline{\rho}, g(p_2) \rangle = \langle \overline{\rho}, (\exp(-s_2 adY)g)(p_2) \rangle \tag{23}$$

Let ρ be the solution of the adjoint equation with initial condition $\rho(t_2) = \overline{\rho}$. We claim that this $\rho(t)$ satisfies $\rho(t)b(t) \equiv 0$. By definition

$$0 = \langle \overline{\rho}, (\exp(-s_2 adY)g)(p_2) \rangle = \langle \rho(t_1), g(p_1) \rangle. \tag{24}$$

Since $f(p_2) \in V$ we also have that $0 = \langle \rho(t_2), f(p_2) \rangle$ and furthermore

$$\begin{aligned}\langle \rho(t_1), f(p_1) \rangle &= \langle \rho(t_2), (\exp(-s_2 adY)f)(p_2) \rangle = \\ &= \langle \rho(t_2), Y(p_2) - (\exp(-s_2 adY)g)(p_2) \rangle = 0.\end{aligned} \tag{25}$$

Combining (23), (24) and (25) gives us that we have at every switching point

$$0 = \langle \rho(t_i), X(p_i) \rangle = \langle \rho(t_i), Y(p_i) \rangle \qquad i = 1, 2. \tag{26}$$

Thus the expression $\rho(t)b(t)$ vanishes at the swiching times t_1 and t_2. Furthermore, for any vector field Z

$$\frac{d}{dt} \langle \rho, Z(x_*) \rangle = \langle \rho, [f + uq, Z](x_*) \rangle. \tag{27}$$

If we apply this property to our problem with $Z = X$ on $[0, t_1]$ and $[t_2, T]$ and with $Z = Y$ on $[t_1, t_2]$, then we obtain on these intervals respectively that

$$\frac{d}{dt} \langle \rho, X(x_*) \rangle = \langle \rho, [X, X](x_*) \rangle \equiv 0 \quad \text{and} \quad \frac{d}{dt} \langle \rho, Y(x_*) \rangle = \langle \rho, [Y, Y](x_*) \rangle \equiv 0.$$

Thus the function ρb is constant on the subintervals $[0, t_1]$, $[t_1, t_2]$ and $[t_2, T]$, and, since it vanishes at the switching times, it therefore vanishes identically. This verifies the construction of the multiplier ρ satisfying (21). Note that this is also the multiplier as it is defined in (8) in Theorem 1.

The compatability condition (ii) takes the form

$$(-\int_0^{(.)}(D^2f(x_*)+u_*D^2g(x_*))(h,h)+2z(Df(x_*)+u_*Dg(x_*))hdt,0)\in \text{Im } B. \quad (28)$$

It can be shown that

$$\text{Im } B = \{(a,\int_0^T \Psi(T,s)\alpha(s)ds + R): \quad a\in \overline{W}_{11}^n(0,T), \quad a(t)=\int_0^t \alpha(s)ds\}, \quad (29)$$

where $\Psi(t,s)$ is the fundamental matrix for the equation $\dot{h}= A(t)h$ and R is the reachable subspace at time T of the variational equation $\dot{h}= A(t)h + b(t)z$ with initial condition $h(0) = 0$. Combining (28) and (29) it follows that the compatibility condition is equivalent to

$$\int_0^T \Psi(T,\cdot)\left[(D^2f(x_*)+u_*D^2g(x_*))(h,h)+2z(Df(x_*)+u_*Dg(x_*))h\right]dt \in R \quad (30)$$

The reachable set R can be calculated explicitly: It follows from the variation of parameters formula that

$$h(T) = \int_0^T \Psi(T,s)b(s)z(s)ds.$$

Splitting $[0,T]$ into the the three intervals where the control is constant ± 1, we obtain

$$h(T) = \Psi(T,s_1)\int_0^{s_1} \Psi(s_1,s)b(s)z(s)ds$$

$$+\Psi(T,s_1+s_2)\int_{s_1}^{s_1+s_2} \Psi(s_1+s_2,s)b(s)z(s)ds)ds +$$

$$+\int_{s_1+s_2}^T \Psi(T,s)b(s)z(s)ds.$$

Notice that for s,t in each of the intervals $[0,t_1],[t_1,t_2],[t_2,T]$ we have $\Psi(t,s)b(s) = b(t)$ since Ψ simply moves the vector $X(s)$ respectively $Y(s)$ along its own flow. Thus

$$h(T) = \Psi(T,s_1)X(p_1)\left(\int_0^{s_1} z(s)ds\right) + \Psi(T,s_1+s_2)Y(p_2)\left(\int_{s_1}^{s_1+s_2} z(s)ds\right) +$$

$$+X(q)\left(\int_{s_1+s_2}^T z(s)ds\right) \quad (31)$$

$$= \exp(-s_3 adX)\left\{\exp(-s_2 adY)X(p_2)(\int_0^{s_1} z(s)ds) + Y(p_2)(\int_{s_1}^{s_1+s_2} z(s)ds) +\right.$$

$$\left.+X(p_2)(\int_{s_1+s_2}^T z(s)ds)\right\} \quad (32)$$

where we use the specific structure of the trajectory in the last step. But the vectors $\exp(-s_2 adY)X(p_2)$, $Y(p_2)$ and $X(p_2)$ span the subspace V given by (16) and thus we have

Proposition 1
$$R = \exp(-s_3 adX)(V). \tag{33}$$

The third condition (iii) determining the set of critical directions concerns the integrand and takes the form
$$\int_0^T z(t)dt = 0. \tag{34}$$

Summarizing we have

Proposition 2 *The set of critical directions $\mathcal{C}(x_*, u_*)$ consists of all $(h, z) \in \ker B$ which satisfy conditions (30) and (34).*

For each of the directions $(h, z) \in \mathcal{C}(x_*, u_*)$ we now have the following necessary conditions for optimality from Theorem 1.

Theorem 2 (Extended second-order Maximum Principle for time-optimal control) *Suppose the admissible process (x_*, u_*) defined on $[0, T]$ is time-optimal. Then for every $(h, z) \in \mathcal{C}(x_*, u_*)$ there exist a number $\nu_0 \geq 0$ and functions $\rho(\cdot) = \rho(h, z)(\cdot)$ and $\psi(\cdot) = \psi(h, z)(\cdot)$ from $[0, T]$ into $(\mathbb{R}^n)^*$, not all identically zero, such that the following conditions are satisfied:*
$$\begin{aligned}\dot{\psi} &= -\psi\left(Df(x_*) + u_* Dg(x_*)\right) \\ &\quad -\rho\left(\left(D^2 f(x_*) + u_* D^2 g(x_*)\right)h + z\left(Df(x_*) + u_* Dg(x_*)\right)\right);\end{aligned} \tag{35}$$

the function $\rho(\cdot)$ satisfies the adjoint equation
$$\dot{\rho} = -\rho\left(Df(x_*) + u_* Dg(x_*)\right) \tag{36}$$

and for a.e. $t \in [0, T]$ the orthogonality condition
$$\rho\left(f(x_*) + u_* g(x_*)\right) = 0; \tag{37}$$

and the following minimum condition is satisfied along (x_*, u_*): for every $u \in M$ and a.e. $t \in [0, 1]$ we have
$$\begin{aligned}0 &\equiv \nu_0 + \psi\left(f(x_*) + u_*(t)g(x_*)\right) + \rho\left(Df(x_*) + u_* Dg(x_*)\right)h, \tag{38}\\ 0 &\leq \nu_0 + \psi\left(f(x_*) + ug(x_*)\right) + \rho\left(Df(x_*) + uDg(x_*)\right)h. \tag{39}\end{aligned}$$

Theorem 2 provides extended and thus extra conditions for optimality of abnormal bang-bang trajectoreis. For lack of space in this paper we only considered the cone of critical directions for the problem in $\mathbb{R}^3$, but all the arguments given are valid in general. This is carried out in [11] and there also a further analysis of the necessary conditions is given.

References

[1] Agrachev, A.A. and Sarychev, A.V., "On abnormal extremals for Lagrange variational problems", *Journal of Mathematical Systems, Estimation and Control*, Vol. 5, No. 1, 1995, pp. 127-130

[2] Arutyunov, A.V., "Second-order conditions in extremal problems. The abnormal points", *Transactions of the American Mathematical Society*, to appear

[3] Avakov, E.R., Necessary conditions for a minimum for nonregular problems in Banach spaces. Maximum principle for abnormal problems of optimal control, *Trudy Mat. Inst. AN. SSSR*, 185, 1988, pp. 3–29, [in Russian]

[4] Dmitruk, A.V., Quadratic order conditions of a local minimum for abnormal extremals, preprint

[5] Hoffmann, K.H. and Kornstaedt, H.J., Higher-order necessary conditions in abstract mathematical programming, *Journal of Optimization Theory and Applications*, Vol.26, 1978, pp. 533-568

[6] Ledzewicz, U. and Schättler, H., An extended maximum principle, *Nonlinear Analysis, Theory, Methods & Applications*, Vol. 29, No. 2, 1997, pp. 159-183,

[7] Ledzewicz, U. and Schättler, H., A high-order generalization of the Lyusternik theorem, *Nonlinear Analysis, Theory, Methods & Applications*, 1998, to appear

[8] Ledzewicz, U. and Schättler, H., High-order approximations and generalized necessary conditions for optimality, *SIAM J. on Control and Optimization*, to appear

[9] Ledzewicz, U. and Schättler, H., A high-order generalization of the Lyusternik theorem and its application to optimal control problems, Proceedings of the International Conference on Dynamical Systems and Differential Equations, Springfield, Missouri, 1998

[10] Ledzewicz, U. and Schättler, H., On a high-order maximum principle, in: *Optimal Control: Theory, Algorithms and Applications*, W.W. Hager and P.M. Pardalos, Eds., Kluwer Academic Publ., 1998

[11] Ledzewicz, U. and Schättler, H., On abnormal bang-bang trajectories, preprint

[12] Milyutin, A.A., Quadratic conditions of an extremum in smooth problems with a finite-dimensional image, in: *Methods of the Theory of Extremal Problems in Economics*, Nauka, Moscow, Russia, 1997, pp. 138-177, [in Russian]

[13] Páles, Z., and Zeidan, V., First- and second-order necessary conditions for control problems with constraints, Tranactions of the American Mathematical Society, Vol. 346, No.2, 1994, pp. 421-453

Urszula Ledzewicz, Dept. of Mathematics and Statistics, Southern Illinois University, Edwardsville, Il 62025, USA

Heinz Schättler, Dept. of Systems Science and Mathematics, Washington University, St. Louis, Mo, 63130, USA

ROBERT BAIER AND ELZA FARKHI
Directed sets and differences of convex compact sets

1 Introduction

A linear normed and partially ordered space is introduced, in which the convex cone of all nonempty convex compact sets in $\mathbb{R}^n$ is embedded. This space of so-called "directed sets" is a Banach and a Riesz space for dimension $n \geq 2$ and a Banach lattice for $n = 1$.

We use essentially the specific parametrization of convex compact sets via their support functions and consider the supporting faces as lower dimensional convex sets. Extending this approach, we define a directed set as a pair of mappings that associate to each unit direction a $(n-1)$-dimensional directed set ("directed supporting face") and a scalar function determining the position of this face in $\mathbb{R}^n$. This method provides recursive definitions, constructions and inductive proofs as well as a visualization of differences of general convex sets with oriented boundary parts.

The basic differences of our approach to other existing embeddings are that there are no equivalence classes (as in [13], [15]) and secondly, that differences of directed convex sets in $\mathbb{R}^n$ are not real-valued functions of n arguments as in [5], but higher-dimensional maps representable as oriented manifolds, e.g. oriented curves/surfaces in the cases $n = 2, 3$. For nonconvex polygons in $\mathbb{R}^2$ see [3] in which an interesting computational-geometric method of polygonal tracings is presented (this approach has been recently extended to polyhedrals in $\mathbb{R}^3$).

The approach is based on the notions of generalized ([6], [11]) or directed intervals ([8], [9]) in the one-dimensional case. In the n-dimensional case, there are essential differences, namely a mixed type part appears which does not exist in the case $n = 1$.

As an application we give an example of set-valued interpolation where nonconvex visualizations of directed sets appear as results.

Basic Notations

Let $\mathcal{C}(\mathbb{R}^n)$ be the set of all convex, compact, nonempty subsets of $\mathbb{R}^n$. The following operations in $\mathcal{C}(\mathbb{R}^n)$ are well-known:

$$A + B := \{a + b \,|\, a \in A,\ b \in B\} \quad \text{(Minkowski addition)}$$
$$\lambda \cdot A := \{\lambda \cdot a \,|\, a \in A\} \quad \text{(scalar multiplication for } \lambda \in \mathbb{R}) \quad (1)$$

Each convex, compact, nonempty set A could be described via its support function $\delta^*(l, A) := \max_{a \in A} <l, a>$ and reconstructed via the intersection of half-spaces with outer normal $l \in S_n$ (S_n is the unit sphere in $\mathbb{R}^n$):

$$A = \bigcap_{l \in S_n} \{x \in \mathbb{R}^n \,|\, <l, x> \leq \delta^*(l, A)\}$$

The support function for $A \in \mathcal{C}(\mathbb{R}^n)$ is Lipschitz-continuous and fulfills
$$\delta^*(l, A + B) = \delta^*(l, A) + \delta^*(l, B), \qquad \delta^*(l, \lambda \cdot A) = \lambda \cdot \delta^*(l, A) \quad (\lambda \geq 0).$$
The Hausdorff-distance between two sets in $\mathcal{C}(\mathbb{R}^n)$ could be expressed via the difference of support functions:
$$\mathrm{d_H}(A, B) = \max_{l \in S_n} |\delta^*(l, A) - \delta^*(l, B)|$$
The supporting face (the set of supporting points) for the direction $l \in S_n$ is
$$Y(l, A) := \{y(l, A) \in A \mid \, <l, y(l, A)> \, = \delta^*(l, A)\}.$$
Some of the definitions of differences of sets which are known in the literature and which are not discussed in the beginning are listed below:

- algebraic difference $A - B := \{a - b \mid a \in A,\ b \in B\}$
  It is not useable in our context, since in general $A - A \supsetneq \{0_{\mathbb{R}^n}\}$.
- differences of intervals
  Classical interval arithmetic uses the algebraic difference (cf. [10], ...), whereas the definition of the subtraction in the space of generalized intervals (cf. [6], [11]) resp. directed intervals (cf. [8], [9]) is specified by the subtraction of the corresponding end points of the intervals.
- Minkowski difference in [4], better known as geometric or star-shaped difference
  $$A \stackrel{*}{-} B := \{x \in \mathbb{R}^n \mid x + B \subset A\}$$
  This difference has the property that $A \stackrel{*}{-} A = \{0\}$, but may often be empty.
- Demyanov's difference in [14]
  $$A \stackrel{.}{-} B := \overline{\mathrm{co}}\{y(l, A) - y(l, B) \mid l \in S_n,\ Y(l, A) \text{ and } Y(l, B) \text{ are singletons}\}$$
  The difference $A \stackrel{..}{-} B$ in [14] is always a superset of Demyanov's difference $A \stackrel{.}{-} B$. There is a close connection between Demyanov's difference and the boundary mapping of the difference of directed sets (cf. Proposition 3.10).

2 Directed Intervals

In interval analysis, $\mathcal{I}(\mathbb{R})$ denotes the set of all real compact intervals
$$[a, b] = \{x \in \mathbb{R} \mid a \leq x \leq b\}.$$
The operations $* \in \{+, -, \cdot, /\}$ known from $\mathbb{R}$ are generalized to the interval case by
$$[a, b] * [c, d] = \{x * y \mid x \in [a, b],\ y \in [c, d]\}.$$
Since the difference is the algebraic difference of intervals, $(\mathcal{I}(\mathbb{R}), +, \cdot)$ is only an Abelian semigroup and not a vector space.

In [6], [11] and [8], [9] generalized resp. directed intervals $\{[\alpha, \beta] \mid \alpha, \beta \in \mathbb{R}\}$ are studied for which the left end point could be greater than the right one. The isomorphism $[\alpha, \beta] \mapsto (\alpha, \beta) \in \mathbb{R}^2$ induces operations/definitions for generalized (directed) intervals. The notion of directed intervals introduced here is in principle equivalent to the generalized intervals of Kaucher ([6]) and the directed intervals of Markov ([8], [9]). Our definition is slightly different, since it is based on support functions and its scalar multiples.

Every interval in $\mathcal{I}(\mathbb{R})$ is convex, compact with support function
$$a_1(l) := \delta^*(l, [a,b]) = \max\{l \cdot a, l \cdot b\} \qquad (l = \pm 1).$$

Definition 2.1 *A directed interval $\overrightarrow{A}$ consists of a function $a_1 : \{\pm 1\} \to \mathbb{R}$, i.e.*
$$\overrightarrow{A} = (a_1(l))_{l=\pm 1} = (a_1(-1), a_1(1)) \in \mathbb{R}^2$$
The notation $\overrightarrow{[\alpha, \beta]} := (-\alpha, \beta)$, where $\alpha = a_1(-1)$, $\beta = a_1(1)$, is often used. Let $\mathcal{D}(\mathbb{R})$ denote the set of all directed intervals. The operations in $\mathcal{D}(\mathbb{R})$ are defined as follows:

$\overrightarrow{A} + \overrightarrow{B}$	$:=$	$(a_1(l) + b_1(l))_{l=\pm 1}$	addition
$\lambda \cdot \overrightarrow{A}$	$:=$	$(\lambda \cdot a_1(l))_{l=\pm 1} \quad (\lambda \in \mathbb{R})$	scalar multiplication
$\overrightarrow{A} - \overrightarrow{B}$	$:=$	$(a_1(l) - b_1(l))_{l=\pm 1}$	subtraction
$\|\overrightarrow{A}\|$	$:=$	$\max_{l=\pm 1} \|a_1(l)\|$	norm
$\overrightarrow{A} \leq \overrightarrow{B}$	$:\iff$	$a_1(l) \leq b_1(l)$ for $l = \pm 1$	partial ordering
$\sup\{\overrightarrow{A}, \overrightarrow{B}\}$	$:=$	$(c_1(l))_{l=\pm 1}$ with $c_1(l) = \max\{a_1(l), b_1(l)\}$	supremum
$\inf\{\overrightarrow{A}, \overrightarrow{B}\}$	$:=$	$-\sup\{-\overrightarrow{A}, -\overrightarrow{B}\}$	infimum

Note that multiplication by negative scalars and subtraction are identical to the corresponding operations on vectors in $\mathbb{R}^2$ and differ from the standard interval operations (as in [10]). The space of directed intervals is isomorphic to the space of generalized intervals, so that according to [6] the following properties of $\mathcal{D}(\mathbb{R})$ could be stated.

Theorem 2.2 *($\mathcal{D}(\mathbb{R}), +, \cdot$) is a vector space with the inverse $-\overrightarrow{A} = (-a_1(l))_{l=\pm 1}$ and the subtraction defined in the table above. Furthermore, it is a a Banach space with the norm $\|\cdot\|$, a Banach lattice with the partial ordering "$\leq$" as well as a Riesz space.*

Example 2.3 *Subtraction of embedded intervals gives $\overrightarrow{[a,b]} - \overrightarrow{[c,d]} = \overrightarrow{[a-c, b-d]}$.*
$$\overrightarrow{[-1, 2]} - \overrightarrow{[-3, 5]} = \overrightarrow{[2, -3]} \quad \text{and} \quad \overrightarrow{[-3, 5]} - \overrightarrow{[-1, 2]} = \overrightarrow{[-2, 3]}$$

The results as well as an improper interval (the inverse of an embedded interval, also called proper interval) and an embedded scalar are visualized in Figure 2.1 resp. 2.2:

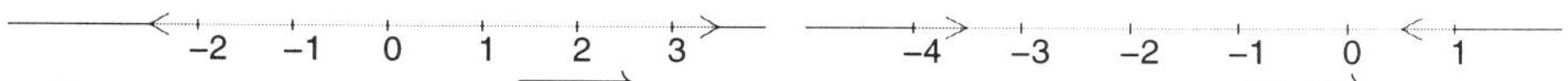

Fig. 2.1: proper interval $\overrightarrow{[-2, 3]} \in \mathcal{D}(\mathbb{R})$ resp. improper interval $\overrightarrow{[1, -4]} \in \mathcal{D}(\mathbb{R})$

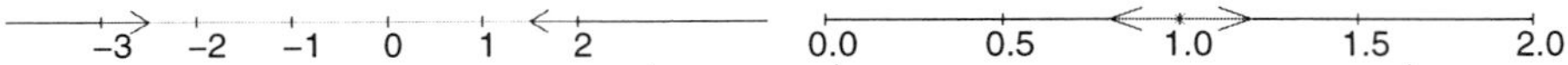

Fig. 2.2: the inverse $\overrightarrow{[2,-3]}$ of $\overrightarrow{[-2,3]}$ resp. degenerate interval $\overrightarrow{[1,1]}$

3 Directed Sets

We construct inductively the linear normed space $\mathcal{D}(\mathbb{R}^n)$ of directed sets in $\mathbb{R}^n$.

Definition 3.1 $\overrightarrow{A}$ is called a directed set

(i) in $\mathbb{R}$, if it is a directed interval and $\|\overrightarrow{A}\|_1 := \max\limits_{l=\pm 1} |a_1(l)|$,

(ii) in $\mathbb{R}^n$, $n \geq 2$, if there exists a continuous function $a_n : S_n \to \mathbb{R}$ and a uniformly bounded function $\overrightarrow{A_{n-1}} : S_n \to \mathcal{D}(\mathbb{R}^{n-1})$ with respect to $\|\cdot\|_{n-1}$.

Then, we denote $\overrightarrow{A} = (\overrightarrow{A_{n-1}(l)}, a_n(l))_{l \in S_n}$ and define

$$\|\overrightarrow{A}\| := \|\overrightarrow{A}\|_n := \max\{\sup_{l \in S_n} \|\overrightarrow{A_{n-1}(l)}\|_{n-1}, \max_{l \in S_n} |a_n(l)|\}, \qquad \|\overrightarrow{A}\| := \max_{l \in S_n} |a_n(l)|$$

The set of all directed sets in $\mathbb{R}^n$ is denoted by $\mathcal{D}(\mathbb{R}^n)$.

The definition above is motivated by describing the convex, compact, nonempty set A for each direction $l \in S_n$ as a pair

("$Y(l, A)$ as $(n-1)$ dimensional (directed) set", $\delta^*(l, A))_{l \in S_n}$.

Each operation is defined recursively and works separately on both components.

Definition 3.2 Let $\overrightarrow{A} = (\overrightarrow{A_{n-1}(l)}, a_n(l))_{l \in S_n}$, $\overrightarrow{B} = (\overrightarrow{B_{n-1}(l)}, b_n(l))_{l \in S_n}$.

$$\begin{aligned}
\overrightarrow{A} + \overrightarrow{B} &:= (\overrightarrow{A_{n-1}(l)} + \overrightarrow{B_{n-1}(l)}, a_n(l) + b_n(l))_{l \in S_n} \\
\lambda \cdot \overrightarrow{A} &:= (\lambda \cdot \overrightarrow{A_{n-1}(l)}, \lambda \cdot a_n(l))_{l \in S_n} \qquad (\lambda \in \mathbb{R}). \\
\overrightarrow{A} - \overrightarrow{B} &:= \overrightarrow{A} + (-\overrightarrow{B}) = (\overrightarrow{A_{n-1}(l)} - \overrightarrow{B_{n-1}(l)}, a_n(l) - b_n(l))_{l \in S_n} \\
\overrightarrow{A} \leq \overrightarrow{B} &:\iff \begin{cases} \text{(i) } \forall l \in S_n: & a_n(l) \leq b_n(l) \\ \text{(ii) if } \exists l \in S_n \text{ with} & a_n(l) = b_n(l), \\ \quad \text{then} & \overrightarrow{A_{n-1}(l)} \leq \overrightarrow{B_{n-1}(l)} \end{cases}
\end{aligned}$$

$$\sup\{\overrightarrow{A}, \overrightarrow{B}\} := (\overrightarrow{S_{n-1}(l)}, \max\{a_n(l), b_n(l)\})_{l \in S_n}$$

$$\overrightarrow{S_{n-1}(l)} := \begin{cases} \overrightarrow{B_{n-1}(l)} & \text{if } a_n(l) < b_n(l) \\ \sup\{\overrightarrow{A_{n-1}(l)}, \overrightarrow{B_{n-1}(l)}\} & \text{if } a_n(l) = b_n(l) \\ \overrightarrow{A_{n-1}(l)} & \text{if } a_n(l) > b_n(l). \end{cases}$$

$$\inf\{\overrightarrow{A}, \overrightarrow{B}\} := -\sup\{-\overrightarrow{A}, -\overrightarrow{B}\}$$

Proposition 3.3 $(\mathcal{D}(\mathbb{R}^n), +, \cdot)$ *is a vector space with the zero element* $\overrightarrow{0_{\mathcal{D}(\mathbb{R}^n)}}$ $= (\overrightarrow{0_{\mathcal{D}(\mathbb{R}^{n-1})}}, 0_{\mathbb{R}})_{l \in S_n}$ *and the inverse of* $\overrightarrow{A}$, $-\overrightarrow{A} = (-\overrightarrow{A_{n-1}(l)}, -a_n(l))_{l \in S_n}$.

Proposition 3.4 $(\mathcal{D}(\mathbb{R}^n), \|\cdot\|)$ *is a Banach space and* $\|\cdot\|$ *is a semi-norm. It is even a lattice and a Riesz space with the ordering and supremum/infimum in Definition 3.2.*

One may interpret the supporting face $Y(l, A)$ as $(n-1)$-dimensional (directed) set, e.g. by the following procedure:

- translate the hyperplane which is orthogonal to l and contains $Y(l, A)$ to the origin by the vector $\delta^*(l, A)l$

- rotate the result into the plane $\{x_n = 0\}$ until the attached orthogonal vector l coincides with e^n

- project the rotated image of $Y(l, A)$ into $\mathbb{R}^{n-1}$

- embed the result in the space $\mathcal{D}(\mathbb{R}^{n-1})$

Definition 3.5 *The set* $A \in \mathcal{C}(\mathbb{R}^n)$ *is embedded into the set* $\mathcal{D}(\mathbb{R}^n)$ *via* $J_n : \mathcal{C}(\mathbb{R}^n) \to \mathcal{D}(\mathbb{R}^n)$:

(i) $J_1([a, b]) := \overrightarrow{[a,b]} = (-a, b)$ *for* $n = 1$

(ii) $J_n(\overrightarrow{A}) := (J_{n-1}(P_{n-1,l}(Y(l, A))), \delta^*(l, A))_{l \in S_n}$ *for* $n \geq 2$
$P_{n-1,l}(x) := \pi_{n-1,n} R_{n,l}(x - \delta^*(l, A)l)$ *and* $\pi_{n-1,n}$ *is the projection from* $\mathbb{R}^n$ *to* $\mathbb{R}^{n-1}$, $R_{n,l}$ *is a rotation matrix which satisfies for the unit vectors* $e^1, \ldots, e^n$

$$R_{n,l}(l) = e^n, \qquad R_{n,l}(span\{l\}^\perp) = span\{e^1, e^2, \ldots, e^{n-1}\} \tag{2}$$

and must be uniquely defined for the embedding. A possible construction is skipped due to the lack of space, only the properties in (2) are used in the proofs.

To define the visualization of a directed set, the convex and the concave part of a directed set are defined.

Definition 3.6 *Let* $\overrightarrow{A} \in \mathcal{D}(\mathbb{R}^n)$. *The definition of its convex and concave part are:*

$$P_n(\overrightarrow{A}) := \{x \in \mathbb{R}^n \,|\, \text{for every } l \in S_n : \,<l, x> \leq a_n(l)\},$$
$$N_n(\overrightarrow{A}) := \{-x \in \mathbb{R}^n \,|\, \text{for every } l \in S_n : \,<l, x> \leq -a_n(l)\}$$

At least one of the convex and the concave part of $\overrightarrow{A}$ is empty, except the case that both are equal and contain only one point. It could happen that both of them are empty and the set coincides with the mixed type part defined in Definition 3.7, but in the one-dimensional case, exactly one of $P_1(\overrightarrow{[a,b]})$ and $N_1(\overrightarrow{[a,b]})$ is empty, if $a \neq b$.

Definition 3.7 Let $\vec{A} \in \mathcal{D}(\mathbb{R}^n)$. The visualization $V_n : \mathcal{D}(\mathbb{R}^n) \Rightarrow \mathbb{R}^n$ consists of three parts, the convex and concave part as well as the mixed type part $M_n(\vec{A})$. $M_n(\vec{A})$ collects all reprojected points from the visualization of the boundary parts $\overrightarrow{A_{n-1}(l)}$ which are not elements of the other two parts. Both sets are defined simultaneously:

$$M_1(\vec{A}) := \emptyset, \quad V_1(\vec{A}) := P_1(\vec{A}) \cup N_1(\vec{A}) \qquad (n = 1)$$
$$M_n(\vec{A}) := \bigcup_{l \in S_n} \{x \in Q_{n,l}(V_{n-1}(\overrightarrow{A_{n-1}(l)})) \mid x \notin P_n(\vec{A}) \cup N_n(\vec{A})\} \quad (n \geq 2)$$
$$V_n(\vec{A}) := P_n(\vec{A}) \cup N_n(\vec{A}) \cup M_n(\vec{A}) \qquad (n \geq 2)$$

with the reprojection $Q_{n,l}(y) := R_{n,l}^{-1} \pi_{n,n-1}(y) + a_n(l)l$, $y \in \mathbb{R}^{n-1}$. $\pi_{n,n-1}$ is the natural embedding of $\mathbb{R}^{n-1}$ into $\mathbb{R}^n$.
The boundary mapping $B_n : \mathcal{D}(\mathbb{R}^n) \Rightarrow \mathbb{R}^n$ is defined as

$$B_n(\vec{A}) := \partial P_n(\vec{A}) \cup \partial N_n(\vec{A}) \cup M_n(\vec{A}).$$

The "boundary" of a directed set consists of the boundary of the convex or concave part and the additional part of mixed type, which is outside the convex and the concave part. This mixed type part is always empty in the case $n = 1$ or if the set is an embedded convex set and usually nonempty otherwise. Each point $x \in \mathbb{R}^n$ from the reprojected image of the visualization of the $(n-1)$-dimensional boundary part $\overrightarrow{A_{n-1}(l)}$ for some "normal" direction $l \in S_n$ is a "boundary" point, i.e. $x \in B_n(\vec{A})$. All these directions l are attached to x and form its directions bundle $\mathcal{O}_n(x, \vec{A})$.

Each part of the boundary of the inverse of a directed set is the (pointwise) negative (according to (1)) of the boundary part of the directed set itself. The convex part of the inverse is the (pointwise) negative of the concave part of the original set. Therefore, the visualization of $-\vec{A}$ is the (pointwise) negative of the visualization of $\vec{A}$.

Proposition 3.8 $\vec{A} \in \mathcal{D}(\mathbb{R}^n)$. It follows with the convention $-\emptyset = \emptyset$:

$$P_n(-\vec{A}) = -N_n(\vec{A}), \quad N_n(-\vec{A}) = -P_n(\vec{A}), \quad V_n(-\vec{A}) = -V_n(\vec{A})$$

Furthermore, the direction bundle of the "negative" points remains the same as of the corresponding "positive" points, i.e.

$$\mathcal{O}_n(-x, -\vec{A}) = \mathcal{O}_n(x, \vec{A}) \quad (x \in B_n(\vec{A}) = -B_n(-\vec{A})). \tag{3}$$

Example 3.9 According to (3) the visualization of the inverse is formed by multiplying all boundary points of the original set with -1 and keeping their corresponding directions l. The outer normals $l \in S_n$ of the directed set $\overrightarrow{[0,2]^2}$ become inner normals of its inverse $-\overrightarrow{[0,2]^2}$ (see Figure 3.1).

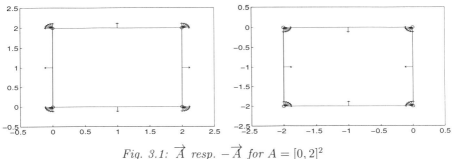

Fig. 3.1: $\overrightarrow{A}$ resp. $-\overrightarrow{A}$ for $A = [0,2]^2$

The visualization of differences of directed sets are strongly related to other differences.

Proposition 3.10 $A, B \in \mathcal{C}(\mathbb{R}^n)$. Then, the following is true:
$$P_n(\overrightarrow{A} - \overrightarrow{B}) = A \stackrel{*}{-} B, \quad N_n(\overrightarrow{A} - \overrightarrow{B}) = -(B \stackrel{*}{-} A), \quad \overline{co}\, B_n(\overrightarrow{A} - \overrightarrow{B}) = A \stackrel{.}{-} B,$$
$$\bigcup_{l \in S_n} \{y(l, A) - y(l, B) \,|\, Y(l, A), Y(l, B) \text{ are singletons}\} \subset B_n(\overrightarrow{A} - \overrightarrow{B})$$

The operations and definitions (addition, scalar multiplication, ordering, norm, ...) are generalizations to the ones known for convex sets.

Proposition 3.11 $A, B \in \mathcal{C}(\mathbb{R}^n)$ and $\lambda \geq 0$. Then, it is valid:

$\overrightarrow{A} + \overrightarrow{B} = \overrightarrow{A+B}$	$V_n(\overrightarrow{A} + \overrightarrow{B}) = A + B$
$\lambda \cdot \overrightarrow{A} = \overrightarrow{\lambda \cdot A}$	$V_n(\lambda \cdot \overrightarrow{A}) = \lambda \cdot A$
$\|\overrightarrow{A}\| = \|\overrightarrow{A}\| = \sup_{a \in A} \|a\|_2$	$\|V_n(\overrightarrow{A})\| = \sup_{a \in A} \|a\|_2$

$\|\cdot\|$ defines a metric on $J_n(\mathcal{C}(\mathbb{R}^n))$ with $\delta(\overrightarrow{A}, \overrightarrow{B}) := \|\overrightarrow{A} - \overrightarrow{B}\| = d_H(A, B)$.

Example 3.12 Let $A = B_2(0)$, $B = [-1,1]^2$. The boundary of $\overrightarrow{A} - \overrightarrow{B}$ consists of the boundary of $A \stackrel{*}{-} B$ (convex part), all other points of $B_n(\overrightarrow{A} - \overrightarrow{B})$ are elements of the mixed type part $M_n(\overrightarrow{A} - \overrightarrow{B})$. All differences of supporting points inside of $A \stackrel{.}{-} B$ are elements of the boundary of $\overrightarrow{A} - \overrightarrow{B}$ (see Figure 3.2 and 3.3).

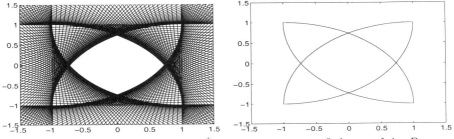

Fig. 3.2: geometric difference $A \stackrel{*}{-} B$ resp. non-convexified part of $A \stackrel{.}{-} B$

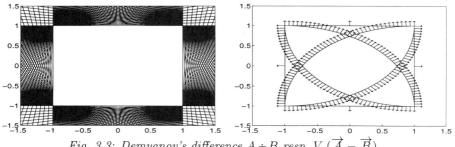

Fig. 3.3: Demyanov's difference $A \stackrel{.}{-} B$ resp. $V_n(\vec{A} - \vec{B})$

4 Applications and Numerical Example

One example of set-valued interpolation is studied to show the visualization of directed sets in applications, especially, if negative weights appear in formulas. Further results and a more detailed research on applications (computation of reachable sets by extrapolation methods in [1], differentiable set-valued mappings in [2], the connection to minimal pairs in [12], error estimates, ...) must be postponed to a forthcoming publication.

Linear interpolation can be done in the space $\mathcal{C}(\mathbb{R}^n)$, but interpolation with a higher polynomial degree creates negative weights.

Example 4.1 *Consider the quadratic interpolation of the set-valued mapping in [7]*

$$F(t) = \begin{pmatrix} (t+1) \cdot (t+2) & 1 \\ 0 & t^2+1 \end{pmatrix} B_1(0), \qquad (t \in [-3,3])$$

with the prescribed sets $F(-3), F(0), F(3)$.

a) geometric difference

$$P_2(t) := \{\, x \in \mathbb{R}^2 \mid \forall l \in S_2 : <l,x> \le p_2(l,t) := \sum_{i=0}^{2} L_i(t) \delta^*(l, F(t_i)) \,\} \qquad (4)$$

with the Lagrange polynomials $L_i(t) = \prod\limits_{\substack{j=0,1,2 \\ j \ne i}} \frac{t-t_j}{t_i-t_j}$ and $t_i = -3 + i \cdot 3$ $(i=0,1,2)$

Although, the prescribed sets are convex sets, $P_2(-1.5)$ is an empty set. $P_2(1)$ is convex and compact, but $p_2(\cdot,1)$ is nonconvex which creates non-supporting hyperplanes in (4).

b) difference of directed sets

$$\vec{P_2}(t) := \sum_{i=0}^{2} L_i(t) \overrightarrow{F(t_i)}$$

$\vec{P_2}(-1.5)$ and $\vec{P_2}(1)$ are "mixed-type" directed sets (see Figure 3.4), $\vec{P_2}(-1.5)$ has an empty convex and concave part, $\vec{P_2}(1)$ has a nonempty convex part $P_2(1)$.

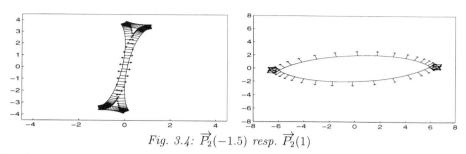

Fig. 3.4: $\vec{P_2}(-1.5)$ resp. $\vec{P_2}(1)$

References

[1] R. Baier. Mengenwertige Integration und die diskrete Approximation erreichbarer Mengen. *Bayreuther Mathematische Schriften* **50** (1995).

[2] T. D. Donchev, E. M. Farkhi. Moduli of Smoothness of Vector Valued Functions of a Real Variable and Applications. *Numer. Funct. Anal. Optim.* **11** (1990), no. 5 & 6, 497–509.

[3] L. Guibas, L. Ramshaw, J. Stolfi. A kinematic framework for computational geometry. In *Proceedings* 24th *annual symposium on foundations of computer science, Nov. 7–9, 1983, Tucson, Arizona*, pp. 100–111, IEEE Computer Press, Los Alamitos, California, 1983.

[4] H. Hadwiger. Minkowskische Addition und Subtraktion beliebiger Punktmengen und die Theoreme von Erhard Schmidt. *Math. Z.* **53** (1950), Heft 3, 210–218.

[5] P. L. Hörmander. Sur la fonction d'appui des ensembles convexes dans un espace localement convexe. *Ark. Mat.* **3** (1954), no. 12, 181–186.

[6] E. Kaucher. Interval Analysis in the Extended Interval Space $\mathbb{IR}$. *Comput. Suppl.* **2** (1980), 33–49.

[7] F. Lempio. Set-Valued Interpolation, Differential Inclusions, and Sensitivity in Optimization. In R. Lucchetti, J. Revalski (eds.), *Recent Developments in Well-Posed Variational Problems*, pp. 137–169, Kluwer Academic, Dordrecht, 1995.

[8] S. Markov. On the presentation of ranges of monotone functions using interval arithmetic. *Interval Comput.* **4** (1992), no. 6, 19–31.

[9] S. Markov. On directed interval arithmetic and its applications. *JUCS* **1** (1995), no. 7, 514–526.

[10] R. E. Moore. *Interval Analysis*. Prentice-Hall, Englewood Cliffs, N.J., 1966.

[11] H.-J. Ortolf. Eine Verallgemeinerung der Intervallarithmetik. *GMD-Bericht* **11** (1969), Bonn.

[12] D. Pallaschke, S. Scholtes, R. Urbański. On Minimal Pairs of Convex Compact Sets. *Bull. Pol. Acad. Sci., Math.* **39** (1989), no. 1–2, 105–109.

[13] H. Rådström. An embedding theorem for spaces of convex sets. *Proc. Amer. Math. Soc.* **3** (1952), 165–169.

[14] A. Rubinov, I. Akhundov. Difference of compact sets in the sense of Demyanov and its application to non-smooth analysis. *Optimization* **23** (1992), 179–188.

[15] K. E. Schmidt. Embedding Theorems for Classes of Convex Sets. *Acta Appl. Math.* **5** (1986), 209–237.

DEAN A CARLSON
Open-loop Nash equilibrium for nonlinear control systems[1]

1 Introduction

Research in the area of differential games has for the most part focused on the problem of obtaining sufficient conditions for the existence of closed-loop or feedback equilibrium for two-player games of either zero-sum type or of min-max type. Notable exceptions to this for p-player games are found in [Başar, T. and Olsder, G., 1982] and the references contained therein. The main tools in establishing these results has been either through the use of dynamic programming or by studying a corresponding system of Hamilton-Jacobi equations in which it can be demonstrated that a viscosity solution exists. In the results reported below we approach the problem somewhat differently by seeking conditions which insure that a p-player game has an open-loop Nash equilibrium. The approach we utilize is to view the game as a static game in an infinite dimensional space and exploit the control-theoretic structure to establish our results with convexity and seminormality conditions approaching those assumed in the classical existence theory of optimal control as outlined in [Cesari, L., 1983]. A typical procedure is to define a so-called "best reply" mapping from the set of feasible points into itself (see for example [Friedman, J. W., 1990]). For a fixed feasible point, say x, the image of this mapping is set of feasible points which represents each players best reply to the situation when all other players play the point x. The existence of a Nash equilibrium is then established by showing that this mapping has a fixed point. The definition of this set-valued mapping, while perhaps useful as a theoretical tool, requires the solution of p optimization problems and therefore may be of limited practical value. This is particularly true if the fixed point is to be obtained via an iterative scheme since at each step it is required to solve p optimization problems.

In the results reported below we define an analogue of the best reply mapping which originates in the work of [Rosen, J., 1965]. What makes this mapping better is that if an iterative scheme is developed it requires solving only one optimization problem. Using a generalization of Kakutani's Fixed Point theorem for set-valued mappings conditions under which a class of p-player differential games has an open-loop Nash equilibrium.

2 The Basic Model and Hypotheses

We consider a control system whose state equation is described by the ordinary differential equation
$$\dot{x}(t) = f(t, x(t), u(t)), \quad \text{a.e.} \quad t_0 \leq t \leq t_1; \tag{2.1}$$

---
[1] This research was supported by the National Science Foundation (INT–9500782)

with boundary conditions,
$$(t_0, x(t_0), t_1, x(t_1)) \in B; \tag{2.2}$$
state constraints,
$$(t, x(t)) \in A \quad \text{for} \quad t_0 \leq t \leq t_1; \tag{2.3}$$
and control constraints
$$u(t) \in U(t, x(t)) \quad \text{a.e.} \quad t_0 \leq t \leq t_1. \tag{2.4}$$

Here we assume that $A \subset [t_0, t_1] \times \mathbb{R}^n$ is a closed set; $U : A \to 2^{\mathbb{R}^m} \setminus \emptyset$ is a set-valued mapping with closed graph,
$$M = \{(t, x, u) : \quad (t, x) \in A \quad \text{and} \quad u \in U(t, x)\}; \tag{2.5}$$
the function $f : M \to \mathbb{R}^n$ is an n-vector valued function in which each component is Lebesgue measurable in t for each fixed (x, u) and continuous in (x, u) for each fixed t (i.e., each component of f is a Carathéodory integrand); and $B \subset \mathbb{R} \times \mathbb{R}^n \times \mathbb{R} \times \mathbb{R}^n$ is a closed set. Clearly this is a standard control system with the usual general hypotheses satisfied.

To introduce the dynamic game we assume that a state trajectory $x : [t_0, t_1] \to \mathbb{R}^n$ can be partitioned into $p \in \mathbb{N}$ components $x = (x_1, x_2, \ldots, x_p)$ in which for $j = 1, 2, \ldots, p$ each component $x_j : [t_0, t_1] \to \mathbb{R}^{n_j}$ represents the state of the jth player. Here, of course, we assume that $n = \sum_{j=1}^{p} n_j$. We notice that in a specific application this decomposition occurs naturally, but it possible that one can select a parirition in such a way that it is possible to exploit some distinguishing features of the dynamical system. We also assume a similar decomposition of the control functions, $u = (u_1, u_2, \ldots, u_p)$ in which for $j = 1, 2, \ldots, p$ each component is a function from $[t_0, t_1]$ into $\mathbb{R}^{m_j}$ with $m = \sum_{j=1}^{p} m_j$. Again, this is often given naturally but perhaps another decomposition could be exploited in a particular application.

The performance of each player will be measured by a performance criterion described by a Lagrange-type functional; J_j for $j = 1, 2, \ldots, p$; given by a definite integral of the form
$$J_j(t_0, t_1, x, u) = \int_{t_0}^{t_1} g_j(t, x(t), u_j(t)) \, dt, \tag{2.6}$$
in which $g_j : A \times \mathbb{R}^{m_j} \to [0, +\infty)$ is a Lebesgue normal integrand. That is, for each fixed t, $g_j(t, \cdot, \cdot)$ is lower semicontinuous and g_j is measurable with respect to the σ-algebra of sets generated by the products of Lebesgue measurable subsets in t with the Borel measurable subsets of $\mathbb{R}^n \times \mathbb{R}^{m_j}$.

For each fixed $j = 1, 2, \ldots, p$ we observe that what is described above is a "standard" Lagrange-type optimal control problem and so, in some sense at least, each player is going to attempt to solve an optimal control problem. What makes this problem difficult of course is that what may be optimal for one player will not necessarily be optimal for another. Thus we must seek a non-cooperative or Nash equilibrium. To do this we must first introduce some additional notations and definitions.

Definition 2.1 *A pair of functions $\{x, u\}$ defined on an interval $[t_0, t_1]$ is an admissible pair for the dynamic game if $x : [t_0, t_1] \to \mathbb{R}^n$ is absolutely continuous, $u : [t_0, t_1] \to \mathbb{R}^m$ is Lebesgue measurable, the optimal control system described by Equations (2.1)-(2.4) is satisfied, and if for each $j = 1, 2, \ldots, p$ the map $t \to g_j(t, x(t), u_j(t))$ is Lebesgue integrable on $[t_0, t_1]$.*

Following standard terminology for an admissilbe pair $\{x, u\}$ we will call x an admissible trajectory and u an admissible control. We let Ω denote the set of all admissible pairs and assume that $\Omega \neq \emptyset$.

The basic idea of a Nash equilibrium is that each player is providing its "best response" to the controls played by the other players. Consequently we must introduce a notation that allows us to focus specifically on a single player's contribution to the state of the control system. Therefore, for $x = (x_1, x_2, \ldots, x_p) \in \mathbb{R}^{n_1} \times \mathbb{R}^{n_2} \ldots \mathbb{R}^{n_p}$ and $y \in \mathbb{R}^{n_j}$ we define the notation,

$$[x^j, y] = (x_1, x_2, \ldots, x_{j-1}, y_j, x_{j+1}, \ldots, x_p),$$

(i.e. $x \in \mathbb{R}^n$ remains the same except in the n_j-vector component where we replace x_j by y). With this notation we introduce the following notion of admissibility.

Definition 2.2 *For a fixed admissible pair $\{x, u\} : [t_0, t_1] \to \mathbb{R}^n$, we say that a trajectory-control pair $\{y, u\} : [t_0, t_1] \to \mathbb{R}^{n_j} \times \mathbb{R}^{m_j}$ is an admissible response to the admissible pair $\{x, u\}$ for player $j = 1, 2, \ldots, p$ the trajectory-control pair $\{[x^j, y], [u^j, v]\}$ is an admissible pair for the dynamic game.*

With this definition we now can give the definition of a Nash equilibrium.

Definition 2.3 *An admissible pair $\{x^*, u^*\}$ for the dynamic game is called a Nash equilibrium if for each player $j = 1, 2, \ldots p$ we have*

$$J_j(t_0, t_1, x^*, u^*) \leq J_j(t_0, t_1, [x^{*j}, y], [u^{*j}, v]) \tag{2.7}$$

holds for all pairs $\{y, v\} : [t_0, t_1] \to \mathbb{R}^{n_j}$ that are admissible responses to the admissible pair $\{x^*, u^*\}$ for player j.

Remark 2.1 *The above definition states that an admissible pair $\{x^*, u^*\}$ is a Nash equilibrium whenever x_j^* is the best reply among all possible admissible responses, y, to the trajectory x^* for player j.*

3 Rosen's "trick" and the Associated Family of Optimal control Problems

With the basic model described above we are ready to begin addressing the problem of providing sufficient conditions for the existence of Nash equilibrium. In a seminal paper [Rosen, J., 1965] conditions for existence and uniqueness for Nash equilibria in static games in $\mathbb{R}^n$ were given. Since we can view the problem described above as a static game in an infinite dimensional setting it is possible to extend the results given by Rosen. Indeed in [Carlson, D. A., 1998] such an extension was given for the case of a Hilbert space. As noted in that work, Rosen's existence results are not directly applicable since they require continuity of the objective functionals. This regularity is not available without assuming more stringent hypotheses than those considered here. On the other hand, as we shall see, we can adopt Rosen's approach to the setting considered here and exploit the control-theoretic structure to achieve the desired results. We begin by letting Ω_T denote the set of all admissible trajectories for the dynamic game. That is,

$$\Omega_T = \{x : [t_0, t_1] \to \mathbb{R}^n \mid \text{there exists } u : [t_0, t_1] \to \mathbb{R}^m \text{ such that } \{x, u\} \in \Omega\}. \quad (3.1)$$

We now fix an admissible trajectory $x \in \Omega_T$; let $r_j > 0$, $j = 1, 2, \ldots p$ be a set of fixed positive weights; and define the combined cost functional defined $J[x, r] : \Omega \to \mathbb{R}$ by

$$J[x, r](t_0, t_1, y, v) = \int_{t_0}^{t_1} G_r(t, x(t), y(t), v(t)) \, dt, \quad (3.2)$$

in which G_r is defined on $\mathbb{R} \times \mathbb{R}^n \times \mathbb{R}^n \times \mathbb{R}^m$ is the combined cost integrand given by

$$G_r(t, x, y, v) = \sum_{j=1}^{p} r_j \, g_j(t, [x^j, y_j], v_j]) \quad (3.3)$$

With this notation we define the family of optimal control problems described briefly by

$$\min_{\{y,v\} \in \Omega(x)} J[x, r](t_0, t_1, y, v). \quad (3.4)$$

This family of control problem is parametrized by the set of admissible trajectories and the weights r_j. We focus our attention here on the admissible trajectories and regard the weights as fixed. For a fixed $x \in \Omega_T$ we let $\Gamma(x)$ denote the set of admissible trajectories that solve the optimal control problem given by equation (3.4). That is,

$$\Gamma(x) = \{y : \text{ there exists } v \text{ such that } \{y, v\} \in \Omega \quad (3.5)$$
$$\text{and } J[x, r](t_0, t_1, y, v) \le J[x, r](t_0, t_1, z, w) \text{ for all } \{z, w\} \in \Omega(x)\}.$$

The significance of the set-valued mapping $x \to \Gamma(x)$ is given by the following proposition.

Proposition 3.1 *If $x^* : [t_0, t_1] \to \mathbb{R}^n$ is an admissible trajectory for the dynamic game that is a fixed point of the set-valued mapping Γ, then it is a Nash equilibrium.*

Proof: Let $\{x^*, u^*\}$ be an admissible pair for the dynamic game such that $x^* \in \Gamma(x^*)$ and suppose that it is not a Nash equilibrium. Then there exists an index k and an admissible response $\{y, v\} : [t_0, t_1] \to \mathbb{R}^{n_k}$ to the trajectory x^* such that

$$J_k(t_0, t_1, x^*, u^*) > J_k(t_0, t_1, [x^{*k}, y], [u^{*k}, v]) \quad (3.6)$$

holds. This means that,

$$\int_{t_0}^{t_1} G_r(t, x^*(t), [x^{*k}, y](t), [u^{*k}, v](t)) \, dt = \sum_{j \neq k} r_j \int_{t_0}^{t_1} g_j(t, x^*(t), u_j^*(t)) \, dt$$

$$+ r_k \int_{t_0}^{t_1} g_k(t, [x^{*k}, y](t), v(t)) \, dt$$

$$< \sum_{j \neq k} r_j \int_{t_0}^{t_1} g_j(t, x^*(t), u_j^*(t)) \, dt$$

$$+ r_k \int_{t_0}^{t_1} g_k(t, [x^{*k}, y](t), v(t)) \, dt$$

$$= \int_{t_0}^{t_1} G_r(t, x^*(t), x^*(t), u^*(t)) \, dt,$$

where the above inequality follows from (3.6) and the non-negativity of the integrands g_j. However, this gives rise to a contradiction since x^* is a fixed point of Γ which necessarily means that

$$\int_{t_0}^{t_1} G_r(t, x^*(t), x^*(t), u^*(t)) \, dt \leq \int_{t_0}^{t_1} G_r(t, x^*(t), [x^{*k}, y](t), [u^{*k}, v](t)) \, dt$$

$$< \int_{t_0}^{t_1} G_r(t, x^*(t), x^*(t), u^*(t)) \, dt.$$

□

The above proposition indicates the direction we pursue to derive conditions under which the differential game has an open-loop Nash equilibrium. Specifically we must provide conditions which insure that the mapping Γ has a fixed point. Many results concerning the existence of a fixed point for a set-valued mapping exist in the literature and we will appeal directly to one of these. The most well known of these is Kakutani's fixed point theorem. Two difficulties arise when utilizing this theorem. The first of these is that it is required that the set-valued mapping Γ must be convex valued. The other difficulty arises since the mapping must also be compact valued (i.e., the sets $\Gamma(x)$ must be compact). Compactness conditions in an infinite dimensional space can be quite restrictive and therefore at first glance this may be the wrong approach. However, a generalization of Kakutani's theorem was proven in [Bohnenblust, H. F. and Karlin, S., 1950] which weakens the compactness hypothesis. In the next section we will provide conditions under which both of this generalization can be used.

4 Existence of Open-Loop Nash Equilibria

In this section we conditions under which an open-loop Nash equilibrium exists for the class of differential games considered here. To affect our results we will appeal to the generalization of Kakutani's fixed point theorem mentioned above.

In establishing both our existence result we require a compactness condition to be satisfied. Specifically we regard the set of admissible trajectories as a subset of absolutely continuous functions endowed with the weak topology. For brevity we denote this space as $AC([t_0, t_1]; \mathbb{R}^n)$ For those unfamiliar with this topology wedirect the reader to [Cesari, L., 1983].

Sufficient conditions, given in the form of growth conditions,under which a set of admissible trajectories is relatively sequentially weakly compact in this topology are well known in optimal control theory and the calculus of variations. A thorough and complete discussion of these results is given in [Cesari, L., 1983] and we refer the interested reader to that work for further information. The condition we choose to use here is defined in the following.

Definition 4.1 *We say that the control system described by (2.1) to (2.4) satisfies the growth condition (γ) if there exists an auxiliary Lebesgue normal integrand $h : M \to \mathbb{R}$ such that for each $\epsilon > 0$ there exists an integrable function $\psi_\epsilon : [t_0, t_1] \to [0, +\infty)$ such that*

$$|f(t, y, v)| \leq \psi_\epsilon(t) + \epsilon h(t, y, v) \tag{4.1}$$

for almost all $t \in [t_0, t_1]$ with $(t, y, v) \in M$.

The growth condition (γ) defined above is apparently one of the most general and it is well known to be equivalent to several other growth conditions used in the work of Rockafellar and others (see [Cesari, L., 1983]). In most existence results in optimal control it is the case that a suitable comparison function $h(\cdot, \cdot, \cdot)$ is the given by the cost integrand. For the game theory case considered here, it is not apparent which of the cost integrands would be a suitable candidate and we must look for another comparison function. One appropriate candidate to replace the cost integrand would be to use the function $h : M \to [0, +\infty$ defined by the formula

$$h(t, y, v) = \inf_{x \in A(t)} G_r(t, x, y, v) \tag{4.2}$$

in which $A(t) = \{x \in \mathbb{R}^n \mid (t, x) \in A\}$ and the function G_r is the combined cost integrand defined in (3.3). Under the assumptions made here the function $h(\cdot, \cdot, \cdot)$ is easily seen to be a Lebesgue normal integrand. The compactness result we now require is given by the following theorem.

Theorem 4.1 *Assume that the control system described by equation (2.1) to (2.4) satisfies the growth condition (γ), let $K \in \mathbb{R}$ be a fixed constant, and let Ω_0 denote any set of admissible pairs for satisfying*

$$\int_{t_0}^{t_1} h(t, y(t), v(t)) \, dt \leq K$$

$$y(t_0) \in X,$$

in which $h(\cdot, \cdot, \cdot)$ is as described in Definition 4.1 and X is a compact subset of $\mathbb{R}^n$. Then the set of admissible trajectories is a relatively weakly compact subset of $AC([t_0, t_1]; \mathbb{R}^n)$.

Proof: See [Cesari, L., 1983, Section 10.4]. □

The remaining hypotheses will place certain convexity and seminormality assumptions on the dynamic game. In optimal control theory these conditions are most easily defined through the use of set-valued mappings. For the discussion here we introduce two such sets. The first of these is $S: A \to 2^{\mathbb{R}^{1+2n}}$, where $2^{\mathbb{R}^{1+2n}}$ denotes the nonempty subsets of $\mathbb{R}^{1+2n}$, defined by the equation

$$S(t,x) = \{(y, z^0, z) : (t,y) \in A,\ z^0 \geq G_r(t,x,y,v),\ z = f(t,y,v),\ v \in U(t,y)\}. \quad (4.3)$$

We will assume that that for each $(t,x) \in A$ that the set $S(t,x)$ is closed and convex. The second set-valued mapping is $\tilde{Q} : \tilde{A} \to 2^{\mathbb{R}^{1+n}}$, in which $\tilde{A} = \{(t,x,y) : \quad (t,x) \in A \text{ and } (t,y) \in A\}$ given by the equation

$$\tilde{Q}(t,x,y) = \{(z^0, z) : \quad z^0 \geq G_r(t,x,y,v), \quad z = f(t,y,v), \quad v \in U(t,y)\}. \quad (4.4)$$

These sets will also be supposed to be closed and convex. In addition we will assume that these sets satisfy the uppersemicontinuity condition (K) with respect to y. That is we assume that

$$\tilde{Q}(t,x,y) = \bigcap_{\delta > 0} \left[\bigcup \{\tilde{Q}(t,x,z) : \quad |z - y| \leq \delta\} \right]. \quad (4.5)$$

Finally, we need the following fixed point theorem to establish our results is the following.

Theorem 4.2 *Let X be a weakly separable Banach space with S a convex, weakly closed set in E. Let $\mathcal{B} : S \to 2^S \setminus \{\emptyset\}$ be a set-valued mapping satisfying the following:*

1. *$\mathcal{B}(\mathbf{x})$ is convex for each $\mathbf{x} \in S$.*

2. *The graph of $\mathcal{B}$, $\{(\mathbf{x}, \mathbf{y}) \in S \times S : \mathbf{y} \in \mathcal{B}(\mathbf{x})\}$, is weakly closed in $X \times X$. That is, if $\{\mathbf{x}_n\}$ and $\{\mathbf{y}_n\}$ are two sequences in S such that $\mathbf{x}_n \to \mathbf{x}$, $\mathbf{y}_n \to \mathbf{y}$, weakly in X with $\mathbf{x}_n \in \mathcal{B}(\mathbf{y_n})$, then necessarily we have $\mathbf{x} \in \mathcal{B}(\mathbf{y})$.*

3. *$\bigcup_{\mathbf{x} \in S} \mathcal{B}(\mathbf{x})$ is contained in a sequentially weakly compact set T.*

Then there exists $\mathbf{x}^* \in S$ such that $\mathbf{x}^* \in \mathcal{B}(\mathbf{x}^*)$.

Proof: See [Bohnenblust, H. F. and Karlin, S., 1950, Theorem 5]. □

With these assumptions we are able to establish the following theorem.

Theorem 4.3 *Let $A \subset [t_0, t_1] \times \mathbb{R}^n$; $U : A \to 2^{\mathbb{R}^m} \setminus \emptyset$; M, $f : M \to \mathbb{R}^n$; $g_j : A \times \mathbb{R}^{m_j}$, $j = 1, 2, \ldots p$ satisfy the general hypotheses outlined in section 2. Further let $r = (r_1, r_j \ldots r_p)$ be a positive set of weights; that the growth condition (γ) holds; that the sets $S(t,x)$, $(t,x) \in A$, are closed and convex; and that the sets $Q(t,x,y)$, $(t,x,y) \in \tilde{A}$,* *are convex and enjoy property (K) with respect to y. Finally assume that there exists a Lebesgue integrable function $\phi : [t_0, t_1] \to \mathbb{R}$ such that*

$$G_r(t, x, y, v) \leq \phi(t)$$

holds for almost all $t \in [t_0, t_1]$ $(t, x, y, v) \in D$, where

$$D = \{(t, x', y', v') : \quad (t, x') \text{ and } (t, y') \in A, v' \in U(t, y')\}.$$

Then there exists an open-loop Nash equilbrium for the differential game.

Proof: Our goal is to apply the above fixed point theorem to the set-valued mapping Γ defined in section 3. To do this we must show that $\Gamma(x)$ is nonempty and convex for each admissible trajectory, that Γ has a weakly closed graph, and that the union of all the sets $\Gamma(x)$ is contained in a sequentially weakly compact set. For brevity we indicate only which hypotyheses are required to show each of these hold. First, the conditions given above are sufficient to insure that for each fixed admissible trajectory x the optimization problem

$$\min_{\{y,v\} \in \Omega} J[x, r](t_0, t_1, y, v)$$

has a solution. This is easily seen by appealing to the known existence theorems found in [Cesari, L., 1983] because the growth condition (γ) holds, and for fixed x the sets $Q(t, x(t), y)$ enjoy property (K) with respect to y. Thus $\Gamma(x)$ is non-empty. Further, the convexity and closedness of the sets $S(t, x)$ are sufficient to insure that $\Gamma(x)$ is a convex set with a weakly closed graph. This is a consequence of standard closure theorems provided in [Cesari, L., 1983]. Finally we observe that because of the integral bound on G_r, the set of all admissible trajectories is relatively weakly compact by Theorem 4.1. Thus we have demonstrated that $\Gamma(\cdot)$ enjoys all of the conditions required to apply Theorem 4.2 and so there exists an admissible trajectory $x^* : [t_0, t_1] \to \mathbb{R}^n$ satisfying $x^* \in \Gamma(x^*)$. The desired conclusion now follows through an application of Proposition 3.1. □

5 Conclusions

In the above presentation we gave explicit conditions under which a class of differential games has an open-loop Nash equilibrium. The assumptions made in establishing these results were described in terms of the control structure and these assumptions we conditions approaching those found in the existence theory of optimal control. The most stringent hypothesis made in the above presentation was, in the opinion of this author, concerns the convexity of the sets $S(t, x)$. This assumption is a convexity condition that is stronger than usually imposed in the existence theory of optimal control and it is made solely for the purpose of applying the fixed point theorem found in [Bohnenblust, H. F. and Karlin, S., 1950]. Consequently one direction for further research is to weaken this convexity condition. Results in this direction are under preparation by this author. Another avenue of pursuit is to consider the question of feedback or closed-loop equilibrium for dynamic games using this framework. General conditions for such equilibrium in a p-player ($p > 2$) game are apparently unknown.

References

[Başar, T. and Olsder, G., 1982] *Dynamic Noncooperative Game Theory*, Academic Press, London.

[Bohnenblust, H. F. and Karlin, S., 1950] On a theorem of Ville. In Kuhn, H. W. and Tucker, A. W., editors, *Contributions to the Theory of Games, Vol. 1*, pages 155–160. Princeton University Press, Princeton, New Jersey.

[Carlson, D. A., 1998] Uniqueness in open-loop dynamic games. preprint.

[Cesari, L., 1983] *Optimization-Theory and Applications: Problems with Ordinary Differential Equations*, volume 17 of *Applications of Applied Mathematics*. Springer-Verlag, New York.

[Friedman, J. W., 1990] *Game theory with applications to economics*, Oxford University Press, New York.

[Rosen, J., 1965] Existence and uniqueness of equilibrium points for concave n-person games. *Econometrica*, 33(3):520–534.

SILVIA C DI MARCO AND ROBERTO L V GONZALEZ
Relaxed minimax optimal control problems. Infinite horizon case

1 Introduction

1.1 General discussion

In this work, we consider some issues concerning the minimax optimal control problem with continuous time and infinite horizon.

The minimax optimal control problem has been analyzed in recent publications from different points of view. For the theoretical analysis of the finite horizon problem (*fhp*), we can see [6] and [7, 10]; the numerical analysis of the *fhp* has been done in [12, 13]. Extensions has been considered in [1, 2, 3, 8, 9, 17]. The open loop necessary conditions of optimality has been presented in [4].

In the infinite horizon problem (*ihp*), the analysis of the optimal cost function and its approximate computation present considerable difficulties. Particularly, characterizing it through the Hamilton-Jacobi-Bellman (HJB) equation requires a careful treatment because the optimal cost function has poor properties of regularity and, even in its integral form, the HJB equation has not unique solution. In [11, 14, 15] we have considered the *ihp* and we have analyzed some properties of the optimal cost function, among them, the issue of regularity and the approximation with finite horizon problems. We have also proved in [14] that the optimal cost function is characterized as the lowest supersolution and as the maximum element of a special class of subsolutions of the associated HJB equation. In this work, other issues concerning the infinite horizon case are analyzed.

To be more specific, let us describe briefly the problem under analysis. We consider a dynamic system which evolves according to the ordinary differential equation

$$\begin{cases} \dfrac{dy}{dt}(t) = g(y(t), \alpha(t)) \quad \text{a.e. } t \in [0, \infty), \\ y(0) = x, \qquad \qquad x \in \Omega \subset \mathbb{R}^m, \ \Omega \text{ an open domain.} \end{cases} \quad (1)$$

The admissible control functions set is $\mathcal{A} = L^\infty([0,\infty); A)$, where $A \subset \mathbb{R}^\nu$.

The problem consists in minimizing (in the set $\mathcal{A}$) the functional

$$J(x, \alpha(\cdot)) = \operatorname{ess\,sup}\{f(y(t), \alpha(t)) : t \in [0, \infty)\}.$$

Here, we want to study the optimal cost function

$$u(x) = \inf\{J(x, \alpha(\cdot)) : \alpha(\cdot) \in \mathcal{A}\}, \quad (2)$$

and to obtain certain results related to its approximation with finite horizon problems. So, let us denote u_T the optimal cost function corresponding to the finite horizon problem and
$$\underline{u} = \lim_{T \to \infty} u_T. \tag{3}$$

In [14] we have proved that $\underline{u} \le u$ and that in general there exists a gap between u and $\underline{u}$.

Several questions arise from this phenomenon, for instance, *how $\underline{u}$ can be interpreted? under which conditions the equality $\underline{u} = u$ holds?*

In this paper we deal with these issues in the following form:

(a) In section 2 we describe a relaxation procedure for the controls. The relaxation procedure is the same one presented in [5, 7]. In that work, Barron and Jensen have proved that in the case of finite horizon, the original control problem and the relaxed control problem has the same optimal cost, i.e. $u_T^r = u_T$.

(b) We show that the function $\underline{u}$ can be interpreted as the optimal cost corresponding to the minimax optimal control problem where the relaxed controls mentioned above are used. Let us denote this cost with u^r. We prove that $\underline{u} = u^r$. Besides, we show the existence of optimal relaxed control for the *ihp* and that $\lim_{T \to \infty} u_T^r = u^r$. We prove this result using properties of weak-$\star$ compactness of $L^\infty([0,T]; \mathcal{M}(A))$.

(c) In section 3, we prove the absence of gap between $\underline{u}$ and u when suitable conditions of convexity of some subset of "velocities" of the system hold, more specifically, when
$$\mathcal{L}(x, \gamma) = \{\eta \in \mathbb{R}^m : \exists\, a \in A\,,\, \eta = g(x,a), f(x,a) \le \gamma\}$$
is convex for each $(x, \gamma) \in \mathbb{R}^{m+1}$.

(d) We show that the existence of an optimal control does not imply, by itself, the property $\underline{u} = u$.

(e) We also show that when the convexity hypothesis concerning $\mathcal{L}(x, \gamma)$ is released, it is possible to find examples where $\underline{u} < u$.

1.2 Technical issues

1.2.1 General assumptions

Let $BUC(\Omega \times A)$ be the family of bounded and uniformly continuous on $\Omega \times A$ and Lips(Ω) the set of uniform Lipschitz functions on Ω. We assume that the following hypotheses are satisfied

(H$_1$) $g : \Omega \times A \mapsto \mathbb{R}^m$, $g \in BUC(\Omega \times A)$, $\|g(x,a)\| \leq M_g$; $g(\cdot, a) \in \text{Lips}(\Omega)$, $\forall\, a \in A$, with Lipschitz constant L_g.

(H$_2$) $f : \Omega \times A \mapsto \mathbb{R}$, $f \in C(\Omega \times A)$.

(H$_3$) The control set A is compact in $\mathbb{R}^\nu$.

(H$_4$) The trajectory $y(t)$ remains in Ω, $\forall\, t \in [0, \infty)$, $\forall\, \alpha \in \mathcal{A}$.

Note 1.1 To get the last hypothesis, we will suppose that the following sufficient condition holds:
$$\langle g(x,a), \eta(x) \rangle < 0, \quad \forall\, x \in \partial\Omega, \ \forall\, a \in A,$$
being $\eta(x)$ the exterior normal vector to $\partial\Omega$ in x.

1.2.2 Auxiliary definitions and results

We give here some definitions and results which will be used in the next sections.

Let $\mathcal{M}(A)$ be the set of probability measures on A and let $M(A)$ be the bounded measures on A endowed with the weak-$\star$ topology of $C(A)'$. The following properties hold.

Lemma 1.1 *The set $\mathcal{M}(A)$ is weakly-$\star$ compact.*

Theorem 1.1 *(Dunford-Pettis Theorem) There exists an algebraic isomorphism between $L^\infty([0,\infty); M(A))$ and $[L^1([0,\infty); C(A))]^*$.*

The proofs of Lemma 1.1 and Theorem 1.1 are given in [18] pp. 265 and pp. 268 respectively.

2 Relaxed problem

2.1 Relaxation procedure

Let the space of relaxed controls be given by $\widehat{\mathcal{A}} = L^\infty([0,\infty); \mathcal{M}(A))$, endowed with the weak-$\star$ topology (the topology induced by $L^1([0,\infty); C(A))$). In other words, $\mu \in \widehat{\mathcal{A}}$ iff $\mu(\tau)$ is a probability measure a.e. $\tau \in [0,\infty)$ and the mapping $t \mapsto \mu(t)$ is measurable.

For any $\mu \in \widehat{\mathcal{A}}$, we consider the relaxed trajectories
$$\hat{y}(\tau) = x + \int_0^\tau \int_A g(\hat{y}(s), a)\, \mu(s, da)\, ds. \tag{4}$$

In consequence, we have that a.e. $t \in [0, \infty)$,

$$\frac{d\hat{y}}{dt}(t) = \hat{g}(\hat{y}(t), \mu(t)), \tag{5}$$

where the relaxed dynamic is given by $\hat{g} : \Omega \times \mathcal{M}(A) \mapsto \mathbb{R}^m$

$$\hat{g}(x, \mu) = \int_A g(x, a) \mu(da). \tag{6}$$

We also define the instantaneous cost, $\hat{f} : \Omega \times \mathcal{M}(A) \mapsto \mathbb{R}$,

$$\hat{f}(x, \mu) = \|f(x, \cdot)\|_{L^\infty(A, \mu)}, \tag{7}$$

where $L^\infty(A, \mu)$ denotes the space of real valued functions with domain A, essentially bounded with respect to the measure $\mu \in \mathcal{M}(A)$.

The relaxed functional to be minimized is $\widehat{J} : \Omega \times \widehat{\mathcal{A}} \mapsto \mathbb{R}$

$$\widehat{J}(x, \mu(\cdot)) = \operatorname{ess\,sup}\left\{\hat{f}(\hat{y}(t), \mu(t)) : t \in [0, \infty)\right\}, \tag{8}$$

and the relaxed optimal cost is

$$u^r(x) = \inf\{\widehat{J}(x, \mu(\cdot)) : \mu \in \widehat{\mathcal{A}}\}. \tag{9}$$

Analogously, we define the finite horizon problem corresponding to this relaxation as follows. The space of relaxed controls is $\widehat{\mathcal{A}}_T = L^\infty([0, T]; \mathcal{M}(A))$. For any $\mu \in \widehat{\mathcal{A}}_T$, $\hat{g}$ and $\hat{f}$ are defined as in (6) and (7). The relaxed functional to be minimized is $\widehat{J}_T : \Omega \times \widehat{\mathcal{A}}_T \mapsto \mathbb{R}$

$$\widehat{J}_T(x, \mu(\cdot)) = \operatorname{ess\,sup}\left\{\hat{f}(\hat{y}(t), \mu(t)) : t \in [0, T)\right\}, \tag{10}$$

and the relaxed optimal cost is

$$u_T^r(x) = \inf\{\widehat{J}_T(x, \mu(\cdot)) : \mu \in \widehat{\mathcal{A}}_T\}. \tag{11}$$

From definitions (6), (7) and hypotheses H_1-H_2 for g and f (data of the original problem) it can be proved that

$$\hat{g} : \Omega \times \mathcal{M}(A) \mapsto \mathbb{R}^m, \; \hat{g}(\cdot, \cdot) \in BUC(\Omega \times \mathcal{M}(A)), \; \frac{\partial \hat{g}}{\partial x} \in L^\infty(\Omega \times \mathcal{M}(A); \mathbb{R}^{m \times m}),$$

$\hat{f} : \Omega \times \mathcal{M}(A) \mapsto \mathbb{R}$, $\hat{f}(\cdot, \mu) \in C(\Omega)$, $\forall \mu \in \mathcal{M}(A)$, $\hat{f}(x, \cdot)$ is sequentially weakly lower semicontinuous in $\mathcal{M}(A)$.

Barron and Jensen have proved in [7] that this relaxation is appropriate for the minimax finite horizon problem in the sense that there exists an optimal control and the relaxed problem has the same optimal cost that the original one.

2.2 Convergence of the finite horizon relaxed problems

The relaxation presented in the previous subsection and used by Barron and Jensen in the *fhp* will be here used in the infinite horizon case. We will prove that $u_T^r \to u^r$. Since it is known that $u_T^r = u_T$ and we have defined $\underline{u} = \lim_{T \to \infty} u_T$, we will have an interpretation for $\underline{u}$: it is the cost of the infinite horizon relaxed problem.

Theorem 2.1 u_T^r *converges to* u^r *when* $T \to \infty$. *There exists an optimal control for the relaxed* ihp.

We give a proof which essentially uses the compactness of the set $L^\infty([0,T); \mathcal{M}(A))$ and properties of the weak convergence of measures of $\mathcal{M}(A)$.

Proof. Since $u_T^r = u_T$ (see [7]) and from definition (3), $\lim_{T \to \infty} u_T^r = \lim_{T \to \infty} u_T = \underline{u}$. In consequence, $\underline{u} \leq u^r$. To prove that $\underline{u} \geq u^r$, let $\mu_n \in \widehat{\mathcal{A}}$ be a sequence of minimizing controls for the relaxed infinite horizon problem; i.e. a sequence such that:

$$\widehat{J}(x, \mu_n(\cdot)) \leq \underline{u}(x) + \frac{1}{n}, \tag{12}$$

and let $y_n(\cdot) \in W^{1,\infty}([0,T); \mathbb{R}^m)$ be the trajectory associated to $\mu_n(\cdot)$.

From the hypothesis (H$_1$), $\{y_n(\cdot)\}_{n \in \mathbb{N}}$ are uniformly bounded and equi-Lipschitz continuous in any interval $[0,T]$. Then there exists a subsequence $\{y_n(\cdot)\}_{n \in \mathbb{N}}$ which converges to $y(\cdot)$ uniformly on any $[0,T]$.

In consequence, there exists $\eta(n) > 0$, such that $\lim_{n \to \infty} \eta(n) = 0$ and

$$\operatorname*{ess\,sup}_{t \in [0,T)} \hat{f}(y(t), \mu_n(t)) \leq \underline{u}(x) + \frac{1}{n} + \eta(n). \tag{13}$$

Moreover, since $\mathcal{A}_T$ is sequentially weakly-$\star$ compact (see [18], pp. 272), there exists a subsequence of $\mu_n(\cdot)$ and $\overline{\mu}(\cdot) \in \mathcal{A}_T$ such that μ_n converges weakly to $\overline{\mu}$. By the diagonal procedure, there exists a subsequence of $\mu_n(\cdot)$ and $\overline{\mu}(\cdot) \in \mathcal{A}$ such that μ_n converges to $\overline{\mu}$ in any interval $[0,T]$. From (4), we have that $y(\cdot)$ is the trajectory associated to the control $\overline{\mu}$.

Let us prove that

$$\operatorname*{ess\,sup}_{t \in [0,T)} \hat{f}(y(t), \overline{\mu}(t)) \leq \underline{u}(x). \tag{14}$$

We denote $\phi(t) = \hat{f}(y(t), \overline{\mu}(t))$ and we assume that (14) does not occur. Then there exists $\varepsilon > 0$ and $T_\varepsilon \subseteq [0,T]$ such that $m(T_\varepsilon) > \varepsilon$ ($m(\cdot)$ is Lebesgue measure) and $\phi(t) \geq \underline{u}(x) + \varepsilon$, $\forall\, t \in T_\varepsilon$. This inequality and the definition of $\hat{f}$ implies that there exists $A_{t,\varepsilon} \subseteq A$ with $\overline{\mu}(t, A_{t,\varepsilon}) > 0$ such that

$$f(y(t), a) \geq \underline{u}(x) + \varepsilon, \quad \forall\, a \in A_{t,\varepsilon},\ \forall\, t \in T_\varepsilon. \tag{15}$$

By defining

$$\Psi(s) = \begin{cases} 0 & \text{if } s \in [0, \underline{u}(x) + \varepsilon/2] \\ 1 & \text{if } s \in [\underline{u}(x) + \varepsilon, \infty) \\ \dfrac{s - \underline{u}(x) - \varepsilon/2}{\varepsilon/2} & \text{if } s \in (\underline{u}(x) + \varepsilon/2, \underline{u}(x) + \varepsilon), \end{cases}$$

from (15) we have that $\Psi(f(y(t), a)) = 1, \ \forall \, a \in A_{t,\varepsilon}, \ \forall \, t \in T_\varepsilon$. Then,

$$\xi(t) = \int_A \Psi(f(y(t),a))\overline{\mu}(t,da) \geq \int_{A_{t,\varepsilon}} \Psi(f(y(t),a))\overline{\mu}(t,da) = \overline{\mu}(t, A_{t,\varepsilon}) > 0.$$

It is clear from the previous statement that

$$\int_{[0,T]} \xi(t) dt \geq \int_{T_\varepsilon} \xi(t) dt > 0. \tag{16}$$

By the weak convergence of μ_n, we have

$$\int_{[0,T]} \int_A \Psi(f(y(t),a))\overline{\mu}(t,da) dt = \lim_{n \to \infty} \int_{[0,T]} \int_A \Psi(f(y(t),a))\mu_n(t,da) dt.$$

From (13), we have $\forall \, n \geq n_\varepsilon$ that $\hat{f}(y(t), \mu_n(t)) \leq \underline{u}(x) + \varepsilon/2$. Thus,

$$f(y(t), a) \leq \underline{u}(x) + \varepsilon/2 \ \forall \, a \in A \setminus E_t,$$

with $\mu_n(t)(E_t) = 0$. In consequence,

$$\int_{[0,T]} \int_A \Psi(f(y(t),a)) \mu_n(t, da) dt = 0$$

and then, $\int_{[0,T]} \xi(t) dt = 0$ which contradicts (16). Then, (14) holds. As T is arbitrary we have

$$\operatorname*{ess\,sup}_{t \in [0,\infty)} \hat{f}(y(t), \overline{\mu}(t)) \leq \underline{u}(x). \tag{17}$$

And so, the obtained control $\overline{\mu}$ is optimal and $u^r = \underline{u}$. □

Remark 2.1 Since, in general there is a gap between $\underline{u}$ and u, so is there between u^r and u. In other words, for the infinite horizon problem the optimal cost with ordinary or relaxed controls are not necessarily equal.

3 Sufficient condition for inexistence of gap

As it was proved in [11, 14], it is valid that $\underline{u} \leq u$. In that work it was shown that there are examples where $\underline{u} < u$ (see also Example 3.2 below). Since $\underline{u}$ is more regular than u (in particular, $\underline{u}$ is lower semicontinuous), it is important to give sufficient conditions which ensures that $\underline{u} = u$. This property allows us to study the problem as the limit of a sequence of finite horizon problems, procedure which is crucial for the analysis of the numerical approximation of the function u.

3.1 A sufficient convexity condition

In the following theorem, we will prove that there are conditions of convexity type which ensure that $\underline{u} = u$. Specifically, we have:

Theorem 3.1 *Let*

$$\mathcal{L}(x,\gamma) = \{\eta \in \mathbb{R}^m : \exists\, a \in A \text{ such that } \eta = g(x,a), f(x,a) \leq \gamma\}$$

be convex and closed for each $(x,\gamma) \in \mathbb{R}^m \times \mathbb{R}$, then there exists optimal control for the infinite horizon problem and $\underline{u} = u$.

The proof of this theorem is based on the Fillipov's lemma (see [16]).

3.2 The existence of optimal control does not imply $u_T \to u$

The convexity of $\mathcal{L}(x,\gamma)$ implies that there exists optimal control and that there does not exist gap between u and $\underline{u}$. However, the existence of optimal control is not enough to ensure that $u_T \to u$, as we show in the following example.

Example 3.1 We consider a system which evolves in $\mathbb{R}^3$ according to the following differential equation

$$\begin{cases} x_1' = x_1(1-x_1^2), \\ x_2' = a, \\ x_3' = (x_2^2 + x_3)(1-x_3^2). \end{cases} \quad a \in A, \tag{18}$$

The set A is given by $A = \{-1, 0, 1\}$ and the instantaneous cost function is f:

$$f(x,a) = \begin{vmatrix} (-|x_1(t)| + 888\,(x_3^2(t) - 2\,x_1^2(t)))^+, & \text{if } a = -1, 1, \\ 8, & \text{if } a = 0. \end{vmatrix} \tag{19}$$

Let $(x_1(0), x_2(0), x_3(0)) = (0,0,0)$ the initial condition of the system. It is easy to see that for this initial condition, $a(t) \equiv 0$ is an optimal control and so, $u(0,0,0) = 8$. Let us consider the finite horizon problem. We introduce the integer parameter ν, and we consider the control subset given by

$$\{\alpha_\nu(\cdot) \in \mathcal{A} : \alpha_\nu(s) = (-1)^i \text{ if } s \in [i/\nu, (i+1)/\nu), i \in \mathbb{N}_\circ\}.$$

By properties of $x_2(\cdot)$ and $x_3(\cdot)$, it easy to check that $J_T(x, \alpha_\nu(\cdot)) \to 0$ when $\nu \to \infty$. Consequently, $u_T(0,0,0) = 0 \ \forall T$ and so $\underline{u}(0,0,0) = \lim_{T\to\infty} u_T(0,0,0) = 0 \neq 8 = u(0,0,0)$.

3.3 The role of the convexity hypothesis

The lack of convexity of the set $\mathcal{L}(x,\gamma)$ may produce the loss of the convergence of the sequence $\{u_T\}_{T>0}$ to u, as it is shown in the following example.

Example 3.2 *A problem where* $\lim_{T\to\infty} u_T \neq u$. *Let the dynamic of the system be given by*

$$\begin{cases} x_1' = a, & a \in A = \{-1,1\}, \\ x_2' = (1 - x_2^2)(x_2 + x_1^2), \\ (x_1(0), x_2(0)) = x, & x \in \mathbb{R}^2. \end{cases} \quad (20)$$

We define the instantaneous cost $f(x_1(t), x_2(t), \alpha(t)) = x_2(t)$ and the control set $\mathcal{A} = L^\infty((0,\infty); \{-1,1\})$.

Obviously, the set $\mathcal{L}(x,\gamma) = \{\eta \in \mathbb{R}^2 : \exists a \in \{-1,1\} \text{ such that } \eta = g(x,a), f(x,a) \leq \gamma\}$ is not convex.

We consider the initial condition $x = (0,0)$. From (20), $x_2(\cdot)$ is a non-decreasing function which verifies $\lim_{t\to\infty} x_2(t) = 1$. Therefore, $\forall\, \alpha(\cdot) \in \mathcal{A}$ it results $J(0,\alpha(\cdot)) = 1$, which implies that $u(0) = 1$. By considering the finite horizon problem and a subset of controls similar to those used in the previous example, it easy to check that

$$J_T(x, \alpha_\nu(\cdot)) = \max_{t\in[0,T]} x_{\nu,2}(t) = x_{\nu,2}(T) \to 0$$

when $\nu \to \infty$. Then, $u_T(0,0) = 0$ and, so $\underline{u}(0,0) = \lim_{T\to\infty} u_T(0,0) = 0 \neq 1 = u(0,0)$.

References

[1] Barles G., Daher Ch., Romano M., *Optimal control on the L^∞ norm of a diffusion process*, SIAM J. on Control and Optimization, Vol. 32, N°3, pp. 612-634, 1994.

[2] Barles G., Daher Ch., Romano M., *Convergence of numerical schemes for parabolic equations arising in finance theory*, Math. Models and Methods in Applied Sc., Vol. 5, N°1, pp. 125-143, 1995.

[3] Barron E.N., *Differential games with maximum cost*, Nonlinear Analysis, T. M. & A., Vol. 14, N°11, pp. 971-989, 1990.

[4] Barron E.N., *The Pontryagin maximum principle for minimax problems of optimal control*, Nonlinear Analysis, T. M. & A., Vol. 15, N°12, pp. 1155-1165, 1990.

[5] Barron E.N., *Optimal control and calculus of variations in L^∞*, in Optimal control in differential equations, N.H. Pavel (Ed.), Marcel Dekker, New York, 1994.

[6] Barron E.N., Ishii H., *The Bellman equation for minimizing the maximum cost*, Nonlinear Analysis, T. M. & A., Vol. 13, N°9, pp. 1067-1090, 1989.

[7] Barron E.N., Jensen R., *Relaxed minimax control*, SIAM J. Control and Optimization, Vol. 33, N°4, pp. 1028-1039, 1995.

[8] Barron E.N., Jensen R., *Relaxation of constrained control problems*, SIAM J. Control and Optimization, 1996.

[9] Barron E.N., Jensen R., Menaldi J.L., *Optimal control and differential games with measures*, Nonlinear Analysis, T. M. & A., Vol. 21, N°4, pp. 241-268, 1993.

[10] Barron E.N., Liu W., *Semicontinuous solutions for Hamilton-Jacobi equations and the L^∞ control problem*, Applied Math. Optim., Vol 34, pp. 325-360, 1996.

[11] Di Marco S.C., *Sobre la optimización minimax y tiempos de detención óptimos*, Thesis, University of Rosario, Argentine, 1996.

[12] Di Marco S.C., González R.L.V., *A numerical procedure for minimizing the maximum cost*, in System modelling and optimization, J. Dolezal and J. Fidler (Eds.), Chapman & Hall, Chap. 33, pp. 285-291, 1996, (extended version: Rapport de Recherche N°2454, INRIA, Rocquencourt, 1995).

[13] Di Marco S.C., González R.L.V., *Une procédure numérique pour la minimisation du coût maximum*, Comptes Rendus Acad. Sc. Paris, Série I, Tome 321, pp. 869-874, 1995.

[14] Di Marco S.C., González R.L.V., *A minimax optimal control problem with infinite horizon*, Rapport de Recherche N°2945, INRIA, Rocquencourt, 1996.

[15] Di Marco S.C., González R.L.V., *A finite state stochastic minimax optimal control problem with infinite horizon*, in Numerical Analysis and Its Applications, L. Vulkov, J. Waśniewski, P. Yalamov (Eds.), Lectures Notes of Computer Science N°1196, pp. 134-141, Springer-Verlag, Berlin, 1997.

[16] Friedman A., *Differential games*, Wiley-Interscience, New York, 1971.

[17] Motta M., Rampazzo F., *Dynamic programming for nonlinear systems driven by ordinary and impulsive controls*, SIAM J. Control and Optimization, Vol. 34, N°1, pp. 199-225, 1996.

[18] Warga J., *Optimal control of differential and functional equations*, Academic Press, New York, 1972.

CONICET – Inst. Beppo Levi, Dpto. Matemática, FCEIA,
Universidad Nacional de Rosario, Argentine.

M d R DE PINHO[1] AND HARRY ZHENG
Conjugate points for optimal control problems with state-dependent control constraints

1 Introduction

Consider the following optimal control problem:

$$(P) \begin{cases} \text{Minimise } J(x,u) = l(x(b)) + \int_a^b L(t,x(t),u(t))dt \\ \text{subject to} \\ \dot{x}(t) = f(t,x(t),u(t)) \quad \text{a.e.} \quad t \in [a,b] \\ 0 = b(t,x(t),u(t)) \quad \text{a.e.} \quad t \in [a,b] \\ 0 \geq g(t,x(t),u(t)) \quad \text{a.e.} \quad t \in [a,b] \\ x(a) = 0 \\ h(x(b)) = 0 \end{cases}$$

where $l : \Re^n \to \Re$, $L : [a,b] \times \Re^n \times \Re^k \to \Re$, $f : [a,b] \times \Re^n \times \Re^k \to \Re^n$, $b : [a,b] \times \Re^n \times \Re^k \to \Re^{m_1}$, $g : [a,b] \times \Re^n \times \Re^k \to \Re^{m_2}$ and $h : \Re^n \to \Re^r$ with $k \geq m_1 + m_2$ and $r \leq n$.

(P) is an optimal control problem involving equality and inequality state dependent control constraints, also known as mixed state-control constraints. For (P) we seek second-order necessary conditions for a weak minimiser. We define a "generalised conjugate point" for (P) and we assert that the nonexistence of such points is a necessary condition of optimality. As it is commonly done in such cases we associate with (P) an auxiliary problem, the so-called "accessory problem", which is to minimise a second variation $J_2(v)$ over all solution (y,v) for some linearised system. If (x,u) solves the original problem we show that the optimal solution of the accessory problem produces a nonnegative cost. Then second-order necessary conditions of optimality to (P) are derived.

Results on necessary conditions in terms of conjugate points are well established in the classical calculus of variation. In the last decades such results have been successfully transferred to the optimal control setting. In this context it is of relevance, among others, the work of Zeidan and Zezza ([4]) and Loewen and Zheng ([1]), both treating standard optimal control problems. Optimal control problems with inequality state-dependent control constraints were then covered by Zeidan in [5]. More recently problems with equality constraints were treated by Stefani and Zezza in [3].

Here we consider optimal control problems with both equality and inequality constraints. In the absence of equality constraints our results are comparable to Zeidan's in [5]. In fact, as in [5] we assume strong normality and our accessory problem reduces

---
[1]This author was supported by J.N.I.C.T. Project PICT/CEG/2438/95, Portugal.

to the one the author defines. However our definition of conjugate point is different. We extend the notion of "generalised conjugate point", introduced by Loewen and Zheng in [1], to cover problems with state-dependent control constraints. Based on our definition of conjugate point we obtain second-order necessary condition of optimality that may detect nonoptimal extremals whereas Zeidan's do not as we illustrate through an example.

2 Preliminaries

For (P) a process is a pair $(x, u) \in W^{1,1}([a, b]; \Re^n) \times L^\infty([a, b]; \Re^n)$ satisfying

$$\begin{aligned}\dot{x}(t) &= f(t, x(t), u(t)) \quad a.e. \quad t \in [a, b] \\ 0 &= b(t, x(t), u(t)) \quad a.e. \quad t \in [a, b] \\ 0 &\geq g(t, x(t), u(t)) \quad a.e. \quad t \in [a, b]\end{aligned}$$

A process is *admissible* if it satisfies the end point constraints.

An admissible process $(\bar{x}, \bar{u})$ for (P) is a *weak local minimiser* if for some $\epsilon > 0$, $(\bar{x}, \bar{u})$ minimises $J(x, u)$ over all admissible processes (x, u) satisfying $\| x - \bar{x} \|_\infty < \epsilon$ and $\| u - \bar{u} \|_\infty < \epsilon$.

Let $I(t)$ be a set of indexes defined by

$$I(t) = \{i \in \{1, \ldots, m_2\} : g_i(t, \bar{x}(t), \bar{u}(t)) = 0\} \tag{2.1}$$

$\bar{\phi}(t)$ will denote the evaluation of a function ϕ at $(t, \bar{x}(t), \bar{u}(t))$, whereas ϕ may be f, b, g, H or its derivatives.

Let $\bar{g}_u^{I(t)}(t)$ denote the matrix we get after removing from $\bar{g}_u(t)$ all the rows of index $i \in \{1, \ldots, m_2\}$ such that $i \notin I(t)$. $\bar{g}_x^{I(t)}(t)$ is similarly defined.

We shall invoke the following hypotheses on (P), which refers to some process $(\bar{x}(\cdot), \bar{u}(\cdot))$ and parameter $\epsilon > 0$:

H1. $L(\cdot, x, u)$, $f(\cdot, x, u)$ and $\tilde{b}(\cdot, x, u)$, where $\tilde{b} = (b, g)$, are measurable for each (x, u).

For almost every $t \in [a, b]$, $f(t, \cdot, \cdot)$, $L(t, \cdot, \cdot)$ and $\tilde{b}(t, \cdot, \cdot)$ are twice continuously differentiable on $(\bar{x}(t), \bar{u}(t)) + \epsilon B$ and their derivatives are integrably bounded at $(t, \bar{x}(t), \bar{u}(t))$.

There exists $k_{\tilde{b}} > 0$ such that for almost every $t \in [a, b]$

$$|\tilde{b}_x(t, \bar{x}(t), \bar{u}(t))| + |\tilde{b}_u(t, \bar{x}(t), \bar{u}(t))| \leq k_{\tilde{b}}.$$

h an l are C^2 on $\{x : | x - \bar{x}(b) | < \epsilon\}$, and $h'(\bar{x}(b))$ (the Jacobian of h) is of full rank.

H2. There exists $K > 0$ such that for almost every $t \in [a, b]$

$$\det \{\Upsilon(t)\Upsilon^*(t)\} \geq K$$

where $\Upsilon(t) = \begin{bmatrix} b_u(t, \bar{x}(t), \bar{u}(t)) \\ g_u^{I(t)}(t, \bar{x}(t), \bar{u}(t)) \end{bmatrix}$

Define the Hamiltonian:

$$H(t, x, p, q, r, u) = p \cdot f(t, x, u) + q \cdot b(t, x, u) + r \cdot g(t, x, u) - \lambda L(t, x, u) \quad (2.2)$$

Theorem 2.1 *(Weak Maximum Principle [2]) Let $(\bar{x}(\cdot), \bar{u}(\cdot))$ be a weak local minimiser for (P). Assume that, for some $\epsilon > 0$, hypotheses H1 and H2 are satisfied. Then there exist $p \in W^{1,1}([a, b]; \Re^n)$ and $\lambda \geq 0$, not both zero, such that:*

(i) $-\dot{p}(t) = H_x(t, \bar{x}(t), p(t), q(t), r(t), \bar{u}(t))$

(ii) $0 = H_u(t, \bar{x}(t), p(t), q(t), r(t), \bar{u}(t))$

(iii) $r(t) \cdot \bar{g}(t) = 0$ *a.e.* $t \in [a, b]$ *and* $r(t) \leq 0$.

(iv) $-p(b) = \lambda \nabla l(\bar{x}(b)) + h'(\bar{x}(b))^* \mu$ *for some* $\mu \in \Re^r$.

Furthermore

$$\begin{pmatrix} q(t) \\ r(t) \end{pmatrix} = - \begin{bmatrix} \bar{b}_u(t)\bar{b}_u^*(t) & \bar{b}_u(t)\bar{g}_u^*(t) \\ \bar{g}_u(t)\bar{b}_u^*(t) & \bar{g}_u(t)\bar{g}_u^*(t) + I \end{bmatrix}^{-1} \begin{bmatrix} \bar{b}_u(t) \\ \bar{g}_u(t) \end{bmatrix} \left(\bar{f}_u^*(t)p(t) + \lambda \bar{L}_u^*(t)\right) \quad (2.3)$$

3 Second Order Necessary Conditions

Consider problem (P). An admissible process for (P), (x, u), is called an *extremal* if it satisfies H1 and H2 and there exist multipliers, i.e., an arc p and a scalar $\lambda \geq 0$ such that conclusions (i) through (iv) of Theorem 2.1 together with (2.3) hold. For any extremal (x, u) we define the set of multipliers as $\Lambda = (p(\cdot), q(\cdot), r(\cdot), \lambda, \mu)$.

An extremal is called a *normal* extremal if $\lambda = 1$. It is called *strongly normal* if the only solution to the system:

$$\begin{aligned} -\dot{p}(t) &= H_x(t, \bar{x}(t), p(t), q(t), r(t), \bar{u}(t)) \\ 0 &= H_u(t, \bar{x}(t), p(t), q(t), r(t), \bar{u}(t)) \\ 0 &= r(t) \cdot g(t, \bar{x}(t), \bar{u}(t)) \\ p(b) &= -h'(\bar{x}(b))^* \mu \end{aligned} \quad (3.1)$$

is $p(\cdot) \equiv 0$.

Suppose that $(\bar{x}, \bar{u})$ is a strongly normal extremal for (P) and consider the variational problem:

$$(AC) \begin{cases} \text{Minimise } J_2(v) := l_2(y(b)) + \int_a^b L_2(t, y(t), v(t))dt \\ \text{subject to} \\ \quad \dot{y}(t) = \bar{f}_x(t)y(t) + \bar{f}_u(t)v(t) \quad \text{a.e.} \quad t \in [a,b] \\ \quad 0 = \bar{b}_x(t)y(t) + \bar{b}_u(t)v(t) \quad \text{a.e.} \quad t \in [a,b] \\ \quad 0 = \bar{g}_x^{I(t)}(t)y(t) + \bar{g}_u^{I(t)}(t)v(t) \quad \text{a.e.} \quad t \in [a,b] \\ \quad y(a) = 0 \\ \quad h'(\bar{x}(b))y(b) = 0 \end{cases}$$

where:

$$\begin{aligned} l_2(y(b)) &= y(b)^* l''(\bar{x}(b))y(b) + y(b)^* [h''(\bar{x}(b))^* \mu] y(b) \\ L_2(t, y(t), v(t)) &= -\left(y(t)^* \bar{H}_{xx}(t) y(t) + 2y(t)^* \bar{H}_{xu}(t) v(t) + v(t)^* \bar{H}_{uu}(t) v(t) \right) \end{aligned}$$

Lemma 3.1 *Suppose that $(\bar{y}, \bar{v})$ is a weak local minimiser for (AC). There exist $\psi \in W^{1,1}([a,b]; \Re^n)$ and $\lambda \geq 0$, not both zero, such that:*

(i) $-\dot{\psi}(t) = \bar{f}_x(t)^* \psi(t) + \bar{b}_x(t)^* \zeta(t) + \bar{g}_x(t)^* \xi(t) - 2\lambda (\bar{H}_{xx}(t)\bar{y}(t) + \bar{H}_{xu}\bar{v}(t))$

(ii) $0 = \bar{f}_u(t)^* \psi(t) + \bar{b}_u(t)^* \zeta(t) + \bar{g}_u(t)^* \xi(t) - 2\lambda(\bar{H}_{xu}\bar{y}(t) + \bar{H}_{uu}\bar{v}(t))$

(iii) $\xi_i(t) = 0$ *for* $i \notin I(t)$.

(iv) $-\psi(b) = \lambda l_2'(\bar{y}(b)) + h'(\bar{x}(b))^* \nu$ *for some* $\nu \in \Re^r$.

Proof. (**An outline**)
Define $\delta(\cdot) \in L^\infty([a,b]; \Re_2^m)$ as $\delta_i(t) = 1$ if $i \in I(t)$ and $\delta_i(t) = 0$ if $i \notin I(t)$. Let $\Delta(t) \in M_{m_2 \times m_2}$ be the diagonal matrix $\Delta(t) = diag\{\delta_1(t), \ldots, \delta_{m_2}(t)\}$. Obviously $\Delta(\cdot)$ is a L^∞ function. Define another matrix $\Delta'(t) = I - \Delta(t)$ where $I \in M_{m_2 \times m_2}$ is the identity matrix. Finally let $w(\cdot)$ be a $L^\infty([a,b]; \Re^{m_2})$. It can be checked that $(\bar{y}, \bar{v}, 0)$ is a weak local minimiser to

$$(LQ^*) \begin{cases} \text{Minimise } J_2(v) := l_2(y(b)) + \int_a^b L_2(t, y(t), v(t))dt \\ \text{subject to} \\ \quad \dot{y}(t) = \bar{f}_x(t)y(t) + \bar{f}_u(t)v(t) \quad \text{a.e.} \\ \quad 0 = \bar{b}_x(t)y(t) + \bar{b}_u(t)v(t) \quad \text{a.e.} \\ \quad 0 = \Delta(t)\bar{g}_x(t)y(t) + \Delta(t)\bar{g}_u(t)v(t) + \Delta'(t)w(t) \quad \text{a.e.} \\ \quad y(a) = 0 \\ \quad h'(\bar{x}(b))y(b) = 0 \end{cases}$$

Applying Theorem 2.1 to (LQ^*) we get conditions of the required form. ∎

Lemma 3.2 *If (x, u) be a strongly normal extremal to (P), then any solution (y, v) to (AC) is normal.*

Lemma 3.1 and Lemma 3.2 imply:

Theorem 3.3 *Let $(\bar{x}, \bar{u})$ be a strongly normal extremal. Assume that conditions H1-H2 are satisfied. If $(\bar{x}, \bar{u})$ is a weak local minimiser to (P) then, for any solution (y, v) of (AC), we have*

$$J_2(v) = l_2(y(b)) + \int_a^b L_2(t, y(t), v(t))dt \geq 0$$

Proof. (**outline**) We start by pointing out that $(\bar{y}(\cdot) \equiv 0, \bar{v}(\cdot) \equiv 0)$ is an admissible solution for (AC) with $J_2(0) = 0$ and it is a normal extremal.

Suppose that there exists another extremal (y, v) with lower cost, i.e., $J_2(v) < 0$. Since any extremal of (AC) is normal, conditions (i)-(iv) of Lemma 3.1 are satisfied with $\lambda = 1$. After possibly rescaling of all multipliers we deduce that

$$J_2(v) = -y(a)^*\psi(a) < 0$$

This implies that $y(a) \neq 0$ contradicting the admissibility of (y, v). The proof is complete. ∎

4 Generalised Conjugate Points

Definition 4.1 *Let (x, u) be an extremal. A point $c \in (a, b)$ is a generalised conjugate point to b if there exists $(y, v, \psi, \xi, \zeta, \nu)$ on $[c, b]$ such that*

$$\begin{cases} \dot{y}(t) &= \bar{f}_x(t)y(t) + \bar{f}_u(t)v(t) \\ 0 &= \bar{b}_x(t)y(t) + \bar{b}_u(t)v(t) \\ 0 &= \bar{g}_x^{I(t)}(t)y(t) + \bar{g}_u^{I(t)}(t)v(t) \\ y(c) &= 0 \\ h'(\bar{x}(b))y(b) &= 0 \\ -\dot{\psi}(t) &= \bar{f}_x(t)^*\psi(t) + \bar{b}_x(t)^*\zeta(t) + \bar{g}_x(t)^*\xi(t) + \\ & \quad \bar{H}_{xx}(t)y(t) + \bar{H}_{xu}(t)v(t) \\ 0 &= \xi_i(t) \quad \text{if } i \notin I(t) \\ \psi(c) &\neq 0 \\ -\psi(b) &= g''(\bar{x}(b))y(b) + [h''(\bar{x}(b))^*\mu]y(b) + h'(\bar{x}(b))^*\nu \\ \gamma(t)^*v(t) &\geq 0 \end{cases}$$

where

$$\gamma(t) = \bar{f}_u(t)^*\psi(t) + \bar{b}_u(t)^*\zeta(t) + \bar{g}_u(t)^*\xi(t) + \bar{H}_{xu}(t)y(t) + \bar{H}_{uu}(t)v(t)$$

and in addition, either (a) or (b) below holds:

(a) $\gamma(t)^*v(t) > 0$ *on a set of positive measure.*

(b) there exit a control v_1 and an arc y_1 such that

$$\begin{cases} \dot{y}_1(t) &= \bar{f}_x(t)y_1(t) + \bar{f}_u(t)v_1(t) \\ 0 &= \bar{b}_x(t)y_1(t) + \bar{b}_u(t)v_1(t) \\ 0 &= \bar{g}_x^{I(t)}(t)y_1(t) + \bar{g}_u^{I(t)}(t)v_1(t) \\ y_1(a) &= 0 \\ y_1(c)^*\psi(c) &> 0 \\ h'(\bar{x}(b))y_1(b) &= 0 \\ \gamma(t)^*v_1(t) &\geq 0 \end{cases}$$

Theorem 4.2 *Let (x, u) be a normal extremal. If there is a generalised conjugate point to b on (a, b), then there exists a pair $(\bar{y}, \bar{v})$ admissible for (AC) such that $J_2(\bar{v}) < 0$.*

Proof. In what follows we present the main steps of the proof. For details see [1][Thm 4.3].

Let (x, u) be a normal extremal and suppose that $c \in (a, b)$ is a conjugate point. Consider (y, v, ψ, ξ, ν) of the definition and define:

$$\bar{y}(t) = \begin{cases} 0, & t \in [a, c) \\ y(t), & t \in [c, b] \end{cases} \quad \text{and} \quad \bar{v}(t) = \begin{cases} 0, & t \in [a, c) \\ v(t), & t \in [c, b] \end{cases}$$

Then $(\bar{y}, \bar{v})$ is admissible for (AC). Definition 4.1 implies:

$$\begin{aligned} J_2(\bar{v}) &\leq l_2(y(b)) + \int_c^b \left\{ y(t)^* \dot{\psi}(t) + y(t)^* \left[\bar{f}_x(t)^*\psi(t)\right] + \left[\bar{f}_u(t)^*\psi(t)\right]^* v(t) + \right. \\ &\quad y(t)^* \left[\bar{b}_x(t)^*\zeta(t)\right] + \left[\bar{b}_u(t)^*\zeta(t)\right]^* v(t) + \\ &\quad \left. y(t)^* \left[\bar{g}_x(t)^*\xi(t)\right] + \left[\bar{g}_u(t)^*\xi(t)\right]^* v(t) \right\} dt \\ &= l_2(y(b)) - \int_c^b \left\{ y(t)^* \left[\bar{H}_{xx}(t)y(t) + \bar{H}_{xu}(t)v(t)\right] \right. \\ &\quad \left. - \left[y(t)^*\bar{H}_{xu}(t) + v(t)^*\bar{H}_{uu}(t)\right] v(t) \right\} dt \\ &= l_2(y(b)) - y(b)^* g''(\bar{x}(b)) y(b) - y(b)^* [h''(\bar{x}(b))\lambda] y(b) - y(b)^* h'(\bar{x}(b))^* \nu \\ &= 0 \end{aligned} \quad (4.1)$$

If condition (a) of definition 4.1 holds then $J_2(\bar{v}) = J_2(v) < 0$ proving the theorem. On the other hand, if condition (b) holds define $v_\alpha = v_1 + \alpha \bar{v}$ and $y_\alpha = y_1 + \alpha \bar{y}$ where $\alpha > 0$. (y_α, v_α) is an admissible process of (AC). Then we can show that

$$J_2(v_\alpha) = J_2(v_1) - 2\alpha y_1(c)^* \psi(c)$$

The right side approaches $-\infty$ as $\alpha \to \infty$, so for all α sufficiently large, we have $J_2(v_\alpha) < 0$. ∎

Theorem 3.3 and 4.2 can now be combined to deduce that:

Theorem 4.3 *Let $(\bar{x}, \bar{u})$ be a strongly normal extremal. If $(\bar{x}, \bar{u})$ solves (P), then the interval (a, b) contains no generalised conjugate points to b.*

5 Example

Consider the following problem:

$$(C) \begin{cases} \text{Minimise } \int_0^{\pi/2}(x^2(t)-u_2^2(t))dt \\ \text{subject to} \\ \quad \dot{x}(t) = u_1 + u_2 \\ \quad 0 \geq x^2 - u_1 \\ \quad x(0) = 0 \\ \quad x(\pi/2) = 0 \end{cases}$$

Here the process $(x,u) = (0,0)$ is extremal but not optimal. The results in [4] do not detect this whereas ours do.

Let p, q and λ be multipliers associated with any pair (x,u) and

$$H(x,u,p,q,\lambda) = p \cdot (u_1 + u_2) + q \cdot (x^2 - u_1) - \lambda(x^2(t) - u_2^2(t))$$

Then

$$\begin{array}{lll} \bar{f}_x(t) = 0 & \bar{f}_u(t) = (1,1) & \bar{g}_x(t) = 2x \\ \bar{g}_u(t) = (-1,0) & \bar{H}_{xx}(t) = 2q - 2\lambda & \bar{H}_{xu}(t) = 0 \\ \bar{H}_{uu} = \begin{pmatrix} 0 & 0 \\ 0 & 2\lambda \end{pmatrix} & & \end{array}$$

Evidently $(x,p,q,u,\lambda) = (0,0,0,0,1)$ is a solution. So $(x,u) = (0,0)$ is an extremal. By definition 4.1, $c \in (0,\pi/2)$ is a conjugate point to $\pi/2$ if there exists (y,ψ,ξ,v) such that

$$\begin{cases} \dot{y}(t) = v_1(t) + v_2(t) \\ 0 = v_1(t) \\ y(c) = 0 \\ y(\pi/2) = 0 \\ \dot{\psi}(t) = 2y(t) \\ \psi(\pi/2) = -\nu \text{ for some } \nu \in \Re \\ \psi(c) \neq 0 \\ (\psi(t) - \xi(t))v_1(t) + (v_2(t)\psi(t) + 2v_2^2(t)) \geq 0 \end{cases}$$

Take any $c \in (0,\pi/2)$ and let $a = \frac{2\pi}{\pi-2c}$. Now define $y(t) = \sin(a(t-c))$, $v_1(t) = 0$ and $v_2(t) = a\cos(a(t-c))$. It is easy to check that

$$\left(y(t), \psi(t) = -\frac{2}{a}\cos(a(t-c)), \xi(t) \equiv 0, v_1(t), v_2(t), \nu = -\psi(\frac{\pi}{2})\right)$$

satisfies the above conditions. In particular,

$$(\psi(t) - \xi(t))v_1(t) + (v_2(t)\psi(t) + 2v_2^2(t)) = 2\cos^2(a(t-c))\left(a^2 - 1\right) \geq 0$$

for almost all $t \in [c, \pi/2]$. This means that any $c \in (0, \pi/2)$ is a generalised conjugate point. Theorem 4.2 implies that $(x, u) = (0, 0)$ is not optimal. In fact, if we let

$$\begin{cases} u_1(t) &= -t^2 + \frac{\pi}{2}t \\ u_2(t) &= 2t - \frac{\pi}{2} + t^2 - \frac{\pi}{2}t \end{cases}$$

then the corresponding trajectory is

$$x(t) = t^2 - \frac{\pi}{2}t$$

and $\int_0^{\pi/2}(x^2(t) - u_2^2(t))dt < 0$.

Let us now consider Zeidan's definition of conjugate point for C and $(x, u) = (0, 0)$. By Zeidan's definition $c \in [0, \psi/2)$ is a conjugate point to $\pi/2$ if there exists a nonzero (y, ψ, ξ, v) with $y \neq 0$ on $[0, c]$ and satisfying

$$\begin{cases} \dot{y}(t) &= v_1 + v_2 \\ 0 &= -v_1 \\ y(c) &= 0 \\ y(\pi/2) &= 0 \\ \dot{\psi}(t) &= 2y(t) \\ \psi(\pi/2) &= \nu \text{ for some } \nu \in \Re \\ \psi - \xi &= 0 \\ \psi + 2v_2 &= 0 \end{cases}$$

We then have

$$\begin{pmatrix} \dot{y} \\ \dot{\psi} \end{pmatrix} = \begin{bmatrix} 0 & -1/2 \\ 2 & 0 \end{bmatrix} \begin{pmatrix} y \\ \psi \end{pmatrix}$$

The solution of the above system is

$$\begin{pmatrix} y \\ \psi \end{pmatrix} = \begin{pmatrix} K_1 cos(t) - K_2 sin(t) \\ 2K_1 sin(t) + 2K_2 cos(t) \end{pmatrix}$$

We consider now the end point constraints. Since $y(\pi/2) = 0$ we have $K_2 = 0$ which implies that $y(t) = K_1 cos(t)$ and $y(c) = K_1 cos(c) = 0$. Since $c \in [0, \pi/2)$ we deduce that $y(t) = \psi(t) = 0$ for $t \in [0, c]$ for any $c \in [0, \pi/2]$. But this means that there is no conjugate point to $\pi/2$ along the extremal $(x, u_1, u_2) = (0, 0, 0)$.

References

[1] P. D. Loewen and H. Zheng, *Generalised Conjugate Points for Optimal Control Problems*, Nonlinear Analysis, Theory, Meth. and Applic., 22, 1994, pp. 771-791.

[2] M. d. R. de Pinho *A Maximum Principle for Optimal Control Problem with Equality and Inequality Constraints*, Proceeding of the 1997 IEEE CDC, San Diego, USA, 1997, pp.1403-1404.

[3] G. Stefani and P. L. Zezza, *Constrained Regular LQ-Control Problems*, SIAM J. Control and Optim., 35, 1997, pp.876-900.

[4] Vera Zeidan and P. L. Zezza, *Necessary Conditions for Optimal Control Problems: Conjugate Points*, SIAM J. Control and Optimization, 26, 1988, pp. 592-608.

[5] Vera Zeidan, *The Riccati Equation for Optimal Control Problems with Mixed State-Control Constraints: Necessity and Sufficiency*, SIAM J. Control and Optimization, 32, 1994, pp. 1297–1321.

JOHN NOBLE AND HEINZ SCHÄTTLER[1]
On sufficient conditions for a strong local minimum of broken extremals in optimal control

1. Introduction

Following classical ideas from field theory, we study sufficient conditions for strong optimality of broken extremals in optimal control theory by embedding the reference trajectory into a field of extremals and establishing the optimality of the corresponding flow over the region R which is covered by the field. It is well-known that the necessary conditions for optimality given in the Pontryagin Maximum Principle also give the characteristic equations for the Hamilton-Jacobi-Bellman equation. Thus, if the flow of extremals covers an open region R of the state-space diffeomorphically, then a smooth solution to the Hamilton-Jacobi-Bellman equation can be constructed on R (see, e.g. [3]). We construct local value-functions by adapting the method of characteristics to the optimal control problem. In earlier papers the construction below was already used for smooth extremals to analyze the value near fold and cusp-singularities [1, 2]. Here we extend this construction to broken extremals. Our aim is to establish a framework which can be used to derive simple and easily applicable criteria for local optimality (amongst all trajectories which lie in R). An immediate application is to bang-bang trajectories. Our results will be less general than for instance those of Sarychev [6] who establishes an index theory for the optimality of bang-bang trajectories based on the second variation specifically adapted to bang-bang trajectories. But our criteria have the advantage that they are easily verifiable and thus applicable. Also, our results are sufficiently general to cover the least degenerate and in this sense typical scenarios. Due to length restrictions we only give the main results indicating steps in the proofs.

2. Parametrized Families of Broken Extremals

Let U be a subset of $\mathbb{R}^m$, the *control set*, and denote by $\mathcal{U}$ the class of all locally bounded Lebesgue measurable maps defined on some interval $I \subset \mathbb{R}$ with values in U, $u : I \to U$, the *space of (admissible) controls*. Suppose $f : \mathbb{R} \times \mathbb{R}^n \times \mathbb{R}^m \to \mathbb{R}^n$, $(t, x, u) \mapsto f(t, x, u)$, *the dynamics of the control system*, is a continuous map which for fixed $t \in \mathbb{R}$ is r-times continuously differentiable in (x, u) and let $N = \{(t, x) : \Psi(t, x) = 0\}$ be a k-dimensional embedded submanifold of (t, x)-space $\mathbb{R} \times \mathbb{R}^n$, the *terminal manifold*. We assume that the components ψ_i, $i = 0, \ldots, n - k$, of Ψ have linearly independent

---
[1] supported in part by NSF Grant DMS-9503356

gradients $\nabla\psi_i$ everywhere on N.

We consider the problem to minimize over $\mathcal{U}$ a cost functional given in Bolza form as

$$\mathcal{J}(u;t,\xi) = \int_t^T L(s,x,u)ds + \varphi(T,x(T)) \tag{1}$$

subject to the dynamics $\dot{x} = f(t,x,u)$ with initial condition $x(t) = \xi$ and terminal condition $(T,x(T)) \in N$. The terminal time T is free.

The Maximum Principle gives necessary conditions for a controlled trajectory (x,u) to be optimal. The definition of parametrized families of extremals below formalizes these necessary conditions of the Maximum Principle while requiring additional regularity on control and smoothness of the parametrization. In our notation we distinguish between tangent vectors which we write as column vectors (such as x, $f(t,x,u)$ etc.) and cotangent vectors which we write as row vectors (like the multipliers λ and ν in the statement of the Maximum principle below). We denote the space of n-dimensional row vectors by $(\mathbb{R}^n)^*$. Define the Hamiltonian function H, $H: \mathbb{R} \times [0,\infty) \times (\mathbb{R}^n)^* \times \mathbb{R}^n \times \mathbb{R}^m \to \mathbb{R}$, by

$$H(t,\lambda_0,\lambda,x,u) = \lambda_0 L(t,x,u) + \lambda f(t,x,u). \tag{2}$$

We parametrize extremals through their endpoints in the terminal manifold N (k-dimensional) and the vector ν in the transversality condition for the terminal condition for the multiplier λ ($n+1-k$-dimensional). However, we also need to enforce the transversality condition on H which pins down the terminal time. Hence the parameter space is n-dimensional.

Definition 1 *A C^r-parametrized family $\mathcal{E}$ of extremals is an 8-tuple $(P; \mathcal{T} = (t_0, t_f); \xi, \nu; x, u, \lambda_0, \lambda)$ consisting of*

- *an open set P in some n-dimensional manifold and a pair $\mathcal{T} = (t_0, t_f)$ of r times continuously differentiable functions $t_0 : P \to \mathbb{R}$ and $t_f : P \to \mathbb{R}$ defined on P which satisfy $t_0(p) < p < t_f(p)$ for all $p \in P$. They define the domain of the parametrization as $D = \{(t,p) : p \in P, t \in I_p = [t_0(p), t_f(p)]\}$.*

- *r times continuously differentiable functions $\xi : P \to N$ and $\nu : P \to (\mathbb{R}^{n+1-k})^*$ which parametrize the terminal conditions for the states and costates respectively.*

- *extremal lifts consisting of controlled trajectories $(x,u) : D \to \mathbb{R}^n \times U$ and corresponding adjoint vectors $\lambda_0 : P \to [0,\infty)$ and $\lambda : D \to (\mathbb{R}^n)^*$. Specifically, we assume*

    1. the multipliers $(\lambda_0(p), \lambda(t,p))$ are nontrivial for all $t \in I_p$,

2. the controls $u = u(\cdot, p)$, $p \in P$, parametrize admissible controls which are continuous in (t, p) and for t fixed depend r-times continuously differentiable on p with the derivatives continuous in (t, p),

3. the trajectories $x = x(t, p)$ solve

$$\dot{x}(t,p) = f(t, x(t,p), u(t,p)), \quad x(t_f(p), p) = \xi(p) \in N \qquad (3)$$

4. the multiplier λ_0 is r-times continuously differentiable on P and the costate $\lambda = \lambda(t, p)$ solves the corresponding adjoint equation

$$\dot{\lambda}(t,p) = -\lambda_0(p) L_x(t, x(t,p), u(t,p)) - \lambda(t,p) f_x(t, x(t,p), u(t,p)), \qquad (4)$$

with terminal conditions

$$\lambda(t_f(p), p) = \lambda_0(p) \varphi_x(t_f(p), \xi(p)) + \nu(p) D_x \Psi(t_f(p), \xi(p)) \qquad (5)$$

5. the controls $u(t, p)$ solve the minimization problem

$$H(t, \lambda_0(p), \lambda(t,p), x(t,p), u(t,p)) = \min_{v \in U} H(t, \lambda_0(p), \lambda(t,p), x(t,p), v) \qquad (6)$$

6. the transversality condition on the terminal time,

$$\begin{aligned} 0 = {} & H(t_f(p), \lambda_0(p), \lambda(t_f(p), p), \xi(p), u(t_f(p), p)) + \\ & \lambda_0(p) \varphi_t(t_f(p), \xi(p)) + \nu(p) D_t \Psi(t_f(p), \xi(p)), \end{aligned} \qquad (7)$$

holds.

It follows from standard results about differentiable dependence on parameters of solutions to ordinary differential equations that the trajectories $x = x(t, p)$ and their time-derivatives $\dot{x}(t, p)$ are r-times continuously differentiable in p for fixed t and that these derivatives are continuous jointly in (t, p). We denote this class of functions by $C^{1,r}$. These partial derivatives can be calculated as solution to the corresponding variational equations which are obtained by interchanging the time derivative with the p-derivatives. The costate $\lambda(t, p)$ has identical smoothness properties because of our smoothness condition on λ_0. Note that the conditions are linear in the multipliers (λ_0, λ) and thus it is possible to normalize this vector. In particular, if $\lambda_0 > 0$, then we can divide by λ_0 and thus assume $\lambda_0 = 1$. We assume that λ_0 is C^r, so that this division does not destroy the required smoothness properties. These kind of extremals are called *normal* while extremals with $\lambda_0 = 0$ are called *abnormal*. Our definition allows to make abnormal extremals part of the parametrized family. The existence of optimal abnormal extremals can in general not be ruled out. For instance, the synthesis for time-optimal control to the origin for the harmonic oscillator contains abnormal trajectories.

The parametrizations (ξ, ν) are typically injective, but note that it is not assumed that the corresponding trajectories do not overlap.

In this paper our interest is to generalize the relevant structures to parametrized families of broken extremals where the controls have discontinuities on manifolds described by r-times continuously differentiable functions of p.

Definition 2 *A C^r-parametrized family $\mathcal{E}$ of broken extremals is an 8-tuple $(P; \mathcal{T}; \xi, \nu; x, u, \lambda_0, \lambda)$ where $\mathcal{T} = t_0 < t_1 < \ldots < t_\ell < t_f$ is a finite family of r-times continuously differentiable functions $t_j : P \to \mathbb{R}$ with the property that the conditions of Definition 1 are satisfied with the following modification on the smoothness of the controls: if*

$$D_j = \{(t,p) \in D: \quad p \in P, \quad t_j(p) \leq t \leq t_{j+1}(p)\} \tag{8}$$

and

$$D_j^* = \{(t,p) \in D: \quad p \in P, \quad t_j(p) < t < t_{j+1}(p)\}, \tag{9}$$

then $u_j = u_j(t,p)$ is an admissible control on D_j^* which for t fixed is r-times continuously differentiable in p and which, as a function of (t,p) has a continuous extension onto D_j. The other conditions remain in effect.

On the domain D we define the parametrized cost $C : D \to \mathbb{R}$ along a C^r-parametrized family $\mathcal{E}$ of (broken) extremals as

$$C(t,p) = \int_t^{t_f(p)} L(s, x(s,p), u(s,p))ds + \varphi(t_f(p), \xi(p)), \tag{10}$$

i.e. $C(t,p)$ denotes the cost for the optimal control problem with initial condition $(t,x) = (t, x(t,p))$ corresponding to the control $u = u(t,p)$. It follows from our assumptions and the above smoothness properties that for $(t,p) \in D_j^*$ the cost C is continuously differentiable in t with time derivative $\frac{\partial C}{\partial t} = -L(t, x(t,p), u(t,p))$ and that both C and its time-derivative $\frac{\partial C}{\partial t}$ are r-times continuously differentiable in p.

Notation: For a function like C we denote the gradient with respect to p (which we consider a row vector) by $\frac{\partial C}{\partial p}$. Consequently, for a column vector like $x = (x_1, \ldots, x_n)^T$ we denote by $\frac{\partial x}{\partial p}$ the matrix whose rows are given by the gradients of the components of x, i.e. $\frac{\partial x}{\partial p} = \left(\frac{\partial x_i}{\partial p_j}\right)_{1 \leq i,j \leq n}$ with row index i and column index j. However, to be consistent, for a row-vector like $\lambda = (\lambda_1, \ldots, \lambda_n)$ we denote the matrix of the partial derivatives $\left(\frac{\partial \lambda_j}{\partial p_i}\right)_{1 \leq i,j \leq n}$ with row index i and column index j by $\frac{\partial \lambda}{\partial p}$. In this sense, we have $\frac{\partial \lambda}{\partial p} = \left(\frac{\partial \lambda^T}{\partial p}\right)^T$. This will allow us to write most formulas without having to use transposes.

The following relation is crucial to the whole construction:

Lemma 1 (Shadow Prices) *Let $\mathcal{E}$ be a C^1-parametrized family of broken extremals. Then the identity*

$$\lambda_0(p)\frac{\partial C}{\partial p}(t,p) = \lambda(t,p)\frac{\partial x}{\partial p}(t,p) \tag{11}$$

holds on each open domain D_j^*, $j = 0, 1, \ldots, \ell$.

SKETCH OF THE PROOF: Without loss of generality consider the case $\ell = 1$ of one switching and let $T_1 = \{(t,p) : t = t_1(p)\}$ denote the switching surface. Adjoining the terminal condition with multiplier $\nu = \nu(p)$ to C and then using the transversality conditions of the maximum principle after differentiating in p, it can be shown that both sides of (11) have identical values at the terminal time $t_f(p)$. It therefore suffices to show that both sides have the same time-derivative. Using the adjoint equation and the variational equation for $\frac{\partial x}{\partial p}$ it follows that for $(t,p) \in D_i^*$, $i = 1, 2$,

$$\frac{d}{dt}\left\{\lambda(t,p)\frac{\partial x}{\partial p}(t,p)\right\} = \lambda_0(p)\frac{\partial^2 C}{\partial t \partial p}(t,p) + H_u \frac{\partial u}{\partial p}(t,p) \tag{12}$$

where H_u is evaluated along the parametrized extremal. But

$$H_u(t, \lambda_0(p), \lambda(t,p), x(t,p), u(t,p))\frac{\partial u}{\partial p}(t,p) \equiv 0 \quad \text{on } D_i \tag{13}$$

follows from the minimization property of the extremal control $u(t,p)$ and thus the left and right hand side in (11) have the same derivatives in t for p fixed. This already implies (11) on D_1^*.

In order to show that (11) also holds on D_0^*, we still need to argue that the derivatives adjust at the switching surface. Denote by C_i the cost on the domains D_i and also index the trajectories and multipliers accordingly. Since u_0 and u_1 extend continuously onto a neighborhood of T_1, the trajectories and multipliers can be extended as solutions of the corresponding differential equations (not necessarily as extremals) onto a neighborhood of T_1. It follows from the definition that C_0 and C_1 agree on T_1 and the same also holds trivially for the trajectories. It is then elementary to see that for every parameter p_0 there exist continuous functions κ_0 and $\kappa = (\kappa_1, \ldots, \kappa_r)^T$ defined near p_0 such that

$$\nabla C_0(t,p) = \nabla C_1(t,p) + \kappa_0(p)(1, -\frac{\partial t_1}{\partial p}(t,p)) \tag{14}$$

and

$$\nabla x_0(t,p) = \nabla x_1(t,p) + \kappa(p)(1, -\frac{\partial t_1}{\partial p}(t,p)) \tag{15}$$

be calculated. For instance, from (10)

$$\kappa_0(p) = L(t_1(p), x(t_1(p), p), u_1(t_1(p), p)) - L(t_1(p), x(t_1(p), p), u_0(t_1(p), p)). \tag{16}$$

Using these values it follows from the fact that the Hamiltonian stays continuous in t also at the switching point that

$$\lambda_0(p)\kappa_0(p) - \lambda(t_1(p), p)\kappa(p) = 0. \tag{17}$$

Hence, and using the continuous extension of (11) from D_1^* onto T_1, it follows that

$$\begin{aligned}
\lambda_0(p)\frac{\partial C_0}{\partial p}(t_1(p), p) &= \lambda_0(p)\frac{\partial C_1}{\partial p}(t_1(p), p) - \lambda_0(p)\kappa_0(p)\frac{\partial t_1}{\partial p}(t, p) \\
&= \lambda(t_1(p), p)\left(\frac{\partial x_1}{\partial p}(t_1(p), p) - \kappa(p)\frac{\partial t_1}{\partial p}(t, p)\right) \\
&= \lambda(t_1(p), p)\frac{\partial x_0}{\partial p}(t_1(p), p). \tag{18}
\end{aligned}$$

This implies (11) on D_0^*. □

3. The Value for a Parametrized Family of Broken Extremals

In this section we show that a diffeomorphic parametrization gives rise to solutions of the Hamilton-Jacobi-Bellman equation

$$V_t(t, x) + \min_{u \in U}\{V_x(t, x)f(t, x, u) + L(t, x, u)\} \equiv 0. \tag{19}$$

$$V(t, x) = \varphi(t, x) \quad \text{for } (t, x) \in N. \tag{20}$$

Because of lack of space here we only consider the simple situation when the flow of extremals is piecewise a diffeomorphism away from the terminal manifold. More generally, fields where extremals merge can be considered as well [4].

Definition 3 *Given a C^r-parametrized family $\mathcal{E}$ of broken extremals, denote by*

$$D_* = \{(t, p) : p \in P, \quad t_0(p) < t < t_f(p)\} \tag{21}$$

the open domain of the parametrization and let

$$\sigma : D_* \to \mathbb{R} \times \mathbb{R}^n, \quad (t, p) \mapsto (t, x(t, p)). \tag{22}$$

We call $\mathcal{E}$ a C^r-parametrized field of broken extremals if the map σ is one-to-one and if the retrictions σ_j of σ to D_j^* are $C^{1,r}$-diffeomorphisms.

For a C^r-parametrized family $\mathcal{E}$ of broken extremals and $j = 0, 1, \ldots, \ell$, let $R_j^* = \sigma(D_j^*)$, $N_j = \sigma(T_j)$, where $T_j = \{(t, p) : t = t_j(p)\}$, and set $R_* = \sigma(D_*)$.

Theorem 2 Let $\mathcal{E}$ be a C^r-parametrized field of normal broken extremals, $r \geq 1$. Then the corresponding value-function $V : R_* \to \mathbb{R}$, $V = C \circ \sigma^{-1}$ and the admissible feedback-control $u^* : R_* \to U$, $u^* = u \circ \sigma^{-1}$ together solve the Hamilton-Jacobi-Bellman equation (19) on R. The value-function V is continuous R_* and has a continuous extension to the terminal manifold $N = N_\ell$. On the regions R_j^* the value is $C^{1,r+1}$. Furthermore, the following identities hold on D_j^*:

$$V_t(t, x(t, p)) = -H(t, \lambda(t, p), x(t, p), u(t, p)) \quad (23)$$
$$V_x(t, x(t, p)) = \lambda(t, p), \quad (24)$$
$$V_{xx}(t, x(t, p)) = \frac{\partial \lambda^T}{\partial p}(t, p) \left(\frac{\partial x}{\partial p}(t, p)\right)^{-1} \quad (25)$$

SKETCH OF THE PROOF: Again assume $\ell = 1$. By assumption σ is injective with $C^{1,r}$ inverse on R_1^* and R_2^*. Thus V and u^* are well-defined and a priori $V \in C^{1,r}$ on R_1^* and R_2^*. By construction V remains continuous at the switching surface and V also extends continuously onto the terminal manifold N: although injectivity of σ is not postulated on N, this follows since the cost $C(t, p)$ depends only on the terminal point $(t_f(p), \xi(p))$, but not directly on the parameter p. Hence we can extend the definition of $V \circ \sigma^{-1}$ to $(t, x) \in N$ by taking any of the pre-images of σ.

Since $C = V \circ \sigma$, we have on D_1^* and D_2^* that

$$\frac{\partial C}{\partial p}(t, p) = V_x(t, x(t, p)) \frac{\partial x}{\partial p}(t, p) \quad (26)$$

and thus, in view of Lemma 1 and the fact that $\frac{\partial x}{\partial p}$ is nonsingular, equation (24) follows. Equations (19) and (23) then follow directly from the conditions of the maximum principle. Furthermore, since also $V_x = \lambda \circ \sigma^{-1}$ on R_1^* and R_2^*, we still have $V_x \in C^{1,r}$. In particular, and observing that we need to take a transpose in λ to keep the notation consistent, we also get

$$V_{xx}(t, x(t, p)) \frac{\partial x}{\partial p}(t, p) = \frac{\partial \lambda^T}{\partial p}(t, p)$$

which implies equation (25). □

Extremals

In this section we argue that the solution V to the Hamilton-Jacobi-Bellman equation constructed from a parametrized field of broken extremals is indeed optimal over the region R. Define the *value function* V_R over a region R of the state-space, $V : R \to \mathbb{R}$, as

$$V_R(t,x) = \inf_{u \in \mathcal{U}_R} \mathcal{J}(u;t,x), \qquad (27)$$

where the infimum of the values of the cost functional $\mathcal{J}(u;t,x)$ (with initial conditions given by (t,x)) is taken over all controls $u \in \mathcal{U}_R$ for which the corresponding trajectory remains in the region R.

Theorem 3 *Let $\mathcal{E}$ be a C^r-parametrized field of normal broken extremals, $r \geq 1$, with corresponding value-function $V : R_* \to \mathbb{R}$, $V = C \circ \sigma^{-1}$, and admissible feedback-control $u^* : R_* \to U$, $u^* = u \circ \sigma^{-1}$. Then u^* is an optimal feedback control with respect to any other admissible control for which the corresponding trajectory lies in R. The value over R is given by $V_R = V$.*

SKETCH OF THE PROOF: Here it actually makes a difference whether we have a field of smooth or of broken extremals. In fact, if there are no switchings the result is classical, well-known and elementary. For, let $(t_0, p_0) \in D_*$ and let v be an admissible control defined over $[t_0, T]$ which steers $(t_0, x_0) = (t_0, x(t_0, p_0))$ into N. If there are no switchings, then V is a differentiable solution to (19) on R_*. Hence, differentiating along the trajectory $x(\cdot)$ corresponding to v, we obtain

$$\frac{dV}{dt}(t,x(t)) = V_t(t,x(t)) + V_x(t,x(t))f(t,x(t),v(t)).$$

But then by (19) $\frac{dV}{dt}(t,x(t)) + L(t,x(t),v(t)) \geq 0$, and thus, integrating from t_0 to T it follows that

$$V(t_0, x_0) \leq \int_{t_0}^{T} L(s, x(s), v(s))ds + \varphi(T, x(T)). \qquad (28)$$

Thus the cost of v is not better than the cost of using u_*.

This simple argument, however, does not generalize directly to the situation when the parametrized controls have switchings. The reason simply is that it cannot be excluded a priori that the trajectory x corresponding to the control v will not lie in the switching surface N_1 for an uncountable set of times. However, this difficulty can be resolved by perturbation arguments. We only indicate how this is done, but the details are too technical to be included here. The argument is due to Sussmann and can be found in a more general version in [5]. First choose a sequence of piecewise constant controls $\{v_n\}_{n \in \mathbb{N}}$ which take values in a compact subset K of the control set U, such that v_n converges to v a.e. and such that the corresponding trajectories $\{x_n\}_{n\in\mathbb{N}}$ converge uniformly to x. Then perturb the terminal point of the trajectories $\{x_n\}_{n\in\mathbb{N}}$ so that the perturbed trajectories corresponding to a fixed, but arbitrary constant control $u \equiv u_0$ intersect the switching surface N_1 at most for a discrete set of times. This is possible since the set of end-points for which the corresponding trajectory meets N_1 in an uncountable set of points has Lebesgue measure zero. This, essentially, is a consequence of Sard's theorem about the image of singular sets. Then taking the limit in all these approximations again the inequality (28) can be obtained thus proving the result. □

Space does not permit to include results on how to localize these arguments and derive respective sufficient conditions for a strong local minimum of broken extremals in optimal control. For this we refer the reader to [4].

References

[1] M. Kiefer and H. Schättler, "Parametrized families of extremals and singularities in solutions to Hamilton-Jacobi-Bellman equations", submitted for publication

[2] M. Kiefer and H. Schättler (1998), "Cut-loci and cusp singularitites in parametrized families of extremals", in: Optimal Contol: Theory, Algorithms and Applications, W.W. Hager and P.M. Pardalos, Eds., Kluwer Academic Publisher B.V., The Netherlands, pp. 250-277

[3] H.W. Knobloch, A. Isidori and D. Flockerzi (1993), Topics in Control Theory, Birkhäuser, Basel, Switzerland

[4] J. Noble (1998), "Parametrized families of broken extremals and sufficient conditions for strong local minima", D.Sc. Thesis, Washington University, St. Louis, Missouri, to appear

[5] B. Piccoli and H.J. Sussmann (1998), Regular synthesis and sufficiency conditions for optimality, to appear

[6] A. Sarychev (1997), First- and second-order sufficient conditions for bang-bang controls, *SIAM J. Control,* Vol. 35, No.1, pp. 315-340

John Noble, The Boeing Company, St. Louis, Missouri, 63166, USA
Heinz Schättler, Dept. of Systems Science and Mathematics, Washington University, St. Louis, Mo, 63130, USA

PATRICK SAINT-PIERRE
Dynamical system with state constraints: theory and applications

1 Introduction

Let us consider the following controlled dynamical system in $\mathbb{R}^N$

(1) $$\dot{x} = f(x, u)), \quad u \in U(x), \quad \text{for a.e. } t \geq 0, \quad x(0) = z$$

where $U : \mathbb{R}^N \to \mathcal{C}$ is an upper semicontinuous set-valued map from $\mathbb{R}^N$ to the set $\mathcal{C}$ of compact subsets of a finite dimensional vector space U.

An absolutely continuous function $t \to x(t)$ starting from x is a solution for (1) if there exists a control $t \to u(t) \in U(x(t))$ measurable such that

(2) $$\dot{x}(t) = f(x(t), u(t))), \quad \text{for a.e. } t \geq 0, \quad x(0) = x$$

We denote by $S_F(z)$ the set of all absolutely continuous function $t \to x(t)$ starting from z solution to (1) and by $\mathcal{U}_K(z)$ the set of all measurable functions $u(\cdot) \in \mathcal{U}$ such that there exists at least one solution $x(\cdot) \in S_F(z)$ remaining forever in a given closed set $K \subset \mathbb{R}^N$:

(3) $$x(t) \in K \text{ for all } t \geq t_0.$$

It is well known that Viability Theory allows to answer with a full generality to the following *qualitative* question: is it possible to find a solution starting from a given point to remain forever in K ? It is clear that this amounts to control the Viability of the system without specifying any intertemporal criteria.

But following the recent studies of optimal control problem with state constraints through Viability approach (see namely [7], [8],[3]), we know that *quantitative* problems, as for instance problems involving Minimal Time, optimal control, with uncertainties, stability and equilibria, can also be solved thanks to Viability Theory and

Set-Valued Numerical Analysis. Then this approach appears fully adapted to take into account the inherent difficulties due to the presence of constraints without any controllability assumption on the boundary of the constraint set. Moreover one can also introduce some rigidity in the constraints through the evaluation of crisis levels.

2 Definitions and preliminary results

Let us consider a set-valued map $F : K \rightsquigarrow {\rm I\!R}^N$ and the differential inclusion:

(4) $$\dot{x}(t) \in F(x(t)), \text{ for almost all } t \geq 0,$$

(5) $$x(0) = z \in K$$

Let us set $F(x) := \{f(x,u),\ u \in U(x)\}$. We assume that $f : {\rm I\!R}^N \times U \to {\rm I\!R}^N$ is continuous and that $\forall x \in K, \exists L_F(x) > 0$ s.t. $\forall y \in \mathcal{B}(x,1)$ we have $F(y) \subset F(x) + L_F(x)\|y - x\|\mathcal{B}_X$ where $\mathcal{B}_X$ is the unit ball in $X = {\rm I\!R}^N$.

Referring for instance to [2], we know that if the set-valued map $x \rightsquigarrow U(x)$ is upper semicontinuous and if $f(\cdot,\cdot)$ is continuous, with any solution to the system (4) one can associate a measurable selection $u(t) \in U(x(t))$ solution to (1).

Let F be a Marchaud map and K be a closed subset of X. The Viability Theorem (see [1]) states that the following properties are equivalent:

(6) $$\begin{array}{ll} i) & \forall x \in K, \exists x(\cdot) \in S_F(x) \cap \mathcal{K} \\ ii) & \forall x \in K,\ F(x) \cap T_K(x) \neq \emptyset \end{array}$$

where $\mathcal{K}$ denotes the set of functions defined on ${\rm I\!R}^+$ onto K and $T_K(x)$ the contingent cone to K at x.

A subset $M \subset \text{Dom}\ (F)$ is a called a *Viability Domain* of F if and only if $\forall x \in M,\ F(x) \cap T_M(x) \neq \emptyset$.

Given K and F, there is no reason that this condition is fulfilled. So we have to consider the *Viability Kernel* of K for F which is the largest closed subset of initial states x from which starts at least one viable solution to (4). If F is a Marchaud map and if K is closed, Aubin's Viability Theorem [1] states that the Viability Kernel of K exists - possibly empty - and that is equal to the largest closed Viability Domain of K for F. We denote it $Viab_F(K)$.

2.1 The Viability Kernel Algorithm

We know already how to approach the Viability Kernel by constructing a sequence of Discrete Viability Kernels, denoted $\overrightarrow{Viab}_{G_\rho}(K)$, associated with constrained discrete dynamical systems of the form

$$x^{n+1} \in G_\rho(x^n), \ x^n \in K \subset X, \ \forall n \in \mathbb{N}$$

where $G_\rho : X \rightsquigarrow X$ is the set-valued map corresponding to the finite difference inclusion $G_\rho(x) := x + \rho F_\rho(x)$. We also know how to construct approximations of the Discrete Viability Kernel using the Viability Kernel Algorithm (see [9]).

Let us first recall the two main results of set-valued numerical analysis.

Let F_ρ be an approximation of F which satisfies

(7) $$\limsup_{\rho \to 0} Graph(F_\rho) \subset Graph(F)$$

(8) $$\forall \rho \in]0, \frac{1}{M(x)}], \ F(x + \rho M(x) \mathcal{B}_X) \subset F_\rho(x)$$

where $M(x) := \max_{y \in \mathcal{B}(x,1) \cap K} \|F(y)\|$ is a local bound of F at x. The limits of sets are taken at the sense of Painlevé-Kuratovski lower limit or upper limit.

Let us define $G_\rho(x) := \{x\} + \rho F_\rho(x)$. From the Convergence Theorem of discrete Viability Kernel under assumptions (7) and (8), we have

$$\limsup_{\rho \to 0} \overrightarrow{Viab}_{G_\rho}(K) = Viab_F(K) = \liminf_{\rho \to 0} \overrightarrow{Viab}_{G_\rho}(K) \subset \overrightarrow{Viab}_{G_\rho}(K)$$

Let us now consider the sequence of sets K^n defined by the recursive relation:

$$K_\rho^0 := K_\rho, \quad K_\rho^{n+1} := \{x \in K_\rho^n, \ | \ G_\rho(x) \cap K_\rho^n \neq \emptyset\}$$

From the Viability Kernel Approximation Theorem we have

$$\overrightarrow{\mathrm{Viab}}_{G_\rho}(K) = \bigcap_{n>0} K_\rho^n = K_\rho^\infty$$

The inclusion $\limsup_{\rho \to 0} \overrightarrow{\mathrm{Viab}}_{G_\rho}(K) \subset Viab_F(K)$ derives from assumption (7)). The difficulty comes in finding a "good" approximation F_ρ of F - that is to say as "small" as possible - which satisfies assumption (8) implying inclusion $Viab_F(K) \subset \overrightarrow{\mathrm{Viab}}_{G_\rho}(K)$. For this task, the rule which leads to the choice of F_ρ is to check that the following property holds true:

$\forall x_0 \in Viab_F(K)$, $\forall x(\cdot) \in S_F(x_0)$ viable in K, $x(\rho) \in G_\rho(x_0)$

so that the sequence $x_n := x(n\rho)$, which belong to $Viab_F(K)$ is a viable solution to the discrete dynamical system and then $x_0 \in \overrightarrow{\mathrm{Viab}}_{G_\rho}(K)$.

Assume that for any $x \in K$ there exists an upper bounded positive function $M_F(x)$ such that for all $y \in \mathcal{B}(x,1)$ we have $|F(y)| \leq M_F(x)$

Proposition 2.1 *Let us consider* $G_\rho(x) := \{x\} + \rho F(x) + \frac{M_F(x)L_F(x)\rho}{2}\mathcal{B}_X$. *Then* $Viab_F(K) \subset \overrightarrow{\mathrm{Viab}}_{G_\rho}(K)$.

3 Some Applications

3.1 The Target Problem with State Constraints

Let C be a target and K the constraint set. A first question which arises when studying target problems is to find the set of points of K from which at least one solution starts, reaching C in a finite time while remaining in K until it reaches K. We denote this set $Vict_F(K,C)$. Let us define $\tilde{F}$ the set-valued map which coincides with F everywhere except on C and $\forall x \in C$, $\tilde{F}(x) = \overline{Co}(F(x) \cup \{0\})$.

Proposition 3.1 *Let F be a Marchaud map and let K and C be closed subsets of X. Let us assume that $Viab_F(K) = \emptyset$. Then*

$$Vict_F(K,C) = Viab_{\tilde{F}}(K)$$

A second question is to determine the Minimal Time function with values in $\mathbb{R}^+ \cup \{+\infty\}$ defined on X by

$$\vartheta_C^K(x_0) := \inf_{x(\cdot) \in S_F(x_0)} \{\tau \mid x(\tau) \in C, \ x(t) \in K, \forall t \leq \tau\}$$

Let us set $\mathcal{H} := \{(x,y) \in K \times \mathbb{R}^+\}$ and consider the auxiliary dynamical system

$$\Phi(x,y) = \begin{cases} F(x) \times \{-1\} \text{ if } x \in X \backslash C \\ \overline{Co}((F(x) \times \{-1\}) \cup (0,0)) \text{ otherwise} \end{cases}$$

The Minimal Time function enjoys the following properties (see [3])

Proposition 3.2 *If F is a Marchaud map and if K and C are closed, then*

a) the function $\vartheta_C^K(\cdot)$ is lower semicontinuous,

b) $\forall x_0 \in Dom(\vartheta_C^K), \ \exists x(\cdot) \in S_F(x_0) \cap \mathcal{K}$ such that $x(\vartheta_C^K(x_0)) \in C$ and $Epigraph(\vartheta_C^K) = Viab_\Phi(\mathcal{H})$

This allows to compute the Minimal Time function in the lack of regularity or controllability assumptions on the boundary of the target or of the constraint set. The only condition is that K is closed. Let be suitable $\tau > 0$ and $\rho > 0$, ρ depending on τ (and on the state discretization step that we do not mention here) and let us consider the sequence of functions defined by $\vartheta_\tau^0 \equiv \mathcal{I}_K$ (the indicator function of K) and

$$\vartheta_\tau^n(x) := (1-\rho)\tau + \inf_{v \in V, |w| \leq 1} \vartheta_\tau^{n-1}(x + \tau(f(x,v) + \rho w)).$$

Proposition 3.3 *Under the previous assumptions, the functions $\vartheta_\tau^n(\cdot)$ are upper bounded by $\vartheta_C^K(\cdot)$ and the sequence $\vartheta_\tau^n(\cdot)$ converges pointewisely to $\vartheta_C^K(\cdot)$.*

This Proposition is a consequence of the theorem of convergence of Discrete Viability Kernels.

3.2 Application to the Time of Crisis problem

Now we admit that the state of the system can violate "temporarily" the constraint K. In such situation we say that a "crisis" occurs (see [6]). We want to determine the

Minimal Time of Crisis function defined by

$$C_{K,F}(x) = \inf_{x(\cdot)\in\mathcal{S}_F(x)} \mu(t \mid x(t) \notin K) = \inf_{x(\cdot)\in\mathcal{S}_F(x)} \int_0^{+\infty} \mathcal{X}_{K^c}(x(s))ds,$$

where μ denotes the Lebesgue measure in $\mathbb{R}$ and $\mathcal{X}_{K^c}$ stands for the characteristic function of $X\backslash K$.

The main mathematical results concern the characterization of the epigraph of the crisis map in terms of a viability kernel of an extended problem and the obtention of an equivalent Hamilton-Jacobi formulation.

Let us introduce the upper semicontinuous set-valued map:

$$\mathcal{X}^{\natural}_{K^c}(x) := \begin{cases} [0,1] & \text{if } x \in \partial K \\ 1 & \text{if } x \notin K \\ 0 & \text{otherwise} \end{cases}$$

and consider the auxiliary differential inclusion system

$$x'(t) \in F(x(t)) \quad y'(t) \in -\mathcal{X}^{\natural}_{K^c}(x).$$

Theorem 3.4 *Let K be a closed subset of X. Let $F: X \rightsquigarrow X$ be a Marchaud set-valued map.*

a) The map $C_{K,F}(\cdot)$ is lower semicontinuous and $\forall x \in \text{Dom}(C_{K,F})$,

$$\exists\, x^*(\cdot) \in \mathcal{S}_F(x) \text{ such that } C_{K,F}(x) = \int_0^{+\infty} \mathcal{X}_{K^c}(x^*(s))ds,$$

b) We have $\text{Epi}(C_{K,F}) = Viab_{\widetilde{F}}(X \times \mathbb{R}^+)$, where $\widetilde{F}(x,y) = (F(x), -\mathcal{X}^{\natural}_{K^c}(x))$.

The following theorem provides an algorithm of approximation of the Minimal Time of crisis function which convergence derives from the Viability Kernel Algorithm.

Let us define $\widetilde{F}_\rho$ by $\widetilde{F}_\rho(x,y) := (x + \rho F(x) + \frac{Mk}{2}\rho^2 B,\; y - \rho\Gamma(\rho,x))$, with $\Gamma(\rho,x) = [0,1]$ if $d_{\partial K}(x) \leq M\rho$, 0 if $d_{K^c}(x) > M\rho$ and 1 if $d_K(x) > M\rho$.

Theorem 3.5 *Under the previous assumptions and if F is k-Lipschitz and bounded by some M on X. Then $\text{Lim}_{\rho \to 0^+} Viab_{\widetilde{F}_\rho}(X \times \mathbb{R}^+) = \text{Epi}(C_{K,F}(.))$.*

In particular, as for the Minimal Time function, the Minimal Time of Crisis function can be approached by an increasing sequence of functions defined on successively refined grids X_h of X.

3.3 The Problem of Games against the Nature

We present now a result in the frame of differential games with constraints ([4]).

Let us consider the differential game which dynamic is given by

(9) $$x'(t) = f(x(t), u(t), v(t)), \quad u(t) \in U, \; v(t) \in V$$

where $f : X \times U \times V \to X$ and where U denotes uncertainty and V denotes the control set.

Let $\mathcal{O}$ be an open target contained in X. The aim is to avoiding $\mathcal{O}$ forever, that is to say to remain in the complement of $\mathcal{O}$ whatever is the evolution of the uncertainty.

We can compute the set of initial positions from which one can avoid the target in the worst situation and without knowledge of the future. This set is the Victory Domain. In the frame of two players differential games it is defined thanks to *nonanticipative strategies* (cf [5]).

Let us consider the set-valued map F defined on $X \times U$ by

$$\forall x \in X, \; \forall v \in V, \; F(x,u) := \bigcup_{v \in V} f(x,u,v).$$

Let us assume that U and V are compact, $f : X \times U \times V \to X$ is continuous and ℓ−Lipschitz and $\forall x \in X, \forall u \in U$, $F(x,u)$ is nonempty, convex and compact. A locally compact set D is a **discriminating domain for** f if it satisfies:

$$\forall u \in U, \; \exists v \in V, \; f(x,u,v) \in T_D(x).$$

Then we can characterize the Victory Domain as the largest discriminating domain, denoted $Disc_F(\mathcal{K})$, associated with the system (9)

Let F_ε be a suitable set-valued map approaching F. Let us define $G_\varepsilon(x,u) := x + \varepsilon F_\varepsilon(x,u)$ and let us denote $\overrightarrow{Disc}_{G_\varepsilon}(K)$ the discrete discriminating kernel associated with G_ε. Then we can compute the Victory Domain as the limit of the sequence of sets $\{K_\varepsilon^n\}_n$ defined by an extention of the Viability Kernel Algorithm

$$K_\varepsilon^0 = K, \quad K_\varepsilon^{n+1} = \{x \in K_\varepsilon^n \,|\, \forall u \in U,\ G_\varepsilon(x, u) \cap K_\varepsilon^n \neq \emptyset\}.$$

Proposition 3.6 *The sequence K_ε^n converges to $\overrightarrow{Disc}_{G_\varepsilon}(K)$ when $n \to \infty$ and $\overrightarrow{Disc}_{G_\varepsilon}(K)$ converges to $Disc_F(\mathcal{K})$ when $\varepsilon \to 0$.*

3.4 Equilibria and Stability

Let us consider a set-valued map $F : X \rightsquigarrow Y$ and the set $Equi_F(K) := \{x \in K \mid 0 \in F(x)\}$ of all equilibria contained in a given set K.

Proposition 3.7 *Let F be a Marchaud map satisfying assumptions (7), (8) and let us consider the set-valued map $\Phi : X \times \mathbb{R} \rightsquigarrow X \times \mathbb{R}$ defined by $\Phi(x, y) := (0, \inf_{u \in F(x)} \|u\|_Y)$. Then Φ is a Marchaud map and*

$$Equi_F(K) \times 0 = Viab_\Phi(K \times 0)$$

From this result we can deduce a numerical method for finding for instance all the roots of a polynomial $P(x) = 0$ as well as of a piecewise lipschitz function.

Let us mention an important application to the calculus of Lyapunov functions allowing, for instance, the stabilization of constrained system near some given point as equilibria or some given domain. This result is due to J.P. Aubin and R. Wets.

Let be $V(\cdot) : X \to \mathbb{R}^+ \cup \{+\infty\}$ a lower semicontinuous function, $\mathcal{V} := Epigraph(V)$ and consider the differential inclusion

(10) $\qquad x'(t) \in F(x(t)),\ \ y'(t) = -ay(t)$ a.e. $t \in [0, +\infty)$,

Let us denote $\Phi(x, y) := (F(x), -ay)$ and $\mathcal{H} := \mathcal{V} \cap K \times \mathbb{R}^+$.

Proposition 3.8 *Let us assume that $F : X \rightsquigarrow X$ is a Marchaud map and that $K \subset X$ is closed. Then*

the Viability Kernel $Viab_\mathcal{H}(\Phi)$ is the epigraph of a function φ) and

$\varphi(\cdot)$ is the lowest lower semicontinuous Lyapunov function greater than V.

Applying the convergence Theorem we prove that the sequence of functions $\varphi^n(\cdot)$ defined by :

$\varphi^0(\cdot) = V(\cdot)$ and $\varphi^n(x) = \max\left[\varphi^{n-1}(x), \inf_{z \in F_\rho(x)} \frac{\varphi^{n-1}(x+\rho z)}{1-a\rho}\right]$

converges pointwisely to $\varphi(\cdot)$ when $n \to +\infty$ and $\rho \to 0^+$.

References

[1] AUBIN J.-P. (1991) *Viability Theory*, Birkhäuser Verlag, Basel

[2] AUBIN J.-P. & CELLINA A. (1984) DIFFERENTIAL INCLUSIONS, Springer-Verlag, Berlin

[3] CARDALIAGUET P., QUINCAMPOIX M. & SAINT-PIERRE P. (1997) *Optimal times for constrained non-linear control problems without local controllability* Applied Mathematics & Optimisation

[4] CARDALIAGUET P., QUINCAMPOIX M. & SAINT-PIERRE P. (1997) NUMERICAL METHODS FOR DIFFERENTIAL GAMES "Stochastic and Differential Games: Theory and Numerical Methods" Annals of the International Society of Dynamic Games, Birkaüser

[5] CARDALIAGUET P. (1996) *A differential game with two players and one target.* SIAM J. Contr. Opti., Vol. 34, No. 4, pp. 1441-1460

[6] DOYEN L. & SAINT-PIERRE P. (1997) SCALE OF VIABILITY AND MINIMAL TIME OF CRISIS Set Valued Analysis, **5**: 227-246

[7] FRANKOWSKA H. (1989) *Optimal trajectories associated to a solution of contingent Hamilton-Jacobi Equations*, Appl. Math. Optim. **19**: 291-311

[8] FRANKOWSKA H. (1993) *Lower semicontinuous solutions of Hamilton-Jacobi-Bellman equations*, SIAM J. Control and Optimization, **31**: 257-272.

[9] SAINT-PIERRE P. (1994) *Approximation of the Viability Kernel*, Applied Mathematics & Optimisation, **29**: 187-209.

[10] SAINT-PIERRE P. (1996) *Equilibria and Stability in Set-Valued Analysis : a Viability Approach.* - Topology in nonlinear Analysis, Banach Center Publications,**35**: 243-255, Warsaw

Centre de Recherche Viabilité, Jeux, Contrôle, ERS CNRS 644
Université Paris IX -Dauphine, Place de Lattre de Tassigny, 75775 Paris cedex 16 - France.

BORIS S MORDUKHOVICH
On variational analysis in infinite dimensions[1]

1. Introduction

Variational analysis has been recognized as an active area in mathematics that, on one hand, studies constrained optimization and related problems and, on the other hand, applies optimization, perturbation, and approximation ideas to the analysis of a broad range of problems which may not be of a variational nature. Nonsmooth functions, sets with nonsmooth boundaries, and set-valued mappings appear naturally and frequently in the framework of variational analysis and require the development of appropriate tools of generalized differentiation. We refer the reader to the book [17] for a systematic exposition and thorough development of the key features of variational analysis in finite dimensions.

This paper concerns with some aspects of variational analysis in infinite dimensional spaces. We present recent results on basic principles in variational analysis, on generalized differentiation of nonsmooth and set-valued mappings, and on dual differential characterizations of Lipschitzian stability.

Our notation is basically standard; see, e.g., [13]. Recall that $\limsup_{x \to \bar{x}} \Phi(x)$ for multifunctions $\Phi : X \rightrightarrows X^\star$ means the *sequential* Kuratowski-Painlevé upper limit with respect to the norm topology in X and the weak-star topology in $X^\star$.

2. Generalized Differentiation

Let X be a Banach space. We consider a nonempty set $\Omega \subset X$ and an extended-real-valued function $\varphi : X \to \bar{\mathbf{R}}$ and define the following constructions

$$\hat{N}_\varepsilon(x; \Omega) := \{ x^\star \in X^\star |\ \limsup_{u \stackrel{\Omega}{\to} x} \frac{\langle x^\star, u - x \rangle}{\|u - x\|} \leq \varepsilon \}, \qquad (2.1)$$

$$\hat{\partial}_\varepsilon \varphi(x) := \{ x^\star \in X^\star |\ \liminf_{u \to x} \frac{\varphi(u) - \varphi(x) - \langle x^\star, u - x \rangle}{\|u - x\|} \geq -\varepsilon \}, \ \varepsilon \geq 0, \qquad (2.2)$$

called, respectively, the set of ε-*normals* to Ω at $x \in \Omega$ and ε-*subgradients* of φ at x with $|\varphi(x)| < \infty$. We put $\hat{N}_\varepsilon(x; \Omega) := \emptyset$ for $x \notin \Omega$ and $\hat{\partial}_\varepsilon \varphi(x) := \emptyset$ if $|\varphi(x)| = \infty$. When $\varepsilon = 0$, the sets (2.1) and (2.2) are the *Fréchet normal cone and subdifferential* denoted by $\hat{N}(x; \Omega)$ and $\hat{\partial}\varphi(x)$.

Based on (2.1) and (2.2), we define the (sequential) *limiting normal cone and subdifferential*

$$N(\bar{x}; \Omega) := \limsup_{x \to \bar{x},\ \varepsilon \downarrow 0} \hat{N}_\varepsilon(x; \Omega), \qquad (2.3)$$

---
[1]This research was partly supported by the National Science Foundation under grants DMS-9404128 and DMS-9704751.

$$\partial \varphi(\bar{x}) := \limsup_{x \xrightarrow{\varphi} \bar{x},\, \varepsilon \downarrow 0} \hat{\partial}_\varepsilon \varphi(x) \qquad (2.4)$$

introduced in [7] as extensions of the corresponding finite dimensional constructions of [8]. In contrast to (2.1) and (2.2), the limiting normal cone and subdifferential are often *nonconvex*; they may not be even closed in infinite dimensions. Nevertheless, these constructions possess a rich calculus and important applications to optimization, control theory, and related areas; see [9, 17] in finite dimensions and [13] in the framework of *Asplund spaces*.

The latter subclass of Banach spaces is sufficiently broad, in particular, it contains reflexive spaces [16]. Although the normal cone (2.3) and subdifferential (2.4) have some useful properties in general Banach spaces, it turns out that the class of Asplund spaces provides a proper framework for their development and applications at the same level of perfection as in finite dimensions.

When X is finite dimensional, it is well known that one can always take $\varepsilon = 0$ in (2.3) and (2.4) for any closed sets and lower semicontinuous (l.s.c.) functions. It is remarkable that these properties hold true when X is Asplund and, moreover, they provide *characterization* of Asplund spaces.

Theorem 1. *Let X be an arbitrary Banach space. Then the following are equivalent:*
(a) *For every closed set $\Omega \subset X$ and every point $\bar{x} \in \Omega$ one has*

$$N(\bar{x}; \Omega) = \limsup_{x \to \bar{x}} \hat{N}(x; \Omega). \qquad (2.5)$$

(b) *For every function $\varphi : X \to \bar{\mathbf{R}}$ l.s.c. around $\bar{x}$, $|\varphi(\bar{x})| < \infty$, one has*

$$\partial \varphi(\bar{x}) = \limsup_{x \xrightarrow{\varphi} \bar{x}} \hat{\partial} \varphi(x). \qquad (2.6)$$

(c) *X is an Asplund space.*

When X is Asplund, representations (2.5) and (2.6) are proved in [13]. The opposite statements are established in [4].

Next we consider a multifunction $\Phi : X \rightrightarrows Y$ between Banach spaces and define for it coderivative objects. Given a normal cone N to the graph of Φ, we say that the set-valued mapping $D_N^* \Phi(\bar{x}, \bar{y}) : Y^* \rightrightarrows X^*$ defined by

$$D_N^* \Phi(\bar{x}, \bar{y})(y^*) := \{x^* \in X^* \mid (x^*, -y^*) \in N((\bar{x}, \bar{y}); \text{gph } \Phi)\} \qquad (2.7)$$

is the *normal coderivative* of Φ at $(\bar{x}, \bar{y})$ generated by the normal cone N. In what follows we consider the case of the normal cone (2.3) in (2.7). Note that this coderivative cannot be treated as a dual object to tangentially generated derivatives since the normal cone (2.3) is nonconvex.

It easily follows from (2.3) and (2.7) that the normal coderivative (2.7) admits the limiting representation

$$D_N^\star \Phi(\bar{x},\bar{y})(\bar{y}^\star) = \limsup_{\substack{(x,y)\to(\bar{x},\bar{y}),\, y^\star \xrightarrow{w^\star} \bar{y}^\star \\ \varepsilon \downarrow 0}} \hat{D}_\varepsilon^\star \Phi(x,y)(y^\star) \quad \forall \bar{y} \in Y^\star \qquad (2.8)$$

through the ε-*coderivatives*

$$\hat{D}_\varepsilon^\star \Phi(x,y)(y^\star) := \{x^\star \in X^\star |\, (x^\star, -y^\star) \in \hat{N}_\varepsilon((x,y); \text{gph } \Phi)\}.$$

This means that $D_N^\star \Phi(\bar{x},\bar{y})(\bar{y}^\star)$ is the collection of all $\bar{x}^\star$ for which there are sequences $\varepsilon_k \downarrow 0$, $(x_k, y_k) \to (\bar{x},\bar{y})$, $(x_k^\star, y_k^\star) \xrightarrow{w^\star} (\bar{x}^\star, \bar{y}^\star)$ with $(x_k, y_k) \in \text{gph } \Phi$ and $x_k^\star \in \hat{D}_{\varepsilon_k}^\star \Phi(x_k, y_k)(y_k^\star)$. Replacing above the weak* convergence of $y_k^\star$ with the norm convergence in $Y^\star$ but keeping the weak* convergence of $x_k^\star$, we arrive at the following construction

$$D_M^\star \Phi(\bar{x},\bar{y})(\bar{y}^\star) := \limsup_{\substack{(x,y,y^\star)\to(\bar{x},\bar{y},\bar{y}^\star) \\ \varepsilon \downarrow 0}} \hat{D}_\varepsilon^\star \Phi(x,y)(y^\star) \qquad (2.9)$$

called the *mixed coderivative* of Φ at $(\bar{x},\bar{y})$. It is crucial for various results involving (2.9) that we consider the norm convergence in $Y^\star$ but not in $X^\star$; see [11, 15] for more details and discussions.

3. Equivalent Principles in Variational Analysis

Fundamental tools of variational analysis and its applications are related to so-called *variational principles*. Roughly speaking, a variational principle means that a bounded below l.s.c. function admits a small perturbation from a given class such that the resulting function attains its minimum at some point. The first variational principle in complete metric spaces was established by Ekeland with nonsmooth perturbations of distance type. When a space is Banach with some additional smoothness properties (the existence of an equivalent smooth norm or bump function), there are *smooth* variational principles of Borwein-Preiss and Deville-Godefroy-Zizler which allow us to choose smooth perturbations; see [16].

Another fruitful approach to variational analysis consists of using so-called *extremal principles* which can be treated as extremal extensions of the classical separation theorem to nonconvex sets and go back to the beginning of dual-space geometric methods in nonsmooth analysis and optimization; see [8, 9]. For Banach spaces with Fréchet differentiable renorms, a version of the extremal principle in terms of ε-normals (2.1) was established in [7].

Recall that $\bar{x} \in \Omega_1 \cap \Omega_2$ is a *locally extremal point* of the set system $\{\Omega_1, \Omega_2\}$ in a Banach space X if for any $\varepsilon > 0$ there are $b \in X$ with $\|b\| < \varepsilon$ and a neighborhood U of $\bar{x}$ such that $(\Omega_1 + b) \cap \Omega_2 \cap U = \emptyset$. This geometric concept of extremality covers conventional notions of optimal solutions to general constrained optimization problems.

Theorem 2. *Let X be an arbitrary Banach space. Then the following are equivalent:*

(a) *The extremal principle holds in X, i.e., for any locally extremal point $\bar{x} \in \Omega_1 \cap \Omega_2$ of the closed set system $\{\Omega_1, \Omega_2\}$ and for any $\varepsilon > 0$ there exist $x_i \in \Omega_i \cap B_\varepsilon(\bar{x})$ and $x_i^* \in \hat{N}(x_i; \Omega_i) + \varepsilon B^*$ such that*

$$\|x_1^*\| + \|x_2^*\| = 1 \quad \text{and} \quad x_1^* + x_2^* = 0.$$

(b) *The fuzzy sum rule holds in X, i.e., for any l.s.c. functions $\varphi_i : X \to \bar{\mathbf{R}}$, $i = 1, 2$, one of which is Lipschitz continuous around $\bar{x}$, for any $\varepsilon > 0$, and for any $x^* \in \hat{\partial}(\varphi_1 + \varphi_2)(\bar{x})$ there exist $x_i \in B_\varepsilon(\bar{x})$ and $x_i^* \in \hat{\partial}\varphi_i(x_i)$ such that*

$$|\varphi_i(x_i) - \varphi_i(\bar{x})| \leq \varepsilon \quad \text{and} \quad \|x^* - x_1^* - x_2^*\| \leq \varepsilon.$$

(c) *X is an Asplund space.*

The equivalences (b)$\Longleftrightarrow$(c) and (a)$\Longleftrightarrow$(c) are proved in [3] and [12], respectively. Due to Theorem 1 the extremal principle in (a) is equivalent to the one in terms of ε-normals [7]. Recently the list of equivalent results in Theorem 2 was expanded in [18].

Next we present another characterization of Asplund spaces through a nonconvex analog of the Bishop-Phelps density theorem [16].

Theorem 3. *Let X be an arbitrary Banach space. Then the following are equivalent:*

(a) *X is an Asplund space.*

(b) *For every proper closed subset Ω of X, the set of points*

$$x \in \text{bd}\, \Omega \quad \text{with} \quad \hat{N}(x; \Omega) \neq \{0\}$$

is dense in the boundary bd Ω *of Ω.*

The implication (a)$\Longrightarrow$(b) is proved in [12] based on the extremal principle of Theorem 2(a). The opposite implication (b)$\Longrightarrow$(a) is proved in [4].

Let us mention that Ekeland's variational principle is actually equivalent to the Bishop-Phelps theorem in any Banach spaces; see [16]. The next result establishes the equivalence between the extremal principle and an appropriate version of the smooth variational principle in a subclass of Asplund spaces.

Theorem 4. *The following variational principle holds and is equivalent to the extremal principle of Theorem 2(a) on the class of Banach spaces admitting Lipschitzian and Fréchet differentiable bump functions:*

For any l.s.c. function $f : X \to (-\infty, \infty]$ bounded below, for any $\varepsilon > 0$ and $\bar{x}$ satisfying

$$f(\bar{x}) < \inf_X f + \varepsilon,$$

and for any $\lambda > 0$ there exist $x_0 \in X$, a neighborhood U of x_0, and a function $g : U \to \mathbf{R}$, which is Lipschitz continuous with modulus ε/λ and Fréchet differentiable with norm-to-norm continuous derivative $\nabla g : X \to X^\star$ on U, such that
  (i) $\|x_0 - \bar{x}\| < \lambda$;
  (ii) $f(x_0) < \inf_X f + \varepsilon$;
  (iii) $f + g$ attains a local minimum at x_0.
Moreover, condition (iii) implies the existence of a Lipschitzian and Fréchet differentiable bump function on any Banach space X admitting a locally uniformly rotund (LUR) renorm.

Sketch of the Proof. Let us show that the extremal principle implies the smooth variational principle of the theorem. The opposite implication is more or less standard.

First we apply Ekeland's variational principle to f and find $\tilde{x} \in X$ with $\|\tilde{x} - \bar{x}\| \leq \lambda$ so that

$$f(\tilde{x}) \leq \inf_X + \varepsilon \quad \text{and} \quad f(\tilde{x}) < f(x) + (\varepsilon/\lambda)\|x - \tilde{x}\| \quad \forall x \in X \setminus \{\tilde{x}\}.$$

Then we observe that $(\tilde{x}, f(\tilde{x}))$ is a locally extremal point of the closed sets

$$\Omega_1 := \operatorname{epi} f \quad \text{and} \quad \Omega_2 := \{(x, \mu) \in X \times \mathbf{R} |\ \mu \leq f(\tilde{x}) - (\varepsilon/\lambda)\|x - \tilde{x}\|\} \qquad (3.1)$$

Now we apply the extremal principle to (3.1) in the product space $X \times \mathbf{R}$ endowed with the sum norm $\|(x, \mu)\| := \|x\| + |\mu|$. Taking $\tilde{\varepsilon} > 0$, we find $(x_i, \mu_i) \in \Omega_i \cap B_\varepsilon((\tilde{x}, f(\tilde{x})))$, $i = 1, 2$, and $(x^\star, \nu) \in X^\star \times \mathbf{R}$ so that

$$(x^\star, -\nu) \in \hat{N}((x_1, \mu_1); \operatorname{epi} f) + \tilde{\varepsilon} B^\star, \quad (-x^\star, \nu) \in \hat{N}((x_2, \mu_2); \Omega_2) + \tilde{\varepsilon} B^\star, \qquad (3.2)$$

and $\max\{\|x^\star\|, |\nu|\} = 1/2$. It follows from (3.1) and (3.2) that $\mu_2 = f(\tilde{x}) - (\varepsilon/\lambda)\|x_2 - \tilde{x}\|$ and $\nu > \tilde{\varepsilon}$ for $\tilde{\varepsilon}$ sufficiently small.

Let us pick $(x_1^\star, -\nu_1) \in \hat{N}((x_1, \mu_1); \operatorname{epi} f)$ with $\|(x_1^\star, \nu_1) - (x^\star, \nu)\| \leq \tilde{\varepsilon}$. Then we have $\nu_1 > 0$ which implies $\mu_1 = f(x_1)$. Thus $x_1^\star / \nu_1 \in \hat{\partial} f(x_1)$. Using (3.2) and the structure of Ω_2 in (3.1), we can get, by a proper choice of $\tilde{\varepsilon}$, the estimate $\|x^\star / \nu\| < \varepsilon/\lambda + \eta$ for any given number $\eta > 0$. Since ε, λ, and η were chosen arbitrarily, we justify the existence of x_0, satisfying (i) and (ii), and $x^\star \in \hat{\partial} f(x_0)$ with $\|x^\star\| < \varepsilon/\lambda$. Finally we apply [2, Prop. VIII.1.2] and find a function $g : U \to \mathbf{R}$ which is Lipschitzian and Fréchet differentiable in a neighborhood U of x_0 such that (iii) holds, $\nabla g(x_0) = x^\star$, and $\nabla g : X \to X^\star$ is norm-to-norm continuous at x_0. The latter yields $\|\nabla g(x)\| < \varepsilon/\lambda$ around x_0 that ensures the local Lipschitzness of g with modulus ε/λ due to the classical mean value theorem. This completes the proof of the smooth variational principle. The last statement of the theorem is proved in [4]. □

For spaces with Fréchet differentiable renorms, the equivalence between both principles in Theorem 4 follows from [1] where equivalence results are established in general bornologically smooth Banach spaces for so-called viscosity β-normals and subdifferentials of controlled rank. Note also that the existence of a smooth bump function is a necessary condition for a smooth variational principle involving property (iii) with no need of the LUR assumption if the minimum in (iii) is global [4]. On the contrary, the extremal principle holds in any Asplund space.

4. Nonconvex Calculus

In this section we present basic sum and chain rules for both normal and mixed coderivatives of Section 2 which generate other calculus results for nonconvex coderivatives, normal cones, and subdifferentials under consideration.

Recall that $\Phi : X \rightrightarrows Y$ is *inner semicontinuous* at $(\bar{x}, \bar{y}) \in$ gph Φ if for any sequence $x_k \to \bar{x}$ with $\Phi(x_k) \neq \emptyset$ there is a sequence $y_k \in \Phi(x_k)$ converging to $\bar{y}$. We say that Φ is *partially sequentially normally compact (p.s.n.c.)* at $(\bar{x}, \bar{y})$ if for any sequence $(x_k, y_k, x_k^\star, y_k^\star)$ satisfying

$$(x_k^\star, y_k^\star) \in \hat{N}((x_k, y_k); \text{gph } \Phi),\ (x_k, y_k) \to (\bar{x}, \bar{y}),\ \|y_k^\star\| \to 0,\ \text{and } x_k^\star \xrightarrow{w^\star} 0$$

one has $\|x_k^\star\| \to 0$ as $k \to \infty$. See [5, 14] for the genesis of the latter property and its comparison with related compactness properties of sets and set-valued mappings.

Given two multifunctions $\Phi_i : X \rightrightarrows Y$ of closed graph, let us define

$$S(x, y) := \{(y_1, y_2) \in Y^2 |\ y_1 \in \Phi_1(x),\ y_2 \in \Phi_2(x),\ y_1 + y_2 = y\} \tag{4.1}$$

and introduce the *exact qualification condition* in terms of the mixed coderivatives:

$$D_M^\star \Phi_1(\bar{x}, \bar{y}_1)(0) \cap (-D_M^\star \Phi_2(\bar{x}, \bar{y}_2)(0)) = \{0\} \tag{4.2}$$

where $(\bar{x}, \bar{y}) \in$ gph $(\Phi_1 + \Phi_2)$ and $(\bar{y}_1, \bar{y}_2) \in S(\bar{x}, \bar{y})$.

Theorem 5. *Let Φ_1 and Φ_2 be multifunctions of closed graph between Asplund spaces, let $\bar{y} \in \Phi_1(\bar{x}) + \Phi_2(\bar{x})$, and let $(\bar{y}_1, \bar{y}_2) \in S(\bar{x}, \bar{y})$. Assume that S in (4.1) is inner semicontinuous at $(\bar{x}, \bar{y}, \bar{y}_1, \bar{y}_2)$, the exact qualification condition (4.2) holds, and either Φ_1 or Φ_2 is p.s.n.c. at $(\bar{x}, \bar{y}_1)$ and $(\bar{x}, \bar{y}_2)$, respectively. Then one has*

$$D^\star(\Phi_1 + \Phi_2)(\bar{x}, \bar{y})(y^\star) \subset D^\star \Phi_1(\bar{x}, \bar{y}_1)(y^\star) + D^\star \Phi_2(\bar{x}, \bar{y}_2)(y^\star)\ \forall y^\star \in Y^\star \tag{4.3}$$

for both normal ($D^\star = D_N^\star$) and mixed ($D^\star = D_M^\star$) coderivatives.

The proof of this theorem is based on the *extremal principle* of Section 3 and can be found in [15]. Therein this result is used to establish the following *chain rule* for coderivatives of general compositions

$$(F \circ G)(x) := F(G(x)) = \bigcup \{F(y) |\ y \in G(x)\} \tag{4.4}$$

where $G : X \rightrightarrows Y$ and $F : Y \rightrightarrows Z$ are set-valued mappings between Banach spaces. Note that although the same chain rule (5.6) holds for both normal and mixed coderivatives of $F \circ G$, the normal coderivative of G is used in both cases.

Theorem 6. *Let all the spaces in (4.4) be Asplund and let the multifunction*

$$M(x,z) := G(x) \cap F^{-1}(z) = \{y \in G(x) | \, z \in F(y)\}$$

be inner semicontinuous at $(\bar{x}, \bar{z}, \bar{y})$ for some $\bar{y} \in M(\bar{x}, \bar{z})$. Assume that

$$D_M^\star F(\bar{y}, \bar{z})(0) \cap (-D_M^\star G^{-1}(\bar{y}, \bar{x})(0)) = \{0\} \tag{4.5}$$

and either F is p.s.n.c. at $(\bar{y}, \bar{z})$ or G^{-1} is p.s.n.c. at $(\bar{y}, \bar{x})$. Then one has

$$D^\star(F \circ G)(\bar{x}, \bar{z})(z^\star) \subset D_N^\star G(\bar{x}, \bar{y}) \circ D^\star F(\bar{y}, \bar{z})(z^\star) \tag{4.6}$$

for both coderivatives $D^\star = D_N^\star$ and $D^\star = D_M^\star$.

The main assumptions in the calculus rules of Theorems 5 and 6 are the qualification conditions (4.2) and (4.5) together with the p.s.n.c. property of the corresponding multifunctions. It follows from the results presented in the next section that these assumptions are automatically fulfilled for a broad class of pseudo-Lipschitzian multifunctions.

5. Lipschitzian Stability

This section contains results related to dual coderivative characterizations of Lipschitzian stability of multifunctions which are significant for coderivative calculus and applications. We mainly consider the *pseudo-Lipschitzian* property of $\Phi : X \rightrightarrows Y$ around $(\bar{x}, \bar{y})$ introduced by Aubin (see [17]): there are neighborhoods U of $\bar{x}$ and V of $\bar{y}$ such that

$$\Phi(x) \cap V \subset \Phi(u) + l\|x - u\|B \quad \forall x, u \in U \tag{5.1}$$

with some modulus $l \geq 0$. We denote by $(\text{plip }\Phi)(\bar{x}, \bar{y})$ the infimum of all such moduli l and call it the *pseudo-Lipschitzian bound* of Φ at $(\bar{x}, \bar{y})$. One can see that (5.1) reduces to the classical local Lipschitzian property when $V = Y$. It is also well known that (5.1) is equivalent to the fundamental properties of metric regularity/openness at a linear rate for the inverse mapping Φ^{-1}.

In the case of finite dimensional spaces X and Y, complete dual characterizations of the mentioned fundamental properties of set-valued mappings are obtained in [10] in terms of the coderivative $D^\star$ which coincides with both constructions (2.8) and (2.9) in finite dimensions. In particular, we prove there that a multifunction $\Phi : \mathbf{R}^n \rightrightarrows \mathbf{R}^m$ of closed graph is pseudo-Lipschitzian around $(\bar{x}, \bar{y})$ *if and only if* each of the following equivalent conditions holds:

$$\|D^\star\Phi(\bar{x}, \bar{y})\| := \sup\{\|x^\star\| \text{ s.t. } x^\star \in D^\star\Phi(\bar{x}, \bar{y})(y^\star), \|y^\star\| \leq 1\} < \infty, \tag{5.2}$$

$$D^\star \Phi(\bar{x}, \bar{y})(0) = \{0\}. \tag{5.3}$$

Moreover, in this case we establish the exact formula for computing the pseudo-Lipschitzian bound:

$$(\text{plip } \Phi)(\bar{x}, \bar{y}) = \|D^\star \Phi(\bar{x}, \bar{y})\|.$$

To extend these results to infinite dimensional settings, we need appropriate compactness conditions. The next theorem is proved in [14].

Theorem 7. *Let $\Phi : X \rightrightarrows Y$ be a multifunction of closed graph between Banach spaces and let $(\bar{x}, \bar{y})$. Consider the properties:*
  (a) Φ *is pseudo-Lipschitzian around* $(\bar{x}, \bar{y})$;
  (b) *one has (5.2) for* $D^\star = D_N^\star$;
  (c) *one has (5.3) for* $D^\star = D_N^\star$.
Then (b)$\Longrightarrow$(c) *and the following hold:*
  (i) *Let Y be finite dimensional. Then* (a)$\Longrightarrow$(b), *and one has*

$$(\text{plip } \Phi)(\bar{x}, \bar{y}) \geq \|D_N^\star \Phi(\bar{x}, \bar{y})\|.$$

  (ii) *Let both X and Y be Asplund and let Φ be p.s.n.c. at* $(\bar{x}, \bar{y})$. *Then* (c)$\Longrightarrow$(a). *Moreover,*

$$(\text{plip } \Phi)(\bar{x}, \bar{y}) \leq \|D_N^\star \Phi(\bar{x}, \bar{y})\|$$

when X is finite dimensional.

One can see that, in contrast to finite dimensional spaces, the normal coderivative (2.7) does not provide unified necessary and sufficient conditions for the pseudo-Lipschitzian property in infinite dimensions. The next theorem shows that the usage of the mixed coderivative (2.9) allows us to improve results in this direction. We refer the reader to [15] for the proof and comments.

Theorem 8. *Let $\Phi : X \rightrightarrows Y$ be a multifunction of closed graph between Banach spaces and let $(\bar{x}, \bar{y}) \in$ gph Φ. Consider the properties:*
  (a) Φ *is pseudo-Lipschitzian around* $(\bar{x}, \bar{y})$;
  (b) Φ *is p.s.n.c. at* $(\bar{x}, \bar{y})$ *and one has (5.2) for* $D^\star = D_M^\star$;
  (c) Φ *is p.s.n.c. at* $(\bar{x}, \bar{y})$ *and one has (5.3) for* $D^\star = D_M^\star$.
Then (a)$\Longrightarrow$(b)$\Longrightarrow$(c) *and one has*

$$(\text{plip } \Phi)(\bar{x}, \bar{y}) \geq \|D_M^\star \Phi(\bar{x}, \bar{y})\|.$$

If both spaces X and Y are Asplund, then in addition (c)$\Longrightarrow$(a), *i.e., each of the equivalent conditions* (b) *and* (c) *is necessary and sufficient for the pseudo-Lipschitzian property of Φ.*

Due to criterion (b) in Theorem 8, we establish the sum and chain rules (4.3) and (4.6) for pseudo-Lipschitzian multifunctions.

Corollary. *In the setting of Theorem 5, the sum rule (4.3) holds when either Φ_1 or Φ_2 is pseudo-Lipschitzian around $(\bar{x}, \bar{y}_1)$ and $(\bar{x}, \bar{y}_2)$, respectively. Similarly, the chain rule (4.6) holds in the setting of Theorem 6 when either F or G^{-1} is pseudo-Lipschitzian around the corresponding points.*

References

1. J.M. Borwein, B.S. Mordukhovich and Y. Shao, On the equivalence of some basic principles in variational analysis, CECM Research Report 97:098 (1997).
2. R. Deville, G. Godefroy and V. Zizler, *Smoothness and renormings in Banach spaces*, Wiley, New York, 1993.
3. M. Fabian, Subdifferentiability and trustworthiness in the light of a new variational principle of Borwein and Preiss, *Acta Univ. Carolinae* **30** (1989), 51–56.
4. M. Fabian and B.S. Mordukhovich, Nonsmooth characterizations of Asplund spaces and smooth variational principles, Dept. of Math., Wayne State Univ., Research Report # 38 (1997).
5. A.D. Ioffe, Codirectional compactness, metric regularity and subdifferential calculus, preprint.
6. A. Jourani and L. Thibault, Qualification conditions for calculus rules of coderivatives of multivalued mappings, *J. Math. Anal. Appl.*, to appear.
7. A.Y. Kruger and B.S. Mordukhovich, Extremal points and the Euler equation in nonsmooth optimization, *Dokl. Akad. Nauk BSSR* **24** (1980), 684–687.
8. B.S. Mordukhovich, Maximum principle in problems of time optimal control with nonsmooth constraints, *J. Appl. Math. Mech.* **40** (1976), 960–969.
9. B.S. Mordukhovich, *Approximation Methods in Problems of Optimization and Control*, Nauka, Moscow, 1988.
10. B.S. Mordukhovich, Complete characterization of openness, metric regularity, and Lipschitzian properties of multifunctions, *Trans. Amer. Math. Soc.* **340** (1993), 1–35.
11. B.S. Mordukhovich, Coderivatives of set-valued mappings: calculus and applications, *Nonlinear Anal.* **30** (1997), 3059–3070.
12. B.S. Mordukhovich and Y. Shao, Extremal characterizations of Asplund spaces, *Proc. Amer. Math. Soc.* **124** (1996), 197–205.
13. B.S. Mordukhovich and Y. Shao, Nonsmooth sequential analysis in Asplund spaces, *Trans. Amer. Math. Soc.* **348** (1996), 1235–1280.
14. B.S. Mordukhovich and Y. Shao, Stability of set-valued mappings in infinite dimensions: point criteria and applications, *SIAM J. Control Optim.* **35** (1997), 285–314.
15. B.S. Mordukhovich and Y. Shao, Coderivatives of set-valued mappings in variational analysis, Dept. of Math., Wayne State Univ., Research Report # 35 (1997).
16. R.R. Phelps, *Convex Functions, Monotone Operators and Differentiability*, 2nd edition, Springer, Berlin, 1993.
17. R.T. Rockafellar and R. J-B Wets, *Variational Analysis*. Springer, Berlin, 1998.
18. Q.J. Zhu, The equivalence of several basic theorem for subdifferentials, *Set-Valued Anal.*, to appear.

ZSOLT PÁLES[1] AND VERA ZEIDAN[2]
On L^1-closed decomposable sets in L^∞

1. Introduction

Throughout the paper let $(\Omega, \mathcal{A}, \mu)$ denote a finite complete measure space and, for $1 \leq p \leq \infty$ let $L^p(\Omega) = L^p(\Omega, \mathbf{R}^m)$ be the space of equivalence classes of L^p-integrable functions equipped with the usual norm $\|\cdot\|_p$. For the sake of brevity, we write a.e. for "μ-almost everywhere".

The notion of decomposable sets in L^p spaces appeared in the work of Hiai and Umegaki [7]: A set $\mathbf{Q} \subseteq L^p(\Omega)$ is called *decomposable* if, for $x, y \in \mathbf{Q}$ and for $A \in \mathcal{A}$, we have
$$1_A x + 1_{\Omega \setminus A} y \in \mathbf{Q}.$$

It is easily seen that if $Q : \Omega \to 2^{\mathbf{R}^m}$ is an arbitrary set-valued function, then the set of L^p-measurable selections
$$\sigma_p(Q) = \{x \in L^p(\Omega) \mid x(t) \in Q(t) \text{ for a.e. } t \in \Omega\} \tag{1}$$

is always a decomposable subset of $L^p(\Omega)$. Therefore, it is natural to ask if the converse of this statement is also true. In the case $1 \leq p < \infty$, Hiai and Umegaki [7] proved that the closed decomposable sets can be derived from set-valued maps. More precisely they proved:

Theorem A ([7, Theorem 3.1]). *Let $1 \leq p < \infty$ and $\mathbf{Q} \subset L^p(\Omega)$ be a nonempty closed decomposable set. Then there exists a measurable nonempty closed set valued map $Q : \Omega \to 2^{\mathbf{R}^m}$ such that $\mathbf{Q} = \sigma_p(Q)$, where $\sigma_p(Q)$ is defined in (1).*

For the case $p = \infty$, there is no similar result available in the literature. In fact, if we consider a nonempty, closed and decomposable subset $\mathbf{Q}$ of $L^\infty(\Omega)$, it is not in general represented by a measurable set-valued map Q with nonempty and closed images.

As we shall show in Lemma 1 below, the image $\sigma_\infty(Q)$ of such set-valued map Q is not only closed in $L^\infty(\Omega)$, but also L^1-closed. Therefore, in $L^\infty(\Omega)$, the L^1-closed decomposable sets are the proper candidates for the representability via set-valued functions.

---
[1] Research supported by the Hungarian Soros Foundation Grant No. 022-2998/95 and by the Hungarian National Foundation for Scientific Research (OTKA), Grant No. T-016846.
[2] Research supported by the National Science Foundation, under grant NSF–DMS-9404591.

A set $\mathbf{Q} \subseteq L^\infty(\Omega)$ is called L^p-*closed* if, whenever a sequence $\{x_n\}$ in $\mathbf{Q}$ converges to $x_0 \in L^\infty(\Omega)$ in the L^p-norm, then $x_0 \in \mathbf{Q}$. The L^p-closure of a set $\mathbf{Q} \in L^\infty(\Omega)$ will be denoted by $cl_p\,\mathbf{Q}$.

In Theorem 2 below, we describe the class of set-valued maps whose members are in a bijective correspondence with L^1-closed decomposable sets. The further results of the paper offer analogous description for L^1-closed decomposable sets $\mathbf{Q}$ with the following additional properties: (i) $\mathbf{Q}$ has nonempty interior, (ii) $\mathbf{Q}$ is convex, (iii) $\mathbf{Q}$ is convex and has nonempty interior. In the last two cases, the corresponding set-valued map has convex images, therefore it can also be described via its supporting functional.

In the last section, we develop the notions of convex analysis for L^1-closed decomposable sets. These results are essential for the investigations of extremum problems, where constraints of the form

$$x(t) \in Q(t) \quad \text{for a.e. } t \in \Omega \quad (x \in L^\infty(\Omega))$$

are present. The need for investigating constraints of this form stems from control theory. Analogous problems with lower semicontinuous set-valued constraints are investigated in [9].

The results presented in this paper are proved in details in [10].

2. Decomposable and L^1-closed subsets in $L^\infty(\Omega)$.

The notion of equality between measurable set-valued maps over Ω is defined as:

$$Q_1 = Q_2 \quad \text{means} \quad Q_1(t) = Q_2(t) \quad \text{for a.e. } t \in \Omega.$$

A characterization of the nonemptiness of the image $\sigma_\infty(Q)$ is given by the following.

Lemma 1. *Let $Q : \Omega \to 2^{\mathbf{R}^m}$ be a measurable set-valued map with nonempty closed images. Then, $\sigma_\infty(Q)$ is L^1-closed and decomposable. Furthermore, $\sigma_\infty(Q) \neq \emptyset$ if and only if*

$$\exists r > 0 : Q(t) \cap B_r \neq \emptyset \quad \text{for a.e. } t \in \Omega. \tag{2}$$

(Here B_r denotes the ball of radius r in $\mathbf{R}^m$ centered at the origin.)

Remark 1. Condition (2) is equivalent to the essential boundedness of the distance function from 0 to $Q(t)$. For the case when $p \in [1, \infty)$, this fact is given in [1].

Consider a nonempty and decomposable subset $\mathbf{Q}$ of $L^\infty(\Omega)$. We furnish first a characterization of the L^p-closure of $\mathbf{Q}$. As a consequence, we will obtain that, for each $p \in [1, \infty)$, the L^1-closure of $\mathbf{Q}$ coincides with its L^p-closure.

Theorem 1. Let $\mathbf{Q} \subset L^\infty(\Omega)$ be a nonempty decomposable set. Then, for $p \in [1, \infty)$, $x \in cl_p \mathbf{Q}$ if and only if there exist a sequence of measurable sets $A_n \in \mathcal{A}$ and $\xi \in \mathbf{Q}$ such that

$$\mu(A_n) \to \mu(\Omega) \quad \text{as } n \to \infty \quad \text{and} \quad 1_{A_n} x + 1_{\Omega \setminus A_n} \xi \in cl_\infty \mathbf{Q} \quad \text{for all } n \in \mathbf{N}.$$

Corollary 1. If $\mathbf{Q} \subseteq L^\infty(\Omega)$ is nonempty and decomposable, then

(i) For all $p \in [1, \infty)$, $cl_1 \mathbf{Q} = cl_p \mathbf{Q}$;

(ii) For all $p \in [1, \infty)$,

$$cl_p(\mathbf{Q}) = \{x \in L^\infty(\Omega) | \exists \{\xi_n\}_{n=1}^\infty \subset \mathbf{Q} : \lim_{n \to \infty} \|\xi_n - x\|_p = 0 \text{ and } \sup_{n \in \mathbf{N}} \|\xi_n\|_\infty < \infty\}.$$

Let us introduce the following classes:

$$\Gamma(\Omega) := \left\{ Q : \Omega \to 2^{\mathbf{R}^m} | Q \text{ is measurable with closed nonempty images satisfying (2)} \right\}$$

and

$$\Sigma(\Omega) := \{\mathbf{Q} \subset L^\infty(\Omega) \mid \mathbf{Q} \text{ is } L^1\text{-closed nonempty and decomposable}\}.$$

From Lemma 1, we know that these two classes are related. The exact nature of their relationship is established in the next theorem.

Theorem 2. The map σ_∞ is a bijection between $\Gamma(\Omega)$ and $\Sigma(\Omega)$.

As we shall prove in the result below, the nonemptiness of the interior of $\sigma_\infty(Q)$ is characterized by the following property of Q:

$$\exists r \geq \rho > 0 \text{ and, for a.e. } t \in \Omega, \exists x_t \in \mathbf{R}^m \text{ such that } B_\rho(x_t) \subseteq Q(t) \cap B_r. \quad (3)$$

Therefore, set:
$$\Gamma_0(\Omega) := \{Q \in \Gamma(\Omega) \mid Q \text{ satisfies (3)}\}$$

and
$$\Sigma_0(\Omega) := \{\mathbf{Q} \in \Sigma(\Omega) \mid \text{int } \mathbf{Q} \neq \emptyset\}.$$

Theorem 3. The map σ_∞ is a bijection between $\Gamma_0(\Omega)$ and $\Sigma_0(\Omega)$.

Now, consider an element Q of $\Gamma(\Omega)$, and $\mathbf{Q} = \sigma_\infty(Q)$. We are interested in investigating the connection between the tangent cones associated with each of $\mathbf{Q}$ and Q. For this purpose we focus on two types of tangent cones to a subset M of a normed vector space X.

Namely: The adjacent cone $T(x_0 \mid M)$ and the Clarke tangent cone $C(x_0 \mid M)$, (where x_0 belongs to the closure of M) are defined by

$$T(x_0 \mid M) := \left\{v \in X \mid \forall \varepsilon_n \to 0^+, \exists v_n \to v \text{ such that } x_0 + \varepsilon_n v_n \in M, \forall n \in \mathbf{N}\right\}$$
$$C(x_0 \mid M) := \{v \in X \mid \forall \varepsilon_n \to 0^+, \forall x_n \in M : x_n \to x_0,$$
$$\exists v_n \to v \text{ such that } x_n + \varepsilon_n v_n \in M, \forall n \in \mathbf{N}\}.$$

Both tangent cones are nonempty and closed, the cone $C(x_0 \mid M)$ is convex, and $C(x_0 \mid M)$ is a subset of $T(x_0 \mid M)$.

Let Q in $\mathbf{\Gamma}(\Omega)$ and $x_0 \in \sigma_\infty(Q)$. To Q, we can associate two set-valued maps $T(x_0 \mid Q)$ and $C(x_0 \mid Q)$ defined, for t in Ω, by

$$T(x_0 \mid Q)(t) := T(x_0(t) \mid Q(t)), \quad \text{and} \quad C(x_0 \mid Q)(t) := C(x_0(t) \mid Q(t)).$$

It is shown by Giner in [5] that these set-valued maps are measurable. Thus, they are elements of $\mathbf{\Gamma}(\Omega)$. On the other hand, we can define these tangent cones for the set $\sigma_\infty(Q)$, that is, $T(x_0 \mid \sigma_\infty(Q))$ and $C(x_0 \mid \sigma_\infty(Q))$. These cones are decomposable and closed subsets in $L^\infty(\Omega)$. However, they are not necessarily L^1-closed, in general. Hence, by Theorem 2, they cannot be represented via the set-valued map $T(x_0 \mid Q)$ and $C(x_0 \mid Q)$, respectively. As we shall show below, these set-valued maps correspond via σ_∞ to the L^1-closures of those tangent cones. The proof of the results is based on the work by Giner [5].

Theorem 4. *Let $Q \in \mathbf{\Gamma}(\Omega)$. Then, for all $x_0 \in \sigma_\infty(Q)$, we have*

$$cl_1\, T(x_0 \mid \sigma_\infty(Q)) = \sigma_\infty(T(x_0 \mid Q)), \quad \text{and} \quad cl_1\, C(x_0 \mid \sigma_\infty(Q)) = \sigma_\infty(C(x_0 \mid Q)).$$

Remark 2. In [5], it is shown, for $p \in [1, \infty)$, that

$$\sigma_p(T(x_0 \mid Q)) \subseteq T(x_0 \mid \sigma_p(Q)) \quad \text{and} \quad C(x_0 \mid \sigma_p(Q)) \subseteq \sigma_p(C(x_0 \mid Q)),$$

and these inclusions are strict in general. Hence, Theorem 4 states that in $L^\infty(\Omega)$, certain equalities occur yielding no gap between the concerned sets.

3. L^1-closed decomposable and convex sets in $L^\infty(\Omega)$

This section is devoted to the study of convexity properties of decomposable sets.

Let $Q : \Omega \to 2^{\mathbf{R}^m}$ be a set-valued map on Ω with convex images. Then $\sigma_\infty(Q)$ is decomposable and convex. However, this subset of $L^\infty(\Omega)$ enjoys a richer property than the two combined that we call the $\mathcal{L}$-convexity.

A set $\mathbf{Q} \subset L^\infty(\Omega)$ is called $\mathcal{L}$-*convex* if for all $x, y \in \mathbf{Q}$, and for all $\lambda : \Omega \to [0, 1]$ measurable, we have

$$\lambda x + (1 - \lambda) y \in \mathbf{Q}.$$

This notion defined for sets in $L^\infty(\Omega)$ is the analogue of the C-convexity defined in [8] for sets of continuous functions over a compact Hausdorff space.

The next result shows how this notion is related to convexity for closed sets in $L^\infty(\Omega)$.

Lemma 2. *Let $\mathbf{Q}$ be a closed subset of $L^\infty(\Omega)$. Then, the following are equivalent:*

(i) $\mathbf{Q}$ *is convex and decomposable;*

(ii) $\mathbf{Q}$ *is $\mathcal{L}$-convex;*

(iii) *For all $k \in \mathbf{N}$, for all $\lambda_1, \ldots, \lambda_k \in L^\infty(\Omega, [0,1])$ such that $\sum_{i=1}^k \lambda_i = 1$, and for all $x_1, \ldots, x_k \in \mathbf{Q}$, the $\mathcal{L}$-convex combination $\sum_{i=1}^k \lambda_i x_i$ belongs to $\mathbf{Q}$.*

Now let Q be in $\mathbf{\Gamma}(\Omega)$. We can associate to Q another set-valued map defined by:

$$(\overline{co}\, Q)(t) = \overline{co}\, Q(t).$$

In [7], it is shown that for $p \in [1, \infty)$,

$$\sigma_p(\overline{co}\, Q) = c\ell_p \operatorname{co}(\sigma_p(Q)).$$

The case when $p = \infty$ has been an open question. Our goal now is to solve this open question. As we shall see below, the above identity is not true when $p = \infty$. However, it is known (cf. [1]) that $\overline{co}\, Q$ is measurable and hence is in $\mathbf{\Gamma}(\Omega)$. The associated set $\sigma_\infty(\overline{co}\, Q)$ is not only L^∞-closed but also L^1-closed and $\mathcal{L}$-convex. Thus, similarly to Theorem 4, we shall show that replacing, in the seeked identity, the L^∞-closure by the L^1-closure, we obtain a valid statement.

Theorem 5. *Let Q be in $\mathbf{\Gamma}(\Omega)$. Then*

$$\sigma_\infty(\overline{co}\, Q) = c\ell_1 \operatorname{co}(\sigma_\infty(Q)).$$

Let us introduce the following subsets of $\mathbf{\Gamma}(\Omega)$ and $\mathbf{\Sigma}(\Omega)$:

$$\Gamma(\Omega) := \{Q \in \mathbf{\Gamma}(\Omega) \mid Q(t) \text{ is convex for a.e. } t \in \Omega\},$$

$$\Gamma_0(\Omega) := \{Q \in \mathbf{\Gamma}_0(\Omega) \mid Q(t) \text{ is convex for a.e. } t \in \Omega\},$$

$$\Sigma(\Omega) := \{\mathbf{Q} \in \mathbf{\Sigma}(\Omega) \mid \mathbf{Q} \text{ is convex}\},$$

$$\Sigma_0(\Omega) := \{\mathbf{Q} \in \mathbf{\Sigma}_0(\Omega) \mid \mathbf{Q} \text{ is convex}\}.$$

The connection between these sets is given by

Corollary 2. *The map σ_∞ is a bijection between $\Gamma(\Omega)$ and $\Sigma(\Omega)$, and between $\Gamma_0(\Omega)$ and $\Sigma_0(\Omega)$.*

4. Support functionals for set-valued maps

Let X be a normed vector space and X^* be its dual. For a nonempty set $Q \subseteq X$ we define the support functional on X^* as:

$$\delta^*(x^* \mid Q) = \sup\{\langle x^*, x\rangle \mid x \in Q\}.$$

Let Q be a measurable set-valued map on Ω with nonempty images in $\mathbf{R}^m$. Then, we can associate to Q the function $q : \Omega \times \mathbf{R}^m \to]-\infty, +\infty]$ defined by

$$q(t, \xi) = \delta^*(\xi \mid Q(t)).$$

Set $\Delta^*(Q) := q$. Hence, one can easily see that

(i) For all $t \in \Omega$, $q(t, \cdot)$ is lower semicontinuous and sublinear on $\mathbf{R}^m$;

(ii) q is $(\mathcal{A} \times \mathcal{B})$-measurable, where $\mathcal{B}$ denotes the collection of Borel subsets of $\mathbf{R}^m$.

On the other hand, for Q in $\mathbf{\Gamma}(\Omega)$, we have the support functional of $\sigma_\infty(Q)$ defined on $(L^\infty(\Omega))^*$. As we shall see below, this support functional, restricted to elements from $L^1(\Omega)$, can be calculated via $\Delta^*(Q)$.

Theorem 6. *Let $\mathbf{Q}$ be a nonempty decomposable subset of $L^\infty(\Omega)$ such that $c\ell_1 \operatorname{co} \mathbf{Q} = \sigma_\infty(Q)$ for some $Q \in \mathbf{\Gamma}(\Omega)$, and let $\varphi \in L^1(\Omega)$. Then, $\int_\Omega \delta^*(\varphi(t) \mid Q(t)) \, d\mu(t)$ is well-defined and*

$$\delta^*(\varphi \mid \mathbf{Q}) = \int_\Omega \delta^*(\varphi(t) \mid Q(t)) \, d\mu(t).$$

The following result is an obvious consequence of the above theorem.

Corollary 3. *Let $Q \in \mathbf{\Gamma}(\Omega)$. Then, for $\varphi \in L^1(\Omega)$,*

$$\delta^*(\varphi \mid \sigma_\infty(Q)) = \int_\Omega \delta^*(\varphi(t) \mid Q(t)) \, d\mu(t).$$

In order to completely characterize set-valued maps Q in $\mathbf{\Gamma}(\Omega)$ in terms of their support functional $\Delta^*(Q)$, it is only natural to consider the images of Q to be convex, that is, Q must be in $\mathbf{\Gamma}(\Omega)$. The following result is a characterization, in terms of $\Delta^*(Q)$, of the nonemptiness of $\sigma_\infty(Q)$.

Lemma 3. *Let Q in $\mathbf{\Gamma}(\Omega)$, and $q = \Delta^*(Q)$, then, $\sigma_\infty(Q) \neq \emptyset$ iff*

$$\exists r > 0 : \text{ for a.e. } t \in \Omega, \ q(t, \xi) + r|\xi| \geq 0 \quad \forall \xi \in \mathbf{R}^m. \tag{4}$$

The above Lemma inspires the consideration of the following sets for support functionals to elements of $\Gamma(\Omega)$.

Set
$$\Lambda(\Omega) := \{q : \Omega \times \mathbf{R}^m \to]-\infty, \infty] \mid q(t, \cdot) \text{ is sublinear and lsc,}$$
$$q \text{ is } (\mathcal{A} \times \mathcal{B})\text{-measurable and satisfies condition (4)}\}.$$

The equality $q_1 = q_2$ in $\Lambda(\Omega)$, means that, for a.e. $t \in \Omega$, $q_1(t, \xi) = q_2(t, \xi)$ for all $\xi \in \mathbf{R}^m$. The intimate connection between $\Lambda(\Omega)$ and $\Gamma(\Omega)$ is presented below.

Theorem 7. *The map Δ^* is a bijection from $\Gamma(\Omega)$ onto $\Lambda(\Omega)$.*

Remark 3. By combining Theorem 7 with Corollary 2, we obtain the equivalence between the three sets: $\Gamma(\Omega)$, $\Sigma(\Omega)$ and $\Lambda(\Omega)$.

The rest of this section is devoted to the characterization of $\Gamma_0(\Omega)$ in terms of elements of $\Lambda(\Omega)$. As we shall see in the result below, the nonemptiness of the interior of $\sigma_\infty(Q)$ is characterized by this property of $q \in \Lambda(\Omega)$

$$\exists r \geq \rho > 0 : \text{ for a.e. } t \in \Omega,$$
$$q(t, \xi) + q(t, \eta) + r|\xi + \eta| \geq \rho[|\xi| + |\eta| + |\xi + \eta|] \quad \forall \xi, \eta \in \mathbf{R}^m. \tag{5}$$

Set
$$\Lambda_0(\Omega) := \{q \in \Lambda(\Omega) \mid q \text{ satisfies (5)}\}.$$

The following result shows that there is a one-to-one correspondence between $\Gamma_0(\Omega)$ and $\Lambda_0(\Omega)$.

Theorem 8. *The map Δ^* is a bijection from $\Gamma_0(\Omega)$ onto $\Lambda_0(\Omega)$.*

5. Normal and polar cones of decomposable sets in $L^\infty(\Omega)$.

Let X be a normed vector space and Q is a nonempty subset of X. For $x_0 \in Q$ we define the normal cone $N(x_0 \mid Q)$ to Q at an element $x_0 \in cl\,Q$ by:

$$N(x_0 \mid Q) := \{x^* \in X^* \mid \langle x^*, x \rangle \leq \langle x^*, x_0 \rangle \, \forall x \in Q\}.$$

Clearly, $N(x_0 \mid Q)$ is a closed convex nonempty cone. When Q is a cone and $x_0 = 0$, $N(0 \mid Q)$ is called the polar cone of Q, denoted by Q^0.

For the rest of this section, we take $X = L^\infty(\Omega)$. We first focus on the investigation of the normal cone of decomposable sets $\mathbf{Q}$. For, we introduce the multiplication in $(L^\infty(\Omega))^*$ by essentially bounded real valued function f as follows: for $x^* \in (L^\infty(\Omega))^*$, the linear functional fx^* is defined by

$$\langle fx^*, x \rangle = \langle x^*, fx \rangle \quad \forall x \in L^\infty(\Omega).$$

Hence, the notions of decomposability and $\mathcal{L}$-convexity extend naturally to subsets Φ of $(L^\infty(\Omega))^*$.

We say that $\Phi \subseteq (L^\infty(\Omega))^*$ is an $\mathcal{L}$-cone if for any $f \in L^\infty(\Omega, \mathbf{R}^+)$ and $x^* \in \Phi$, $fx^* \in \Phi$.

The characterization obtained in Lemma 2 for the $L^\infty(\Omega)$ setting remains valid for the $(L^\infty(\Omega))^*$ case. Furthermore, similar arguments show the following.

Lemma 4. *Let Φ be a closed cone of $(L^\infty(\Omega))^*$. The following are equivalent.*

(i) *Φ is decomposable and convex;*

(ii) *Φ is a convex $\mathcal{L}$-cone;*

(iii) *$\sum_{i=1}^k f_i x_i^* \in \Phi$ whenever, for all i, $x_i^* \in \Phi$ and $f \in L^\infty(\Omega, \mathbf{R}^+)$.*

Lemma 5. *Let $\mathbf{Q} \subset L^\infty(\Omega)$ be a decomposable set. Then, for $x_0 \in \mathbf{Q}$, $N(x_0 \mid \mathbf{Q})$ is a closed convex $\mathcal{L}$-cone.*

Let Q be a set-valued map in $\Gamma(\Omega)$, and let $\mathbf{Q} = \sigma_\infty(Q)$, and x_0 in Q. Then, to $\mathbf{Q}$ we can associate a set-valued map $N(x_0 \mid Q)$ defined via the pointwise normal cones to $Q(t)$ at $x_0(t)$, that is, $N(x_0 \mid Q)(t) = N(x_0(t) \mid Q(t))$. This map is measurable, since $\delta^*(\xi \mid Q(\cdot))$ is measurable and

$$N(x_0(t) \mid Q(t)) = \{\xi \in \mathbf{R}^m \mid \delta^*(\xi \mid Q(t)) = \langle \xi, x_0(t) \rangle\} \qquad \text{for a.e. } t \in \Omega.$$

The goal is to describe the L^1-elements of $N(x_0 \mid \mathbf{Q})$ in terms of the set-valued map $N(x_0 \mid Q)$. However, we shall show that such a characterization in terms of $N(x_0 \mid Q)$ is true for subsets $\mathbf{Q}$ of $L^\infty(\Omega)$ which are not necessarily L^1-closed, but rather their L^1-closures are decomposable. This property is met by both the Clarke and the adjacent tangent cones used in Section 2.

Theorem 9. *Let $\mathbf{Q}$ be a nonempty decomposable subset of $L^\infty(\Omega)$ such that $c\ell_1\, \mathbf{Q} = \sigma_\infty(Q)$ for some $Q \in \Gamma(\Omega)$. Then, for $x_0 \in \mathbf{Q}$,*

$$N(x_0 \mid \mathbf{Q}) \cap L^1(\Omega) = \sigma_1(N(x_0 \mid Q)).$$

Corollary 4. *If $Q \in \Gamma(\Omega)$, then*

$$N(x_0 \mid \sigma_\infty(Q)) \cap L^1(\Omega) = \sigma_1(N(x_0 \mid Q)).$$

The characterization of the polar cones of the Clarke and the adjacent tangent cones are presented in the sequel.

Corollary 5. *Let $Q \in \Gamma(\Omega)$ and $x_0 \in \sigma_\infty(Q)$, then*

$$(T(x_0 \mid \sigma_\infty(Q))^0 \cap L^1(\Omega) = \sigma_1(T^0(x_0 \mid Q))$$

and

$$(C(x_0 \mid \sigma_\infty(Q))^0 \cap L^1(\Omega) = \sigma_1(C^0(x_0 \mid Q)),$$

where, for $t \in \Omega$,

$$T^0(x_0 \mid Q)(t) := (T(x(t) \mid Q(t)))^0 \quad \text{and} \quad C^0(x_0 \mid Q)(t) := (C(x_0(t) \mid Q(t)))^0.$$

References

[1] Aubin, J.-P. and Frankowska, H. (1990) *Set-valued analysis*, Birkhäuser Verlag, Boston–Basel.

[2] Castaing, C. and Valadier, M. (177) *Convex-analysis and measurable multifunctions*, Lecture Note in Mathematics, 580, Editors: A. Dold and B. Eckmann, Springer Verlag, Berlin–Heidelberg–New York.

[3] Clarke, F. H. (1983) *Optimization and nonsmooth analysis*, John Wiley, New York.

[4] Ekeland, I. and Temam R. (1974) *Analyse convexe et probleèmes variationelles*, Hermann, Paris. (English translation: (1977) *Convex Analysis and variational problems*, North-Holland, Amsterdam.)

[5] Giner, E. (1985) *E'tudes des fonctionnelles integrables*, Thesis, Université de Pau, France.

[6] Giner, E. *Sous-differentiabilité des fonctionnelles integrables (II)*, Thesis, sém. Anal. Num. No. 6, 83-84.

[7] Hiai, F. and Umegaki, H. *Integrals, conditional expectations, and martingales of multivalued functions*, J. Multivariate Analysis 7 (1977), 149-182.

[8] Páles, Zs. and Zeidan, V. *Characterization of closed and open C-convex sets in $C(T, \mathbf{R}^r)$*, preprint.

[9] Páles, Zs. and Zeidan, V. *Optimum problems with lower semicontinuous set-valued constraints*, SIAM J. Contr. Opt., to appear.

[10] Páles, Zs. and Zeidan, V. *Characterization of L^1-closed decomposable sets in L^∞*, preprint.

[11] Roberts, A. W. and Varberg, D. E. (1973) *Convex functions*, Academic Press, New York–London.

[12] Rockafellar, R. T. (1970) *Convex analysis*, Princeton University Press, Princeton, NJ.

[13] Rockafellar, R. T. *Integrals which are convex functionals*, Pac. J. Math. 24 (1968), 525-539.

[14] Rockafellar, R. T. *Integrals which are convex functionals, II*, Pac. J. Math. 39 (1971), 439-469.

ZSOLT PÁLES, Institute of Mathematics and Informatics, L. Kossuth University, H-4010 Debrecen, Pf. 12, Hungary, e-mail: pales@math.klte.hu

VERA ZEIDAN, Department of Mathematics, Michigan State University, East Lansing, MI 48824, e-mail: zeidan@math.msu.edu

MARCIN STUDNIARSKI
Characterizations of weak sharp minima of order one in nonlinear programming

1. Introduction

The notion of a weak sharp minimum was introduced by Burke and Ferris in [2]. It is an extension of a strict (or strongly unique [4]) minimum to include the possibility of a nonunique solution set. Weak sharp minima play an important role in the convergence analysis of iterative numerical methods (see Section 4 of [2]). Some results concerning characterizations of such minimizers for constrained optimization problems have been obtained in [7], with special attention given to weak sharp local minimizers of order two. In this paper, we derive a new characterization of weak sharp local minimizers of order one for a standard nonlinear programming problem with constraints of both inequality and equality type. We show that some additional conditions can be attached to the Kuhn-Tucker conditions, giving a set of conditions which characterize this sort of local minimizers. We assume that the functions defining equality constraints are differentiable, while the objective function and the functions defining inequality constraints are locally Lipschitzian and have convex directional derivatives (i.e., are regular in the sense of Clarke).

2. A characterization of the contingent and Ursescu tangent cones to a set of feasible points

In this section, we review a characterization of the contingent cone to a set defined by equality and inequality constraints. It was obtained by the author in [6]. We assume that the functions defining the constraints are locally Lipschitzian and directionally differentiable, so that our characterization can be stated in terms of directional derivatives. The corresponding theorem also includes the statement that the Ursescu tangent cone to such a set is equal to the contingent cone.

Throughout this section, X will be a real Banach space. For $x_0 \in X$ and $\delta > 0$, we denote $B(x_0, \delta) := \{x \in X \mid \|x - x_0\| \leq \delta\}$. For a given subset A of X, the distance function to A is denoted by d_A. Given a locally Lipschitzian function $f : X \to \mathbf{R}$, we denote by $\partial f(x_0)$ the generalized gradient of f at x_0 (see [3], p. 27).

Our characterization is based on the following sufficient condition for the metric regularity of a set of feasible points. It is a particular case of Theorem 3.2(a) in [1].

Theorem 2.1. *Let* $g = (g_1, ..., g_q) : X \to \mathbf{R}^q$ *be a locally Lipschitzian function. Denote* $I := \{1, ..., r\}, J := \{r+1, ..., q\}$, *(where* $1 \leq r \leq q$ *). Define the following*

sets:
$$B := \{y \in \mathbf{R}^q \mid y_i \leq 0, \forall i \in I; y_j = 0, \forall j \in J\}, \tag{2.1}$$
$$C := \{x \in X \mid g(x) \in B\}. \tag{2.2}$$

Suppose that $x_0 \in C$ and define $I(x_0) := \{i \in I \mid g_i(x_0) = 0\}$. Suppose further that the following assumption is satisfied:

(A1) For each $y \in \mathbf{R}^q$ satisfying the conditions
$$y_i = 0, \forall i \in I \setminus I(x_0); y_i \geq 0, \forall i \in I(x_0)$$
the following implication holds:
$$z_i^* \in \partial g_i(x_0) \ (\forall i \in I \cup J), \ \sum_{i=1}^{q} y_i z_i^* = 0 \implies y = 0.$$

Then there exist $K > 0, \delta > 0$ such that
$$d_C(x) \leq K \max\{g_i^+(x), \mid g_j(x) \mid \mid i \in I, j \in J\} \tag{2.3}$$
for all $x \in B(x_0, \delta)$, where $g_i^+ := \max\{g_i, 0\}$.

The following two notions of "tangent" cones (see [8]) will be used in the sequel: the *contingent cone* to C at x, defined by
$$K(C, x) := \{y \mid \exists (t_n, y_n) \to (0^+, y) \text{ such that } x + t_n y_n \in C, \forall n\}, \tag{2.4}$$
and the *Ursescu tangent cone* to C at x, defined by
$$k(C, x) := \{y \mid \forall (t_n) \to 0^+, \exists (y_n) \to y \text{ such that } x + t_n y_n \in C, \forall n\}. \tag{2.5}$$

We can now formulate the main result of this section.

Theorem 2.2. *Let $x_0 \in C$ where C is defined by (2.2). Suppose that the functions g_i, $i \in I \cup J$, are locally Lipschitzian and possess (one-sided) directional derivatives $g_i'(x_0; y)$ for all y. Suppose also that assumption (A1) is satisfied. Then*
$$K(C, x_0) = k(C, x_0) = C(x_0) \tag{2.6}$$
where
$$C(x_0) := \{y \in X \mid g_i'(x_0; y) \leq 0, \forall i \in I(x_0); g_j'(x_0; y) = 0, \forall j \in J\}. \tag{2.7}$$

For the proof, see Section 2 of [6]. In order to compare Theorem 2.2 with the corresponding result in [5], let us observe that, for Fréchet differentiable functions, assumption (A1) is weaker than the assumption used in Theorem 2.3 of [5] that the gradients of all equality constraint functions and all active inequality constraint functions are linearly independent. In fact, if C is the constraint set in a standard nonlinear programming problem for which x_0 is a stationary point, then (A1) is equivalent to the statement that there are no nonzero abnormal Lagrange multipliers at x_0 (see [3], p. 235), that is, the Kuhn-Tucker necessary optimality conditions hold at x_0.

3. Problems with an abstract set constraint

In this section, we consider the following nonlinear programming problem in a finite-dimensional space $\mathbf{R}^p$:
$$\min\{f(x) \mid x \in C\}, \tag{3.1}$$
where $f: \mathbf{R}^p \to \mathbf{R}$ and C is a nonempty closed subset of $\mathbf{R}^p$.

Let $m \geqslant 1$ be an integer. Suppose that f is constant on some closed set $S \subset C$, and that $x_0 \in S$. We say that x_0 is a *weak sharp local minimizer of order m* for (3.1) if there exist $\varepsilon > 0, \beta > 0$ such that
$$f(x) - f(x_0) \geqslant \beta (d_S(x))^m, \qquad \forall x \in C \cap B(x_0, \varepsilon). \tag{3.2}$$

In order to formulate necessary and sufficient conditions for x_0 to be a weak sharp local minimizer of order one for (3.1), we now introduce two concepts of normal cones to S at x_0. First, for any $x \in \mathbf{R}^p$, call
$$P(S, x) := \{w \in S \mid \| x - w \| = d_S(x)\}.$$

Now, let $x_0 \in S$. The *normal cone* to S at x_0 (in the sense of Mordukhovich) is defined by
$$\begin{aligned} N(S, x_0) \ : \ &= \{y \in \mathbf{R}^P \mid \exists \{y_n\} \to y, \{x_n\} \to x_0, \{t_n\} \subset (0, +\infty), \{s_n\} \subset \mathbf{R}^p \\ \text{with } s_n \ &\in \ P(S, x_n) \text{ and } y_n = (x_n - s_n)/t_n\}. \end{aligned} \tag{3.3}$$

We also introduce a variation of the normal cone that takes the set C into account. The *normal cone* to S at x_0 relative to C is defined by
$$\begin{aligned} N_C(S, x_0) \ : \ &= \{y \in \mathbf{R}^P \mid \exists \{y_n\} \to y, \{x_n\} \to x_0, \{t_n\} \subset (0, +\infty), \{s_n\} \subset \mathbf{R}^p \\ \text{with } x_n \ &\in \ C, s_n \in P(S, x_n) \text{ and } y_n = (x_n - s_n)/t_n\}. \end{aligned} \tag{3.4}$$

Comparing (3.3) and (3.4), we see that
$$N_C(S, x_0) \subset N(S, x_0). \tag{3.5}$$

The following theorem is an extension of Theorem 2.2 in [7] (for $m = 1$) to the case of constrained optimization problem (3.1).

Theorem 3.1. *Let S, C be nonempty closed subsets of $\mathbf{R}^p$, such that $S \subset C$. Let $f: \mathbf{R}^p \to \mathbf{R}$ be constant on S, and let $x_0 \in S$. The following conditions are equivalent:*

(a) x_0 is a weak sharp local minimizer of order one for (3.1), that is, condition (3.2) is satisfied for $m = 1$ and for some $\varepsilon > 0, \beta > 0$;

(b) for all $y \in N_C(S, x_0)$ with $\| y \| = 1$, for all sequences $C \supset \{x_n\} \to x_0, \{s_n\} \subset P(S, x_n)$ with $\{(x_n - s_n)/ \| x_n - s_n \|\} \to y$, we have
$$\liminf_{n \to \infty} (f(x_n) - f(s_n))/ \| x_n - s_n \| > 0. \tag{3.6}$$

Proof. (a) $\Longrightarrow$ (b): Suppose that (a) holds. Let $y \in N_C(S, x_0)$ with $\|y\| = 1$, and let $C \supset \{x_n\} \to x_0, \{s_n\} \subset P(S, x_n)$ be such that
$$\{(x_n - s_n)/\|x_n - s_n\|\} \to y.$$
Since $s_n \in S$ and $x_n \in C$, it follows by (a) that there exists $\beta > 0$ such that, for n large enough,
$$f(x_n) - f(s_n) = f(x_n) - f(x_0) \geq \beta d_S(x_n) = \beta \|x_n - s_n\|.$$
Thus
$$\liminf_{n \to \infty} (f(x_n) - f(s_n))/\|x_n - s_n\| \geq \beta > 0,$$
and (b) holds.

(b) $\Longrightarrow$ (a) (by contraposition): Suppose that x_0 is not a weak sharp minimizer of order one for (3.1). Then there exists a sequence $C \supset \{x_n\} \to x_0$ such that
$$f(x_n) - f(x_0) < d_S(x_n)/n. \tag{3.7}$$
For each n, let $s_n \in P(S, x_n)$. Inequality (3.7) implies that $x_n \notin S$, and so $x_n \neq s_n$. Taking a subsequence if necessary, we may assume without loss of generality that the sequence $\{(x_n - s_n)/\|x_n - s_n\|\}$ converges to some y. Then $y \in N_C(S, x_0)$, $\|y\| = 1$, and by (3.7),
$$f(x_n) - f(s_n) = f(x_n) - f(x_0) < d_S(x_n)/n = \|x_n - s_n\|/n.$$
Hence
$$\liminf_{n \to \infty} (f(x_n) - f(s_n))/\|x_n - s_n\| \leq 0.$$
Therefore (b) does not hold, and the proof is complete. $\square$

Condition (b) of Theorem 3.1 gives a general characterization of weak sharp local minimizers of order one for (3.1). However, it is rather difficult to apply it in practice. Therefore, we need some easier conditions stated in terms of certain directional derivatives of f. Following [8], we will use the notation
$$f^K(x; y) := \liminf_{(t,v) \to (0^+, y)} (f(x + tv) - f(x))/t.$$

We will also use the following modification of the previous concept, introduced in [7]. Let A be a nonempty subset of $\mathbf{R}^p$, and let $x \in \operatorname{bd} A$. For $y \in \mathbf{R}^p$, define
$$f_A^K(x; y) := \liminf_{\operatorname{bd} A \ni s \to x; (t,v) \to (0^+, y)} (f(s + tv) - f(s))/t.$$

Now, consider again two closed sets S, C such that $S \subset C$. For $x_0 \in S$, we define
$$S(x_0) := \{x_0\} \cup \bigcup_{x \in C \setminus S} P(S, x).$$

Using these concepts, we can prove a sufficient condition for weak sharp local minimality of order one.

Theorem 3.2. *Let S, C be nonempty closed subsets of $\mathbf{R}^p$, such that $S \subset C$. Let $f : \mathbf{R}^p \to \mathbf{R}$ be constant on S, and let $x_0 \in S$. If*

$$f^K_{S(x_0)}(x_0; y) > 0, \quad \forall y \in N_C(S, x_0) \setminus \{0\}, \tag{3.8}$$

then x_0 is a weak sharp local minimizer of order one for (3.1).

Proof. Let $y \in N_C(S, x_0)$ with $\| y \| = 1$. Then there exist sequences $\{y_n\} \to y, \{x_n\} \to x_0, \{t_n\} \subset (0, +\infty), \{s_n\} \subset \mathbf{R}^p$ such that $x_n \in C, s_n \in P(S, x_n)$ and $y_n = (x_n - s_n)/t_n$. Since $\| y \| = 1$, we may assume without loss of generality that $t_n = \| x_n - s_n \|$. By the definition of $P(S, x_n)$, we have $\| x_n - s_n \| \leqslant \| x_n - x_0 \|$, and so $t_n \to 0^+$. Moreover, $s_n \to x_0$ and $s_n \in \text{bd } S(x_0)$ for all n. Now by (3.8), it follows that

$$\liminf_{n \to \infty}(f(x_n) - f(s_n))/ \| x_n - s_n \| = \liminf_{n \to \infty}(f(s_n + t_n y_n) - f(s_n))/t_n$$
$$\geqslant f^K_{S(x_0)}(x_0; y) > 0.$$

Hence x_0 is a weak sharp local minimizer of order one for (3.1) by Theorem 3.1. □

Necessary conditions for weak sharp local minimizers of of order one for (3.1) have already been established in [8]. We will derive another form of them which, under some additional assumptions, turns out to characterize this sort of minimizers. To simplify the presentation, we assume here that the objective function is locally Lipschitzian.

Theorem 3.3. *Let S, C be nonempty closed subsets of $\mathbf{R}^p$, such that $S \subset C$, and let $f : \mathbf{R}^p \to \mathbf{R}$ be locally Lipschitzian. Suppose that f is constant on S, and $x_0 \in S$. Suppose further that $K(S, x_0) \cap N(S, x_0) = \{0\}$. If x_0 is a weak sharp local minimizer of order one for (3.1), then*

$$f^K(x_0; y) > 0, \quad \forall y \in (N(S, x_0) \cap k(C, x_0)) \setminus \{0\}. \tag{3.9}$$

Proof. Suppose that x_0 is a weak sharp local minimizer of order one for (3.1). We apply the first part of Corollary 4.1 in [8] (note that assumption (8) used there is always satisfied if f is locally Lipschitzian). It follows that there exists $\beta > 0$ such that

$$f^K(x_0; y) \geqslant \beta d_{K(S, x_0)}(y), \quad \forall y \in k(C, x_0). \tag{3.10}$$

Now, take any $y \in (N(S, x_0) \cap k(C, x_0)) \setminus \{0\}$. Since $K(S, x_0) \cap N(S, x_0) = \{0\}$, we have $y \notin K(S, x_0)$, and so $f^K(x_0; y) > 0$ by (3.10) (note that $K(S, x_0)$ is closed). We have thus verified condition (3.9). □

By comparing Theorems 3.2 and 3.3, we can determine assumptions under which condition (3.9) characterizes weak sharp local minima for (3.1).

Theorem 3.4. *Let S, C be nonempty closed subsets of $\mathbf{R}^p$, such that $S \subset C$, and let $f : \mathbf{R}^p \to \mathbf{R}$ be locally Lipschitzian. Suppose that f is constant on S, and $x_0 \in S$. Suppose further that the following conditions are satisfied:*

(b) $N_C(S, x_0) \subset k(C, x_0)$,
(c) $f^K(x_0; y) = f^K_{S(x_0)}(x_0; y)$, $\forall y \in N_C(S, x_0)$.

Then x_0 is a weak sharp local minimizer for (3.1) if and only if condition (3.9) holds.

Proof. If x_0 is a weak sharp local minimizer of order one for (3.1), then condition (3.9) holds by Theorem 3.3. On the other hand, suppose that (3.9) is satisfied. To verify condition (3.8), let us take any $y \in N_C(S, x_0) \setminus \{0\}$. Then, assumption (b) and condition (3.5) imply that $y \in (N(S, x_0) \cap k(C, x_0)) \setminus \{0\}$. Hence, $f^K_{S(x_0)}(x_0; y) = f^K(x_0; y) > 0$ by assumption (c) and condition (3.9). The proof is complete. □

4. Nonlinear programming problems with constraints of both inequality and equality type

In this section, we give characterizations of strict local minimizers of order one for (3.1) with C defined by (2.2), in two versions: the first one is without, and the second one — with Lagrange multipliers. Combining Theorems 2.2 and 3.4, we easily get the following.

Theorem 4.1. *Suppose that the constraint set C in problem (3.1) is given by (2.1), (2.2), where $g : \mathbf{R}^p \to \mathbf{R}^q$ is a locally Lipschitzian function which possesses one-sided directional derivatives at $x_0 \in C$. Suppose also that assumption (A1) is satisfied. Let $f : \mathbf{R}^p \to \mathbf{R}$ be locally Lipschitzian. Suppose further that f is constant on some closed set $S \subset C$ such that $x_0 \in S$, and that assumptions (a)-(c) of Theorem 3.4 are fulfilled (where $k(C, x_0)$ may be replaced by $C(x_0)$). Then x_0 is a weak sharp local minimizer of order one for (3.1) if and only if*

$$f^K(x_0; y) > 0, \qquad \forall y \in (N(S, x_0) \cap C(x_0)) \setminus \{0\}. \tag{4.1}$$

Example 4.2. *Let $f : \mathbf{R}^2 \to \mathbf{R}$ be defined by $f(x_1, x_2) = \max\{0, x_1^2 + x_2^2 - 1\}$. Consider the problem of minimizing f subject to one inequality constraint $g(x_1, x_2) = x_1 \leqslant 0$. Then the point $x_0 = (0, 1)$ is a weak sharp local minimizer of order one for this problem, where the set S is given by*

$$S = \{(x_1, x_2) \mid x_1^2 + x_2^2 \leqslant 1, x_1 \leqslant 0\}.$$

It is easy to verify that
$N(S, x_0) = \{(x_1, x_2) \mid x_1 \geqslant 0, x_2 \geqslant 0\}$,
$N_C(S, x_0) = \{(0, x_2) \mid x_2 \geqslant 0\}$,
$K(S, x_0) = \{(x_1, x_2) \mid x_1 \leqslant 0, x_2 \leqslant 0\}$ *and*

$k(C, x_0) = C(x_0) = \{(x_1, x_2) \mid x_1 \leq 0\}$.

Therefore, assumptions (a) and (b) of Theorem 3.4 are satisfied. Also, assumption (c) holds since $f^K(x_0; (0,1)) = f^K_{S(x_0)}(x_0; (0,1)) = 2$. Moreover, $N(S, x_0) \cap C(x_0) = N_C(S, x_0) = \{(0, x_2) \mid x_2 \geq 0\}$. Hence, condition (4.1) is satisfied which is equivalent to the weak sharp local minimality of x_0. Let us also note that, for this example, the sufficient condition given in Theorem 3.1 of [7] is not fulfilled since the cone $K_S(C, x_0)$ used there is equal to the whole space $\mathbf{R}^2$, and $f^K_S(x_0; (1,0)) = 0$ for $(1,0) \in N(S, x_0)$.

In order to prove a characterization of weak sharp local minimizers in terms of Lagrange multipliers, we will need some additional regularity assumptions on the functions describing the problem. We remind that a locally Lipschitzian function $f : \mathbf{R}^p \to \mathbf{R}$ is *regular* at x if the one-sided directional derivative $f'(x; y)$ exists for all y and satisfies the equality

$$f'(x; y) = f^\circ(x; y) := \limsup_{(t,u) \to (0^+, x)} (f(u + ty) - f(u))/t.$$

Theorem 4.3. *Consider problem (3.1) where C is given by (2.1), (2.2). Suppose that the functions f and $g_i, i \in I$, are locally Lipschitzian and regular at x_0, and the functions $g_j, j \in J$, are strictly differentiable at x_0 (for the definition, see [3], p. 30). Suppose also that assumption (A1) holds. Suppose further that f is constant on some closed set $S \subset C$ such that $x_0 \in S$, and that assumptions (a)-(c) of Theorem 3.4 are fulfilled (where $k(C, x_0)$ may be replaced by $C(x_0)$). Then x_0 is a weak sharp local minimizer of order one for (3.1) if and only if the following conditions are satisfied:*

(a) $N(S, x_0) \cap C(x_0) \cap \{y \in \mathbf{R}^p \setminus \{0\} \mid f'(x_0; y) = 0\} = \emptyset$;

(b) *(Kuhn-Tucker conditions) there exist numbers $\lambda_i, i \in I \cup J$, such that $\lambda_i \geq 0$ for $i \in I$, and*

$$0 \in \partial(f + \sum_{i=1}^{q} \lambda_i g_i)(x_0), \tag{4.2}$$

$$\lambda_i g_i(x_0) = 0 \text{ for } i \in I. \tag{4.3}$$

Proof. (i) *Necessity.* Suppose that x_0 is a weak sharp local minimizer of order one for problem (3.1). Then it is also a local minimizer for this problem; therefore, Theorem 6.1.1 in [3] implies the existence of numbers $\lambda_i, i \in \{0\} \cup I \cup J$, not all equal to zero, such that $\lambda_i \geq 0$ for $i \in \{0\} \cup I$,

$$0 \in \partial(\lambda_0 f + \sum_{i=1}^{q} \lambda_i g_i)(x_0), \tag{4.4}$$

and condition (4.3) holds. Conditions (4.3), (4.4) and the calculus of generalized gradients imply

$$0 \in \lambda_0 \partial f(x_0) + \sum_{i \in I(x_0) \cup J} \lambda_i \partial g_i(x_0).$$

Now, if $\lambda_0 = 0$, then assumption (A1) implies $\lambda_i = 0$ for $i \in I \cup J$ — a contradiction. Hence, $\lambda_0 > 0$, and we may assume $\lambda_0 = 1$, so that (4.2) holds. This completes the proof of (b). Condition (a) follows from Theorem 4.1 and the equality $f^K(x_0; y) = f'(x_0; y)$.

(ii) *Sufficiency.* By Theorem 4.1, it suffices to show that $f'(x_0; y) > 0$ for all $y \in (N(S, x_0) \cap C(x_0)) \setminus \{0\}$. Take any such y. Our assumptions on f and g_i imply that the function $h := f + \sum_{i \in I(x_0) \cup J} \lambda_i g_i$ is regular at x_0 (cf. [3], Proposition 2.3.6). Then, using conditions (4.2), (4.3) and the definition of the generalized gradient, we obtain

$$0 \leqslant h^\circ(x_0; y) = h'(x_0; y) = f'(x_0; y) + \sum_{i \in I(x_0) \cup J} \lambda_i g'_i(x_0; y) \leqslant f'(x_0; y), \qquad (4.5)$$

where the last inequality is a consequence of (2.7). Now, the desired inequality $f'(x_0; y) > 0$ follows from (4.5) and (a). □

5. Conclusions

In this paper, we have presented several theorems characterizing weak sharp local minimizers of order one for nonlinear programming problems. Among them, Theorem 4.3 seems to be most promising as far as practical applications are concerned. However, this requires further investigation with the aim of obtaining easily verifiable criteria for assumptions (a)-(c) of Theorem 3.4 to hold. This is connected with the natural (but difficult) question whether it is possible to obtain characterizations of the cones $K(S, x_0), N(S, x_0)$ and $N_C(S, x_0)$ in terms of functions describing the problem.

References

[1] BORWEIN J.M., Stability and regular points of inequality systems, *J. Optim. Theory Appl.* **48**, 9-52 (1986).

[2] BURKE J.V. & FERRIS M.C., Weak sharp minima in mathematical programming, *SIAM J. Control Optim.* **31**, 1340-1359 (1993).

[3] CLARKE F.H., *Optimization and Nonsmooth Analysis*, John Wiley & Sons, New York (1983).

[4] CROMME L., Strong uniqueness: a far reaching criterion for the convergence of iterative procedures, *Numer. Math.* **29**, 179-193 (1978).

[5] DI S. & POLIQUIN R., Contingent cone to a set defined by equality and inequality constraints at a Fréchet differentiable point, *J. Optim. Theory Appl.* **81**, 469-478 (1994).

[6] STUDNIARSKI M., Characterizations of strict local minima for some nonlinear programming problems, *Nonlinear Analysis Th. Meth. Appl.* **30**, 5363-5367 (1997) (*Proc. 2nd World Congress of Nonlinear Analysts*).

[7] STUDNIARSKI M. & WARD D.E., Weak sharp minima: characterizations and sufficient conditions (to appear).

[8] WARD D.E., Characterizations of strict local minima and necessary conditions for weak sharp minima, *J. Optim. Theory Appl.* **80**, 551-571 (1994).

Faculty of Mathematics, University of Łódź, ul. S. Banacha 22, 90-238 Łódź, Poland. E-mail: marstud@imul.uni.lodz.pl

DOUG WARD
Second-order necessary conditions in nonsmooth programming

1. Introduction

We consider the inequality-constrained program

$$\min \{ f_0(x) \mid f_i(x) \leq 0, \ i \in I \}, \tag{P}$$

where $I = \{1,\ldots,m\}$ and each $f_i : \mathbb{R}^n \to \overline{\mathbb{R}} := [-\infty, +\infty]$. Over the past twenty years, research in nonsmooth analysis has led to first-order necessary optimality conditions for (P) which are valid for large classes of functions that properly contain the class of locally Lipschitzian functions (see e.g. [4-5, 7, 10]). In this paper, we present second-order necessary optimality conditions for (P) in a similar "non-Lipschitzian" setting. These conditions are stated in terms of the generalized Clarke subdifferential [2, 4-5, 7] and the upper parabolic second-order epiderivative [8-9]. We establish these conditions in §3 after reviewing the definitions of the relevant directional derivatives--and stating a key inequality--in §2.

Throughout this paper, we will let $\|\cdot\|$ and $\langle\cdot,\cdot\rangle$ denote, respectively, the Euclidean norm and inner product on $\mathbb{R}^n$, and let $B(x,\varepsilon) := \{y \in \mathbb{R} \mid \|x - y\| \leq \varepsilon\}$ be the closed ball of radius $\varepsilon > 0$ about the point $x \in \mathbb{R}^n$. We will also use the usual notation and terminology of convex and extended-real analysis. In particular, the *epigraph* of a function $f:\mathbb{R}^n \to \overline{\mathbb{R}}$ is defined by $\operatorname{epi} f := \{(x,r) \in \mathbb{R}^n \times \mathbb{R} \mid f(x) \leq r\}$, its *effective domain* is the set $\operatorname{dom} f := \{x \in \mathbb{R} \mid f(x) < +\infty\}$, and f is said to be *proper* if $\operatorname{dom} f$ is nonempty and f never takes on the value $-\infty$. In addition, f is said to be lower semi-continuous (abbreviated l.s.c.) if $\operatorname{epi} f$ is closed, and f is said to be *strictly l.s.c.* at x if there exists $\varepsilon > 0$

such that epi $f \cap B((x,f(x)),\varepsilon)$ is closed. For a convex set $S \subset \mathbb{R}$, the *recession cone of S* is the set
$$0^+S := \{y \in \mathbb{R}^n | x + y \in S \quad \forall\, x \in S\}.$$
Finally, if $C_i \subset \mathbb{R}^n$, $i = 1,\ldots,p$, are convex cones, we say these cones are in *strong general position* ([7, 10, 11]) if
$$\bigcap_{i=1}^{m} C_i - C_{m+1} = \mathbb{R}^n \quad \forall\, m = 1,\ldots,p-1.$$
As is shown in [11], this property does not depend upon the order in which the cones are listed.

2. Directional Derivatives and Subdifferentials

Let $f: \mathbb{R}^n \to \overline{\mathbb{R}}$ be finite at $x \in \mathbb{R}^n$. Our optimality conditions for (P) involve two first-order directional derivatives: the *upper epiderivative* of f at x in the direction $y \in \mathbb{R}^n$, defined by

$$f^+(x;y) := \limsup_{\substack{t \to 0^+ \\ w \to y}} \inf (f(x+tw) - f(x))/t$$
$$:= \sup_{\varepsilon > 0} \inf_{\lambda > 0} \sup_{t \in (0,\lambda)} \inf_{w \in B(y,\varepsilon)} (f(x+tw) - f(x))/t;$$

and the *upper subderivative* of f at x in the direction y, which for l.s.c. functions is defined by

$$f^\uparrow(x;y) := \limsup_{\substack{t \to 0^+ \\ z \to_f x}} \inf_{w \to y} (f(z+tw) - f(z))/t,$$

where "$z \to_f x$" stands for "$z \to x$ with $f(z) \to f(x)$". (A general definition, valid for arbitrary functions, can be found in [4-5,7].) We will also make use of the *generalized Clarke subdifferential*

$$\partial f(x) := \{x^* \in \mathbb{R}^n | \langle x^*, y \rangle \leq f^\uparrow(x;y) \quad \forall\, y \in \mathbb{R}^n\}.$$

The calculus and applications of $f^+(x;\cdot)$, $f^\uparrow(x;\cdot)$, and ∂f are discussed in [1, 4-5, 7] and their references. We note that $f^+(x;\cdot)$ is l.s.c., while $f^\uparrow(x;\cdot)$ is l.s.c. and sublinear and $\partial f(x)$ is the usual subdifferential of $f^\uparrow(x;\cdot)$ at 0. In addition, if f is Lipschitzian near x, then dom $f^+(x;\cdot)$ = dom $f^\uparrow(x;\cdot) = \mathbb{R}^n$ (in fact, $f^+(x;\cdot)$ and $f^\uparrow(x;\cdot)$ are themselves Lipschitzian) and these directional derivatives may be written (see e.g., [1, Chap. 6]) as

$$f^+(x;y) := \limsup_{t \to 0^+} (f(x + ty) - f(x))/t$$

and

$$f^\uparrow(x;y) = f^o(x;y) := \limsup_{\substack{t \to 0^+ \\ z \to x}} (f(z + tw) - f(z))/t,$$

where f^o is the *Clarke directional derivative* [2].

If $v \in \mathbb{R}^n$ is such that $f^+(x;v)$ is finite, we define the *upper parabolic second-order directional derivative of f* by

$$f^{++}(x,v;y) :=$$
$$= \limsup_{t \to 0^+} \inf_{w \to y} 2(f(x + tv + t^2w/2) - f(x) - tf^+(x;v))/t^2.$$

The properties and calculus of $f^{++}(x,v;y)$ are detailed in [1, 8-9] and their references. In particular, if f is twice Fréchet differentiable at x with derivative $\nabla f(x)$ and second derivative $\nabla^2 f(x)$, then it follows from Taylor's Theorem that

$$f^{++}(x,v;y) = \langle \nabla f(x), y \rangle + \nabla^2 f(x)(v,v); \quad (2.1)$$

and if f is Lipschitzian near x, then as in [6, Prop. 3.4],

$$f^{++}(x,v;y)$$
$$= \limsup_{t \to 0^+} 2(f(x + tv + t^2y/2) - f(x) - tf^+(x;v))/t^2. \quad (2.2)$$

In general, it does not seem possible to split $f^{++}(x,v;y)$ into first- and second-order terms. However, we do have the following inequality:

Proposition 2.1. Let $f: \mathbb{R}^n \to \overline{\mathbb{R}}$ be finite at $x \in \mathbb{R}^n$, and suppose that $f^+(x;v)$ is finite. Then for all $y_1, y_2 \in \mathbb{R}^n$,

$$f^{++}(x,v;y_1 + y_2) \le f^{++}(x,v;y_1) + f^\uparrow(x;y_2). \quad (2.3)$$

In particular, for all $y \in \mathbb{R}^n$,

$$f^{++}(x,v;y) \le f^{++}(x,v;0) + f^\uparrow(x;y). \qquad (2.4)$$

Proof. Apply [8, Lemma 2.4] to the set $S := \text{epi } f$ with $x := (x, f(x))$ and $v := (v, f^+(x;v))$ to obtain (2.3). Inequality (2.4) then follows from (2.3) when we set $y_1 := 0$, $y_2 := y$. ∎

Inequality (2.4) provides a sort of "upper convex approximate" for $f^{++}(x,v;\cdot)$. It will play an important role in the proof of our optimality conditions.

Remark 2.2. If f is continuously Fréchet differentiable near x_0, then one can use the mean value theorem to show that equality holds in (2.4). (See [9, Proposition 4.2].)

3. Second-Order Neccessary Optimality Conditions

Let $x_0 \in \mathbb{R}^n$ be feasible for problem (P). Define
$$I(x_0) := \{i \in I \mid f_i(x_0) = 0\}$$
and $J(x_0) := \{0\} \cup I(x_0)$. For $v \in \mathbb{R}^n$ with $f_i^+(x_0;v) \le 0$ for all $i \in J(x_0)$, set
$$J(x_0, v) := \{i \in J(x_0) \mid f_i^+(x_0;v) = 0\}.$$

In this section, we will use the calculus of f^+ and f^{++}, inequality (2.4), and the sublinearity of $f^\uparrow$ to prove some second-order necessary optimality conditions for problem (P).

Theorem 3.1. Let x_0 be a local minimizer for (P). Suppose that f_i, $i \in J(x_0)$, are strictly l.s.c. at x_0 with $f_i^+(x_0;\cdot)$ proper; f_i, $i \in I \setminus I(x_0)$, are continuous at x_0; and dom $f_i^\uparrow(x_0;\cdot)$, $i \in J(x_0)$, are in strong general position. Let $v \in \mathbb{R}$ be such that $f_i^+(x_0;v) \le 0$ for all $i \in J(x_0)$, dom $f_i^{++}(x_0,v;\cdot) = \mathbb{R}^n$ for all $i \in J(x_0) \setminus J(x_0,v)$, and $f_i^{++}(x_0,v;0)$ is finite for all $i \in J(x_0,v)$. Then there exist $\lambda_i \ge 0$, $i \in J(x_0,v)$, not all equal zero, such that
$$0 \in \sum_{i \in J(x_0,v)} \lambda_i \, \partial f_i(x_0) \qquad (3.1)$$
and

$$0 \le \sum_{i \in J(x_0,v)} \lambda_i f_i^{++}(x_0,v;0), \qquad (3.2)$$

where "$\lambda_i \partial f_i(x_0)$" means $0^+ \partial f_i(x_0)$ whenever $\lambda_i = 0$. Moreover, if $0 \in J(x_0,v)$ and

$$\operatorname{dom} f_0^\uparrow(x_0;\cdot)$$
$$\cap \{y \mid f_i^\uparrow(x_0;y) < 0 \ \forall\, i \in J(x_0,v)\setminus\{0\}\} \ne \emptyset, \qquad (3.3)$$

then $\lambda_0 \ne 0$ in (3.1) and (3.2).

Proof. Define $F(x) := \max\{f_0(x) - f_0(x_0), f_i(x), i \in I\}$. Then x_0 is a local minimizer of F. Since $f_i^+(x_0;\cdot)$ is proper for each $i \in J(x_0)$, $f_i^+(x_0;0) = 0$, and so [8, Theorem 2.8, Proposition 4.1] imply that

$$0 \le F^+(x_0;y) = \max_{J(x_0)} f_i^+(x_0;y) \quad \forall\, y \in \mathbb{R}^n. \qquad (3.4)$$

Let $v \in \mathbb{R}^n$ satisfy $f_i^+(x_0;v) \le 0$ for all $i \in J(x_0)$, with $\operatorname{dom} f_i^{++}(x_0,v;\cdot) = \mathbb{R}^n$ for all $i \in J(x_0)\setminus J(x_0,v)$ and $f_i^{++}(x_0,v;0)$ finite for all $i \in J(x_0,v)$. Then by (3.4), we have $F^+(x_0;v) = 0$, and we may apply [8, Theorem 2.8, Proposition 4.1] and Proposition 2.1 to obtain the inequalities

$$0 \le F^{++}(x_0,v;y)$$
$$= \max_{J(x_0,v)} f_i^{++}(x_0,v;y) \qquad (3.5)$$
$$\le \max_{J(x_0,v)} \{f_i^{++}(x_0,v;0) + f_i^\uparrow(x_0;y)\} \quad \forall\, y \in \mathbb{R}^n.$$

Now let p be the cardinality of $J(x_0,v)$, and define the sets

$$S_1 := \{z \in \mathbb{R}^p \mid z \le 0\}$$

and

$$S_2 := \{w \in \mathbb{R}^p \mid (f_i^{++}(x_0,v;0) + f_i^\uparrow(x_0;y))_{i \in J(x_0,v)} \le w$$
$$\text{for some } y \in \mathbb{R}^n\}.$$

Observe that S_1 and S_2 are convex and $\operatorname{int} S_1 \cap S_2 = \emptyset$, so there exist a nonzero $\lambda \in \mathbb{R}^p$, $\lambda = (\lambda_i)_{i \in J(x_0,v)}$, and $\alpha \in \mathbb{R}$ with

$$\langle \lambda, z \rangle \leq \alpha \leq \langle \lambda, w \rangle \quad \forall z \in S_1, \forall w \in S_2. \tag{3.6}$$

By (3.6), each $\lambda_i \geq 0$ and $\alpha \geq 0$. Hence for all $y \in \mathbb{R}^n$,

$$0 \leq \sum_{i \in J(x_0,v)} \lambda_i \{f_i^{++}(x_0,v;0) + f_i^{\uparrow}(x_0;y)\}. \tag{3.7}$$

Letting $y = 0$ in (3.7), we obtain (3.2). Because $f_i^{\uparrow}(x_0;\cdot)$ is positively homogeneous and $f_i^{++}(x_0,v;0)$ is finite, (3.7) also implies that

$$0 \leq \sum_{i \in J(x_0,v)} \lambda_i f_i^{\uparrow}(x_0;y) \quad \forall y \in \mathbb{R}^n. \tag{3.8}$$

The subgradient sum formula [3, Theorem 23.8] then yields

$$0 \in \sum_{i \in J(x_0,v)} \partial(\lambda_i f_i)(x_0),$$

from which (3.1) follows, provided we interpret $\lambda_i \partial f_i(x_0)$ as $0^+ \partial f_i(x_0)$ whenever $\lambda_i = 0$. Finally, if $0 \in J(x_0,v)$ and (3.3) holds, then $\lambda_0 \neq 0$ by (3.8). ∎

Remark 3.2. In Theorem 3.1, the constraint qualification "$0 \in J(x_0,v)$ and (3.3) holds" may be replaced by

$$\text{dom } f_0^{\uparrow}(x_0;\cdot)$$
$$\cap \{y \mid f_i^{\uparrow}(x_0;y) < 0 \quad \forall i \in I(x_0)\} \neq \emptyset. \tag{3.9}$$

To see this, observe that (3.9) implies (3.3). In addition, (3.9) and the hypotheses of Theorem 3.1 also imply (by [8, Theorem 4.7], e.g.) that

$$f_0^+(x_0;y) \geq 0 \quad \forall y \text{ such that } f_i^+(x_0;y) \leq 0 \quad \forall i \in I(x_0),$$

from which it follows that $0 \in J(x_0,v)$.

We illustrate Theorem 3.1 with a simple example:

Example 3.3. In problem (P), let $m = 1$ and define f_0, $f_1: \mathbb{R} \to \mathbb{R}$ by $f_0(x_1,x_2) := |x_1|^{1/2} - x_2^2$, $f_1(x_1,x_2) = -x_2$. Let $x_0 = (0,0)$. In this example, dom $f_0^{\uparrow}(x_0;\cdot) = \{0\} \times \mathbb{R}$ and dom $f_1^{\uparrow}(x_0;\cdot) = \mathbb{R}^2$, so these sets are in strong general position. In addition, f_0 and f_1 are strictly l.s.c. and $f_0^+(x_0;\cdot)$ and $f_1^+(x_0;\cdot)$ are proper.

Let $v = (0,1)$. Then $J(x_0,v) = \{0,1\}$, $f_0^{++}(x_0,v;0) = -2$, and $f_1^{++}(x_0,v;0) = 0$. The only nonzero (λ_0, λ_1) for which (3.1) holds are $\lambda_0 > 0$, $\lambda_1 = 0$, and (3.2) does not hold for any of these (λ_0, λ_1). We therefore conclude that x_0 is not a local minimizer in this example.

In the case in which each f_i is Lipschitzian near x_0, the statement of Theorem 3.1 can be substantially simplified. Note first that if f is Lipschitzian near x, then by (2.2),
$$f^{++}(x,v;0) = d_+^2(x;v) := \limsup_{t \to 0^+} 2(f(x+tv) - f(x) - tf^+(x;v))/t^2.$$
As a result, the locally Lipschitzian case of Theorem 3.1 has the following form:

Corollary 3.4. Let f_i, $i \in I \cup \{0\}$, be Lipschitzian near x_0, a local minimizer of (P). Let $v \in \mathbb{R}^n$ be such that $f_i^+(x_0;v) \leq 0$ for all $i \in J(x_0)$ and $d_+^2 f_i(x_0;v)$ is finite for all $i \in J(x_0)$. Then there exist $\lambda_i \geq 0$, $i \in J(x_0,v)$, not all equal zero, such that

$$0 \in \sum_{i \in J(x_0,v)} \lambda_i \partial f_i(x_0) \tag{3.10}$$

and

$$0 \leq \sum_{i \in J(x_0,v)} \lambda_i d_+^2 f_i(x_0;v). \tag{3.11}$$

Moreover, if $0 \in J(x_0,v)$ and

$$\{y| f_i^\circ(x_0;y) < 0 \ \forall \ i \in J(x_0,v)\setminus\{0\}\} \neq \emptyset, \tag{3.12}$$

then $\lambda_0 \neq 0$ in (3.10) and (3.11).

Proof. Since each f_i is Lipschitzian near x_0, f_i is continuous at x_0 and $f_i^+(x_0;\cdot)$ and $f\uparrow(x_0;\cdot) = f^\circ(x_0;\cdot)$ are finite-valued. In particular, then, each $f_i^+(x_0;\cdot)$ is proper and dom $f_i^\uparrow(x_0;\cdot)$, $i \in I(x_0)$, are in strong general position. Since each $d_+^2 f_i(x_0;v)$ is finite, (2.4) implies that dom $f_i^{++}(x_0,v;\cdot) = \mathbb{R}^n$. By Theorem 3.1, we obtain (3.10) and (3.11). Finally, (3.3) reduces to (3.12) in this case, so the final assertion also follows from Theorem 3.1. ∎

Remark 3.5. (a) In the case in which each f_i is continuously Fréchet differentiable near x_0, Theorem 3.1 gives the necessary optimality conditions in [8, Corollary 5.5].
(b) The assumption in Corollary 3.4 that $d_\pm^2 f_i(x_0;v)$ is finite is satisfied, in particular, when f is $C^{1,1}$ at x_0 (see [9]).

4. Conclusion

We have derived second-order necessary optimality conditions that are valid for inequality-constrained programs whose objective and constraint functions belong to a large class of possibly nonsmooth functions. These conditions should have applications elsewhere in optimization theory--in particular, in the calculation of first- and second-order directional derivatives of marginal functions of mathematical programs with perturbations. We plan to investigate these applications in a future paper.

References

1. J.-P. Aubin and H. Frankowska, *Set-Valued Analysis* (Birkhäuser, Boston, 1990).

2. F.H. Clarke, *Optimization and Nonsmooth Analysis* (Wiley, New York, 1983).

3. R.T. Rockafellar, *Convex Analysis* (Princeton University Press, Princeton, N.J., 1970).

4. R.T. Rockafellar, "Directionally Lipschitzian functions and subdifferential calculus," *Proceedings of the London Mathematical Society* 39 (1979) 331-355.

5. R.T. Rockafellar, *The Theory of Subgradients and its Applications to Problems of Optimization: Convex and Nonconvex Functions* (Heldermann, Berlin, 1981).

6. M. Studniarski, "Second-Order Necessary Conditions for Optimality in Nonsmooth Nonlinear Programming," *Journal of Mathematical Analysis and Applications* 154 (1991) 303-317.

7. D.E. Ward and J.M. Borwein, "Nonsmooth calculus in finite dimensions," *SIAM Journal on Control and Optimization* 25 (1987) 1312-1340.

8. D.E. Ward, "Calculus for parabolic second-order derivatives," *Set-Valued Analysis* 1 (1993) 213-246.

9. D.E. Ward, "A comparison of second-order epiderivatives: calculus and optimality conditions," *Journal of Mathematical Analysis and Applications* 193 (1995) 465-482.

10. D.E. Ward, "Dini derivatives of the marginal function of a non-Lipschitzian program," *SIAM Journal on Optimization* 6 (1996) 198-211.

11. C. Zalinescu, "On convex sets in general position," *Linear Algebra and its Applications* 64 (1985) 191-198.

ROXIN ZHANG
Upper Hölder continuity of inverse subdifferentials and the proximal point algorithm

1. Introduction

Let f be an extended real-valued, lower-semicontinuous (l.s.c.) function on $\mathbb{R}^n$. We are concerned with two important properties of f: the local growth rate of f from a given subset in $\mathbb{R}^n$ and the upper Hölder continuity of the inverse subdifferential multifunction of f at 0. It has been well observed that these two properties play important roles in many areas of optimization such as the convergence of numerical algorithms, sensitivity analysis, optimality conditions and so on[1]-[12].

Let us begin with some of the standard notations and definitions.

Definition 1.1. *The set $\partial f(\bar{z}) \subset \mathbb{R}^n$ is said to be a subdifferential of f at $\bar{z} \in \mathbb{R}^n$ in the sense of Mordukhovich [7] if*

$$\partial f(\bar{z}) := \limsup_{z \xrightarrow{f} \bar{z}} \{v | \langle v, d \rangle \leq f^k(z;d), \forall d \in \mathbb{R}^n\},$$

where $f^k(z;d)$ is the contingent directional derivative of f at z

$$f^-(z;d) := \liminf_{\substack{d' \to d \\ \tau \downarrow 0}} \frac{f(z + \tau d') - f(z)}{\tau},$$

and limsup is the Kuratowski-Painlevé upper limit for multifunctions.

When f is convex, ∂f reverts to the subdifferential in the convex analysis. Therefore, through the rest of this paper, we simply refer ∂f defined above as the subdifferential of f. The Euclidian distance from a point x to the set C in $\mathbb{R}^n$ is denoted by $\text{dist}(x, C)$ and the corresponding closed ball of x with radius δ by $\mathbb{B}(x, \delta)$. The notation $\mathbb{B}$ by itself stands for the unit ball at the origin.

Definition 1.2. *A multifunction $F : \mathbb{R}^m \rightrightarrows \mathbb{R}^n$ is said to be upper Hölder continuous of order κ at the point $\bar{v} \in \mathbb{R}^m$ if there exist $\tau > 0$ and $a \geq 0$ such that*

$$F(v) \subset F(\bar{v}) + a||v - \bar{v}||^\kappa \mathbb{B}, \qquad \forall v \in \mathbb{B}(\bar{v}, \tau)$$

where κ is referred to as the upper Hölder modulus of F.

Equivalently, one can also state the above definition as follows: F is upper Hölder continuous at $\bar{v}$ if there exist $\tau > 0$ and $a > 0$ such that for all $v \in \mathbb{B}(\bar{v}, \tau)$

and $z \in F(v)$ one has $\text{dist}(z, F(\bar{v})) \leq a||\bar{v}||$. The latter has been extensively used by Luque [5] in his paper to show the convergences of a proximal point algorithm.

When F is upper Hölder continuous of order $\kappa = 1$ at $\bar{v}$, F is said to be upper Lipschitz continuous at the point in the sense of Robinson [8]. Without loss of generality, we may always assume $\tau < 1$ in the above definition. Therefore, for a multifunction F at a point $\bar{v}$, the upper Hölder continuity of order κ implies the upper Hölder continuity of order κ' if $\kappa > \kappa'$.

It should be pointed out that, in the definition 1.2, we do not exclude the possibility that κ takes the values beyond the interval $(0, 1)$, which differs in this aspect from the definition given by Klatte [4]. Especially, κ can be either 0 or ∞. Refer to Robinson [7], Zhang and Treiman [9] and the references therein for further details on the upper Lipschitz continuity of a multifunction.

Definition 1.3. *A l.s.c. function f is said to have a local growth rate of p from the set C if there exist $c > 0$ and $\delta > 0$ such that*

$$f(z) \geq f^* + a[\text{dist}(z, C)]^p, \qquad \forall z \in C + \delta \mathbb{B}$$

where $f^* := \inf\{f(z) | z \in C\}$.

Once again, without loss of generality, we can always assume $\delta < 1$ in the above definition, therefore, a function with growth rate of p must have a growth rate of p' if $p' < p$.

Functions with a linear ($p = 1$) local growth rate from their argmin sets have been referred to as having *weak sharp minima* [2], [3]. In general, functions with a local growth rate p from their argmin set are referred to as having *local weak sharp minima of order p* [11].

A function with linear growth rate from its argmin set shares many nice properties with the polyhedral functions and so does a function with quadratic growth rate with strongly convex functions. Many classical results regarding functions with isolated minimizers of order p have been extended to the cases with nonunique minimizers, especially since the introduction and the extensive study of the concept of weak sharp minima by M. C. Ferris and J. Burke [2], [3]. Their work is mainly concentrated on the functions with linear growth rate. D. Ward has extended the study to the areas concerning functions with general growth rate p and defined them as having weak sharp minima of order p. D. Ward [11] has contributed work along the line of A. Auslender [1] and M. Studniarski [10] on the necessary conditions for the weak sharp minima of general order.

On the other hand, many results have been obtained by utilizing the upper Lipschitz or upper Hölder continuity properties of the inverse subdifferential multifunctions. The prominent work regarding the convergence of the Proximal Point Algorithm (PPA) includes that of M. C. Ferris [3] and F. J. Luque [5]. M. C. Ferris has shown that the PPA for $0 \in \partial f(x)$ has finite convergence if f has *sharp minima* (of order 1) while a result of F. J. Luque [5] can be interpreted as that the PPA converges linearly if the inverse subdifferential multifunction of f is upper Lipschitz continuous at 0.

The upper Hölder continuity is most suitable for the study of a subdifferential or a inverse subdifferential multifunction. Other Lipschitz-type continuities such as the Pseudo Lipschitz continuity and Lipschitz continuity for multifunctions will most likely fail to hold in these cases. The same assertion can be made to the solution multifunction of a parametric mathematical programming problem. As a simple example, consider the problem of minimizing $f(\cdot) + \langle \cdot, v \rangle$ where f is an extended real-valued proper convex function. One has that the solution multifunction $S(v)$ coincides with the inverse subdifferential multifunction. In a much more general setting, recently D. Klatte has shown that for a parametric mathematical programming problem, under certain conditions, the solution multifunction is upper Hölder continuous of order κ with respect to the parameter if the objective function has a growth rate of $1/\kappa$ [4].

One of the main purposes of this paper is to establish the bridges between the two important properties: the growth rate of a l.s.c. function from the set of all minimizers or the set of stationary points $\{x|0 \in \partial f(x)\}$ and the upper Hölder continuity of the inverse subdifferential multifunction of the under line function.

2. Functions with Upper Hölder Continuous Inverse Subdifferentials

In this section, we will study the following questions: What growth properties does f have if its inverse subdifferential multifunction satisfies an upper Hölder continuity condition at $\bar{v} = 0$ and vise versa? What are the specific relationships between the growth rate of f and the upper Hölder continuity modulus of its inverse subdifferential multifunction.

Let f be a l.s.c. function on $\mathbb{R}^n$, we denote the inverse of the subdifferential multifunction of f by $(\partial f)^{-1}$ and the set of all minimizers of f by S. Intuitively, when $(\partial f)^{-1}$ is upper Hölder continuous at 0, any point z with $\partial f(z)$ containing small elements must be "close" to the set of the stationary points $(\partial f)^{-1}(0)$ at the rate of κ.

In the following theorem and its corollaries, we will show that:

(a) The upper Hölder continuity of $(\partial f)^{-1}$ at 0 of order κ implies that the function f yields a growth rate of $(1+\kappa)/\kappa$ from the set $(\partial f)^{-1}(0)$;
(b) In the above result, One can replace $(\partial f)^{-1}(0)$ by S if certain "separation condition" holds.
(c) If f is a closed proper convex function, then the upper Hölder continuity of $(\partial f)^{-1}$ at 0 of order κ and the fact that the function f yields a local growth rate of $(1+\kappa)/\kappa$ from the argmin set S are equivalent.

Theorem 2.1. *If $(\partial f)^{-1}$ is upper Hölder continuous of order κ at 0, then for any $C \subset (\partial f)^{-1}(0)$ there exist $c > 0$ and $\delta > 0$ such that*

$$f(z) \geq \inf f + c[\operatorname{dist}(z, (C + 2\delta \mathbb{B}) \cap (\partial f)^{-1}(0))]^{(1+\kappa)/\kappa}, \qquad \forall z \in C + \delta \mathbb{B}. \quad (2.1)$$

In particular, by setting $C := (\partial f)^{-1}(0)$, one has

$$f(z) \geq \inf f + c[\mathrm{dist}(z, (\partial f)^{-1}(0))]^{(1+\kappa)/\kappa}, \qquad \forall z \in (\partial f)^{-1}(0) + \delta \mathbb{B}. \tag{2.2}$$

Proof. If $\inf f = -\infty$ or $(\partial f)^{-1}(0) = \emptyset$, then (2.1) holds trivially. So let us assume $\inf f > -\infty$, $(\partial f)^{-1}(0) \neq \emptyset$ and $(\partial f)^{-1}$ is upper Hölder continuous of order κ at 0, then there exist $a > 0$ and $\tau > 0$ such that for all $v \in \tau \mathbb{B}$ and $z \in (\partial f)^{-1}(v)$, one has

$$\mathrm{dist}(z, (\partial f)^{-1}(0)) \leq a \|v\|^\kappa. \tag{2.3}$$

Let $E := (\partial f)^{-1}(0)$. We first prove (2.2). In fact, if (2.2) were not true, then for arbitrary $c_1 > 0$ and $\tau_1 > 0$ there exists $z_1 \in E + \tau_1 \mathbb{B}$ such that

$$f(z_1) < \inf f + c_1 [\mathrm{dist}(z_1, E)]^{(1+\kappa)/\kappa}. \tag{2.4}$$

By Ekeland's variational principle, one has that for all $\lambda > 0$ there exists y_1 such that $\|z_1 - y_1\| \leq \lambda$ and

$$f(y_1) \leq f(z) + \frac{c_1}{\lambda}[\mathrm{dist}(z_1, E)]^{(1+\kappa)/\kappa} \|z - y_1\|, \qquad \forall z.$$

Namely, y_1 minimizes the function $z \mapsto f(z) + (c_1/\lambda)[\mathrm{dist}(z_1, E)]^{(1+\kappa)/\kappa} \|z - y_1\|$. Consequently by the elementary properties of ∂f and the fact that $\partial \|\cdot\| = \mathbb{B}$, one has

$$0 \in \partial f(y_1) + (c_1/\lambda)[\mathrm{dist}(z_1, E)]^{(1+\kappa)/\kappa} \mathbb{B}.$$

Now for arbitrary $\tau' > 0$ and $a > 0$, selecting λ, c_1 and τ_1 such that $\lambda := \theta \, \mathrm{dist}(z_1, E)$ with $\theta \in (0, 1)$, $c_1 \leq \theta[(1-\theta)/2a]^{1/\kappa}$ and $\tau_1 < (\theta \tau'/c_1)^\kappa$, one obtains that

$$\|z_1 - y_1\| \leq \theta \, \mathrm{dist}(z_1, E)$$

and there exists $v_1 \in \partial f(y_1)$ with $v_1 \neq 0$ such that

$$\|v_1\| \leq \frac{c_1}{\lambda}[\mathrm{dist}(z_1, E)]^{(1+\kappa)/\kappa} = \frac{c_1}{\theta}[\mathrm{dist}(z_1, E)]^{1/\kappa} \leq \frac{c_1}{\theta}\tau_1^{1/\kappa} < \tau'. \tag{2.5}$$

It follows that $v_1 \in \tau' \mathbb{B}$ and

$$\mathrm{dist}(z_1, E) \geq (\theta/c_1)^\kappa \|v_1\|^\kappa. \tag{2.6}$$

On the other hand,

$$\mathrm{dist}(z_1, E) \leq \|z_1 - y_1\| + \mathrm{dist}(y_1, E) \leq \theta \, \mathrm{dist}(z_1, E) + \mathrm{dist}(y_1, E). \tag{2.7}$$

Combining (2.6) and (2.7), one has

$$\mathrm{dist}(y_1, E) \geq (1-\theta) \, \mathrm{dist}(z_1, E) \geq (1-\theta)(\frac{\theta}{c_1})^\kappa \|v_1\|^\kappa \geq 2a \|v_1\|^\kappa. \tag{2.8}$$

Thus we have shown that for arbitrary $\tau' > 0$ and $a > 0$, there exist $v_1 \in \tau'\,\mathbb{B}$ and $y_1 \in (\partial f)^{-1}(v_1)$ such that $\mathrm{dist}(z_1, (\partial f)^{-1}(0)) \geq 2a||v_1||^\kappa$. This contradicts with the upper Hölder continuity of $(\partial f)^{-1}$ of order κ as assumed in the theorem. Therefore, (2.2) must be true.

To verify (2.1), it suffices to show that for all $z \in C + \delta\,\mathbb{B}$ one has $\mathrm{dist}(z, (C + 2\delta\,\mathbb{B}) \cap E) \leq \mathrm{dist}(z, E)$. In fact, let $z' \in E$ such that $||z - z'|| = \mathrm{dist}(z, E)$, then

$$\mathrm{dist}(z', C) \leq ||z' - z|| + \mathrm{dist}(z, C) = \mathrm{dist}(z, E) + \mathrm{dist}(z, C) \leq 2\,\mathrm{dist}(z, C) \leq 2\delta$$

since $C \subset E$. It follows that $z' \in (C + 2\delta\,\mathbb{B}) \cap E$ and therefore $\mathrm{dist}(z, E) = ||z - z'|| \geq \mathrm{dist}(z, (C + 2\delta\,\mathbb{B}) \cap E)$. This completes the proof. □

Notice that when $\kappa = 1$, the Theorem 2.1 claims the existence of $c > 0$ and $\delta > 0$ such that

$$f(z) \geq \inf f + c[\mathrm{dist}(z, (C + 2\delta\,\mathbb{B}) \cap (\partial f)^{-1}(0))]^2, \qquad \forall z \in C + \delta\,\mathbb{B}$$

provided that $(\partial f)^{-1}$ is upper Lipschitz continuous at 0. This is the result given in Theorem 4.2 of Zhang and Treiman [9] which becomes a special case of the Theorem 2.1.

We must point out that, in general, one can not replace $(\partial f)^{-1}(0)$ in (2.2) by the set of minimizers S. An counter example is given in the Example after the next two corollaries. However, this can be done if certain conditions that "separates" the set S from the rest of $(\partial f)^{-1}(0)$ are satisfied. To be specific, we have the following Corollary.

Corollary 2.2. *If $(\partial f)^{-1}$ is upper Hölder continuous of order κ at 0 and either $S = (\partial f)^{-1}(0)$ or there exists $\epsilon > 0$ satisfying*

$$(S + \epsilon\,\mathbb{B}) \cap (\partial f)^{-1}(0) \subset S, \tag{2.9}$$

then there exist $c > 0$ and $\delta > 0$ such that

$$f(z) \geq \inf f + c[\mathrm{dist}(z|S)]^{(1+\kappa)/\kappa}, \qquad \forall z \in S + \delta\,\mathbb{B}. \tag{2.10}$$

Proof. In (2.1), by setting $C := S$ and observing that

$$\mathrm{dist}(z, (S + \delta\,\mathbb{B}) \cap (\partial f)^{-1}(0)) \leq \mathrm{dist}(z, S),$$

one obtains (2.10) immediately. □

As a special case of the above theorem, we have a necessary condition for a function to have weak sharp minima (of order 1) as stated in the following Corollary.

Corollary 2.3. *If $(\partial f)^{-1}$ is upper Lipschitz continuous with constant 0, i.e., there exists $\tau > 0$ such that*

$$(\partial f)^{-1}(v) \subset (\partial f)^{-1}(0), \qquad \forall v \in \tau \mathbb{B} \tag{2.11}$$

and either $S = (\partial f)^{-1}(0)$ or (2.9) is satisfied, then there exist $c > 0$ and $\delta > 0$ such that

$$f(z) \geq \inf f + c\,\mathrm{dist}(z, S), \qquad \forall z \in S + \delta \mathbb{B}. \tag{2.12}$$

Proof. If (2.11) holds, then for an arbitrary $\kappa \geq 0$ and all $v \in \tau \mathbb{B}$ one has $(\partial f)^{-1}(v) \subset (\partial f)^{-1}(0) + \|v\|^\kappa \mathbb{B}$. By Corollary 2.2, there exist $c > 0$ and $\delta > 0$ such that

$$f(z) \geq \inf f + c[\mathrm{dist}(z, S)]^{(1+\kappa)/\kappa}, \qquad \forall z \in S + \delta \mathbb{B}.$$

Let $\kappa \to \infty$, one obtains (2.12). □

From the calculus of the subdifferential ∂f, it is necessary that $0 \in \partial f(z)$ if z minimizes f. Therefore one always has $S \subset (\partial f)^{-1}(0)$. In this extend, the essence of (2.9) demands the property that the sets S and $(\partial f)^{-1}(0)\backslash S := \{z | z \in (\partial f)^{-1}(0), z \notin S\}$ are "separated". One obvious example of functions with this property is a convex function since $S = (\partial f)^{-1}(0)$. On the other hand, the examples that fail the condition (2.9) can be constructed as the staircase functions in which the set of minimizing points can not be "separated" from the set $(\partial f)^{-1}(0)\backslash S$.

Example 2.4. *For a given constant $r > 0$, consider the function*

$$f(z) := \begin{cases} (1/2)^{r(n+1)} & \text{if } (1/2)^{n+1} < |z| \leq (1/2)^n \\ 0 & \text{otherwise} \end{cases}.$$

This function fails the condition (2.9) because $S = \{0\}$ and $(\partial f)^{-1}(0) = \mathbb{R}$. On the other hand, f has a growth rate of r since it is minorized by the power function z^r while $(\partial f)^{-1}$ is upper Hölder continuous of any order at 0. Thus the specific relationship between the order of the upper Hölder continuity of $(\partial f)^{-1}$ and the growth rate in (2.10) ceases to exist.

It also should be noticed that even under the condition (2.9), the property (2.10) dose not imply the upper Hölder continuity of $(\partial f)^{-1}$ of order κ. In other words, the converse of the Corollary 2.2 is not true in general. However, the sufficient part the Corollary dose hold if f happens to be closed proper convex. The details are shown in the next theorem.

Theorem 2.5. *If f is a closed proper convex function with $S \neq \emptyset$, then $(\partial f)^{-1}$ is upper Hölder continuous of order κ at 0 if and only if there exist $\delta > 0$ and $c > 0$ such that*

$$f(z) \geq \inf f + c[\mathrm{dist}(z, S)]^{(1+\kappa)/\kappa}, \qquad \forall z \in S + \delta \mathbb{B}. \tag{2.13}$$

Proof. If $(\partial f)^{-1}$ is upper Hölder continuous of order κ at 0, the growth property (2.13) follows from the Corollary 2.2 since the condition (2.9) holds for convex functions. To prove the converse, let us assume f satisfies (2.13). Since f is closed proper convex, the graph of ∂f is closed from the convex analysis. Therefore both ∂f and $(\partial f)^{-1}$ are upper semicontinuous everywhere. Specifically, one has

$$\limsup_{v \to 0} (\partial f)^{-1}(v) \subset (\partial f)^{-1}(0) = S.$$

It follows that there exists $\tau > 0$ such that

$$\text{dist}(z, S) \leq \delta, \, \forall v \in \mathbb{B}(0, \tau) \text{ and } z \in (\partial f)^{-1}(v).$$

On the other hand, by the assumption, f satisfies (2.13)

$$f(z) - \inf f \geq c[\text{dist}(z, S)]^{1+1/\kappa}, \, \forall z \in S + \delta \mathbb{B}.$$

Let $z_0 \in S$ such that $\text{dist}(z, S) = \|z - z_0\|$. The above inequality becomes

$$c[\text{dist}(z, S)]^{1/\kappa} \|z - z_0\| \leq f(z) - f(z_0),$$

and therefore

$$c[\text{dist}(z, S)]^{1/\kappa} \leq \frac{f(z) - f(z_0)}{\|z - z_0\|}. \tag{2.14}$$

Due to the convexity of f, one has

$$f(z) - f(z_0) \leq \langle v, z - z_0 \rangle \leq \|v\| \cdot \|z - z_0\|. \tag{2.15}$$

Combining (2.14) with (2.15), we have

$$\text{dist}(z, S) \leq c^{\kappa} \|v\|^{\kappa}, \, \forall v \in \mathbb{B}(0, \tau) \text{ and } z \in (\partial f)^{-1}(v).$$

Which concludes that $(\partial f)^{-1}$ is upper Hölder continuous of order κ. □

Remarks. Theorem 2.1 and all of its corollaries are valid if we replace ∂f by any other type of "subdifferential" as long as the following properties hold true:
(a) $\partial(f + g) \subset \partial f + \partial g$,
(b) $0 \in \partial f(y)$ if y minimizes f and
(c) the set $\partial \|\cdot\|$ is bounded.

3. The Convergence of the Proximal Point Algorithms

In this section, we will apply the Theorem 2.5 to some of the known results regarding the convergence rate of the PPA for a convex programming problem. Consequently, the convergence conditions are interpreted in terms of the local growth rate of the underline function.

For a maximal monotone mapping T, the original form of the Proximal Point Algorithm (PPA), introduced by Martinet [4], uses the proximal mapping $(I+cT)^{-1}$ of Moreau to solve the problem $0 \in T(z)$ by generating a sequence $\{z_k\}$ such that

$$z_{k+1} = P_k(z_k) \tag{3.1}$$

where $P_k := (I + \lambda_k T)^{-1}$ which is single-valued and $\lambda_k > 0$.

The convergence rate of the algorithm has been extensively studies by many authors under various conditions, see for example [3], [5] and [9]. Rockafellar [8] has considered the PPA with the generation of the sequence z_k inexact

$$||z_{k+1} - P_k(z_k)|| \leq \delta_k ||z_{k+1} - z_k||, \quad \sum_0^\infty \delta_k < \infty \tag{3.2}$$

and showed, assuming $T^{-1}(0)$ is a singleton, that the PPA has finite convergence if T^{-1} is polyhedral, linear convergence if T^{-1} is *strictly Lipschitz*. The latter property is actually equivalent to T^{-1} being upper Lipschitz continuous at 0 with $T^{-1}(0)$ containing a single point.

Luque [5] later extended the convergence results of Rockafellar by removing the assumption that $T^{-1}(0)$ is a singleton. M. C. Ferris [2] has also considered the convergence issue of the PPA with exact iterations and shown that $\{z_k\}$ has finite convergence if f has *sharp minima* (see [3] for exact definitions).

Let us first summarize some of Luque's results in terms of the upper Hölder or upper Lipschitz continuity of the mapping T_{-1}.

Lemma 3.1. *Let $\{z_k\}$ be the sequence generated by (3.2) and consider the convergence of the sequence of $\{z_k\}$ to a point in $T^{-1}(0)$.*

(a) *The convergence is finite if T^{-1} is upper Lipschitz continuous at 0 with modulus 0, i.e., there exist $\tau > 0$ such that*

$$T^{-1}(v) \subset T^{-1}(0), \forall v \in \tau \mathbb{B}.$$

(b) *The convergence is linear if T^{-1} is upper Lipschitz continuous at 0 with modulus $a > 0$.*

c. *The convergence rate is $o(k^{\kappa/2})$ if T^{-1} is upper Hölder continuous of order κ at 0.*

Applying the PPA to a convex programming problem of minimizing $f(z)$ by letting $T := \partial f$, one has that z minimizes f if and only if z solves $0 \in T(z)$. In this context, $P_k = (I + \lambda_k \partial f)^{-1}$ and z_{k+1} generated from (3.1) can be calculated as a solution to the problem of minimizing the function $f(z_k) + (1/2\lambda_k)||z - z_k||^2$.

A direct application of the Theorem 2.5 to the Lemma 3.1 reveals the fact that the convergence rate of the PPA applied to a convex programming problem mainly depends on the local growth properties of f.

Theorem 3.2. *Let f be a closed proper convex function with $S := \operatorname{argmin} f \neq \emptyset$, $T := \partial f$ and $\{z_k\}$ be generated by (3.2). Consider the convergence of $\{z_k\}$ to a point in S.*
(a) *The convergence is finite if f has a (local) linear growth rate from S.*
(b) *The convergence is linear if f has a (local) quadratic growth rate from S.*
(c) *The convergence rate is $o(k^\kappa)$ if f has a (local) growth rate of $(2\kappa + 1)/\kappa$ from S.*

4. Acknowledgments

The author would like to thank Professor R. T. Rockafellar for pointing out some of the facts regarding the PPA.

REFERENCES

2. BURKE, J. V. & FERRIS, M. C., Weak sharp minima in mathematical programming, *SIAM J. Control and Optimization*, **31**, No. 5, 1340-1359.
3. FERRIS, M. C., Finite termination of the proximal point algorithm, *Math. Programming*, **50**, 359-366, (1991).
4. KLATTE, D., On quantitative stability for non-isolated minima, *Control and Cybernetics*, **23**, No. 1/2, 183-200 (1994).
5. LUQUE, F. J. Asymptotic convergence analysis of the proximal point algorithm, *SIAM J. Control and Optimization*, **22**, No. 2, 277-293, 1984.
6. MARTINET, B., Regularity, d'inéquations variationelles par approximations succesives, *Rev. Francaise d'Inform Recherche Oper.*, **4**, 154-159, (1970).
7. MORDUKHOVICH, B. S., Sensitivity analysis in nonsmooth optimization, *Theoretical Aspects of Industrial Design (Edited by D. A. Field and V. Komkov)*, SIAM, 32-46, (1992).
8. ROBINSON, S. M., Generalized equations and their solutions, Part I: basic theory, *Math. Program. Study*, **10**, 128-141, (1979).
9. ROCKAFELLAR, R. T., Monotone operators and proximal point algorithm, *SIAM J. Control Optim.*, **14**, 877-898, (1976).
10. STUDNIARSKI, M., Necessary and sufficient conditions for isolated local minima of nonsmooth functions, *SIAM j. Control and Optimization*, **24**, 1044-1049, 1986.
11. WARD, D. E., Characterization of strict local minima and necessary conditions for weak sharp minima, *J. Optimization and Appl.*, **80**, No. 3, 551-571, 1994.
12. ZHANG, R. & TREIMAN, J., Upper-Lipschitz multifunctions and inverse subdifferentials, *Nonlinear Analysis, Theory, Methods & Appl.*, **24**, No. 2, 273-286, (1995).

III. Optimization and operations research

V P IL'EV AND I V OFENBAKH
Optimization models and algorithms of constructing reliable networks[*]

1 Introduction

Reliability is one of the most important properties of communication networks. Reliability of a network depends on its topology and the concept of structural reliability characterizes this dependence.

The communication network is modeled adequately by a graph. In this way connectedness of a graph is the natural index of structural reliability of a network. We remind that, according to Whitney's criterion (Harary, 1969), a graph $G = (V, U)$ is k-connected if and only if for all $u, v \in V$ there are at least k vertex-disjoint (u, v)-chains. We consider *the optimal synthesis of the structurally reliable communication network problem* whose mathematical setting can be formulated as follows.

Shortest k-connected spanning subgraph problem. Let $G = (V, U)$ be an undirected graph with the vertex set V and the edge set U, the connectedness of G is greater or equal to k. Let $f : U \to R_+$ be an additive non-negative weight function. One has to find the k-connected spanning subgraph of the graph G with minimal weight.

For $k = 1$ we have the well-known *shortest spanning tree problem*; for $k \geq 2$ the problem becomes much more complicated. We consider this problem as special case of more general minimization problem over an universal combinatorial object called a system.

2 Systems: main notions

Let U be a finite set and $\mathcal{A} \subseteq 2^U$ be a nonvoid family of subsets of U with the following monotonicity property:
$$(A \in \mathcal{A},\ A' \subseteq A) \Rightarrow A' \in \mathcal{A}.$$

Then $\mathcal{A}$ is called an *independence system* on U. Sets $A \in \mathcal{A}$ are called *independent*, all sets $D \in 2^U \setminus \mathcal{A}$ are called *dependent*. We denote by $\mathcal{D}$ the family of all dependent sets. It is easy to see that the family $\mathcal{D}$ satisfies the following condition:
$$(D \in \mathcal{D},\ D \subseteq D') \Rightarrow D' \in \mathcal{D}.$$

---
[*]This work was supported in part by the Russian Fund for Basic Research, Grant No 97-01-00771

Each of the families $\mathcal{A}, \mathcal{D}$ defines the other uniquely. We shall call the pair $\mathcal{S} = (\mathcal{A}, \mathcal{D})$ the *system* on U.

Bases of the system $\mathcal{S}$ are maximal (under inclusion) independent sets and *circuits* of $\mathcal{S}$ are minimal (under inclusion) dependent sets. The families of all bases and all circuits of $\mathcal{S}$ are denoted by $\mathcal{B}$ and $\mathcal{C}$, respectively.

For a system $\mathcal{S} = (\mathcal{A}, \mathcal{D})$ we define the *dual system* $\overline{\mathcal{S}} = (\overline{\mathcal{A}}, \overline{\mathcal{D}})$ as follows:

$$\overline{\mathcal{A}} = \{\overline{A} \subseteq U : U \setminus \overline{A} \in \mathcal{D}\}.$$

It is easy to see that $\overline{\overline{\mathcal{S}}} = \mathcal{S}$.

For $W \subseteq U$ we denote by $\mathcal{B}(W)$ the family of all maximal (under inclusion) sets of $\mathcal{A}$ contained in W and by $\mathcal{C}(W)$ the family of all minimal (under inclusion) sets of $\mathcal{D}$ containing the set W.

There are two rank functions of an independence system. For $W \subseteq U$ we set

$$lr(W) = \min\{|B| : B \in \mathcal{B}(W)\} (lower\ rank\ of\ W),$$

$$ur(W) = \max\{|B| : B \in \mathcal{B}(W)\}\ (upper\ rank\ of\ W).$$

The quantity

$$c_\mathcal{A} = c_\mathcal{A}(\mathcal{S}) = \min_{W \subseteq U} \frac{lr(W)}{ur(W)}$$

is called an *independence curvature* of the system $\mathcal{S}$.

Similarly, we define two new fundamental measures, the *lower girth* and the *upper girth* of W as the cardinality of the smallest and the largest minimal dependent superset of W, respectively:

$$lg(W) = \min\{|C| : C \in \mathcal{C}(W)\}\ (lower\ girth\ of\ W),$$

$$ug(W) = \max\{|C| : C \in \mathcal{C}(W)\}\ (upper\ girth\ of\ W).$$

We also define the *dependence curvature* of the system $\mathcal{S}$ as

$$c_\mathcal{D} = c_\mathcal{D}(\mathcal{S}) = \max_{\substack{W \subseteq U,\\ W \notin \mathcal{D}}} \frac{ug(W) - |W|}{lg(W) - |W|}.$$

Obviously, $0 < c_\mathcal{A} \leq 1$ and $c_\mathcal{D} \geq 1$. A system with $c_\mathcal{A} = 1$ is called a *matroid* and a system with $c_\mathcal{D} = 1$ is called an *upper matroid* on U.

Theorem 2.1. *Let $\mathcal{S} = (\mathcal{A}, \mathcal{D})$ be an arbitrary system on U and $\overline{\mathcal{S}} = (\overline{\mathcal{A}}, \overline{\mathcal{D}})$ be a dual system. Then*

$$\frac{1}{c} \leq \overline{c},$$

where $c = c_\mathcal{A}(\mathcal{S})$, $\overline{c} = c_{\overline{\mathcal{D}}}(\overline{\mathcal{S}})$.

Proof. It is not difficult to show, by taking complements of sets, that
$$\overline{c} = \max_{\substack{W \subseteq U \\ W \notin \mathcal{A}}} \frac{|W| - lr(W)}{|W| - ur(W)}.$$

We suppose that $c < 1$ (in case $c = 1$ the conclusion of the Theorem is trivial). Let W^* be a minimal (under inclusion) set satisfying $c = lr(W^*)/ur(W^*)$. Then $W^* = B_1 \cup B_2$, where $B_1, B_2 \in \mathcal{B}(W^*)$, $|B_1| = lr(W^*)$, $|B_2| = ur(W^*)$. Thus, $|W^*| \leq ur(W^*) + lr(W^*)$. We have

$$|W^*|(ur(W^*) - lr(W^*)) \leq ur^2(W^*) - lr^2(W^*),$$

hence

$$ur(W^*)(|W^*| - ur(W^*)) \leq lr(W^*)(|W^*| - lr(W^*)).$$

Then

$$\frac{1}{c} = \max_{W \subseteq U} \frac{ur(W)}{lr(W)} = \frac{ur(W^*)}{lr(W^*)} \leq \frac{|W^*| - lr(W^*)}{|W^*| - ur(W^*)} \leq \max_W \frac{|W| - lr(W)}{|W| - ur(W)} = \overline{c}.$$

The Theorem follows.

Quantities $c_\mathcal{A}$ and $c_\mathcal{D}$ play an important role in discrete optimization. They can be considered as measures of worst-case behaviour of the greedy algorithms over systems.

3 Systems and the greedy algorithms

Let $\mathcal{S} = (\mathcal{A}, \mathcal{D})$ be a system on a finite set U. We consider the following combinatorial minimization problem

$$\min\{f(X) : X \in \mathcal{D}\}, \tag{1}$$

where $f : U \to R_+$ is an additive non-negative function and $\mathcal{D}$ is the family of all dependent sets of the system $\mathcal{S}$.

We note that the family $\mathcal{D}$ of all k-connected spanning subgraphs of a graph $G = (V, U)$ (considered as subsets of the edge set U) is the family of all dependent sets of some system $\mathcal{S}$. So the above mentioned shortest k-connected spanning subgraph problem is special case of the problem (1).

Just as significant part of combinatorial problems (1) the shortest k-connected spanning subgraph problem is NP–hard (Garey and Johnson, 1979). Apparently there are no algorithms which yield an optimal solution to this problem at the polynomial time. Now we formulate a polynomial-time approximate algorithm of finding a k-connected spanning subgraph of a given graph which is a version of the greedy algorithm over a system.

procedure GD (*Greedy Descent Algorithm*);
begin

let $\{u_1, \ldots, u_m\}$ be an ordering of U with $f(u_i) \geq f(u_{i+1})$;
$Gr := U$;
for $i = 1$ **to** m **do**
    if $Gr \setminus \{u_i\} \in \mathcal{D}$ **then** $Gr := Gr \setminus \{u_i\}$;
end.

We note that as a rule the following combinatorial problems over independence systems are considered:

$$\max\{f(X) : X \in \mathcal{B}\} \quad \text{or} \quad \min\{f(X) : X \in \mathcal{B}\}, \qquad (2)$$

where $\mathcal{B}$ is the family of all bases of the system $\mathcal{S}$.

We formulate the greedy algorithm for the maximization problem (2) over an independence system:

procedure GA (*Greedy Ascent Algorithm*);
begin
    let $\{u_1, \ldots, u_m\}$ be an ordering of U with $f(u_i) \leq f(u_{i+1})$;
    $Gr := \emptyset$;
    for $i = 1$ **to** m **do**
        if $Gr \cup \{u_i\} \in \mathcal{A}$ **then** $Gr := Gr \cup \{u_i\}$;
end.

B.Korte and D.Hausmann obtained the following bound on worst–case behaviour of the greedy algorithm GA for the maximization problem (2) over an independence system in terms of its curvature.

Theorem 3.1 (Korte and Hausman, 1978). *Let $\mathcal{S}$ be an arbitrary system on U, $Opt \in \mathcal{A}$ be an optimal solution to the maximization problem (2) and $Gr \in \mathcal{A}$ be a solution produced by the greedy algorithm* GA. *Then for any additive objective function $f : U \to R_+$*

$$\frac{f(Gr)}{f(Opt)} \geq c_\mathcal{A}.$$

In (Korte and Hausman, 1978) authors proved that there is no bound on worst–case behaviour of similar greedy ascent algorithm for the minimization problem (2) over an independence system, i.e. this algorithm can yield an arbitrary bad solution.

In (Il'ev, 1996) the similar results were obtained for the problem (1).

Theorem 3.2 (Il'ev, 1996). *Let $\mathcal{S}$ be an arbitrary system on U, $Opt \in \mathcal{D}$ be an optimal solution to the problem (1) and $Gr \in \mathcal{D}$ be a solution yielded by the algorithm* GD. *Then for any additive objective function $f : U \to R_+$*

$$\frac{f(Gr)}{f(Opt)} \leq c_\mathcal{D}.$$

In (Il'ev, 1996) the upper bound on the dependence curvature of the family $\mathcal{D}$ of all 2-connected spanning subgraphs of an arbitrary n-vertex graph was obtained: $c_\mathcal{D} \leq n-2$. This immediately implies the bound on worst-case behaviour of the greedy descent algorithm for the shortest 2-connected spanning subgraph problem:

$$\frac{f(Gr)}{f(Opt)} \leq n-2.$$

As shown in (Il'ev, 1996), this bound is tight.

For $k \geq 3$ the problem is open. Therefore, finding bounds on curvatures of a system is of great interest. Some of them will be presented in the next Section.

4 Bounds on curvatures

We remind that system $\mathcal{S} = (\mathcal{A}, \mathcal{D})$ is a *matroid* if $c_\mathcal{A} = 1$ and an *upper matroid* if $c_\mathcal{D} = 1$.

As a corollary of Theorem 3.1 we can obtain the following well-known result (Rado, 1957), (Edmonds, 1971).

Theorem 4.1 (Rado–Edmonds). *The greedy ascent algorithm* GA *yields an optimal solution to the maximization problem* (2) *for any additive objective function* $f : U \to R_+$ *iff the system* $\mathcal{S} = (\mathcal{A}, \mathcal{D})$ *is a matroid.*

Similar result for upper matroids and the algorithm GD is an immediate corollary of Theorem 3.2.

Theorem 4.2. *The greedy descent algorithm* GD *yields an optimal solution to the minimization problem* (1) *for any additive objective function* $f : U \to R_+$ *iff the system* $\mathcal{S} = (\mathcal{A}, \mathcal{D})$ *is an upper matroid.*

It is well known that a system $\mathcal{S}$ with family of independent sets $\mathcal{A}$ is a matroid if and only if for all $A, A' \in \mathcal{A}$

$$|A'| = |A| + 1 \Rightarrow \exists\, a \in A' \setminus A : A \cup a \in \mathcal{A}. \tag{3}$$

It is easy to show that a system $\mathcal{S}$ with family of dependent sets $\mathcal{D}$ is an upper matroid if and only if for all $D, D' \in \mathcal{D}$

$$|D| = |D'| + 1 \Rightarrow \exists\, d \in D \setminus D' : D \setminus d \in \mathcal{D}. \tag{4}$$

We define two new structural characteristics of a system which can be considered as measures of nearness of the system and matroids.

For a system $\mathcal{S} = (\mathcal{A}, \mathcal{D})$ we define $\beta_\mathcal{A} = \beta_\mathcal{A}(\mathcal{S})$ as a minimal positive integer $\beta \geq 1$ with the following property: for all $A, A' \in \mathcal{A}$

$$|A'| = |A| + \beta \Rightarrow \exists\, a \in A' \setminus A : A \cup a \in \mathcal{A}. \tag{5}$$

Similarly, we define an index $\beta_\mathcal{D} = \beta_\mathcal{D}(\mathcal{S})$ as a minimal positive integer $\beta \geq 1$ with the following property: for all $D, D' \in \mathcal{D}$

$$|D| = |D'| + \beta \Rightarrow \exists\, d \in D \setminus D' : D \setminus d \in \mathcal{D}. \tag{6}$$

We note that $\beta_\mathcal{A} < |U|$ ($\beta_\mathcal{D} < |U|$) and $\beta_\mathcal{A} = 1$ ($\beta_\mathcal{D} = 1$) if and only if the system $\mathcal{S} = (\mathcal{A}, \mathcal{D})$ is a matroid (an upper matroid).

Theorem 4.3. *If $\mathcal{S} = (\mathcal{A}, \mathcal{D})$ and $\overline{\mathcal{S}} = (\overline{\mathcal{A}}, \overline{\mathcal{D}})$ are dual systems then*

$$\beta_\mathcal{A}(\mathcal{S}) = \beta_{\overline{\mathcal{D}}}(\overline{\mathcal{S}}).$$

Proof. It is easy to show, by taking complements of sets, that the following propositions are equivalent:
(a) $|A'| = |A| + \beta \Rightarrow \exists\, a \in A' \setminus A : A \cup a \in \mathcal{A}$, for all $A, A' \in \mathcal{A}$;
(b) $|D| = |D'| + \beta \Rightarrow \exists\, d \in D \setminus D' : D \setminus d \in \overline{\mathcal{D}}$, for all $D, D' \in \overline{\mathcal{D}}$.
This equivalence implies the conclusion of Theorem.

Lemma 4.4. *For any system $\mathcal{S} = (\mathcal{A}, \mathcal{D})$ on U*

$$\beta_\mathcal{A} = \max_{W \subseteq U}(ur(W) - lr(W) + 1).$$

Proof. We prove first that

$$\beta_\mathcal{A} \geq ur(W) - lr(W) + 1, \quad \text{for any } W \subseteq U. \tag{7}$$

Suppose $\beta_\mathcal{A} \leq ur(W) - lr(W)$. Let $A, B \in \mathcal{B}(W)$, $|A| = lr(W)$, $|B| = ur(W)$ and let C be an arbitrary subset of $B \setminus A$ with cardinality $ur(W) - lr(W) - \beta$. We set $A' = B \setminus C$. Then $|A'| = |A| + \beta$ and by (5) there is $a \in A' \setminus A : A \cup a \in \mathcal{A}$, a contradiction to the maximality of A.

Obviously, conclusion of Lemma holds for $\beta_\mathcal{A} = 1$ (because in this case $\mathcal{S}$ is a matroid). To prove it for $\beta_\mathcal{A} > 1$ we note that by definition of $\beta_\mathcal{A}$ there exist $A, B \in \mathcal{A}$ with $|B| = |A| + \beta_\mathcal{A} - 1$ and $A \cup b \notin \mathcal{A}$, for all $b \in B \setminus A$. Hence $A \in \mathcal{B}(W^*)$, where $W^* = A \cup B$ and

$$ur(W^*) \geq |B| = |A| + \beta_\mathcal{A} - 1 \geq lr(W^*) + \beta_\mathcal{A} - 1.$$

We have $\beta_\mathcal{A} \leq ur(W^*) - lr(W^*) + 1$ that, together with (7), implies $\beta_\mathcal{A} = ur(W^*) - lr(W^*) + 1$.

The Lemma follows.

Theorem 4.5. *For any system* $\mathcal{S} = (\mathcal{A}, \mathcal{D})$

$$c_{\mathcal{A}} \geq \frac{1}{\beta_{\mathcal{A}}}.$$

Proof. By Lemma 4.4 for $W \subseteq U$ we have

$$\frac{lr(W)}{ur(W)} \geq \frac{1}{ur(W) - lr(W) + 1} \geq \frac{1}{\beta_{\mathcal{A}}},$$

hence

$$c_{\mathcal{A}} = \min_{W \subseteq U} \frac{lr(W)}{ur(W)} \geq \frac{1}{\beta_{\mathcal{A}}}.$$

Corollary. *Let $\mathcal{S}$ be an arbitrary system on U, $Opt \in \mathcal{A}$ be an optimal solution to the maximization problem (2) and $Gr \in \mathcal{A}$ be a solution produced by the greedy algorithm* GA. *Then for any additive objective function* $f : U \to R_+$

$$\frac{f(Gr)}{f(Opt)} \geq \frac{1}{\beta_{\mathcal{A}}}.$$

Bounds on curvature $c_{\mathcal{D}}$ and worst–case behaviour of the greedy descent algorithm for the problem (1) can be proved similarly.

Theorem 4.6. *For any system* $\mathcal{S} = (\mathcal{A}, \mathcal{D})$

$$c_{\mathcal{D}} \leq \beta_{\mathcal{D}}.$$

Corollary. *Let $\mathcal{S}$ be an arbitrary system on U, $Opt \in \mathcal{D}$ be an optimal solution to the minimization problem (1) and $Gr \in \mathcal{D}$ be a solution produced by the greedy algorithm* GD. *Then for any additive objective function* $f : U \to R_+$

$$\frac{f(Gr)}{f(Opt)} \leq \beta_{\mathcal{D}}.$$

Finally, as a consequence of Theorems 4.5 and 4.6 we have

Theorem 4.7. *For any dual systems* $\mathcal{S} = (\mathcal{A}, \mathcal{D})$ *and* $\overline{\mathcal{S}} = (\overline{\mathcal{A}}, \overline{\mathcal{D}})$

$$\frac{1}{\beta} \leq c \leq 1 \leq \bar{c} \leq \beta,$$

where $c = c_{\mathcal{A}}(\mathcal{S})$, $\bar{c} = c_{\overline{\mathcal{D}}}(\overline{\mathcal{S}})$ *and* $\beta = \beta_{\mathcal{A}}(\mathcal{S}) = \beta_{\overline{\mathcal{D}}}(\overline{\mathcal{S}})$.

References

Edmonds J. (1971) *Matroids and the greedy algorithm.* Math. Programming, v.1, No.2, p.127–136.

Garey, M.R. and Johnson D.S. (1979) *Computers and Intractability: a Guide to the Theory of NP-completeness.* W.H.Freeman, San Francisco.

Harary, F. (1969) *Graph Theory.* Addison–Welsey, Reading, MA.

Il'ev V.P. (1996) *Error estimation of the gradient algorithm for independence systems.* Diskret. Analiz i Issledovanie Oper., v.3, No.1, p.9–22 (Russian).

Korte, B. and Hausmann D. (1978) *An analysis of the greedy heuristic for independence systems.* Annals of Discrete Mathematics, v.2, p.65–74.

Rado R. (1957) *Note on independence functions.* Proc. London Math. Soc., v.7, No.26, p.300–320.

Victor P. Il'jev
Omsk State University
department of mathematics
PO Box 8501, Omsk 644070
Russia
Iljev@univer.omsk.su

Igor V. Ofenbakh

230 W. Sumner ave. #70B
Roselle Park, NJ 07204
USA
vlad7@idt.net

JOSEF STOER, MARTIN WECHS AND SHINJI MIZUNO
Higher order methods for solving sufficient linear complementarity problems

1. Introduction

In this paper we consider horizontal linear complementarity problems

(LCP)
$$Px + Qy = q,$$
$$x \circ y = 0,$$
$$x \geq 0, \ y \geq 0$$

Here P, Q are $n \times n$-matrices with

$$\operatorname{rk}[P,Q] = n,$$

and

$$x \circ y := [x_1 y_1, \ldots, x_n y_n]^T$$

is the componentwise product of the vectors x and y. By e we denote the vector $e := [1,\ldots,1]^T$ and by $X := \operatorname{diag}(x)$ the diagonal matrix belonging to a vector x. The vector $z = (x,y)$ is called *feasible* for (LCP) if $Px + Qy = q$ and $z \geq 0$; it is a *solution* of (LCP) if in addition $x \circ y = 0$. The set of all solutions is denoted by S, and Φ is the linear subspace

$$\Phi := \mathcal{N}[P,Q] = \{(x,y) \mid Px + Qy = 0\}.$$

We consider only LCP's that are *monotone* or at least *sufficient*: Here, Φ, resp. (LCP), is called *monotone* if $x^T y \geq 0$ for all $(x,y) \in \Phi$. It is wellknown that all convex quadratic programs and in particular all linear programs can be formulated as LCP's that are monotone. Still more general are *sufficient* LCP's introduced by Cottle Pang and Venkateswaran [1] that can be defined as follows:

Φ, resp. (LCP) is *column-sufficient* if

$$(x,y) \in \Phi, \quad x \circ y \leq 0 \implies x \circ y = 0,$$

it is *row-sufficient* if

$$(x,y) \in \Phi^\perp, \quad x \circ y \geq 0 \implies x \circ y = 0,$$

and it is called *sufficient* if Φ is row- and column-sufficient.

In the following we use only the following rather weak assumption

(V)
1) $\text{rk}\,[P,Q] = n$, i.e. $\dim \Phi = n$,
2) (LCP) is sufficient,
3) $S \neq \emptyset$.

Note that the degeneracy of (LCP) is not excluded, it may have more than one solution.

It is customary to describe the structure of S by the index sets

$$I := \{i \mid \exists (x,y) \in S : x_i > 0\},$$
$$J := \{j \mid \exists (x,y) \in S : y_j > 0\},$$
$$K := \{k \mid x_k = y_k = 0 \;\forall (x,y) \in S\}.$$

For sufficient LCP's it is known (see [1]) that I, J and K form a partition of

$$N := \{1, 2, \ldots, n\} = I \cup J \cup K.$$

Note that $K = \emptyset$ iff (LCP) has strictly complementary solutions. In particular this holds for those LCP's resulting from linear programs.

Interior-point methods for *strictly feasible* LCP's follow the path of solutions $z = (x,y)(r,\eta)$, $r \downarrow 0$, of

$$Px + Qy = q, \quad (x,y) > 0,$$
$$x \circ y = r\eta,$$

where $\eta > 0$ is a positive weight vector.

More generally, if strictly feasible solutions do not exist or are not known, it is customary to choose $r_0 = 1$, $x_0 > 0$, $y_0 > 0$, to put $\bar{q} := Px_0 + Qy_0 - q$, $\eta := x_0 \circ y_0$ and to follow the *infeasible-interior-point path* of solutions $z = (x,y)(r,\eta)$, $r \downarrow 0$, of

$(LCP)_{r,\eta}$
$$Px + Qy = q + r\bar{q}, \quad (x,y) > 0,$$
$$x \circ y = r\eta.$$

Then by $S \neq \emptyset$, the set

$$M_r := \{z = (x,y) > 0 \mid Px + Qy = q + r\bar{q}\}$$

is nonempty for all $0 < r \leq 1$. So it is natural to assume in the following

(V') $\qquad\qquad\qquad$ (V) and $M_1 \neq \emptyset$.

In this paper we describe a simple infeasible-interior-point method of arbitrarily high order for solving linear complementarity problems (LCP) in the difficult case, when (LCP) has no strictly complementary solutions, $K \neq \emptyset$, and this under the assumption (V') only.

2. Analysis of interior-point paths

The analytical properties of infeasible interior-point-paths depend on whether $K = \emptyset$ or not. The following main results were obtained in [4], that we quote without proof:

Lemma 1: *Under assumption (V') the system $(LCP)_{r,\eta}$ has a unique solution $(x,y)(r,\eta)$ for all $0 < r \leq 1$, $\eta > 0$, so that infeasible-interior-point paths are well defined.*

Theorem 1: *Assume (V') and $K = \emptyset$. Then the normalized functions $(\tilde{x}, \tilde{y})(r, \eta)$ defined by*

$$(x_I(r,.), x_J(r,.)) =: (\tilde{x}_I(r,.), r\tilde{x}_J(r,.)),$$
$$(y_I(r,.), y_J(r,.)) =: (r\tilde{y}_I(r,.), \tilde{y}_J(r,.)),$$

and $(x,y)(r,\eta)$ are analytic functions of (r,η) for $0 < r \leq 1$, $0 < \eta$, that can be analytically extended to $r = 0$ and satisfy $\tilde{x}(0,\eta) > 0$, $\tilde{y}(0,\eta) > 0$ for $\eta > 0$.

In particular, the limits of all derivatives of order m with respect to r exist for all $m \geq 0$ and all $\eta > 0$

$$(x^{(m)}, y^{(m)})(0, \eta) = \lim_{r \downarrow 0} \left(\frac{\partial}{\partial r}\right)^m (x,y)(r,\eta). \tag{2.1}$$

Remark: The theorem applies to the LCP's associated with linear programs.

Theorem 2: *Assume (V') and $K \neq \emptyset$. Put $\rho := \sqrt{r}$, $\rho^2 = r$, and define the functions $\tilde{x}, \tilde{y}$ as functions of (ρ, η) by*

$$(x_I(\rho^2,.), x_J(\rho^2,.), x_K(\rho^2,.)) =: (\tilde{x}_I(\rho,.), \rho^2 \tilde{x}_J(\rho,.), \rho \tilde{x}_K(\rho,.)),$$
$$(y_I(\rho^2,.), y_J(\rho^2,.), y_K(\rho^2,.)) =: (\rho^2 \tilde{y}_I(\rho,.), \tilde{y}_J(\rho,.), \rho \tilde{y}_K(\rho,.)).$$

Then $(\tilde{x}, \tilde{y})(\rho, \eta)$ are analytic functions of (ρ, η) for $0 < \rho \leq 1$, $\eta > 0$ that can be analytically extended to $\rho = 0$ so that $\tilde{x}(0,\eta) > 0$, $\tilde{y}(0,\eta) > 0$ for $\eta > 0$.

In particular, the limits of the following derivatives exist for all $m \geq 0$ and $\eta > 0$:

$$\lim_{r \downarrow 0} \left(\sqrt{r} \frac{\partial}{\partial r}\right)^m (x,y)(r,\eta), \tag{2.2}$$

(note that $2\sqrt{r}\partial/\partial r = \partial/\partial\rho$ by $\rho = \sqrt{r}$).

Remark: These theorems generalize the results of Monteiro and Tsuchyia [3] who proved the existence of limits of first order derivatives with respect to r and (2.1) and (2.2) for $m = 0$ and $m = 1$ for *monotone* LCP's.

3. Applications
We outline some applications of Theorem 2 to the difficult case $K \neq \emptyset$ and note only that there are similar results for the case $K = \emptyset$ based on Theorem 1 (see [5] for details).

According to Theorem 2, it is reasonable to consider the path $(x,y)(.,\eta)$ as a function of $\rho := \sqrt{r}$, because it is then analytic for $\rho \geq 0$, $\eta > 0$. Therefore, we redefine $(x,y)(\rho,\eta)$ as the solution (x,y) of

$$Px + Qy = q + \rho^2 \bar{q}, \quad (x,y) \geq 0,$$
$$x \circ y = \rho^2 \eta, \tag{3.1}$$

and define $(\tilde{x}, \tilde{y})(\rho, \eta)$ (compare Theorem 2) by

$$(x_I(\rho,.), x_J(\rho,.), x_K(\rho,.)) =: (\tilde{x}_I(\rho,.), \rho^2 \tilde{x}_J(\rho,.), \rho \tilde{x}_K(\rho,.)),$$
$$(y_I(\rho,.), y_J(\rho,.), y_K(\rho,.)) =: (\rho^2 \tilde{y}_I(\rho,.), \tilde{x}_J(\rho,.), \rho \tilde{y}_K(\rho,.)). \tag{3.2}$$

We first derive some crucial boundedness results for the derivatives of $(x,y)(\rho,\eta)$. Consider the compact set

$$\mathcal{C}(\beta) := \{(\rho, \eta) \mid 0 \leq \rho \leq 1, \, \|\eta - e\| \leq \beta\},$$

where $0 < \beta < 1$. Then, due to the analyticity of $(\tilde{x}, \tilde{y})(\rho, \eta)$ for $0 \leq \rho \leq 1$, $\eta > 0$ and because of $(\tilde{x}, \tilde{y})(0, \eta) > 0$ (see Theorem 2), there are constants $c_m = c_m(\beta) > 0$, $\delta = \delta(\beta) > 0$, so that for all derivatives D^m with respect to ρ, η of any order $m \geq 0$ and all $(r, \eta) \in \mathcal{C}(\beta)$

$$\|D^m(x,y)(\rho,\eta)\| \leq c_m,$$
$$\|D^m(\tilde{x},\tilde{y})(\rho,\eta)\| \leq c_m,$$
$$0 < \delta e \leq \tilde{x}(\rho, \eta) \leq c_0 e, \tag{3.3}$$
$$0 < \delta e \leq \tilde{y}(\rho, \eta) \leq c_0 e.$$

These bounds can be applied to design *high-order* infeasible-interior-point-methods for the case $K \neq \emptyset$.

A typical iteration of these methods has the form

$$(x^{(k)}, y^{(k)}, \rho_k, \eta^{(k)}) \longrightarrow (x^{(k+1)}, y^{(k+1)}, \rho_{k+1}, \eta^{(k+1)}).$$

It will be such that for all $k \geq 0$

$$0 < (x^{(k)}, y^{(k)}) = (x,y)(\rho_k, \eta^{(k)}), \quad \eta^{(k)} = \frac{x^{(k)} \circ y^{(k)}}{\rho_k^2} \tag{3.4}$$
$$0 < \rho_{k+1} < \rho_k, \quad \|\eta^{(k)} - e\| \leq \beta,$$

where $\beta \leq 1/4$ is some positive constant. Hence for all $k \geq 0$

$$(\rho_k, \eta^{(k)}) \in \mathcal{C}(\beta),$$

and the bounds (3.3) apply for all $(\rho, \eta) = (\rho_k, \eta^{(k)})$ and all $m \geq 0$.

To simplify the notation when describing the k-th iteration, we drop the iteration indices and write briefly (x, y, ρ, η) and $(x^+, y^+, \rho_+, \eta^+)$ for $(x^{(k)}, y^{(k)}, \rho_k, \eta^{(k)})$ and $(x^{(k+1)}, y^{(k+1)}, \rho_{k+1}, \eta^{(k+1)})$, respectively.

The simplest such method is a predictor-corrector method, where the predictor-step determines the next parameter value $\rho_+ =: \rho_{k+1}$ as follows:

1. Given (x, y, ρ, η), one approximates $z(\sigma, \eta) := (x, y)(\sigma, \eta)$ near $\sigma = \rho$ by a *Taylor-polynomial* of degree p in σ

$$\hat{z}(\sigma, \eta) \equiv \hat{z}(\sigma) := z(\rho) + (\sigma - \rho)\dot{z}(\rho) + \cdots + \frac{(\sigma - \rho)^p}{p!} z^{(p)}(\rho).$$

Here of course, $\hat{z}(\sigma, \eta) = (\hat{x}, \hat{y})(\sigma, \eta) \equiv \hat{z}(\sigma)$, and

$$z^{(m)}(\rho) := \frac{\partial^m}{\partial \sigma^m} z(\sigma, \eta)\big|_{\sigma = \rho},$$

which can be computed recursively by solving the linear equations obtained by differentiating (3.1) repeatedly with respect to ρ.

Therefore, by Taylor's theorem and (3.4), (3.3) there is a constant $c = c(\beta)$ so that for all $0 < \sigma \leq \rho$,

$$\begin{aligned} \|\hat{x}(\sigma, \eta) - x(\sigma, \eta)\| &\leq c\rho^{p+1}, \\ \|\hat{y}(\sigma, \eta) - y(\sigma, \eta)\| &\leq c\rho^{p+1}, \\ \left\|\frac{\hat{x}(\sigma, \eta) \circ \hat{y}(\sigma, \eta)}{\sigma^2} - \eta\right\| &\leq c\frac{(\rho - \sigma)^{p+1}}{\sigma^2}, \end{aligned} \quad (3.5)$$

Remark: The definition of $(x, y)(\rho, \eta)$ as the solution of (3.1) was also used by Sturm [6], but he considered only monotone LCP's and the case $p = 2$. His proof is rather technical, since it is not based on the analyticity of $(x, y)(\rho, \eta)$ but only on the existence of the limits

$$\lim_{\rho \downarrow 0} (x, y)(\rho, \eta), \quad \lim_{\rho \downarrow 0} \frac{\partial}{\partial \rho}(x, y)(\rho, \eta)$$

proved by Monteiro and Tsuchyia [3].

2. The next parameter $(\rho_{k+1} =)\rho_+ < \rho(= \rho_k)$ is then determined by the standard approach, that is by

$$\rho_+ := \inf\left\{\bar{\rho} \,\Big|\, \left\|\frac{\hat{x}(\sigma) \circ \hat{y}(\sigma)}{\sigma^2} - e\right\| \leq 2\beta, \,\forall\, \bar{\rho} \leq \sigma \leq \rho\right\}.$$

An upper bound $\bar{\rho}$ for ρ_+ is found as follows: Define $\bar{\rho}$ as the unique solution of the equation
$$c\frac{(\rho-\bar{\rho})^{p+1}}{\bar{\rho}^2} = \beta \tag{3.6}$$
satisfying $0 < \bar{\rho} < \rho$. Then by (3.5) for all σ with $\bar{\rho} \leq \sigma \leq \rho$
$$\left\|\frac{\hat{x}(\sigma)\circ\hat{y}(\sigma)}{\sigma^2} - \eta\right\| \leq c\frac{(\rho-\sigma)^{p+1}}{\sigma^2} \leq c\frac{(\rho-\bar{\rho})^{p+1}}{\bar{\rho}^2} = \beta.$$
Therefore, the triangle-inequality and $\|\eta - e\| \leq \beta$ imply for these σ
$$\left\|\frac{\hat{x}(\sigma)\circ\hat{y}(\sigma)}{\sigma^2} - e\right\| \leq \left\|\frac{\hat{x}(\sigma)\circ\hat{y}(\sigma)}{\sigma^2} - \eta\right\| + \|\eta - e\| \leq 2\beta,$$
so that $0 < \rho_+ \leq \bar{\rho} < \rho$. Hence by (3.6)
$$1 \geq \left(\frac{\rho-\bar{\rho}}{\rho}\right)^{p+1} = \frac{\beta}{c}\frac{\bar{\rho}^2}{\rho^{p+1}} \implies \bar{\rho}^2 \leq \frac{c}{\beta}\rho^{p+1},$$
that is
$$\rho_{k+1} = \rho_+ \leq \bar{\rho} \leq \sqrt{\frac{c}{\beta}}\rho_k^{(p+1)/2}.$$
This shows that the path parameters ρ_k (and therefore also the $r_k = \rho_k^2$) converge Q-superlinearly to 0 with order $(p+1)/2$.

3. The *corrector-step* is standard: one computes
$$(x^+, y^+), \quad \eta^+ := \frac{x^+ \circ y^+}{\rho_+^2},$$
and thereby
$$(x^{(k+1)}, y^{(k+1)}, \eta^{(k+1)}) := (x^+, y^+, \eta^+)$$
by the usual *recentering technique*: (x^+, y^+) is obtained as the solution of the linear equations
$$Px^+ + Qy^+ = q + \rho_+^2 \bar{q}$$
$$\hat{Y}(\rho_+)x^+ + \hat{X}(\rho_+)y^+ = \rho_+^2 e + \hat{x}(\rho_+)\circ\hat{y}(\rho_+),$$
that is by one step of Newton's method for solving
$$Px + Qy = q + \rho_+^2 \bar{q}, \quad (x,y) \geq 0,$$
$$x \circ y = \rho_+^2 \eta,$$
using $(\hat{x}(\rho_+), \hat{y}(\rho_+))$ as starting value.

Then it is wellknown for $0 < \beta \leq 1/4$ that, because of
$$\left\|\frac{x(\rho_+) \circ y(\rho_+)}{\rho_+^2} - e\right\| \leq 2\beta \leq 1/2,$$
the following holds for x^+, y^+ and $\eta^+ = x^+ \circ y^+/\rho_+^2$
$$(x^{(k+1)}, y^{(k+1)}) = (x^+, y^+) > 0,$$
$$\|\eta^{(k+1)} - e\| = \|\eta^+ - e\| \leq \beta.$$
Hence, $(x^{(k+1)}, y^{(k+1)}, \rho_{k+1}, \eta^{(k+1)}) := (x^+, y^+, \rho_+, \eta^+)$ again satisfies (3.4).

4. The convergence properties of the vectors $x^{(k)}$, $y^{(k)}$ easily follow from the Q-superlinear convergence of the ρ_k, (3.2) and (3.3) and $(\rho_k, \eta^{(k)}) \in \mathcal{C}(\beta)$ for all $k \geq 0$ because this implies for all $k \geq 0$
$$x_I^{(k)} = \tilde{x}_I^{(k)}, \quad x_J^{(k)} = \rho_k^2 \tilde{x}_J^{(k)}, \quad x_K^{(k)} = \rho_k \tilde{x}_K^{(k)}$$
$$y_I^{(k)} = \rho_k^2 \tilde{y}_I^{(k)}, \quad y_J^{(k)} = \tilde{y}_J^{(k)}, \quad y_K^{(k)} = \rho_k \tilde{y}_K^{(k)}$$
$$0 < \delta e \leq \tilde{x}^{(k)} \leq c_0 e,$$
$$0 < \delta e \leq \tilde{y}^{(k)} \leq c_0 e,$$
$$\rho_{k+1} \leq \sqrt{\frac{c}{\beta}} \rho_k^{(p+1)/2}.$$

Moreover
$$\|\eta^{(k)} - e\| \leq \beta \implies (1-\beta)e \leq \eta^{(k)} \leq (1+\beta)e$$
$$\eta^{(k)} = \frac{x^{(k)} \circ y^{(k)}}{\rho_k^2} \implies \rho_k^2 e^T \eta^{(k)} = (x^{(k)})^T y^{(k)}$$
$$\rho_k^2 n(1-\beta) \leq (x^{(k)})^T y^{(k)} \leq \rho_k^2 n(1+\beta).$$

This proves the main result of this paper:

Theorem: *Assume* (V'), *and, perhaps,* $K \neq \emptyset$. *Then the method generates* ρ_k, $x^{(k)}$, $y^{(k)}$ *with the following properties:*

The "small" quantities ρ_k, $x_J^{(k)}$, $x_K^{(k)}$, $y_I^{(k)}$, $y_K^{(k)}$, *and* $(x^{(k)})^T y^{(k)}$ *converge Q-superlinearly to 0 with order* $(p+1)/2$.

The "large" vectors $x_I^{(k)}$, $y_J^{(k)}$, *remain bounded for all* $k \geq 0$,
$$0 < \delta e \leq x_I^{(k)}, \ y_J^{(k)} \leq c_0 e,$$
but need not converge.

Here of course, a quantity is called "small" if it converges to zero because of a factor ρ_k, and "large" otherwise.

References
[1] R.W. Cottle, J.-S. Pang, V. Venkateswaran, "Sufficient matrices and the linear complementarity problem," *Linear Algebra Appl.* 114/115 (1989) 231 – 249.
[2] S. Mizuno, "A superlinearly convergent infeasible-interior-point algorithm for geometrical LCP's without a strictly complementarity condition," *Mathematics of Operations Research,* 21 (1996) 382 – 400.
[3] R.D.C. Monteiro, T. Tsuchyia, "Limiting behavior of the derivatives of certain trajectories associated with a monotone horizontal linear complementarity problem," *Mathematics of Operations Research,* 21 (1996) 793 – 814.
[4] J. Stoer, M. Wechs, "Infeasible-interior-point paths for sufficient linear complementarity problems and their analyticity," (1996), to appear in *Mathematical Programming.*
[5] J. Stoer, M. Wechs, S.Mizuno, "High order infeasible-interior-point methods for solving sufficient linear complementarity problems", Research Memorandum No. 634 (The Institute of Statistical Mathematics, Tokio), Techn. Report Nr. 222 (Universität Würzburg), 1997, to appear in *Mathematics of Operations Research.*
[6] J.F. Sturm, "Superlinear convergence of an algorithm for monotone linear complementarity problems, when no strictly complementary solution exists," Report 9656/A, Econometric Institute, Erasmus University Rotterdam, Rotterdam, The Netherlands (1996).

Address of the authors:
Josef Stoer, Martin Wechs
Institut für Angewandte Mathematik und Statistik
Universität Würzburg
Am Hubland
D-97074 Würzburg
Germany

Shinji Mizuno
The Institute of Statistical Mathematics
4-6-7 Minami-Azabu, Minato-ku
Tokyo 106
Japan

GIUSEPPE CONFESSORE, PAOLO DELL'OLMO
AND STEFANO GIORDANI
A linear time approximation algorithm for a storage allocation problem

1. Introduction

In a general storage allocation problem it is given a set J of n items to be stored, and, for each item $j \in J$, an arrival time $r(j) \in Z_0^+$, a departure time $d(j) \in Z^+$, and a size $s(j) \in Z^+$ are known. Let $T(j) = [r(j), d(j))$ be the time interval in which item j has to be stored; two items i, j are in conflict if $T(i) \cap T(j) \neq \emptyset$. Let $\sigma: J \to [1,..., D]$ be a function such that for every $j \in J$ the storage interval allocated for j is $I(j) = [\sigma(j), \sigma(j)+1,..., \sigma(j)+s(j)-1]$. A feasible storage allocation for J is a function σ such that, for all $j \in J$, $I(j)$ is contained in $[1,..., D]$, and for all couple of items $i, j \in J$, $I(i) \cap I(j)$ can be non-empty only if $T(i) \cap T(j) = \emptyset$. The objective of the problem is to minimize the required allocation size D.

This problem, known in literature as Dynamic Storage Allocation problem, is $\mathcal{NP}$-hard in strong sense (Garey and Johnson [3]). Approximation algorithms have been proposed by Kierstead [5], with a performance ratio at most 80, and successively improved by the same author [6], with an algorithm with performance ratio at most 6.

Studies on these problems have been motivated by several applications in telecommunication systems. Recent advances in communication technologies have made it possible the integrations of heterogeneous traffics of different bandwidths in a single broadband network, like optical fiber network, achieving the flexibility and the economic advantage of sharing a network [4]. In such a situation, if no traffic access control is exerted, the network resources, such as transmission bandwidth, will be occupied unfairly resulting in a low throughput usage of the network. Thus, it is needed a traffic control strategy that allocates the network bandwidth fairly to achieve maximum throughput. As an example, consider a radio network where the transmitters request some consecutive frequency slots to allocate their data. After the transmission, the slots are released. The goal of a storage algorithm is to allocate the slots so as to minimize the total amount of used frequency space. Another application area of these problems concerns memory management in computer systems [1].

In this paper, we study the storage allocation problem in the case in which the arrival and departure times of items define inclusion-free time intervals, as, for example, in the case in which all time periods have the same length [2]. From the point of view of the mentioned applications, this implies that the transmission period is the same for all the messages. These assumptions are often matched in both such application fields.

To the best of our knowledge, this problem has not been studied before, and its complexity is open. In this paper, we prove that the problem is $\mathcal{NP}$-hard, and we give a linear time 2-approximation algorithm for it. The bound is shown to be tight.

2. The $\mathcal{NP}$-hardness result

The storage allocation problem with inclusion-free time intervals is $\mathcal{NP}$-hard, as it can be shown by reduction from the Subset Sum problem (SSP), which is known to be $\mathcal{NP}$-complete [3]. This problem is defined as follows:

Subset Sum problem: Given a set $X = \{a_1,..., a_z\}$ of z non negative integers, and a positive integer B, is there a subset $X' \subseteq X$ such that the $\sum_{a_i \in X'} a_i = B$?

Theorem 1 *The storage allocation problem with inclusion-free time intervals is $\mathcal{NP}$-hard.*

Proof: Given an instance of SSP with $\sum_{a_i \in X} a_i = 3B$, let us define a corresponding instance I of the storage allocation problem, with inclusion-free time intervals. Let $J = J_1 \cup J_2$ be the set of items to be stored, where $J_1 = \{j_1,..., j_z\}$ and $J_2 = \{b, c, d, e, f, g\}$.

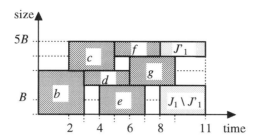

Figure 1: The instance for the proof of $\mathcal{NP}$-hardness

Let the sizes of the items be assigned as follows:

$s(j_i) = a_i$, for $i = 1,..., z$;

$s(b) = 3B$, $s(c) = 2B$, $s(d) = B$, $s(e) = 2B$, $s(f) = B$, $s(g) = 2B$.

Let the arrival and departure times of items be the following ones:

$r(j_i) = 8$ and $d(j_i) = 11$, for $i = 1,..., z$;

$r(b) = 0$, $r(c) = 2$, $r(d) = 3$, $r(e) = 4$, $r(f) = 5$, $r(g) = 6$;

$d(b) = 3$, $d(c) = 5$, $d(d) = 6$, $d(e) = 7$, $d(f) = 8$, $d(g) = 9$.

It is simple to verify that this assignment defines a storage allocation problem instance, with inclusion-free time intervals.

Now, we show that the instance of SSP has a *yes* answer if and only if the corresponding instance of the storage allocation problem, with inclusion-free time intervals, has a solution with allocation size equal to $5B$, that is an optimal solution since a storage of allocation size at least $5B$ is required to store a maximal set of mutually conflicting items.

Let us first assume that the instance of SSP has a *yes* answer. In this case we may refer to $J'_1 \subseteq J_1$, such that $\sum_{j_i \in J'_1} s(j_i) = B$, and it is possible to construct an optimal allocation layout of size $5B$ for the instance I as shown in Figure 1.

Conversely, we show that if there exists an optimal storage allocation of size $5B$, then there exists a *yes* answer to the instance of SSP. Let us examine the possible allocation for the elements of J_2. We can give only two allocation layouts for them such that the size of the required storage is not greater than $5B$. These are the one shown in Figure 1, and its mirror image, which leave two storage intervals of size B and $2B$, respectively, in which items in J_1 have to be stored. If this can be done then there exist a *yes* answer to the instance of SSP. □

3. Item partitioning

In this section, we show how to partition the set J of items, in order to design an approximate algorithm for the storage allocation problem with inclusion-free time intervals.

Given a totally ordered set L, let us define the function $\rho: L \rightarrow Z^+$, such that, for each element $l \in L$, $\rho(l)$ is the position of l in L. Moreover, let $first(L)$ be the first element of L, and $last(L)$ be the last element of L.

In the following, we consider the set $J = \{j_1,..., j_n\}$ of items, ordered according to non decreasing arrival times, that is, $r(j_i) \leq r(j_{i+1})$.

Without loss of generality, we consider the case in which each item is in conflict at least with two different items, except the first and the last item of J, that can be in conflict at least with one different item. On the contrary, the problem instance can be decomposed in non in conflict instances.

Let $\mathcal{P} = (J_1,..., J_p)$ be a set partition of J obtained as follows:

- $J_1 = \{j_i \in J: r(j_i) < d(j_1)\}$
- $J_i = \{j_i \in \{J \setminus (J_1 \cup \cdots \cup J_{i-1})\}: r(j_i) < d(j_{\rho(last(J_{i-1})) + 1})\}$, for $i = 2,..., p$.

The sets J_i are ordered in the same way as J. Note that the partition $\mathcal{P}$ does not depend from the size $s(j)$ of the items $j \in J$. We can show that:

Theorem 2 *Given the partition $\mathcal{P} = (J_1,..., J_p)$ of J, there is no couple (J_i, J_{i+k}), with $k \geq 2$, of partition elements, such that there exist two in conflict items $x \in J_i$ and $y \in J_{i+k}$.*

Proof: By costruction, the arrival times of all the items in J_{i+k}, with $k \geq 2$, are greater than or equal to the departure time of the first item of J_{i+1}, that is, $r(j_i) > d(first(J_{i+1}))$, $\forall j_i \in (J_{i+2} \cup \cdots \cup J_p)$. Due to the order of items in J, we have $d(last(J_i)) < d(first(J_{i+1}))$, then $r(j_i) > d(last(J_i))$, $\forall j_i \in (J_{i+2} \cup \cdots \cup J_p)$. That is, the items belonging to $(J_{i+2} \cup \cdots \cup J_p)$ are not in conflict with the items belonging to J_i. □

For example, let us compute the partition $\mathcal{P}$ for the following istance: $J = \{j_1, j_2, j_3, j_4, j_5, j_6\}$, $\{r(j_i)\} = \{0, 1, 3, 5, 6, 7\}$, $\{d(j_i)\} = \{3, 4, 6, 8, 9, 10\}$. In this case we have $\mathcal{P} = (J_1, J_2, J_3)$, where $J_1 = \{j_1, j_2\}$, $J_2 = \{j_3, j_4\}$, and $J_3 = \{j_5, j_6\}$.

The partition $\mathcal{P} = (J_1,..., J_p)$ can be easily computed in $O(n)$ time starting from the ordered set J. Put one by one in J_i the elements of J such that their arrival times are less than the departure time of the first item inserted in J_i; repeat this step for $i = 1,..., p$, until all items has been considered.

4. An approximation algorithm

Given the partition $\mathcal{P} = (J_1,..., J_p)$, due to the Theorem 2, items in J_i can occupy the same allocation region of items in J_{i+k}, with $k \geq 2$. Hence, we can consider two non overlapping allocation regions A and B, and store the items belonging to $(J_1 \cup J_3 \cup \cdots \cup J_{2\lfloor (p-1)/2 \rfloor+1})$ in region A, and the items belonging to $(J_2 \cup J_4 \cup \cdots \cup J_{2\lfloor p/2 \rfloor})$ in region B.

This allows us to define an heuristic algorithm $\mathcal{H}$ composed by two steps. In step 1 we compute the partition $\mathcal{P}$. In step 2, considering the items in the order in which they are in $J_1, J_3,..., J_{2\lfloor (p-1)/2 \rfloor+1}$, we allocate them in the lowest available area of region A, and we make the same for the items in $J_2, J_4,..., J_{2\lfloor p/2 \rfloor}$ to be allocated in region B.

In this way, the size of each allocation region is not greater than the minimum size $Q = \max_{i=1,...,p} \{\sum_{j \in J_1} s(j)\}$ of storage space required to contain the items in any $J_i \in \mathcal{P}$.

Denoted by $D^{\mathcal{H}}$ the heuristic solution value, which is the sum of the size of regions A and B, we have that $D^{\mathcal{H}} \leq 2Q$.

Moreover, denoted by $\mathcal{J}$ the set of all subsets of mutually in conflict items of J, we define $S = \max_{K \in \mathcal{J}} \{\sum_{j \in K} s(j)\}$ as the minimum size of a storage space required to contain any set of mutually in conflict items. From the definition of S, it follows that S is a lower bound on any feasible allocation size D, that is $S \leq D$. Moreover, since every J_i of $\mathcal{P}$ is clearly a set of mutually in conflict items of J, we have that $J_i \in \mathcal{J}$, and, hence, it results $Q \leq S \leq D$.

Theorem 3 *Denoted by D^* the optimal allocation size of the storage allocation ploblem with inclusion-free time intervals, it results*

$$\frac{D^{\mathcal{H}}}{D^*} \leq 2.$$

Moreover, the bound is tight.

Proof: By definition of $\mathcal{H}$, we have that $D^{\mathcal{H}} \leq 2Q$. Moreover, knowing that $Q \leq S \leq D$, it follows that $Q \leq D^*$; hence, the bound on the performance ratio follows. In order to show that the bound is tight, let us consider the following problem instance: $J = \{j_1, j_2, j_3, j_4, j_5, j_6, j_7\}$, $\{r(j_i)\} = \{0, 1, 2, 3, 4, 5, 7\}$, $\{d(j_i)\} = \{2, 4, 5, 6, 7, 8, 9\}$, $\{s(j_i)\} = \{W-a, a, a, W-2a, a, a, W-a\}$, with $W > 2a$. The heuristic algorithm gives the partition $\mathcal{P} = (J_1, J_2, J_3)$, where $J_1 = \{j_1, j_2\}$, $J_2 = \{j_3, j_4, j_5\}$, $J_3 = \{j_6, j_7\}$, and allocates the items of J_1 and J_3 in region A and the items of J_2 in region B, as shown in Figure 2. Both regions A, B have size equal to W; hence, $D^{\mathcal{H}} = 2W$. As for the optimal solution value D^*, let us consider the optimal item allocation represented in Figure 3 with $D^* = W$. Hence the ratio $D^{\mathcal{H}} / D^*$ is equal to 2. □

As for algorithm time complexity, given an ordered set of n items J which is, for example, the case in which the items are considered while they arrive, we can performed both step 1 and step 2 in $O(n)$ time.

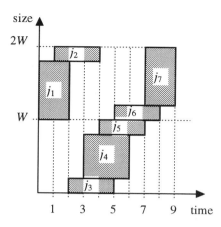

Figure 2: The instance for the tightness: the heuristic solution

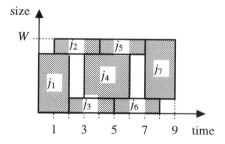

Figure 3: The instance for the tightness: the optimal solution

5. Conclusions

In this paper, we investigated a particular case of a storage allocation problem in which the arrival and departure times of items define inclusion-free time intervals. We have shown that the problem is $\mathcal{NP}$-hard, and proposed a linear time 2-approximation algorithm for it, based on a partition of the set of items. The layout of the heuristic solution suggests further research on looking for new approximation algorithms with better performance ratio.

References

[1] Coffman, E.G. (1983), An Introduction to Combinatorial Models of Dynamic Storage Allocation, *SIAM Review*, **25**, 311–325.

[2] Confessore, G., Dell'Olmo, P. and Giordani S. (1997), Polynomial Time Approximation Algorithms for Dynamic Storage Allocation Problems, Tech. Rep. 448, IASI-CNR, Rome, Italy, January 1997.

[3] Garey, M.R. and Johnson, D.S. (1979), *Computers and Intractability: A Guide to the Theory of NP-Completeness*, Freeman, San Francisco, CA.

[4] Hale, W.K. (1980), Frequency Assignment: Theory and Applications, *Proc. of IEEE*, **68**, 1497–1514.

[5] Kierstead, H.A. (1988), The Linearity of First-Fit Coloring of Interval Graphs, *SIAM J. Disc. Math.*, **1**, 526–530.

[6] Kierstead, H.A. (1991), A polynomial time approximation algorithm for Dynamic Storage Allocation, *Disc. Math.*, **88**, 231–237.

Giuseppe Confessore, Paolo Dell'Olmo, Stefano Giordani
Dipartimento di Informatica, Sistemi e Produzione
Università di Roma "Tor Vergata"
Via di Tor Vergata - I 00133 Roma Italy
e-mail: {confessore,dellolmo,giordani}@iasi.rm.cnr.it

CEDRIC C F FONG[*] JOHN C S LUI, MAN HON WONG[†]
AND EDMUNDO A DE SOUZA E SILVA[‡]

General framework in analyzing mobile terminal tracking protocols

Abstract

The protocols of minimizing the cost of tracking mobile terminals in cellular networks are analyzed in this paper. Recently, many methods such as the time-based method, the movement-based method, and the distance-based method, have been proposed. Mathematical analyses have also been done to show the efficiency of these different schemes. In these analyses, however, the model used are always too simple to represent realistic situations. It is often assumed that a user will follow a random walk motion. Moreover, there is also an assumption that the call interarrival time is much longer compared to user's motion so that steady-state location probabilities can be used in the cost estimation. These assumptions may not be valid in real cases since call interarrival time may be much shorter than the residence time within a cell. Steady analysis is not applicable. In this paper, we develop a general framework that can capture the various user motion patterns. Moreover, we use transient analysis to derive the location probabilities. This model uses a more realistic assumption, allows us to predict the user location and thereby reducing the total cost.

1 Introduction

In a cellular network environment [4], a mobile terminal may turn off its connect with the wired network when no communication is needed. This can release the precious wireless communication bandwidth to other terminals in need. However, this practices causes a problem at the wired network. Since a mobile terminal can be disconnected for a certain time, the wired system no longer has the exact location information of the mobile terminal. When the wired system attempts to connect to the mobile terminal, a *paging* process is necessary to locate a disconnected terminal inside the cellular network.

In the paging process, it would be very costly if all the cells in the cellular network are paged. Instead, it is favorable for the wired network to keep the approximate location information so that, when the wired network wants to connect to the mobile terminal,

---
[*]This work is supported in part by the UGC and CUHK research grants.
[†]This work is supported in part by the UGC research grant.
[‡]The work is partially supported by grants from ProTem-CC and Pronex (CNPq-Brazil).

the system can page the cells that the terminal is most likely to be situated in. In this way, the expected cost can be lowered. The method for the system to keep track of the location information about the mobile terminals if to let the mobile terminals send their location information periodically. This is called the *location update*. Note that the more frequent location update takes place, the lower is the paging cost. On the other hand, location update also requires system cost which is directly related to its frequency.

As a result, the *mobile terminal tracking problem* has been widely studied at the aim of balancing the tradeoffs between the location update and paging so as to minimize the overall system costs. This is very important because the limited system resources are shared by a large number of mobile terminals. An effective utilization of them can therefore increases the system capacity and enhances the performance.

Currently, an approach has been adopted in Global System for Mobile Communication (GSM) in Europe and IS-41 in the United States. In this approach, the cellular network is partitioned into many registration areas. The problem for the above protocol is that the boundaries for the registration areas are static. In this regard, several *update protocols* has been proposed recently. These protocols are based on some other factors to determine the time for a mobile terminal to send an update message. Two common approaches are time-based protocols [2, 7] and the distance-based protocols [1, 5]. In the time-based protocols, a specific time period is set. If a terminal terminal has not sent any message for this time period, an update message will be sent. In the distance-based protocols, a terminal will send an update message if it moves farther than a particular distance from the previous reporting location.

In [1, 2, 5, 7], authors have done mathematical analyses to show the efficiency of their protocols. In [1, 2], the authors assume that the user has a symmetric movement pattern, i.e, the user will move from one cell to any of the neighboring cells with even probability. Moreover, steady analysis is used. These assumptions are not very realistic. A user usually has a unique, skew movement pattern over the cellular network rather than the symmetric one. Moreover, the length of interarrival time of incoming calls may be comparable to that of the expected residence time of the user within a cell. Therefore, using the steady state probabilities for cell locations of a user may not be an accurate estimation.

In this paper, we develop a general framework that enables us to model different kinds of terminal movement pattern inside network. By using this model, predictions on users' time-varying location can be more accurate. This enables us to further minimize the cost in real operations. Moreover, more accurate performance analyses can also be carried out. Our paper addresses on how the time-varying location probabilities can be derived more accurately via the uniformization method [3]. Since many optimization protocols rely on the location probabilities, a more accurate prediction on the user's current location can achieve a better optimization for the tracking protocols. Another important advantage of our model is to provide a unified framework to compare the performance of different tracking protocols. Using this framework, we can also analyze the impacts of different movement patterns to the protocols.

This paper is organized as follows. In Section 2, we introduce a mathematical representation for the cellular system. Using this model, we derive an expression for the transient location probabilities for a mobile terminal. In Section 3, we introduce the time-based method, and derived performance measures for this protocol. Next, we investigate the distance-based protocol in Section 4. Lastly, a conclusion is given in Section 5.

2 System Model

In this paper, we consider a cellular network which consists of a collection $\mathcal{C}$ of N cells, each of which is labelled $c_1, c_2, \ldots, c_N$. Mobile terminals can move freely around the network. Assume that the network fully covers the whole geographic area so that the mobile terminals will not lose connection to the network wherever they are situated inside the region. Since the mobility of a mobile terminal is independent of that of each other, henceforth, we will consider an instance mobile terminal. The mobility of this terminal can be represented by an $N \times N$ *mobility matrix* $\mathbf{Q} = [q_{ij}]$ where

$$q_{ij} = \begin{cases} \text{the } \textit{rate} \text{ at which the terminal moving from } c_i \text{ to } c_j & i \neq j \\ -\sum_{j:j \neq i} q_{ij} & \text{otherwise} \end{cases} \quad (1)$$

This matrix can be input as a user specification, gathered from statistics on the user's past routes, or a combination of the above two. The formulation of this mobility matrix allows us to model practically any kind of movement patterns. Also, because each element in the matrix defines the rate at which the mobile terminal crosses a cell boundary, $q_{ij} > 0$ only if the cells c_i and c_j are next to each other. Under this restriction, the mobility matrix $\mathbf{Q}$ is always a sparse matrix. In addition, this matrix is general enough to model the mobility pattern of a terminal under a network of different topologies. Note that with the definition of q_{ii}, the matrix $\mathbf{Q}$ becomes a transition rate matrix which governs a continuous time Markov process representing the movement of the mobile terminal. Let this process be $\mathcal{X}_m = \{X_m(t) : t \geq 0\}$ whose state space consists of all possible cell locations $\{c_1, c_2, \ldots, c_N\}$ of the cellular network. The subscript m means that $X_m(0) = c_m$; and t is the time elapsed from the last location update at cell c_m.

Now we are going to derive the *transient* location probabilities for that Markov process. It is very important as in many location update schemes, the cost minimization is done on these probabilities. Let the *transient* location probability vector of the process $\mathbf{X}_m$ be $\boldsymbol{\pi}_m(t) = <\pi_{1|m}(t), \pi_{2|m}(t), \ldots, \pi_{N|m}(t)>$, where $\pi_{i|m}(t)$ is the probability that the terminal is located at cell c_i at time t since the last report of location at cell c_m. Note that the initial probability vector $\boldsymbol{\pi}_m(0)$ is a unit vector where $\pi_{i|m}(0) = 0$ for all $i \neq m$ and $\pi_{m|m}(0) = 1$.

Using this notation of transient location probabilities, we mean that the location probabilities of a user only depends on the time t elapsed and the last report of location at cell c_m. This is in fact a simplified model. A more accurate model is to include the moving direction of the mobile terminal. For example, along a two way highway, a car will move straight along nearly the same direction. It is very unlikely that the car will move in opposite direction suddenly. By including the direction of movement into the state space of the process, we can then model this kind of information. However, the size mobility matrix will increase by $O(N_d)$, where N_d is the number of possible directions of movement.

Here we use the uniformization method [3], which is well-known for its efficiency and numerical stability. In our case, we are now having a continuous time Markov process $\mathcal{X}_m$ which is governed by the initial location probability vector $\boldsymbol{\pi}_m(0)$ and transition rate matrix $\mathbf{Q}$. By uniformization, $\mathcal{X}_m$ can be transformed into a discrete time Markov chain with the same state space $\mathcal{C}$ whose transition events are governed by a Poisson process with rate $\Lambda = \max\{-q_{ii} : 1 \leq i \leq N\}$. The step transition matrix for this Markov chain can be computed as $\mathbf{P} = \mathbf{I} + \mathbf{Q}/\Lambda$. By conditioning on the number of transitions over the time interval $(0, t)$, we can establish the equation:

$$\boldsymbol{\pi}_m(t) = \sum_{n=0}^{\infty} e^{-\Lambda t} \frac{(\Lambda t)^n}{n!} \mathbf{p}_m(n) \qquad (2)$$

where $\mathbf{p}_m(n) = \mathbf{p}_m(0)\mathbf{P}^n$ is the probability vector for the discrete Markov chain after n transitions. The element $p_{i|m}(n)$ of the vector $\mathbf{p}_m(n)$ is the probability that the mobile terminal is located at cell c_i after it has crossed n boundaries since its last report of location at cell c_m. Note also that $\mathbf{p}_m(0) = \boldsymbol{\pi}_m(0)$ as the initial setting.

The infinite series in the expression causes difficulty in actual computation. To alleviate this problem, we take the summation up to a particular number, M, of terms. The associated error bound $\varepsilon(M)$ for each element $\pi_{i|m}(t)$ in the probability vector is given by

$$\varepsilon(M) = 1 - \sum_{n=0}^{M} e^{-\Lambda t} \frac{(\Lambda t)^n}{n!} \qquad (3)$$

It is worthwhile to note that the error bound does not depend on the movement pattern of the user. Rather, it just depends on the maximum rate at which the user moves. Moreover, it is easy to precompute the number of terms needed to be summed by specifying the error bound. Therefore, we can optimize for the tradeoff between prediction accuracy and processing time. Lastly, we let $\boldsymbol{\pi} = <\pi_1, \pi_2, \ldots, \pi_N>$ be the steady location probability vector of the mobile terminal in the network, where π_m represents the steady probability that the mobile terminal is at cell c_m. This probability vector can be found by standard iterative methods.

3 The Time-Based Protocol

In the time-based protocol, the time domain is divided into consecutive periods. Each period is initiated by an *update event*. An update event is an event that enables the static network to know the exact cell location of the mobile terminal at the time the event takes place. It can be the event that the mobile terminal sends an update message to the static network. Or it can be the event that the static network pages for the mobile terminal when an incoming call arrives. Lastly, it can also be the event that the mobile terminal wants to disconnect an established connection with the static network. It is worthwhile to note that we do *not* account for the terminal movement during the times the connection between the mobile terminal and the static network has been established. This is out of the scope of this paper and is related to the *handoff* problem. Also, we try to ignore the case that the mobile terminal ends a connection.

It is because such event has incurs no update cost.

Once a new time period is started, it will terminated by another update event. In this time-based protocol, a time limit T is specified *a priori*, and if the period last for this time limit without an incoming paging signal, the terminal will send an update message to the static network. Therefore, an update cost will be incurred in this case. On the other hand, if there is an incoming call waiting at the static network, the static network will page for the mobile terminal. When the terminal receives a paging signal before the time limit for the period, it will reply the static network. The time period will last with a *paging cost*.

Here in our analysis, we assume that the update cost is U per update message; and the paging cost for a terminal in a cell is C. Here we have also made a simplification that these two costs are constant. In actual cases, the update cost may vary with the distance (either topological or geographical) between the mobile terminal and the HLR of the terminal. The paging may also depends on the geographical sizes of the cell and also, the bandwidth available at the corresponding base station. Indeed, our model can be easily extended to accommodate these factors. However, for notational convenience, we insist on assuming that the cost are constant in our analysis.

Now we are going to derive an expression for the expected cost of the time-based protocols. The derivation can be considered in two cases. The first case when there is no incoming paging signal to the terminal during a time period. In this case, an update message will be sent when the time limit expires, which results in an update cost of U for this time period.

In the second case, a call arrives at time t, $0 \le t < T$, of the period; and the static network will page for the mobile terminal. Our paging method is like this: around the last location report of the terminal we define a *location area* which consist of cells that are less than more equal to d apart from the reporting cell. In the hexagonal topology that we have assumed, the number of cells inside the location area is $3(d+1)d+1$. The paging process will proceed in two phases. In the first phase, the static network will poll all the cells inside the location area. If the terminal cannot be found, it will proceed to the second phase. In this phase, all other cells inside the cellular network will be polled. If the mobile terminal still cannot be found, the static network will report failure. Otherwise, the corresponding cell location will be reported.

The reasons for this two-phase searching is two-folded. Firstly, by separate the searching process in the two phases, the system can search inside in the cells with high probability/cost ratio first. This can lower the searching cost greatly. Another advantage is that we can employ some searching methods in the first phase, such as the one in [8].

In this paging case, if the mobile terminal remains inside the location area, the cost of polling is KC, where C is the overhead for polling a single cell, where $K = 3(d+1)d+1$. However, if the terminal is outside the location area, the *overall* cost of polling for the terminal is NC. Let $\mathbf{r}_m = <r_{i|m} : i = 1, 2, \ldots, N>$ be a column vector such that $r_{i|m} = 1$ if cell c_i is inside the location area around cell c_m and $r_{i|m} = 0$ otherwise. Using the location probabilities derived previously, the probability of the terminal staying inside the area is $\boldsymbol{\pi}_m(t)\mathbf{r}_m$ and that of outside the area is $1 - \boldsymbol{\pi}_m(t)\mathbf{r}_m$. Therefore the polling cost when an incoming call arrives at time t is

$$KC[\boldsymbol{\pi}_m(t)\mathbf{r}_m] + NC[1 - \boldsymbol{\pi}_m(t)\mathbf{r}_m] \tag{4}$$

Assume that the arrivals of the incoming forms a Poisson process with parameter μ. The probability density function for the interarrival time is therefore $p(t) = \mu e^{-\mu t}$; and the probability distribution function is $P(t) = 1 - e^{-\mu t}$. Subsequently, the probability that the time period terminates with an incoming call is given by $1 - e^{-\mu T}$, while the probability that the period terminates by the terminal delivering an update message is $e^{-\mu T}$. Unconditioning update and paging cases, we have the expected cost of tracking the mobile terminal for one single period:

$$C_m^T(T) = Ue^{-\mu T} + \int_0^T \{KC[\boldsymbol{\pi}_m(t)\mathbf{r}_m] + NC[1 - \boldsymbol{\pi}_m(t)\mathbf{r}_m]\} \mu e^{-\mu t}\, dt \tag{5}$$

Substituting Equation (2) into the above formula and then integrating the expression, we can simplify it to

$$C_m^T(T) = Ue^{-\mu T} + NC(1 - e^{-\mu T}) - (N-K)C\mu T \sum_{n=0}^{\infty} e^{-(\Lambda+\mu)T} \frac{[(\Lambda+\mu)T]^n}{n!} f(n) \tag{6}$$

where $f(n)$ can be evaluated recursively as:

$$f(n+1) = \frac{n+1}{n+2} f(n) + \left(\frac{\Lambda}{\Lambda+\mu}\right)^{n+1} \frac{\mathbf{p}_m(n+1)\mathbf{r}_m}{n+2} \tag{7}$$

and $f(0) = \mathbf{p}_m(0)\mathbf{r}$.

The distribution of the duration of the time period is exponential truncated at T with rate μ. Therefore, the expected duration of the period is given by:

$$D_m^T(T) = \int_0^T t\mu e^{-\mu t} dt + Te^{-\mu T} = \frac{1 - e^{-\mu T}}{\mu} \tag{8}$$

Note that both the expected cost and the expected duration of each time period depend on the last reporting location, c_m, of the mobile terminal By the weak law of large numbers, we have the *expected average cost* for this time based protocol with parameter T:

$$E_T(T) = \frac{\sum_{m=1}^N C_m^T(T)\pi_m}{\sum_{m=1}^N D_m^T(T)\pi_m} \tag{9}$$

In this protocol, we want to find an optimal parameter T^* such that the average expected cost $E_T(T^*)$ is minimized. However, in our model, we allow a non-uniform terminal movement pattern over the cellular network. Instead of finding a single optimal value for the time limit, it is better to find an optimal vector $\mathbf{T}^* = <T_i^* : 1 \leq i \leq N>$ such that each element T_i^* represents the optimal value for cell c_i. These values can be found for each cell separately because they are independent of each other.

4 Distance-Based Protocol

In the distance-based protocol, the mobile terminal will be aware of the location area. When a new period is started, a location area is created around the reporting cell location. Similar to the above, there are two types of update events. The first type of event is when the an incoming call to the mobile terminal arrives and paging carried out. The second type of the event, however, is when the mobile terminal exits the location area. For the first event, the static system will page over the cellular network. On the other hand, we can see that in this protocol, the mobile terminal will sure be inside the location area. It is because it the mobile terminal exits the area, it will send an update to notify the static system. So it is not necessary for the static system to paging outside the location area. In this regard, the paging cost is always equal to KC, where K is the number of cells in the location area. In the second event, the mobile terminal exits the location area. It will send an update message to the static, resulting in a message cost of U. Here, we want to explore the optimal size of the location area such that the operating cost is minimized.

We are now going to derive the expected cost for a time period. In this protocol, the derivation is more difficult because the probability distribution of the period length is not a simple one (as compared to the truncated exponential distribution in the time-based protocol). This distribution is a joint distribution of the call interarrival time distribution and the distribution of the time at which the mobile terminal leaves the location area. The call interarrival time distribution is an exponential distribution with rate μ. The distribution of the time the mobile terminal leaves the location area is a *phase distribution* [6] that can be derived as follows.

From the system model, we reconstructed a new continuous Markov process of which the state space consists of $K+1$ states, where $K = 3d(d+1)+1$ is the number of cells inside the location area. The first K states correspond to the K possible cell locations in $\mathcal{R} = \{c_{r_1}, c_{r_2}, \ldots, c_{r_K}\}$ of the location area. The last state corresponds to the case that the mobile terminal is outside the registration area. The time at which the mobile terminal leaves location area can then be determined as the first-passage time to the $(K+1)$-th state in this new process.

From the original mobility matrix $\mathbf{Q}$, we construct a transition rate matrix $\mathbf{Q}' = [r_{ij}]$ for this new Markov chain. Without loss of generality, let the first K states be $s_1, s_2, \ldots, s_K$ which correspond to cell locations $c_{r_1}, c_{r_2}, \ldots, c_{r_K}$ inside the location area respectively, and the extra state s_{K+1} be the state of the mobile terminal is outside the location area. The transition rate matrix $\mathbf{R}$ for this chain has the following form:

$$\begin{bmatrix} \mathbf{U} & \mathbf{A} \\ \mathbf{0} & 0 \end{bmatrix} \quad (10)$$

where $\mathbf{U} = [u_{ij}]$ is a $K \times K$ matrix whose elements correspond to the transition rates when the mobile terminal is within the registration area. Therefore, $u_{ij} = q_{r_i,r_j}$ in $\mathbf{Q}$. The matrix $\mathbf{A} = [a_i]$ is a $K \times 1$ matrix whose elements are the rates at which the mobile terminal goes out of the location area from cell c_{r_i}. In this case, $a_i = \sum_{c_k \notin \mathcal{R}} q_{r_i,k}$. In this analysis, we are only interested in the movement of the mobile terminal when it is inside the location area. Therefore, we can construct this Markov chain by regarding the state of outside the location area be an *absorbing state*. Using this construction, the elements in the last row, which corresponds to the outgoing rate of state s_{K+1}, are all zeros.

Using uniformization, we transform the above Markov process into a discrete version and get the step transition matrix $\mathbf{P}' = \mathbf{I} + \mathbf{Q}'/\Lambda'$ for the new discrete chain, where $\Lambda' = \max\{-r_{ii} : i = 1, 2, \ldots, N\}$. $\mathbf{P}'$ will then have the following form:

$$\begin{bmatrix} \mathbf{U}_0 & \mathbf{A}_0 \\ \mathbf{0} & 1 \end{bmatrix} \tag{11}$$

Let $\boldsymbol{\pi}'_m(t) = <\pi'_{i|m}(t) : 1 \leq i \leq K>$ be the probability vector whose element $\pi'_{i|m}$ represents the probability that the mobile terminal is located inside cell c_{r_i} in the location area at time t. This probability is then given by:

$$\boldsymbol{\pi}'_m(t) = \sum_{n=0}^{\infty} e^{-\Lambda' t} \frac{(\Lambda' t)^n}{n!} \boldsymbol{\pi}'_m(0) \mathbf{U}_0^n \tag{12}$$

The probability distribution function $F(t)$ of the time until the mobile terminal leaves the registration area (i.e., going to the absorbing state, s_{K+1}) is then defined by:

$$F(t) = 1 - \boldsymbol{\pi}'_m(t)\mathbf{e} = 1 - \sum_{n=0}^{\infty} e^{-\Lambda' t} \frac{(\Lambda' t)^n}{n!} \boldsymbol{\pi}'_m(0) \mathbf{U}_0^n \mathbf{e} \tag{13}$$

where $\mathbf{e}$ is a column vector whose elements are all 1's. Differentiate Equation (13), we get the probability density function

$$f(t) = \sum_{n=0}^{\infty} e^{-\Lambda' t} \frac{(\Lambda' t)^n}{n!} \boldsymbol{\pi}'_m(0) \mathbf{U}_0^n (\mathbf{I} - \mathbf{U}_0)\mathbf{e} = \sum_{n=0}^{\infty} \Lambda' e^{-\Lambda' t} \frac{(\Lambda' t)^n}{n!} \boldsymbol{\pi}'_m(0) \mathbf{U}_0^n \mathbf{A}_0 \tag{14}$$

with the identity $\mathbf{A}_0 = (\mathbf{I} - \mathbf{U}_0)\mathbf{e}$ for the matrix $\mathbf{R}_0$.

To deduce the expected cost of this distance scheme, we consider two cases. In the first case, there is no call arrived to the mobile terminal before the terminal sends an update message. In this case, the cost incurred is the update cost, U. For another case, there is a call arrive before the update message is sent, in this case the cost is the search cost, KC. By total probability, the expected cost for this scheme can be summarized as follows:

$$C_m^D(d) = \int_0^{\infty} U f(t) e^{-\mu t} dt + \int_0^{\infty} KC \mu e^{-\mu t} [1 - F(t)] dt \tag{15}$$

Substitute Equations (13) and (14) into (15) and do the integration, we can simplify the equation to:

$$C_m^D(d) = \sum_{n=0}^{\infty} \left(\frac{\Lambda'}{\Lambda' + \mu}\right)^n \frac{U\Lambda' \boldsymbol{\pi}'_m(0) \mathbf{U}_0^n \mathbf{A}_0 + KC\mu \boldsymbol{\pi}'_m(0) \mathbf{U}_0^n \mathbf{e}}{\Lambda' + \mu} \tag{16}$$

Using similar approach as the expected cost, the expected period length is given by:

$$D_m^D(d) = \int_0^{\infty} t f(t) e^{-\mu t} dt + \int_0^{\infty} t \mu e^{-\mu t} [1 - F(t)] dt \tag{17}$$

and the result is:

$$D_m^D(d) = \sum_{n=0}^{\infty} \left(\frac{\Lambda'}{\Lambda' + \mu}\right)^n \frac{(n+1)[\boldsymbol{\pi}_m'(0)\mathbf{U}_0^n(\Lambda'\mathbf{A}_0 + \mu\mathbf{e})]}{(\Lambda' + \mu)^2} \tag{18}$$

Lastly, by the weak law of large numbers, the average expected cost for the distance-based protocol is then given by:

$$E_D(d) = \frac{\sum_{m=1}^{N} C_m^D(d)\pi_m}{\sum_{m=1}^{N} D_m^D(d)\pi_m} \tag{19}$$

In this protocol, we manage to find the optimal value d^* such that the average expected cost $E_D(d^*)$ is minimized. Again, because we allow a non-uniform movement pattern of a terminal inside the network, it is advantageous to evaluate an optimal vector $\mathbf{d}^* = <d_i^* : 1 \leq i \leq N>$ where each element d_i^* represents the optimal distance value for the cell c_i. Since the values for each element in the vector is independent, we can derive the optimal value for each cell separately.

5 Conclusion

Recently, the mobile terminal tracking problem has received very much attention. There are efforts made to propose protocols that target at balancing the update cost with the paging cost. To test the performance of their protocols, they devise some simple configurations of the cellular network and user movement. However, these configurations may not be valid. In particular, for easy formulation, they assume that each mobile terminal has a spatially uniform, symmetric movement pattern. Moreover, they assume that the incoming call interarrival time is very long. Therefore, they estimate the expected cost by employing the steady location probabilities. Since these assumptions are not always valid, as we have discussed before. To remove those assumptions, we have proposed a general framework that uses the uniformization method to derive the transient location probabilities accurately and efficiently. This framework can also be employed to compare different update and paging protocols. Having this more accurate prediction, we have shown that the result deviates a lot as compared to the steady analysis. Therefore, the steady state analysis are shown to be inaccurate. Lastly, our framework can model different kinds of user movement. In this regards, we can also compare the performance of paging and update method under different movement patterns.

References

[1] Ian F. Akyildiz and Joseph S. M. Ho. Dynamic Mobile User Location Update for Wireless PCS Networks. In *ACM Journal of Wireless Networks*, volume 1. Baltzer Publishers, 1995.

[2] Amotz Bar-Noy, Ilan Kessler, and Moshe Sidi. Mobile users: To Update or not to Update? In *Proceedings of IEEE INFOCOM*, 1994.

[3] Edmundo de Souza e Silva and H. Richard Gail. Analyzing Scheduled Maintainence Policies for Repairable Computer Systems. In *IEEE Transactions on Computers*, November 1990.

[4] Victor O. K. Li and Xiaoxin Qiu. Personal Communication Systems (PCS). In *Proceedings of IEEE*, volume 83, September 1995.

[5] Upamanyu Madhow, Michael L. Honig, and Ken Steiglitz. Optimization of Wireless Resources for Personal Communications Mobility Tracking. In *Proceedings of IEEE Infocom*, 1994.

[6] Randolph Nelson. *Probability, Stochastic Processes, and Queueing Theory*. Springer-Verlag, 1995.

[7] C. Rose. Minimizing the Average Cost of Paging and Registration: A Timer-Based Method. In *ACM Journal of Wireless Networks*, volume 2. Baltzer Publishers, 1996.

[8] C. Rose and R. Yates. Minimizing the Average Cost of Paging under Delay Constraints. In *ACM Journal of Wireless Networks*, volume 1. Baltzer Publishers, 1995.

Cedric C. F. Fong[*]   John C. S. Lui[*]   Man Hon Wong[*]

Edmundo A. de Souza e Silva[+]

[*] Department of Computer Science & Engineering
The Chinese University of Hong Kong

[+] Federal University of Rio de Janeiro, Brazil

MARIO LEFEBVRE AND RICHARD LABIB[1]
Risk sensitive optimal control of wear processes: the vector case

1. Introduction

The first improvement on the modelling of wear processes was made by Rishel (1991) who rightfully argued that wear increases through time and therefore cannot be modelled as the solution of a scalar Ito equation containing a Wiener process enabling the wear to both increase and decrease on any given interval. In order to correct this imprecision, Rishel modelled the wear as the solution of the system of stochastic differential equations

$$dx(t) = \rho\left[x(t), y(t)\right]dt \tag{1}$$
$$dy(t) = f\left[x(t), y(t)\right]dt + \sigma\left[x(t), y(t)\right]dW(t) \tag{2}$$

where $x(t)$ represents the wear of a certain machine at time t and ρ is non-negative and continuously differentiable in order for the wear to not decrease through time. He introduced the environmental variable $y(t)$ which influences directly the value of the wear. The function σ that accompanies the Brownian motion $W(t)$ must be positive definite.

Next, to introduce controlled wear models, Rishel considered that the functions involved in the stochastic equations depend on a control variable, namely u, in a class of feedback controls which he supposed to be the operating speed, for example, of the machine under consideration. Given the functions $\rho(x,y,u)$, $f(x,y,u)$, $\sigma(x,y,u)$ our aim is to compute the control $u^* = u^*(x,y)$ that maximizes the expected value of the performance criterion

$$J(x,y) = \int_0^{T_B} \left\{ r\left[u(x(t), y(t))\right] - c\left[x(t), y(t)\right] \right\} dt \tag{3}$$

where the functions $r(u)$ and $c(x,y)$ are respectively the return per unit time obtained when the machine is given to operate at speed u and the cost incurred per unit time when the wear is at level x and the environmental variable is at level y. T_B represents the time it

---

[1] The first author was supported by a grant from the Natural Sciences and Engineering Research Council of Canada.

takes for the machine to reach the worn out level B knowing that the wear level was initially x and the value of the environmental variable was y. Thus

$$T_B = T_B(x,y) = \inf\{t \geq 0 : x(t) = B | x(0) = x, y(0) = y\}. \tag{4}$$

For specific values of the functions $\rho\,(x,y,u)$, $f\,(x,y,u)$ and $\sigma\,(x,y,u)$, Rishel was able to solve the above control problem pertaining to wear processes, under certain specific conditions.

The next improvement was to refine the control model obtained by Rishel in order to make it even more realistic. Lefebvre and Gaspo (1996) replaced the performance criterion by the risk sensitive criterion which enables us to take the risk sensitivity of the operator of the machine into account. This criterion given by

$$C(\theta) = C(x,y,\theta) = -\frac{1}{\theta}\log E[\exp(-\theta\, J)] \tag{5}$$

introduces the risk parameter θ which, as suggested by Whittle (1982, p.268), determines if the optimizer is risk seeking ($\theta > 0$) or risk averse ($\theta < 0$). All the tools were now in place for Lefebvre and Labib (1997) to look at how to compute the solution of a special class of wear models where the optimal control can be obtained by considering a related uncontrolled process, as suggested by Kuhn (1985). Lefebvre and Labib considered the following model for the wear $x(t)$ of a certain machine

$$dx(t) = \rho_1[x(t), y(t)]dt \tag{6}$$
$$dy(t) = f_1[x(t), y(t)]dt + bu(t)dt + \sigma_1[x(t), y(t)]dW(t). \tag{7}$$

This particular form for the derivative of the environmental variable is inspired by a result proved by Kuhn, where b is dependent on $x(t)$ and $y(t)$. Now, the specific form for the performance criterion is assumed to be

$$J(x,y) = \int_0^{T_B}\left[\frac{1}{2}r_0 u^2(t) - c_0\right]dt \tag{8}$$

where r_0 and c_0 are non-negative since they represent meaningfully the return and the cost respectively. Once again, the objective is to find the expression for the feedback control u^* that maximizes the risk sensitive criterion established earlier. In order to compute u^*,

Lefebvre and Labib linearized the non-linear dynamic programming equation and transformed the control problem into an uncontrolled diffusion process. All the cases considered so far are defined as scalar since the environmental factor $y(t)$ is defined as a scalar function. The present paper takes into consideration all the improvements and results obtained previously in order to generalize the scalar model to a vectorial wear process. After a relationship has been established linking the multi-dimensional wear model to an uncontrolled multi-dimensional diffusion process, two specific examples are solved explicitly.

2. The Vector Case

We will first consider a three-dimensional wear process. Let

$$w(t) := (x(t), y(t), z(t)) \tag{9}$$

represent the state-space of the system defined by

$$dx(t) = \rho_1[w(t)]dt \tag{10}$$
$$dy(t) = \eta_1[w(t)]dt \tag{11}$$
$$dz(t) = f_1[w(t)]dt + bu(t)dt + \sigma_1[w(t)]dW(t) \tag{12}$$

where $x(t)$ still represents the wear and the environmental factor is now defined by the vector variable $q(t)$

$$q(t) := (y(t), z(t)). \tag{13}$$

This is a case where Brownian motion (or Wiener process) is only present in the second component of the environmental vector variable. For the optimal control problem to be completely defined we need to specify the performance criterion to be used. We have

$$J(x,y,z) = \int_0^{T_B} \left[\frac{1}{2}r_0 u^2(t) - c_0\right]dt \tag{14}$$

where all the quantities are the same than for the scalar case and defined earlier. Taking the risk sensitivity of the operator of the machine into account, we obtain, using Kuhn's result, an expression for the dynamic programming equation in terms of the derivatives of the maximum expected reward starting from the state (x,y,z), namely $V(x,y,z)$. We have

$$\inf_{u}\left[\frac{1}{2}r_0 u^2 - c_0 + \rho_1 V_x + \eta_1 V_y + (f_1 + bu)V_z - \frac{\theta}{2}\sigma_1^2 V_z^2 + \frac{1}{2}\sigma_1^2 V_{zz}\right] = 0. \tag{15}$$

Moreover the dynamic programming equation is subjected to the boundary condition

$$V(B, y, z) = 0. \tag{16}$$

Solving back in terms of $u := u(0)$, we obtain the optimal control to be

$$u^* = -\frac{b}{r_0} V_z. \tag{17}$$

Substituting back in the original equation, we obtain

$$-c_0 + \rho_1 V_x + \eta_1 V_y + f_1 V_z - \left(\frac{b^2}{2r_0} + \frac{\theta}{2}\sigma_1^2\right)V_z^2 + \frac{1}{2}\sigma_1^2 V_{zz} = 0. \tag{18}$$

In order to linearize the preceding equation to get a feasible solution, we will define the following vectorial transformation

$$\frac{1}{\phi}\begin{pmatrix}\phi_x \\ \phi_y \\ \phi_z\end{pmatrix} = \begin{pmatrix}\psi_1(w) & 0 & 0 \\ 0 & \psi_2(w) & 0 \\ 0 & 0 & -K(w)\end{pmatrix}\begin{pmatrix}V_x \\ V_y \\ V_z\end{pmatrix} \tag{19}$$

where $\phi(x, y, z)$ is the new vector-valued function replacing $V(x, y, z)$ in the dynamic programming equation, ψ_1 and ψ_2 are any functions of the state space and $K(w)$ is given by

$$K(w) := \frac{b^2}{r_0 \sigma_1^2} + \theta. \tag{20}$$

It is important to note that in order to obtain a feasible value for $V(x, y, z)$ the three partial differential equations containing the terms V_x, V_y and V_z respectively must be compatible. Thus, the functions ψ_1 and ψ_2 must be chosen with care. We have

$$\begin{pmatrix}V_x \\ V_y \\ V_z \\ V_{zz}\end{pmatrix} = \begin{pmatrix}\psi_1^{-1}(w)\frac{\phi_x}{\phi} \\ \psi_2^{-1}(w)\frac{\phi_y}{\phi} \\ -K^{-1}(w)\frac{\phi_z}{\phi} \\ K^{-2}(w)K_z(w)\frac{\phi_z}{\phi} - K^{-1}(w)\frac{\phi_{zz}}{\phi} + K^{-1}(w)\frac{\phi_z^2}{\phi^2}\end{pmatrix}. \tag{21}$$

The dynamic programming equation becomes

$$-c_0 + \rho_1 \psi_1^{-1}(w)\frac{\phi_x}{\phi} + \eta_1 \psi_2^{-1}(w)\frac{\phi_y}{\phi} - f_1 K^{-1}(w)\frac{\phi_z}{\phi} + \frac{1}{2}\sigma_1^2 \left[K^{-2}(w) K_z(w)\frac{\phi_z}{\phi} - K^{-1}(w)\frac{\phi_{zz}}{\phi} \right] = 0 \quad (22)$$

since the term $\left(\frac{b^2}{2r_0} + \frac{\theta}{2}\sigma_1^2 \right)$ can be replaced by $\frac{1}{2}\sigma_1^2 K(w)$. Rearranging terms and dropping off the argument w for ease of presentation, we get

$$c_0 \phi - \frac{\rho_1 \phi_x}{\psi_1} - \frac{\eta_1 \phi_y}{\psi_2} + \left(\frac{f_1}{K} - \frac{1}{2}\sigma_1^2 \frac{K_z}{K^2} \right)\phi_z + \frac{1}{2}\frac{\sigma_1^2}{K}\phi_{zz} = 0 \quad (23)$$

The above dynamic programming equation is linearized and we recognize the form of the Kolmogorov equation enabling us to interpret $\phi(x,y,z)$ as

$$\phi(x,y,z) = E_{(x,y,z)} \left[\exp\left(\int_0^T c_0 dt \right) \right] \quad (24)$$

where

$$T = T(x,y,z) := \inf\{ t \geq 0 : (X(t), Y(t), Z(t)) \in D \subset \Re^3 | X(0) = x, Y(0) = y, Z(0) = z \} \quad (25)$$

giving us the three-dimensional diffusion process model dependent on the new state-space vector $(X(t), Y(t), Z(t))$

$$dX(t) = -\frac{\rho_1}{\psi_1} dt \quad (26)$$

$$dY(t) = -\frac{\eta_1}{\psi_2} dt \quad (27)$$

$$dZ(t) = \left(\frac{f_1}{K} - \frac{1}{2}\sigma_1^2 \frac{K_z}{K^2} \right) dt + \sigma_1 K^{-1/2} dW(t). \quad (28)$$

This three-dimensional model for wear is therefore degenerate. A result proved, in this case, by Lefebvre and Labib shows that if we can find a continuous function $\phi(x,y,z)$ which is a viscosity solution of the Kolmogorov equation, then it is the unique solution of the system. Moreover if $\phi(x,y,z)$, corresponding to $V(x,y,z)$, satisfies the boundary condition then the optimal control is given by the expression calculated earlier.

In order to improve on our generalization, we will consider another three-dimensional model where this time both components of the vector representing the environmental variable will contain stochastic terms. We have the following form for the wear process

$$dx(t) = \rho_1[w(t)]dt \tag{29}$$
$$dy(t) = f_1[w(t)]dt + b_1 u_1(t)dt + \sigma_1[w(t)]dW_1(t) \tag{30}$$
$$dz(t) = f_2[w(t)]dt + b_2 u_2(t)dt + \sigma_2[w(t)]dW_2(t) \tag{31}$$

where $W_1(t)$ and $W_2(t)$ are independent standard Brownian motions. Again, the model to be controlled is subjected to the following performance criterion

$$J(x,y,z) = \int_0^{T_B} \left[\frac{1}{2}(r_1 u_1^2(t) + r_2 u_2^2(t)) - c_0 \right] dt. \tag{32}$$

For this problem, the dynamic programming equation is given by

$$\inf_u \left[\frac{1}{2}(r_1 u_1^2 + r_2 u_2^2) - c_0 + \rho_1 V_x + (f_1 + b_1 u_1)V_y + (f_2 + b_2 u_2)V_z - \frac{\theta}{2}\sigma_1^2 V_y^2 - \frac{\theta}{2}\sigma_2^2 V_z^2 + \frac{1}{2}\sigma_1^2 V_{yy} + \frac{1}{2}\sigma_2^2 V_{zz} \right] = 0 \tag{33}$$

where the boundary condition remains unchanged. The vector-valued optimal control is

$$\begin{bmatrix} u_1^* \\ u_2^* \end{bmatrix} = \begin{bmatrix} -\dfrac{b_1}{r_1} V_y \\ -\dfrac{b_2}{r_2} V_z \end{bmatrix} \tag{34}$$

Using a similar technique as for the previous model, we use the following transformation

$$\frac{1}{\phi} \begin{pmatrix} \phi_x \\ \phi_y \\ \phi_z \end{pmatrix} = \begin{pmatrix} \psi(w) & 0 & 0 \\ 0 & -K_1(w) & 0 \\ 0 & 0 & -K_2(w) \end{pmatrix} \begin{pmatrix} V_x \\ V_y \\ V_z \end{pmatrix} \tag{35}$$

on the dynamic programming equation

$$-c_0 + \rho_1 V_x + f_1 V_y + f_2 V_z - \left(\frac{b_1^2}{2r_1} + \frac{\theta}{2}\sigma_1^2 \right) V_y^2 - \left(\frac{b_2^2}{2r_2} + \frac{\theta}{2}\sigma_2^2 \right) V_z^2 + \frac{1}{2}\sigma_1^2 V_{yy} + \frac{1}{2}\sigma_2^2 V_{zz} = 0. \tag{36}$$

As previously we have

$$\begin{bmatrix} K_1(w) \\ K_2(w) \end{bmatrix} := \begin{bmatrix} \dfrac{b_1^2}{r_1 \sigma_1^2} + \theta \\ \dfrac{b_2^2}{r_2 \sigma_2^2} + \theta \end{bmatrix}. \tag{37}$$

Proceeding in the same manner as before, we end up with the following Kolmogorov equation containing as expected second derivatives terms in y and z given that they are the stochastic components of the environmental variable

$$c_0\phi - \frac{\rho_1\phi_x}{\psi} + \left(\frac{f_1}{K_1} - \frac{1}{2}\sigma_1^2 \frac{K_{1y}}{K_1}\right) + \left(\frac{f_2}{K_2} - \frac{1}{2}\sigma_2^2 \frac{K_{2z}}{K_2^2}\right)\phi_z + \frac{1}{2}\frac{\sigma_1^2}{K_1}\phi_{yy} + \frac{1}{2}\frac{\sigma_2^2}{K_2}\phi_{zz} = 0. \quad (38)$$

The corresponding three-dimensional diffusion process is given by

$$dX(t) = -\frac{\rho_1}{\psi}dt \quad (39)$$

$$dY(t) = \left(\frac{f_1}{K_1} - \frac{1}{2}\sigma_1^2 \frac{K_{1y}}{K_1^2}\right)dt + \sigma_1 K_1^{-1/2} dW_1(t) \quad (40)$$

$$dZ(t) = \left(\frac{f_2}{K_2} - \frac{1}{2}\sigma_2^2 \frac{K_{2z}}{K_2^2}\right)dt + \sigma_2 K_2^{-1/2} dW_2(t). \quad (41)$$

The functions in front of the Brownian motion terms represent the infinitesimal standard deviations of the stochastic processes and must therefore be positive. It is interesting to note that the first three-dimensional model considered is in fact a particular case of the second wear model where the functions b_1 and σ_1 in the latter model are set equal to zero. Now, given the above procedure, one can generalize the results obtained to compute the optimal feedback control for an n-dimensional model of the same form. The proof of the generalization is straightforward.

3. Explicit solutions

Kuhn established a relationship, between the functions under consideration in a multi-dimensional stochastic model, which enables us to transform the problem into a purely probabilistic one. We are therefore interested to find explicit solutions to problems for which Kuhn's relationship between functions does not hold. We will first consider the initial three-dimensional wear model and let

$$dz(t) = z(t)dt + z(t)u(t)dt + \left[z^2(t)\right]^{1/2} dW(t). \quad (42)$$

Thus, the uncontrolled process for the second component of the environmental vector is a geometric Brownian motion, which has an inaccessible boundary at the origin, and we assume that $z(0) = z$ is negative. Moreover, for practical reasons we will assume that the risk parameter θ is equal to zero. By taking

$$r_0 = -(B-x)yz \tag{43}$$

we simplify $K(w)$ to $(r_0)^{-1}$. If the derivative of $y(t)$ with respect to time given by η_1 is chosen to be positive, then we can take y to be positive. Therefore $r_0 > 0$ as it should be. We now have to solve the linearized partial differential equation (23). In order to obtain a solution satisfying the specific conditions of the problem, we choose the functions ψ_1 and ψ_2 to be

$$\psi_1 = -2\rho_1 y^3 z^3 (B-x)^2 \tag{44}$$

$$\psi_2 = -2\eta_1 y^{-3} z^{-2} (B-x)^{-2}. \tag{45}$$

The cost incurred per unit time which must be positive is taken to be the following

$$c_0 = (B-x)^2 y^2 z^2 \tag{46}$$

thus enabling us to obtain

$$\phi = \exp[(B-x)yz]. \tag{47}$$

The unknowns ρ_1 and η_1 have to be positive and such that the functions ψ_1 and ψ_2 are compatible in order to solve for $V(x,y,z)$ as specified earlier. We thus leave it to the optimizer to specify the nature of these functions which can be chosen according to the type of wear process under consideration (ρ_1 can be proportional to $y(t)$, for example (see Lin and Cheng (1989)).

The same procedure along with the same conditions apply for the second three-dimensional model studied in this paper. The components of the environmental vector are

$$dy(t) = y(t)dt + y(t)u(t)dt + \left[y^2(t)\right]^{1/2} dW(t) \tag{48}$$

and

$$dz(t) = z(t)dt + z(t)u(t)dt + \left[z^2(t)\right]^{1/2} dW(t). \tag{49}$$

Taking $r_1 = r_2 = -(B-x)yz$ as well as $\phi = \exp[(B-x)yz]$, we see that by setting

$$c_0 = 2(B-x)^2 y^2 z^2 + (B-x)^{-2} y^{-2} z^{-2} \tag{50}$$

we obtain

$$\psi = \rho_1 y^{-2} z^{-2} (B-x)^{-3}. \quad (51)$$

Even though the examples are purely mathematical they show the usefulness of the results obtained.

4. Conclusion

The aim of this paper was to consider a generalization of the paper of Lefebvre et Labib. The risk sensitivity of the optimizer was taken into account to establish the ensuing optimal control. These generalized results were obtained by considering two three-dimensional models of specific form. The initial controlled processes were transformed to give degenerate first hitting time problems which are difficult to solve explicitly. Nevertheless explicit solutions were found to provide examples for which Kuhn's relationship does not hold.

5. References

1. Kuhn, J. (1985) *The risk-sensitive homing problem*, J.Appl.Probab. 22, 796-803.
2. Lefebvre, M. and Gaspo, J. (1996) *Controlled wear processes : a risk-sensitive formulation*, Eng. Optimization 26, 187-194.
3. Lefebvre, M. and Labib, R. (1997) *Risk sensitive optimal control of wear processes*, Lectures in Applied Mathematics 33, 163-174.
4. Lin, J.-Y. and Cheng, H.S. (1989) *An analytical model for dynamic wear*, ASME J. Tribology 111, 468-474.
5. Rishel, R. (1991) *Controlled wear processes : Modeling optimal control*, IEEE Trans. Automat. Contr. 36, 1100-1102.
6. Whittle, P. (1982) *Optimization over time*, Vol. I, John Wiley, Chichester.

Département de mathématiques et de génie industriel, École Polytechnique de Montréal, C.P. 6079, Succursale Centre-Ville, Montréal, Québec, Canada H3C 3A7
E-mail address : lefebvre@mathappl.polymtl.ca

Département de mathématiques et de génie industriel, École Polytechnique de Montréal, C.P. 6079, Succursale Centre-Ville, Montréal, Québec, Canada H3C 3A7
E-mail address : richard@mathappl.polymtl.ca

ROBERT LIPSET, PAUL E DEERING AND ROBERT P JUDD
Deadlock in manufacturing systems: application of its necessary and sufficient conditions

1. Introduction

The problem of allocating shared resources to avoid deadlock in a flexible manufacturing system is a major one. Many methods have been developed to deal with this problem. [1] proposed a deadlock avoidance algorithm (DAA) using Petri nets. They developed a restriction policy based upon the production route information to guarantee that no circular wait situations will occur. This method eliminated all the deadlocked states but also prohibited many admissible or safe states. [5] developed a framework which captures the production route information relevant to deadlock avoidance that is less restrictive than in [1]. This improved DAA but it is still not an optimal solution. [10], using Petri nets, developed a deadlock avoidance algorithm which employed a look-ahead policy. The problem with this method is that it is unclear as to how many steps are needed to look ahead to avoid deadlock. This algorithm did not detect all deadlocked states and suggested using a recovery mechanism in case of system deadlock. [11] developed a deadlock prevention technique by introducing the concepts of sequential mutual exclusions (SME) and parallel mutual exclusions (PME). Again, this method eliminated all deadlocked states but is too restrictive and eliminated many admissible states. [2] developed a two-phase deadlock prevention technique to solve a class of FMS called Simple Sequential Processes with Resources (S^3PR). This technique requires locating control places called siphons, which are places that once becoming token free, remain token free. The problem is there exists no efficient algorithm to locate siphons [4]. [3] used system status graphs to develop the concept of simple and non-simple bounded circuits with empty and non-empty shared resources to detect part flow deadlock and impending part flow deadlock. This method introduced the concept of a bounded circuit to detect deadlock. It detected deadlock based on characteristics of this bounded circuit. This method sometimes predicted deadlock when in fact there existed no deadlock. In our work [7,8] we showed that it is possible to precisely quantify both necessary and sufficient conditions for deadlock to occur in a manufacturing system. The approach was to put the system into an evaluation state and then calculate the slack on all the closed paths to determine if deadlock existed.

In [9], we presented a deadlock detection algorithm (DDA), which was polynomial in complexity, that proposed a set of safe part movements to clear the parts from a manufacturing system with unit capacity resources. At every step, all the enabled parts are examined. If the algorithm could find a sequence of safe part movements to clear the manufacturing system, then the original state was not deadlocked. If it could not find a sequence to clear the system then the original state was declared deadlocked. In this paper we expand the DDA in [9] to handle multiple capacity resources. We believe that none of the moves the algorithm proposes will put a non-deadlocked system into a deadlocked state. However, we have not formally proved this conjecture yet.

This paper is organized as follows: Section 2 presents a short description of the construction of the simple static model of a manufacturing system; Section 3 defines the dynamics of part movements within the manufacturing system; Section 4 defines deadlock and explains the order of complexity of the proposed deadlock algorithm; Section 5 presents the deadlock detection algorithm and the set of safe part movements that avoid deadlock; Section 6 is a listing of the deadlock detection algorithm; Section 7 demonstrates an example applying the deadlock detection algorithm.

2. System Model

A manufacturing system consists of a set R of resources, which produce a set P of products. Each resource $r \in R$ has a capacity of cap(r) units. For each product $p \in P$, the process plan, plan(p)=$r_1 r_2 r_3$... defines the sequence of resources that are required to produce p. It is assumed that all process plans are fixed, finite and sequential. The number of steps necessary to produce product p in plan(p) is len(p). The function res(p,n) returns the nth resource in plan(p).

A manufacturing system can be conveniently represented by a graph $G=(R,A)$. G is constructed using the set of resources R and all the process plans in P. The nodes in the graph represent the resources. An arc $a \in A$ is drawn from node r_1 to node r_2 if r_2 is the next resource after r_1 in some process plan $p \in P$. The function input(r) returns the number of input arcs entering resource r and the function output(r) returns the number of output arcs leaving r. A resource with input(r)>1 is called a *converging* resource; a resource with output(r)>1 is called a *diverging* resource.

A *subgraph* $G_1=(R_1,A_1) \subset G$ consists of a subset of the resources and arcs of G such that all of the arcs in G_1 connect resources in G_1. A *path* is a subgraph whose resources can be ordered in a list $r_1 r_2 r_3 ... r_n$ where each resource in the list follows the resource preceding it. A *simple path* is a path with no repeated elements in the ordered list. A *closed path* is a path with $r_1=r_n$. A *simple circuit* is a closed path with no repeated elements. Let circuit(r) be the set of all simple circuits in G that contain r.

3. Model Execution

The graph G describes the static capabilities of the manufacturing system. We now define the model dynamics. At any given time, a manufacturing system is working on a set Q of parts. Each part $q \in Q$ must belong to one of the products P, which are produced by the system. The function class(q) defines the product class for q. At any given time every part q is being processed by one of the resources listed in its process plan. Let step(q) be the position in plan(class(q)) of the resource which is currently processing q. When a new part q_0 is add to the system, then step(q_0)=1. As the part is moved from resource to resource according to its plan, step(q_0) is incremented until it reaches the end of its plan and exits the system.

The state s of a manufacturing system is defined to be the current step(q) for all $q \in Q$.

A state is *feasible* if no resource is allocated more parts then it can process. Let S be the set of all feasible states for the system. A resource is said to be *free* if it is not filled to capacity. The function parts(r) returns the set of parts resource r is currently processing. The function nparts(r) returns the number of parts resource r is currently processing, i.e. $0 \leq \text{nparts}(r) \leq \text{cap}(r)$, $\forall r \in R$. The function freeunits returns the number of free resource units in resource r, that is, freeunits(r) = cap(r) − nparts(r).

A part is *enabled* if either the next resource in its process plan is free or the part is in the last step of its process plan. The system state s can be changed in one of three ways: (1) A new part q enters the system. In this case res(class(q),1) must be free; (2) An enabled part q can exit the system; (3) An enabled part q can advance to the next resource in its process plan. The second two cases will be denoted by the function move(q).

A resource r is said to be *committed* to arc a if resource r is presently processing a product that must follow arc a to the next resource in its process plan.

4. Deadlock

Given a manufacturing system in state s, if a series of part movements cannot be found which clears the system, then the system is in a deadlocked state [6]. One method to determine if a system is in deadlock is to exhaustively search every possible move. If a sequence of moves exists that clears the system, then the system is not deadlocked. The complexity of this process clearly is not scaleable even with simple systems containing few parts.

The algorithm presented in the next two sections limits the search space in the above process. At every step, all the enabled parts are examined. The algorithm will determine which, if any, of the enabled parts can be moved without placing the system into a deadlocked state (assuming it was not already). It will be shown that the decision to move an enabled part can be determined in at most $O(|R|)$ calculations. This means that each iteration takes no more than $O(|E\|R|) \leq O(|Q\|R|)$ calculations. Since each move advances a part along its process plan, then the number of iterations of the algorithm is bounded by

$$N = \sum_{q \in Q} \text{len}(\text{class}(q)) \qquad (4.1)$$

Therefore, deadlock can be determined with less than $O(N|Q\|R|)$ calculations; that is, polynomial in complexity.

The algorithm only determines a single state trajectory to clear the system. This may not be the optimal trajectory based on desired performance characteristics. Also, the algorithm does not directly indicate whether introducing a new part will result in a deadlocked state. Both these problems are easily overcome by examining all the possible advancements of the system state at a given time. These form the set of proposed moves for the system. Before a proposed part movement is allowed to change the system state s, the proposed part movement must be evaluated to determine if it will cause deadlock. If deadlock is detected, then the proposed part movement is not allowed to occur. If deadlock is not detected, then the proposed move is allowed. The optimal allowed proposed move is then executed. It is clear that all proposed moves can be evaluated with

$O(N |Q|^2 |R|)$ calculations. Since the resources have bounded capacity, then $|Q| < \alpha |R|$, for $\alpha = \max_{r \in R}\{\text{cap}(r)\}$. Using this result and (4.1) we conclude that $N < \beta |R|$, for $\beta = \alpha \max\{\text{len}(\text{class}(q))\}$. Hence, determining the optimal move of a manufacturing system in any state using this algorithm has complexity $O(|R|^4)$.

5. Safe Moves

To reiterate a few of the definitions presented in [9], *primary deadlock* occurs when all resources on a simple circuit are committed to arcs on that circuit. *Impending deadlock* is another form of deadlock and occurs when there exist enabled parts in the system but the system must terminate in primary deadlock after a finite number of part movements [3]. A *safe move* is defined to be a part move that will not transition the system from a non-deadlocked state to a deadlocked state. If a system is in impending deadlock, a move may still be identified as safe but deadlock will be determined at a later state. A safe move is never the root cause of deadlock.

5.1 Parts on the Terminal Resource Moving a part that is in its terminal resource will never transition the system from a non-deadlocked state to a deadlocked state. Clearly, it is always safe to move a part that is in its last step of its plan out of the system.

5.2 Simple Circuits Consider a simple circuit with at least one free unit of resource. It is clear that parts on this circuit can be moved until they all exit the system. Therefore, this system is not in deadlock [8].

5.3 Circuit Interation If a part moves into a converging resource, and does not become committed to another circuit, then the move is safe [9]. Here, we consider safe part movements that move into a converging resource and become committed to another circuit. Now consider the case illustrated in Figure 5.1 where two circuits are joined by a single resource r_0. Assume all resources are of single capacity and that when q_1 is moved to r_0 it will be committed to arc a_1. Then clearly this move causes primary deadlock in circuit C_2. This is due to the fact that when part q_1 is moved from r_2 to r_0, it enters circuit C_2, decreasing the number of free resources on this circuit. In this example the decrease was critical, since it eliminated the only free resource on C_2, placing the circuit into primary deadlock. However, it is clear that q_2 can be moved without causing deadlock since C_1 has two free resources. Therefore, special care must be taken to determine if a part can be safely moved into a converging resource with one free resource unit.

5.4 Counter-Flow The situation in Figure 5.1 is generalized in Figure 5.2. In this situation, parts of class a flow to the right and exit the system at resource r_{10}. Parts of class b flow to the left and exit the system at resource r_2. All converging resources are capacity one and all non-converging resources may be capacity one or greater. No arcs are shared by circuits. This situation is called *part counter-flow*. In Figure 5.2, assume all resources are unit capacity. If enabled parts a_0, a_2, or a_3 are moved, deadlock results; however, enabled parts b_1 and b_2 are safe to move. The reason for this is there is a non-converging resource r_1 with at least one free resource unit on C_1. Part b_2 can be moved into this resource, and then part b_1 can be moved into r_7; this allows all the a parts to exit the system, and finally, all the b parts can leave.

To determine if a part q in resource r_0 is safe to move, we try to construct a path from

r_0 to a free non-converging resource. This path is called a *safe counter-flow path* for part q. We construct this path as follows. Let path π be initialized to contain the resource r_0.

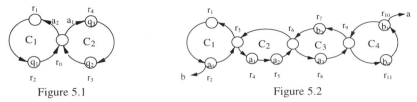

Figure 5.1   Figure 5.2

Trace the flow of q to its next resource r=res(class(q),step(q)+1). There are five cases to consider:

Case 1: If r is a free non-converging resource: append r to π, π is a safe counter-flow path. Exit algorithm.

Case 2: If r contains a part q and step(q)=len(class(q)): append r to π, π is a safe counter-flow path. Exit algorithm.

Case 3: If r is not free and case 2 does not apply: append r to π, let Z be the set of parts in r, for each $z \in Z$, let $q = z$ and call *Find safe counter-flow path*.

Case 4: If r is a free converging resource and r is not on π: append r to π, and call *Find safe counter-flow path*.

Case 5: If r is a free converging resource and r is the last free converging resource already on π, then a safe counter-flow path does not exist for part q; Exit algorithm. ∎

Algorithm 5.1. Find safe counter-flow path.

Applying this algorithm to the situation in Figure 5.2, clearly there are no safe counter-flow paths for any of the a parts. Notice that as the resources on circuit C_4 are appended to π and resource r_9 is encountered for the second time, the situation flagged by case 5 exists. Applying this algorithm to part b_1 results in a safe counter-flow path $\pi = r_{10} r_9 r_7 r_6 r_3 r_1$ and both parts b_1 and b_2 are free to move.

5.5 Counter-Flow Applied to Three or More Circuits Section 5.4 illustrates a bi-directional part flow. Part flows may be n-directional. This situation is characterized by n circuits being connected through a subgraph. Whether the part flows are bi-directional or n-directional the algorithm presented in section 5.4 still applies; it determines if an enabled part is safe to pass through a subgraph from one circuit and into another circuit. Consider the simple system in Figure 5.3 where the subgraph is just a single resource r_1. In this situation the process plan for part a is plan(class(a)) = $r_0 r_1 r_2 r_1 r_3 r_1 r_0$. We illustrate the application of the above algorithm in two cases. First, consider the case where part a_1 is in its initial resource. The algorithm will not identify any safe counter-flow path. Therefore, the algorithm will exit indicating that the system is in impending deadlock. It is easy to verify that this is the case. Next, consider the case where part a_1 is in its terminal resource (i.e., step(a_1)=7). The algorithm will then indicate that it is safe to move a_3. Therefore, this move will not deadlock the system.

5.6 Circuits Intersecting at Multiple Capacity Resources with Two Free Resource Units Assume that in Figure 5.1 resource r_0 is capacity two and all other resources have capacity one. Further assume that when q_1 is moved to r_0 it will be committed to arc a_1 and when q_2 is moved to r_0 it will be committed to arc a_2. Moving part q_1 or q_2 into resource r_0 would be a safe move. This is because both circuits that share r_0 have at least two free resources. Moving parts q_1 or q_2 only decreases the number of free resources by one on either circuit C_1 or C_2, thus this part movement can not cause deadlock. Moving a part into a converging resource with two or more free resource units is a safe move.

5.7 Circuits Intersecting at Multiple Capacity Resource with One Free Resource Unit An interesting phenomenon occurs when two simple circuits are joined by a single capacity resource versus a multiple capacity resource. Consider Figure 5.4 and Figure 5.5 where two simple circuits are joined by a resource r_0 and all parts are committed to their outgoing arcs. Assume that in Figure 5.5, cap(r_0)=2 and q_5 is committed to a_1 and all other resources in Figures 5.4 and 5.5 are single capacity. Further assume that if q_1 is moved to r_0 it will be committed to arc a_1 and if q_2 is moved to r_0 it will be committed to arc a_2. Even though in both manufacturing systems resource r_0 is free with one free resource unit, the system in Figure 5.4 is deadlocked and the system in Figure 5.5 is not deadlocked. In Figure 5.4 if either part q_1 or q_2 were moved into r_0 then primary deadlock would result on circuit C_2 or C_1 respectively. In Figure 5.5 if part q_1 were moved into r_0 then primary deadlock would result on circuit C_2 but moving part q_2 is a safe move. This is because if q_2 were moved into r_0 then resource r_3 would be free and allow the parts to advance along circuit C_2. Special care should be taken to determine if a part might be safely moved into a converging resource that has one free resource unit.

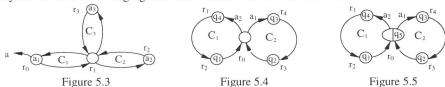

Figure 5.3   Figure 5.4   Figure 5.5

5.8 Evaluating Multiple Capacity Converging Resources Consider the situation in Figure 5.6 in which we want to determine if enabled part q_1 can safely move into resource r_2 where cap(r_2)=2. Assume part b_1 is committed to arc a_1 and if part q_1 moved to r_2 it also would be committed to arc a_1. Also assume arbitrary path p_1 terminates on resource r_3. To conclude whether part q_1 can safely move into r_2 we must determine if path p_1 is a safe path for part q_1 or part b_1. If one of the two paths were safe then moving q_1 would be a safe move. Likewise, if both paths are unsafe then moving q_1 is an unsafe move. It may be necessary to check the paths of the next part that enters and the parts contained in the multiple capacity converging resource to conclude if a move is safe or not.

Figure 5.6   Figure 5.7

Consider again the situation above in Figure 5.6 and assume that if part b_2 moved it would become committed to arc a_2 and that path p_1 is safe for b_1. We could trace two paths to determine if part b_2 is a safe move. Tracing the path $r_3r_2r_1r_2$ for b_2 would be an unsafe path because this would suggest primary deadlock on circuit C_1. But tracing the path $r_3r_2p_1r_3$ for b_2 we can conclude that moving b_2 would be safe. This is because we encountered the original part b_2 after the converging resource r_2 on the path, forming a simple circuit C_2. In essence, moving part b_2 freed space for part b_1. This type of move will be termed safe.

5.9 Circuits with More than One Free Converging Resource Suppose tracing the committed arcs of filled and converging resources results in a closed circuit beginning and terminating on a converging resource r_1. If there is only one converging resource on the circuit, then moving any part into this resource will result in primary deadlock on this circuit. However, a situation like the one illustrated in Figure 5.7 may exist, where the circuit C_2 contains more than one free converging resource. The other free converging resources can be used like a free non-converging resource. Therefore, moving a part into resource r_1 should be considered a safe move.

6. Deadlock Detection Algorithm

The program first advances all the parts in the system until they either exit, fill converging resources until there is one resource unit free in the converging resource, bump up against a converging resource, or are blocked by another part. This is done in function Advance_part. Advance_part returns whether any parts were actually moved. If a part has been moved, then the main algorithm identifies other enabled parts and tries to advance them. From Section 5.1, 5.2 and 5.6, we know that all these moves are safe.

At this point, the next move must place a part on a converging resource. If moving this part does not commit it to another circuit, then it is safe to move, according to section 5.3. Stay_on_circuit checks to see if moving any enabled parts into a converging resource will result in that part still being committed to the same circuits. If it does, then it moves the first part it finds that satisfies this condition. Then Main will advance all the parts again. Function Stay_on_circuit needs to know about all the circuits in the system G. However, the circuit information can be calculated off-line and stored.

Finally, the algorithm must move a part into a converging resource that will commit that part to an arc on another circuit. Find_safe_move looks at each enabled part and determines if it is safe to move according to the conditions presented in sections 5.4, 5.8, and 5.9. If a safe move does exist, Find_safe_move moves the first part it finds and the algorithm continues.

The main algorithm continues until one of three conditions occurs. First, if there are no parts left in the system, then the algorithm found a way to empty the system, and the original state of the system was not deadlocked. Another exit point is when there are still parts in the system, but none are enabled. The system then is in primary deadlock. Finally, the algorithm may exit if there are enabled parts, but no safe moves. In this case the system is in impending deadlock. Since the algorithm only makes safe moves, if the system is in a deadlocked state when it exits the algorithm, then the original state was deadlocked.

Main()
1. **if** $Q==\Phi$ **then**
    Deadlock = F
    exit
2. $E \leftarrow \{\text{enabled parts}\}$
3. **if** $E==\Phi$ **then**
    Deadlock = T
    exit
4. **if** Advance_part(E) **then goto** 1
5. **if** Stay_on_circuit(E) **then goto** 1
6. **if** Find_safe_move(E) **then**
    goto 1
    else
    Deadlock = T
    exit

Figure 6.1 Main algorithm

Advance_part(E)
1. PartMoved ← F
2. if E==Φ then
   return(PartMoved)
3. Remove q from E
4. if (step(q) == len(class(q)) then goto 8
5. if input(res(class(q), step(q)+1)) = 1) then goto 8
6. if (freeunits(res(class(q), step(q)+1))> 1) then goto 8
7. goto 2
8. move(q)
9. PartMoved ← T
10. goto 2

Figure 6.2 Function to advance parts on a circuit

Stay_on_circuit(E)
1. if E==Φ then
   return(F)
2. Remove q from E
3. if circuit(res(class(q),step(q))≠circuit(res(class(q),step(q)+1)) then goto 1
4. move(q)
5. return(T)

Figure 6.3 Function to advance part that stays on circuit

Find_safe_move(E)
1. if E==Φ then return(F)
2. Remove q from E
3. π= res(class(q), step(q))
4. if not(find_safe_path(q, q,π)) then goto 1
5. move(q)
6. return(T)

Figure 6.4 Function to find safe move

Find_safe_path(q_{org},q, π) /* value parameters */
1. r=res(class(q), step(q)+1)
2. if free(r)==T then goto 6
3. if r ∉ π then
      Z←parts(r)
      goto 15
   endif
4. if q_{org} ∈ parts(r) then return(T)
5. return(F)
6. if input(r)==1 then return(T)
7. if freeunits(r)>1 then return(T)
8. if r ∉ π then goto 12
9. Let r* be the last resource on π where input(r*)>1 and freeunits(r*)=1
10. if r==r* then return(F)
11. return(T)
12. if res(class(q),len(class(q))==r then
    return(T)
13. Z←parts(r)
14. append q to Z
15. append r to π
16. for each z ∈ Z do
       if find_safe_path(q_{org}, z, π) then
          return(T)
       endif
17. return(F)

Figure 6.5 Function to find safe path

7. Example

Let a manufacturing system be composed of five resources r_1, r_2, r_3, r_4 and r_5, all with capacity one, except for r_3 which has capacity 2. The system consists of three simple circuits $C_1=r_1r_2r_1$, $C_2=r_2r_3r_2$, and $C_3=r_3r_4r_5r_3$. Suppose the system manufactures two products p_1 and p_2 specified by the following process plans: plan(p_1)=$r_1r_2r_3r_4r_5$ and plan(p_2)=$r_4r_5r_3r_2r_1$. Assume the a parts belong to class p_1 and the b parts belong to class p_2. Table 7.1 shows one sequence of moves that the deadlock detection algorithm may use to clear the system.

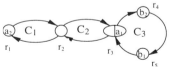

Figure 7.1

Author Note

The authors can be contacted in care of the Department of Industrial and Manufacturing Systems Engineering, Ohio University, Athens, Ohio 45701 USA.

State	Enabled Part	Safe Moves	Comment	Procedure which Moved Part
1	a_2, b_1	b_1	r_5 to r_3	Find_Safe_Move
2	a_2, b_1, b_2	b_2	r_4 to r_5	Advance_Part
3	a_1, a_2, b_1	a_1	r_3 to r_4	Advance_Part
4	a_2, b_1, b_2	a_2	r_1 to r_2	Find_Safe_Move
5	a_2, b_2	a_2	r_2 to r_3	Find_Safe_Move
6	b_1	b_1	r_3 to r_2	Find_Safe_Move
7	b_1, b_2	b_1	r_2 to r_1	Advance_Part
8	b_1, b_2	b_1	b_1 exits	Advance_Part
9	b_2	b_2	r_5 to r_3	Find_Safe_Move
10	a_1, b_2	a_1	r_4 to r_5	Advance_Part
11	a_1, a_2, b_2	a_1	a_1 exits	Advance_Part
12	a_2, b_2	a_2	r_3 to r_4	Advance_Part
13	a_2, b_2	a_2	r_4 to r_5	Advance_Part
14	a_2, b_2	a_2	a_2 exits	Advance_Part
15	b_2	b_2	r_3 to r_2	Find_Safe_Move
16	b_2	b_2	r_2 to r_1	Advance_Part
17	b_2	b_2	b_2 exits	Advance_Part

Table 7.1 State Table for Example

References

[1] Banaszak, Z. and B. Krogh, "Deadlock Avoidance in Flexible Mfg. Sys. with Concurrently Competing Process Flows." *IEEE Trans. on Robotics and Automation*, vol. 6, pp. 724-733.

[2] Barkaoui, K. and I.B. Abdallah, "A Deadlock Method for a Class of FMS." *Proceedings of the 1995 IEEE Int. Conf. On Systems, Man and Cybernetics*, pp. 4119-4124

[3] Cho, H., T.K. Kumaran, and R. Wysk, "Graph-Theoretic Deadlock Detection and Resolution for Flexible Manufacturing Systems." *IEEE Trans. on Robotics and Automation*, vol. 11, no. 3, pp. 550-527

[4] Ezpeleta, J., J. Colom, and J. Martinez, "A Petri Net Based Deadlock Prevention policy for Flexible Manufacturing Systems." *IEEE Trans. on Robotics and Auto.*, vol. 11, pp. 173-184

[5] Guro, D. and H. Muller, "Toward an Optimal Deadlock Avoidance Algorithm for Flexible Manufacturing Systems." *Proceedings of the 1995 IEEE Int. Conf. On Systems, Man and Cybernetics*, pp. 4137-4142

[6] Hsieh, F. and S. Chang, "Dispatching-driven deadlock avoidance controller synthesis for flexible manufacturing systems." *IEEE Trans. Robotics and Auto.*, vol. 10, pp. 196-209

[7] Judd, R.P. and T. Faiz, "Deadlock Detection and Avoidance for a Class of Manufacturing Systems." *Proceedings of the 1995 American Control Conference*, pp. 3637-3641

[8] Lipset, R., P. Deering, and R.P. Judd, "Necessary and Sufficient Conditions for Deadlock in Manufacturing Systems." *Proc. of the 1997 American Control Conf.*, vol. 2, pp. 1022-1026

[9] Judd, R.P., P. Deering, and R. Lipset, "Deadlock Detection in Simulation of Manufacturing Systems." *Proceedings of the 1997 Summer Computer Simulation Conference*, pp.317-322

[10] Viswanadham, N., Y. Narahari, and T. Johnson, "Deadlock Prevention and Deadlock Avoidance in Flexible Manufacturing Systems Using Petri Net Models," *IEEE Trans. on Robotics and Auto.*, vol. 6, no. 6, pp. 713-723

[11] Zhou, M. and F. DiCesare, "Parallel and Sequential Mutual Exclusion for Petri Net Modeling of Manufacturing Systems with Shared Resources," *IEEE Trans. on Robotics and Auto.*, vol. 7, no. 4, pp. 550-527

JAMES R PERKINS AND R SRIKANT
Failure-prone production systems with uncertain demand[1]

1 Introduction

We consider fluid models of failure-prone machines whose buffer evolution $x(t)$ is described by the following equation:

$$x(t) = x(0) + \int_0^t u(s)ds - D(0,s),$$

where $u(t)$ is the controlled production rate and $d(t)$ is the demand rate process. The optimal control problem is to minimize the expected long-run average of the sum of *backlog* and *surplus* inventory costs. Specifically, let c^+ and c^- denote the per-unit surplus and backlog costs, respectively. Then, the objective is to minimize

$$\lim_{T \to \infty} \frac{1}{T} E \int_0^T \left(c^+ x^+(t) + c^- x^-(t) \right) dt, \qquad (1.1)$$

where $x^+ = \max(0,x)$ and $x^- = -\min(0,x)$. Special cases of the above model have been studied earlier. The only known case where an optimal solution has been explicitly found is one in which $D(0,t) = dt$, and $u(t)$ is Markov-modulated, i.e., $0 \leq u(t) \leq \mu$ when the machine is *up* and $u(t) = 0$ when the machine is *down*, and the machine stays in the *up* and *down* states for exponentially-distributed durations [3]. We also point out that related models have been studied in the context of controlled queues. For example, the problem of controlling the rate of service in a queue where the demand arrives according a Poisson process and at each arrival epoch in the Poisson process, the arriving demand is drawn from a fixed distribution has been studied in [4]. Here the server is assumed to be always *up*. Both [4] and [3] obtain explicit optimal solutions.

Generalizations of the above formulation to problems with multiple machines, multiple part-types and discounted costs have also been studied; see [6] for some recent results and references to other work along these lines. A common theme in such problems is the difficulty in obtaining the optimal solution. For the single machine, one part-type problem, an alternative to obtaining optimal solutions is to obtain solutions to related controlled diffusion problems and approximate the solution to the original problem using the diffusion limits. This approach is taken in the recent work by Krichagina et al. [2].

---
[1] This research was supported in part by the National Science Federation under Grant No. ECS-94-10242 and the University of Illinois Research Board under Grant No. 1-2-68075.

In this paper, we generalize the above results to the following problem: We consider the problem in which the demand $D(0,t)$ arrives as in [4], i.e., the demand arrives in bursts where the burst-arrival process is Poisson and the burst size is exponentially distributed. However, unlike [4], the machine can be *up* or *down*, and the buffer content is not restricted to be positive. More precisely, $D(0,t)$ is given by

$$D(0,t) = \sum_0^{N(t)} d_i,$$

where d_i is the size of the ith burst and $N(t)$ is the total number of burst-arrivals up to time t. We assume $N(t)$ is a Poisson process of rate λ and d_i's are i.i.d. exponential random variables with mean $1/\lambda_d$. We assume that the machine *up* and *down* times are independent, and exponentially distributed with means $1/q_d$ and $1/q_u$. To ensure stability, we assume that

$$\frac{q_u}{q_d + q_u}\mu > \frac{\lambda}{\lambda_d}.$$

2 Probability Density Equations

In this section, we assume that the control is a hedging point policy and derive equations for the steady-state buffer distribution. Specifically, we assume that there exists a $z > 0$ that defines the control as follows:

$$u(t) = \begin{cases} \mu & x(t) < z \text{ and machine is up} \\ 0 & \text{else.} \end{cases}$$

We also define

$$\gamma_j \triangleq \lim_{t \to \infty} \text{Prob}(x(t) = z \text{ and in machine state } j).$$

Now, straightforward conservation of probability flow arguments lead to the following equations:

$$0 = q_d\, p_1(x) - (q_u + \lambda)\, p_0(x) + \lambda\, \gamma_0\, p(z-x) + \lambda \int_0^{z-x} p_0(x+y)\, p(y)\, dy$$

$$\mu \frac{dp_1(x)}{dx} = -(q_d + \lambda)\, p_1(x) + q_u\, p_0(x) + \lambda\, \gamma_1\, p(z-x) + \lambda \int_0^{z-x} p_1(x+y)\, p(y)\, dy,$$

where we have used $p(y) = \lambda_d e^{-\lambda_d y}$ to denote the burst-size probability density function. The boundary conditions are

$$q_d\, \gamma_1 = (q_u + \lambda)\, \gamma_0 \quad \mu\, p_1(z) + q_u\, \gamma_0 = (q_d + \lambda)\, \gamma_1 \quad \left(\frac{q_u}{q_d + q_u} - \gamma_1\right)\mu = \frac{\lambda}{\lambda_d}$$

and
$$\int_{-\infty}^{z} p_0(x)\,dx + \gamma_0 = \frac{q_d}{q_d+q_u}, \quad \int_{-\infty}^{z} p_1(x)\,dx + \gamma_1 = \frac{q_u}{q_d+q_u}.$$
Combining the first two of these equations yields
$$p_1(z) = \frac{\lambda}{\mu}(\gamma_0 + \gamma_1).$$
Let $w = z - x$, so that we have a variable that is always positive which would allow us to use Laplace transforms. Also define $f_j(w) := p_j(z-w)$. Then,
$$0 = q_d\, f_1(w) - (q_u + \lambda)\, f_0(w) + \lambda\, \gamma_0\, p(w) + \lambda \int_0^w f_0(w-y)\, p(y)\, dy$$
and
$$-\mu \frac{df_1(w)}{dw} = -(q_d + \lambda)\, f_1(w) + q_u\, f_0(w) + \lambda\, \gamma_1\, p(w) + \lambda \int_0^w f_1(w-y)\, p(y)\, dy.$$
Define
$$F_j(s) = \int_0^\infty f_j(w)\, e^{-sw}\, dw, \quad P(s) = \int_0^\infty p(w)\, e^{-sw}\, dw.$$
Then, taking Laplace transform equations, and using the fact that $f_1(0) := p_1(z) = \frac{\lambda}{\mu}(\gamma_0 + \gamma_1)$, we have
$$0 = q_d\, F_1(s) - (q_u + \lambda)\, F_0(s) + \lambda\, \gamma_0\, P(s) + \lambda\, F_0(s)\, P(s)$$
and
$$-\mu\, s\, F_1(s) + \lambda\, (\gamma_0 + \gamma_1) = -(q_d + \lambda)\, F_1(s) + q_u\, F_0(s) + \lambda\, \gamma_1\, P(s) + \lambda\, F_1(s)\, P(s).$$
Solving these equations yields
$$F_0(s) = \frac{\mu\, q_d\, \gamma_1\, s}{\Delta(s)} - \gamma_0 \quad \text{and} \quad F_1(s) = \frac{\mu\, (q_u + \lambda\, Q(s))\, \gamma_1\, s}{\Delta(s)} - \gamma_1$$
with
$$\gamma_0 = \frac{q_d}{q_u + \lambda}\left[\frac{q_u}{q_d + q_u} - \frac{\lambda}{\lambda_d\, \mu}\right] \quad \text{and} \quad \gamma_1 = \frac{q_u}{q_d + q_u} - \frac{\lambda}{\lambda_d\, \mu}$$
where
$$\Delta(s) \triangleq \mu\, q_u\, s + \lambda\, (\mu\, s - q_d - q_u)\, Q(s) - \lambda^2\, Q^2(s) \quad \text{and} \quad Q(s) \triangleq 1 - P(s).$$
Substituting the exponential density for $p(y)$ yields
$$P(s) = \frac{\lambda_d}{s + \lambda_d} \implies Q(s) = \frac{s}{s + \lambda_d}.$$

Then, inverting the Laplace Transforms, we get

$$p_0(x) = a_0 e^{-\lambda_1(z-x)} + b_0 e^{-\lambda_2(z-x)} \quad \text{and} \quad p_1(x) = a_1 e^{-\lambda_1(z-x)} + b_1 e^{-\lambda_2(z-x)},$$

where

$$\lambda_1 = \frac{1}{2\mu(\lambda + q_u)}\left[2\theta_1 + \lambda\theta_2 + \lambda\sqrt{(\mu\lambda_d - \lambda + q_d - q_u)^2 + 4q_d(\lambda + q_u)}\right]$$

$$\lambda_2 = \frac{1}{2\mu(\lambda + q_u)}\left[2\theta_1 + \lambda\theta_2 - \lambda\sqrt{(\mu\lambda_d - \lambda + q_d - q_u)^2 + 4q_d(\lambda + q_u)}\right]$$

$$a_0 = \frac{\lambda\gamma_0\left[(\lambda_1 - \lambda_d)(\lambda_d\mu + q_d + q_u) + \lambda_1\lambda\right]}{\mu(\lambda + q_u)(\lambda_1 - \lambda_2)}$$

$$b_0 = \frac{\lambda\gamma_0\left[(\lambda_d - \lambda_2)(\lambda_d\mu + q_d + q_u) - \lambda_2\lambda\right]}{\mu(\lambda + q_u)(\lambda_1 - \lambda_2)}$$

$$a_1 = \frac{\lambda\gamma_0\left[(\lambda_1 - \lambda_d)(q_d + q_u) + \lambda_1\lambda\right]}{\mu q_d(\lambda_1 - \lambda_2)}$$

$$b_1 = \frac{\lambda\gamma_0\left[(\lambda_d - \lambda_2)(q_d + q_u) - \lambda_2\lambda\right]}{\mu q_d(\lambda_1 - \lambda_2)}$$

$$\theta_1 = \mu\lambda_d q_u - \lambda(q_d + q_u) \quad \theta_2 = \lambda_d\mu + q_d + q_u - \lambda$$

$$\gamma_1 = \frac{\theta_1}{\mu\lambda_d(q_d + q_u)} \quad \gamma_0 = \frac{q_d\theta_1}{\mu\lambda_d(\lambda + q_u)(q_d + q_u)}.$$

2.1 Example: Quadratic Cost

Our derivation of the steady-state probabilities has not used the linear form of the cost function, although optimality of the hedging-point policy may depend on the explicit form of the cost. Thus, if one is simply interested in implementing a hedging-point policy, the best value of z, denoted by z^*, can be found using the expressions for the steady-state distribution, as we demonstrate below.

Let us consider a cost function of the form:

$$J = \lim_{T\to\infty} \frac{1}{T} E\left[\int_0^T g(x(t))\,dt\right],$$

where we impose the following restrictions on $g(x)$: $g(x)$ is nonnegative, differentiable, strictly convex and $g(0) = 0$. Then, it is easy to see that of z^* satisfies

$$E\left[\frac{dg(x)}{dx}\right]\bigg|_{x=z^*} = 0.$$

Specifically, let us consider the quadratic instantaneous cost function:

$$g(x) = c\,x^2$$

for some constant $c > 0$. The optimal hedging point must be chosen so that $E[x] = 0$. Thus,

$$0 = \int_{-\infty}^{z} x \left(p_0(x) + p_1(x) \right) dx + z \left(\gamma_0 + \gamma_1 \right)$$
$$= \frac{(a_0 + a_1)(\lambda_1 z - 1)}{\lambda_1^2} + \frac{(b_0 + b_1)(\lambda_2 z - 1)}{\lambda_2^2} + z(\gamma_0 + \gamma_1).$$

Then, using the fact that

$$\frac{a_0 + a_1}{\lambda_1} + \frac{b_0 + b_1}{\lambda_2} + \gamma_0 + \gamma_1 = 1,$$

it follows that, the optimal hedging point is given by

$$z^* = \frac{a_0 + a_1}{\lambda_1^2} + \frac{b_0 + b_1}{\lambda_2^2} = \frac{\lambda}{\lambda_d \gamma_1} \left[\frac{1}{\lambda_d \mu} + \frac{q_d}{(q_d + q_u)^2} \right].$$

2.2 Example: Linear, Absolute Value Cost

We return to the cost function defined in (1.1), i.e., the linear, absolute value cost function. From [3, 6], the optimal $z^* > 0$ satisfies

$$P(X < 0) = \frac{c^+}{c^+ + c^-}$$

which implies that

$$\frac{a_0 + a_1}{\lambda_1} e^{-\lambda_1 z} + \frac{b_0 + b_1}{\lambda_2} e^{-\lambda_2 z} = \frac{c^+}{c^+ + c^-}.$$

Note that, unlike the case with quadratic cost, we do not have a closed-form expression for z^*. We have to compute z^* by repeated substitution in the above equation.

In many cases, $\lambda_1 << \lambda_2$. In such cases, a valid approximation is to ignore the first term, and solve analytically for the approximate hedging point. The optimal hedging point under such a dominant eigenvalue approximation is given by

$$z^*_{dom} = \frac{1}{\lambda_2} \log_e \left[\frac{b_0 + b_1}{\lambda_2} \left(1 + \frac{c^-}{c^+} \right) \right].$$

3 Numerical Example

For the purposes of numerical comparison, we compare our results to the diffusion approximation proposed in [1, 2]. Under the diffusion approximation, $p(x) = \theta e^{-\theta(z-x)}$. Thus, the optimal hedging point is given by

$$z^*_d = -\frac{1}{\theta} \log_e \frac{c^+}{c^+ + c^-}.$$

Clearly, the diffusion approximation will be bad when γ is significantly larger than zero. Thus, we will also study the following modification.

Modified Diffusion Approximation: $p(x) = (1-\gamma)\theta e^{-\theta(z-x)}$. In this case,

$$z_m^* = -\frac{1}{\theta}\log_e \frac{c^+}{(1-\gamma)(c^+ + c^-)}.$$

The expression for θ is calculated from the appropriate reflected Brownian motion limit as

$$\theta \equiv \frac{-2m}{\sigma^2},$$

where

$$m = \lambda \bar{d} - \frac{\mu q_u}{q_u + q_d},$$

$$\sigma^2 = 2\lambda \bar{d}^2 + \frac{2 q_d q_u \mu^2}{(q_d + q_u)^3},$$

and $\bar{d} = 1/\lambda_d$ is the mean burst size.

Let $\lambda = q_u = q_d = c^+ = c^- = \bar{d} = 1$. Then,

$$z_d^* = -\frac{2 + 0.25\mu^2}{\mu - 2} \log_e 0.5,$$

and

$$z_m^* = -\frac{2 + 0.25\mu^2}{\mu - 2} \log_e \frac{2\mu}{\mu + 6}.$$

For stability, we require μ to be larger than 2. The optimal hedging point is compared to the various approximations in Table 1.

μ	z_{opt}^*	z_{dom}^*	z_d^*	z_m^*
2.1	20.7565	20.7565	21.5049	20.3766
2.5	4.0258	4.0208	4.9387	3.7807
3.0	1.8595	1.8285	2.9459	1.7232
4.0	0.6932	0.6009	2.0794	0.6694
5.0	0.2506	0.1102	1.9062	0.2621
6.0	0	0	1.9062	0
7.0	0	0	1.9755	0

Table 1: The optimal hedging point and other approximations

The diffusion approximation performs well only under heavy traffic, i.e., μ close to 2. The other approximations all perform fairly well throughout. But, we also point out that, in the dominant eigenvalue method, the optimal hedging point will be $z_{dom}^* = 0$ for $\mu \geq 5.3358$. Thus, the JIT condition is not exactly captured with this approximation.

4 HJB Equations

We have thus far studied the numerical computation of the hedging point. In the rest of this paper, we turn our attention to providing an outline of the proof of the optimality of the policy. We restrict ourselves to the linear cost of the form in (1.1). The Hamilton-Jacobi-Bellman (HJB) equations can be derived along the lines of [5] and are provided below.

$$g(x) + q_u W_1(x) + \lambda \int_0^\infty W_0(x-y)\, p(y)\, dy - (q_u + \lambda) W_0(x) = J^* \quad (4.2)$$

$$\min_{u(x)} \left\{ g(x) + q_d W_0(x) + \lambda \int_0^\infty W_1(x-y)\, p(y)\, dy - (q_d + \lambda) W_1(x) + \frac{dW_1(x)}{dx} u(x) \right\}$$
$$= J^*. \quad (4.3)$$

Now we have to solve these HJB equations for $W_1(x)$ and $W_0(x)$ and show that the optimal hedging point policy solves the above equations. Of course, one has to impose restrictions on the class of allowable policies to ensure the sufficiency of solving the HJB equations to obtain the optimal policy. This can be done as in [3].

5 Solution to the HJB Equations

We divide the HJB equations into three regions. For clarity, we use the following notation:
$$\begin{array}{ll} W^-(x) = W(x), & \text{if } x < 0 \\ W^+(x) = W(x), & \text{if } 0 \leq x < z \\ W^{++}(x) = W(x) & \text{if } x \geq z. \end{array} \quad (5.4)$$

In the region $x < 0$, we have

$$-c^- x + q_u W_1^-(x) + \lambda \lambda_d \int_0^\infty W_0^-(x-y)\, e^{-\lambda_d y}\, dy - (q_u + \lambda) W_0^-(x) = J^*$$

and

$$-c^- x + q_d W_0^-(x) + \lambda \lambda_d \int_0^\infty W_1^-(x-y)\, e^{-\lambda_d y}\, dy - (q_d + \lambda) W_1^-(x) + \mu \frac{dW_1^-(x)}{dx} = J^*.$$

For $0 \leq x < z$,

$$c^+ x + q_u W_1^+(x) + \lambda \lambda_d \int_0^x W_0^+(x-y)\, e^{-\lambda_d y}\, dy + \lambda \lambda_d \int_x^\infty W_0^-(x-y)\, e^{-\lambda y}\, dy$$
$$- (q_u + \lambda) W_0^+(x) = J^*$$

and

$$c^+ x + q_d W_0^+(x) + \lambda \lambda_d \int_0^x W_1^+(x-y)\, e^{-\lambda_d y}\, dy + \lambda \lambda_d \int_x^\infty W_1^-(x-y)\, e^{-\lambda y}\, dy$$
$$- (q_d + \lambda) W_1^+(x) + \mu \frac{dW_1^+(x)}{dx} = J^*.$$

For $x \geq z$,

$$c^+ x + q_u W_1^{++}(x) + \lambda \lambda_d \int_0^{x-z} W_0^{++}(x-y) e^{-\lambda_d y} dy + \lambda \lambda_d \int_{x-z}^x W_0^+(x-y) e^{-\lambda y} dy$$

$$+ \lambda \lambda_d \int_x^\infty W_0^-(x-y) e^{-\lambda_d y} dy - (q_u + \lambda) W_0^{++}(x) = J^*$$

and

$$c^+ x + q_d W_0^{++}(x) + \lambda \lambda_d \int_x^{x-z} W_1^{++}(x-y) e^{-\lambda_d y} dy + \lambda \lambda_d \int_{x-z}^x W_1^+(x-y) e^{-\lambda y} dy$$

$$+ \lambda \lambda_d \int_x^\infty W_1^-(x-y) e^{-\lambda_d y} dy - (q_d + \lambda) W_1^{++}(x) = J^*.$$

We assume the following continuity boundary conditions:

$$W_j^-(0) = W_j^+(0) \quad W_j^+(z) = W_j^{++}(z) \quad \text{for } j = 0, 1.$$

These conditions then imply that

$$\left. \frac{dW_1^-(x)}{dx} \right|_{x=0} = \left. \frac{dW_1^+(x)}{dx} \right|_{x=0} \quad \text{and} \quad \left. \frac{dW_1^+(x)}{dx} \right|_{x=z} = 0.$$

Also, since the differential cost functions are defined only up to a translational constant, we will arbitrarily set $W_1^+(z) = 0$. Substituting the following forms for the W_j's, $j = 0, 1$,

$$W_j^-(x) = a_j^- + b_j^- x + c_j^- x^2$$
$$W_j^+(x) = a_j^+ + b_j^+ x + c_j^+ x^2 + d_j^+ e^{-\lambda_1 x} + g_j^+ e^{-\lambda_2 x}$$
$$W_j^{++}(x) = a_j^{++} + b_j^{++}(x-z) + c_j^{++}(x-z)^2 + d_j^{++} e^{-\lambda_3 (x-z)},$$

we can solve the above equations and show that these expressions, along the optimal hedging point policy, satisfy the HJB equations. The details of these calculations are omitted due to space limitations.

References

[1] E. V. Krichagina, S. X. C. Lou, S. P. Sethi, and M. I. Taksar, "Production control in a failure-prone manufacturing system: Diffusion approximation and asymptotic optimality," *Annals of Applied Probability*, Vol. 3, No. 2, pp. 421–453, 1993.

[2] E. V. Krichagina, S. X. C. Lou, S. P. Sethi, and M. I. Taksar, "Diffusion approximation for a controlled stochastic manufacturing system with average cost minimization," *Mathematics of Operations Research*, Vol. 20, No. 4, pp. 895–922, 1995.

[3] T. Bielecki and P. R. Kumar, "Optimality of zero-inventory policies for unreliable manufacturing systems," *Operations Research*, Vol. 36, pp. 532–541, July-August 1988.

[4] B. T. Doshi, "Optimal control of the service rate in an $M/G/1$ queueing system," *Advances in Applied Probability*, Vol. 10, pp. 682–701, 1978.

[5] M. Caramanis and G. Liberopoulos, "Perturbation analysis of flexible manufacturing system flow controllers," *Operations Research*, vol. 40, pp. 1107–1125, November–December 1992.

[6] J. R. Perkins and R. Srikant, "Scheduling multiple part-types in failure-prone flexible manufacturing system," *IEEE Transactions on Automatic Control*, March 1997.

Address:

Prof. Jim Perkins, Department of Manufacturing Engineering, Boston University, 15 St. Mary's St., Boston, MA 02215; Email: perkins@engc.bu.edu

Prof. R. Srikant, Coordinated Science Laboratory and Department of General Engineering, University of Illinois, 1308 W. Main St., Urbana, IL 61801; Email: rsrikant@uiuc.edu

EDMUNDO DE SOUZA E SILVA[1], H RICHARD GAIL
AND JOÃO CARLOS GUEDES

Transient distributions of cumulative rate and impulse based reward with applications

1 Introduction

Transient distributions of cumulative reward have been found useful in analyzing the performance of various computer system models. Most previous work considered models with rewards associated with states that were rate based or models for which the rewards were associated with transition and were impulse based. In this paper we consider a rate plus impulse based reward model and study a numerical algorithm derived in [3] that is polynomial in both the number of rate based rewards and the number of impulse based rewards. The approach is based on the uniformization technique, also called randomization or Jensen's method, which has been found to be useful in computing transient measures.

A method for calculating the distribution of cumulative rate reward using uniformization was presented in [2], but the algorithm derived there was exponential in the number of rate rewards. Then, using Laplace transform techniques, Donatiello and Grassi [5] were the first to obtain a polynomial algorithm for the cumulative rate reward distribution. Later, the method of [2] was refined in [4] to yield a polynomial algorithm for the rate model with a better computational complexity than [5]. Recently, Nabli and Sericola [6] developed an algorithm for the rate case based on uniformization that involves calculations with real numbers between 0 and 1, which leads to nice computational properties.

The rate plus impulse model was first introduced by Qureshi and Sanders in the paper [7]. They extended the rate methodology of [2] to analyze this new model by uniformization techniques. However, the result was an algorithm to calculate the distribution of cumulative (rate plus impulse) reward that is exponential in both the rate rewards and the impulse rewards. Recently, the approach of [4] has been extended to the rate and impulse model to obtain a polynomial algorithm for this case in [3]. In this paper, we first review this new method for calculating the distribution of the total rate plus impulse based reward accumulated over a time

---
[1]The work of E. de Souza e Silva was partially supported by grants PROTEM-CC and PRONEX from CNPq(Brazil).

interval $(0, t)$ and averaged over the length t. We then discuss some details that are useful when actually implementing the algorithm of [3]. Finally, we present several numerical examples to illustrate the applicability of our approach.

2 Transient Distributions of Cumulative Reward

Consider a time-homogeneous continuous-time Markov chain $\mathcal{X} = \{X(t), t \geq 0\}$ with finite state space $\mathcal{S} = \{1, \ldots, M\}$. We assume that the chain has a reward structure, in that reward rates are associated with states and impulse rewards are associated with transitions, i.e., with pairs of states. Thus a reward per unit time based on the rate for state s is accumulated while the process is in that state, while a reward for transition $s \to s'$ is gained whenever that event occurs.

Specifically, there are $K+1$ rate rewards $r_1 > r_2 > \cdots > r_{K+1}$ associated with the states of the Markov chain. For any state $s \in \mathcal{S}$, we let $c(s)$ be the index of the reward rate for that state, i.e., $c(s) \in \{1, \ldots, K+1\}$ and state s has reward rate $r_{c(s)}$. The random variable $r_{c(X(t))}$ then represents the instantaneous reward rate at the time point t. The cumulative rate based reward during an observation period $(0, t)$ averaged over the length t of the interval is

$$ACR(t) = \frac{1}{t} \int_0^t r_{c(X(\tau))} d\tau. \tag{1}$$

Also, there are $\widehat{K} + 1$ nonnegative impulse rewards $\rho_1 > \rho_2 > \cdots > \rho_{\widehat{K}+1} \geq 0$ associated with the transitions of the Markov chain, i.e., with pairs of states. For any transition $s \to s'$ (any pair (s, s') of states), let $\widehat{c}(s, s')$ be the index of the impulse reward for that transition, i.e., $\widehat{c}(s, s') \in \{1, \ldots, \widehat{K}+1\}$ and transition $s \to s'$ has impulse reward $\rho_{\widehat{c}(s,s')}$. Let τ_n, $n = 1, 2, \ldots$, be the time of the nth transition ($\tau_0 = 0$) of the chain $\mathcal{X}$. The transition itself, say σ_n, can be identified with the pair of states $\sigma_n = (X(\tau_{n-1}), X(\tau_n))$. Let $NT(t)$ be the number of transitions that occur during $(0, t)$. Then the cumulative impulse based reward during $(0, t)$ averaged over the length of time t is

$$ACI(t) = \frac{1}{t} \sum_{n=1}^{NT(t)} \rho_{\widehat{c}(\sigma_n)}. \tag{2}$$

Note that if $NT(t) = 0$, then no transitions occurred during the observation period. This implies that $ACI(t) = 0$, which agrees with the value of the empty sum in (2) in this case. The total reward accumulated during $(0, t)$ averaged over the length t is

$$ACIR(t) = ACI(t) + ACR(t). \tag{3}$$

Our interest is in calculating the distribution $P[ACIR(t) \leq r]$, for a given reward level r. Since the impulse rewards are nonnegative and $ACR(t) \geq r_{K+1}$, we may assume $r \geq r_{K+1}$. Note that $CIR(t)$, the total reward accumulated during $(0,t)$, is simply $tACIR(t)$. Thus calculating the distribution $P[CIR(t) \leq r']$ for a given r' is equivalent to calculating $P[ACIR(t) \leq r]$ where $r = r'/t$.

It is advantageous from a numerical point of view to modify the rate rewards so that they lie between 0 and 1. This normalization may be accomplished as follows. First replace r_i with $r_i^* = (r_i - r_{K+1})/(r_1 - r_{K+1})$, for $i = 1, \ldots, K+1$, which ensures that $r_1^* = 1$, $r_{K+1}^* = 0$. Next replace the distribution reward level r with $r^* = (r - r_{K+1})/(r_1 - r_{K+1})$, and so $r^* \geq 0$. Finally replace the impulse reward ρ_j with $\rho_j^* = \rho_j/(r_1 - r_{K+1})$, for $j = 1, \ldots, \widehat{K}+1$. Let $ACI^*(t)$, $ACR^*(t)$, and $ACIR^*(t)$ be the corresponding quantities for the system with this new reward structure, and note that

$$ACI(t) = (r_1 - r_{K+1})ACI^*(t) \quad (4)$$
$$ACR(t) = (r_1 - r_{K+1})ACR^*(t) + r_{K+1}. \quad (5)$$

Adding these two equations and using (3), we have the equivalence

$$P[ACIR(t) \leq r] = P[(r_1 - r_{K+1})ACIR^*(t) \leq r - r_{K+1}] = P[ACIR^*(t) \leq r^*]. \quad (6)$$

Since the state space $\mathcal{S}$ is finite, $\mathcal{X}$ can be thought of as a discrete-time Markov chain subordinated to a Poisson process. This *uniformization* technique [9], also called randomization or Jensen's method, has been used to calculate a variety of transient measures, and is described as follows. Let $\mathbf{Q} = [q_{s,s'}]$ be the infinitesimal generator of $\mathcal{X}$, and choose $\Lambda \geq \max_{s,s'} |q_{s,s'}|$, so that the matrix $\mathbf{P} = \mathbf{I} + \mathbf{Q}/\Lambda$ is stochastic. Then $X(t) = Z(N(t))$ for all t, where $\mathcal{Z} = \{Z(n) : n = 0, 1, \ldots\}$ is a discrete-time Markov chain with transition matrix $\mathbf{P}$, and $\mathcal{N} = \{N(t) : t \geq 0\}$ is a Poisson process of rate Λ, with $\mathcal{Z}$ and $\mathcal{N}$ independent.

We now review the methodology of [2] for calculating $P[ACIR(t) \leq r]$. First condition on $ACI(t)$, the total impulse reward averaged over t, to obtain

$$P[ACIR(t) \leq r | ACI(t) = \widehat{r}] = P[ACR(t) \leq r - \widehat{r}]. \quad (7)$$

To find the distribution of $ACR(t)$, condition on the number of transitions that occurred during the observation period $(0,t)$. Given n transitions, the interval $(0,t)$ is divided into $n+1$ subintervals. Since the transitions correspond to events from the Poisson process $\mathcal{N}$, the lengths of these subintervals are exchangeable random variables [9]. The state of $\mathcal{X}$ during each subinterval is given by the state of the discrete-time Markov chain $\mathcal{Z}$, i.e., by $Z(0), \ldots, Z(n)$, each with a corresponding reward rate. Again, by exchangeability, the distribution of the total reward does not depend on the order in which the individual rewards are earned, but only on the number of each reward rate. That is, let k_i be the number of subintervals with rate reward r_i, $i = 1, \ldots, K+1$, and define the *rate coloring* $\mathbf{k} = \langle k_1, \ldots, k_{K+1} \rangle$. Here $\|\mathbf{k}\| \triangleq k_1 + \cdots k_{K+1} = n+1$. Then, any two realizations of $Z(0), \ldots, Z(n)$ which yield the same $\mathbf{k}$ give the same cumulative rate reward.

Define the set of rate colorings corresponding to n transitions (i.e., to $n+1$ subintervals), $n = 0, 1, \ldots$, as $\mathcal{K}_n = \{\mathbf{k} : \|\mathbf{k}\| = n+1\}$. Using a result of Weisberg [10], it is shown in [3] that the conditional distribution $P[ACR(t) \leq r|n, \mathbf{k}]$, for n transitions and a rate coloring $\mathbf{k} \in \mathcal{K}_n$, is given in terms of the functions

$$f_i[x, n, \mathbf{k}, r, l] \triangleq \frac{1}{l!} \cdot \frac{d^l}{dx^l} \left\{ \frac{(x-r)^n}{\prod_{\substack{j=1 \\ j \neq i}}^{K+1} (x-r_j)^{k_j}} \right\}. \tag{8}$$

Here, the parameter ranges are $n \geq 0$, $\mathbf{k} \in \mathcal{K}_n$, $l \geq 0$, $i = 1, \ldots, K+1$. In particular, it is shown that

$$P[ACR(t) \leq r|n, \mathbf{k}] = \sum_{i: r_i \leq r} f_i[r_i, n, \mathbf{k}, r, k_i - 1]. \tag{9}$$

Hence from (7)

$$P[ACIR(t) \leq r|n, \mathbf{k}, \hat{r}] = \sum_{i: r_i \leq r - \hat{r}} f_i[r_i, n, \mathbf{k}, r - \hat{r}, k_i - 1]. \tag{10}$$

Unconditioning on n, $\mathbf{k}$, and $\hat{r}$ gives

$$P[ACIR(t) \leq r] = \sum_{n=0}^{\infty} e^{-\Lambda t} \frac{(\Lambda t)^n}{n!} \sum_{\hat{r} \leq r} \sum_{\mathbf{k} \in \mathcal{K}_n} \Theta[n, \mathbf{k}, \hat{r}] \sum_{i: r_i \leq r - \hat{r}} f_i[r_i, n, \mathbf{k}, r - \hat{r}, k_i - 1], \tag{11}$$

where $\Theta[n, \mathbf{k}, \hat{r}]$ is the probability of a rate coloring $\mathbf{k}$ and an average accumulated impulse reward $\hat{r}$, given n transitions.

Because of the sum over all possible rate colorings $\mathbf{k}$, an algorithm based directly on (11) is exponential in the number of rate rewards. However, a polynomial algorithm was obtained in [3] by grouping the rate coloring vectors and finding a recursion on these groups in the following way. First, using the formula for the derivative of a product, the f_i can be written in the form

$$f_i[x, n, \mathbf{k}, r, l] = \sum_{m=0}^{l} \psi_n[x, r, m] \varphi_i[x, \mathbf{k}, l - m], \tag{12}$$

where we have defined the functions

$$\varphi_i[x, \mathbf{k}, m] = \frac{1}{m!} \cdot \frac{d^m}{dx^m} \left\{ \frac{1}{\prod_{\substack{j=1 \\ j \neq i}}^{K+1} (x - r_j)^{k_j}} \right\} \tag{13}$$

and

$$\psi_n[x, r, m] = \frac{1}{m!} \cdot \frac{d^m}{dx^m} \left\{ (x-r)^n \right\}. \tag{14}$$

Next, define the partitions $G_g[i,n] = \{\mathbf{k} \in \mathcal{K}_n : k_i = g\}$, for $i = 1,\ldots,K+1$, $g = 0,\ldots,n+1$. Thus $G_g[i,n]$ is the set of all rate colorings corresponding to n transitions for which there are g subintervals with reward rate r_i.

Using these partitions, define for $s \in \mathcal{S}$, $i = 1,\ldots,K+1$, $n = 0,1,\ldots$, $u = 0,\ldots,n$, $\hat{r} \geq 0$, the functions

$$\Psi_s[i,n,u,\hat{r}] = \sum_{g=n+1-u}^{n+1} \sum_{\mathbf{k}\in G_g[i,n]} \Theta_s[n,\mathbf{k},\hat{r}]\varphi[r_i,\mathbf{k},g+u-(n+1)]. \quad (15)$$

Also define the vector $\boldsymbol{\Psi}[i,n,u,\hat{r}] = \langle \Psi_1[i,n,u,\hat{r}],\ldots,\Psi_M[i,n,u,\hat{r}]\rangle$. It is shown in [3] that

$$P[ACIR(t) \leq r] = \sum_{n=0}^{\infty} e^{-\Lambda t}\frac{(\Lambda t)^n}{n!} \sum_{\hat{r}\leq r}\sum_{i:r_i\leq r-\hat{r}}\sum_{u=0}^{n} \psi_n[r_i, r-\hat{r}, u]\|\boldsymbol{\Psi}[i,n,n-u,\hat{r}]\|. \quad (16)$$

Further, $\Psi_s[i,n,u,\hat{r}]$ can be calculated from the recursion

$$\Psi_s[i,n,u,\hat{r}] = \frac{1}{\omega_{i,c(s)}}\left(\sum_{s'\in\mathcal{S}} p_{s's}\Psi_{s'}[i,n-1,u-1,\hat{r}-\hat{r}_{\widehat{c}(s',s)}] - \Psi_s[i,n,u-1,\hat{r}]\right) \quad (17)$$

for $i \neq c(s)$, and

$$\Psi_s[i,n,u,\hat{r}] = \sum_{s'\in\mathcal{S}} p_{s's}\Psi_{s'}[i,n-1,u,\hat{r}-\hat{r}_{\widehat{c}(s',s)}] \quad (18)$$

for $i = c(s)$. Here $\mathbf{P} = [p_{s's}]$ and $\omega_{i,j} = r_i - r_j$ for $i \neq j$. The initial conditions are

$$\Psi_s[i,0,u,0] = \begin{cases} 0 & i \neq c(s), u = 0 \\ \pi_s(0)/\omega_{i,c(s)} & i \neq c(s), u = 1 \\ \pi_s(0) & i = c(s), u = 0 \\ 0 & i = c(s), u = 1. \end{cases} \quad (19)$$

Also, $\psi_n[x,r,l]$ can be calculated from the recursion

$$\psi_n[x,r,l] = \psi_{n-1}[x,r,l-1] + (x-r)\psi_{n-1}[x,r,l]. \quad (20)$$

The initial conditions are

$$\psi_0[x,r,l] = \begin{cases} 1 & \text{for } l = 0 \\ 0 & \text{for } l > 0. \end{cases} \quad (21)$$

In [4] a scaling of the $\omega_{i,j}$ is described that yields a way to transform the above algorithm so that all multipliers are bounded by 1 in absolute value. Additionally, two near identical rate rewards may be merged to avoid division by small quantities. Collapsing to the larger reward gives a lower bound for $P[ACIR(t) \leq r]$ (since the total accumulated reward is greater in the merged system), while collapsing to the smaller reward yields an upper bound for this distribution.

3 Application Examples

In this section we present several examples to illustrate the applicability of the algorithm of Section 2. The first example is a model of a simple queueing system with a limited buffer. Customers arrive to this system at an exponential rate of λ customers/time unit and are served with exponential rate equal to μ customers/time unit. Let T_1 and T_2 be two threshold values that divide the number of customers q in the system into three regions. A cost function that depends on the cumulative time the system spends in each region and on the number of times the system queue rises above each threshold is defined as follows. If $q \leq T_1$, no cost is incurred by the system. However, if q rises above T_1, then a fee of \$1.0 is paid. Furthermore, there is a penalty of \$0.5 per time unit charged when $T_1 < q \leq T_2$. Similarly, each time q rises above the second threshold T_2, a fee of \$2.0 is paid, and there is a penalty of \$1.0 per time unit charged when $q > T_2$.

Our interest is in calculating the distribution of the total cost during an observation interval $(0,t)$, i.e., $P[ACIR(t) \leq C/t] = P[CIR(t) \leq C]$. In Figure 1 a plot of this distribution for $C = 2$ and $1 \leq t \leq 10$ is presented for a system with $\lambda = 3$, $\mu = 1$, $T_1 = 4$, $T_2 = 8$ and buffer size 10. For comparison, the distribution of total cost is plotted in the case when no fee is paid as the system crosses the thresholds (i.e., no impulse rewards are included). We observe a sharp decrease in the total cost probability as t increases. Furthermore, there is a significant difference in the distribution of the cost when no impulse rewards are present.

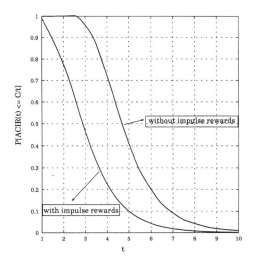

Figure 1: $P[ACIR(t) \leq C/t]$ for $C = 2$.

Our second example is a fluid model of an asynchronous multiplexor queue (i.e., an ATM switch) fed by multimedia sources. Fluid models have been utilized previously to study communication systems, e.g., [1], and are particularly useful in analyzing ATM switches. The multiplexor queue has infinite buffer space and serves cells uniformly by removing them from the queue at a constant rate of C Mbps. The multimedia traffic is modeled by N independent and identical on-off sources. That is, each source alternates between the *active* (or *on*) state, during which traffic arrives continuously at a rate λ Mbps, and the *silent* (or *off*) state, during which no traffic is generated. The length of the active period is assumed to be exponentially distributed with mean α seconds, while the silent period lasts for an exponentially distributed amount of time with mean σ seconds.

Most communication models in the literature deal with steady state solutions, and models which capture transient system behavior are not common. One transient study appears in [8], where the queue length probability as a function of time is characterized by a set of ordinary differential equations. In our example we define a reward model and apply the algorithm of Section 2 to calculate the distribution of the cumulative reward and obtain the dynamics of the system.

Let $q(t)$ be the number of bits in the queue at time t. Our interest is in the probability that $q(t)$ rises above a given threshold value during a busy period. Specifically, we wish to study how this probability varies with time during a busy period that starts with the sources in a given state (for example, all are active). We also wish to determine the largest value that this probability may achieve.

The traffic generated by the sources can be characterized by a birth-death chain with $N+1$ states, where a state represents the number of active sources. The rate from state s to $s-1$ is s/α, while the rate from s to $s+1$ is $(N-s)/\sigma$. Associate a reward to each state equal to the total bit rate into the system minus the output rate (the system capacity). Provided that $tACR(t)$, the cumulative reward up to time t, remains positive, it equals $q(t)$, the number of bits in the multiplexor queue buffer. Now consider a process $q^*(t)$ which represents the queue size during a busy period, i.e., $q^*(t)$ remains at 0 once the queue empties for the first time. It is not difficult to see that $P[ACR(t) > T/t]$ is an upper bound for the probability that $q^*(t)$ is above the threshold T in this busy period case.

For this second example we set $\alpha = 0.518$ sec, $\sigma = 0.510$ sec, $N = 5$ sources, $\lambda = 1.554$ Mbps and $C = 4.875$ Mbps. The reward model has 6 states, and the corresponding rewards are: 5, 2.895; 4, 1.341; 3, -0.213; 2, -1.767; 1, -3.321; 0, -4.875 (before normalizing the rewards to lie between 0 and 1). Figure 2 gives a plot of $P[ACR(t) > T/t]$ (an upper bound to $P[q^*(t) > T]$) for three different values of the threshold T. From this figure we can conclude that only buffers with a size larger than 4M bits have a low probability that the queue length increases beyond the threshold during a busy period starting with all sources active.

The last example is similar to the second, but we consider a system with heterogeneous sources and show how impulse rewards can be incorporated in the

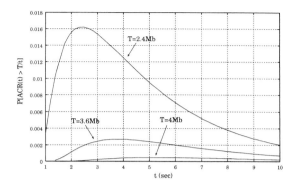

Figure 2: $P[ACR(t) > T/t]$ for model without impulse rewards.

fluid traffic model. We assume that there are two distinct on-off sources. The first source has parameter values $\alpha = 0.4$ sec, $\sigma = 0.6$ sec and $\lambda = 1.5$Mbps. The second source transmits at a rate of 2.0Mbps during an active period, with the mean length of an active period equal to 0.01 sec, which is significantly smaller than that of the first source. The duration of the silent period of the second source is equal to 0.1 sec. The server capacity is 1 Mbps.

The duration of the active period of the second source is very small when compared to that of the first source. Thus we represent the first source with two states as before, but the second source is modeled with impulse rewards associated to an event that triggers at (exponentially distributed) intervals of 0.1 seconds in either of the states of the first source. In Figure 3 we show $P[ACIR(t) > T/t]$ (an upper bound to $P[q^*(t) > T]$) for three different values of the threshold T.

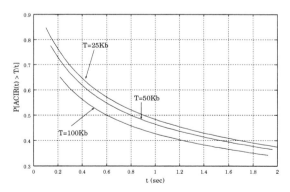

Figure 3: $P[ACIR(t) > T/t]$ for model with impulse rewards.

References

[1] D. Anick, D. Mitra, and M. Sondhi. Stochastic theory of a data-handling system with multiple sources. *Bell System Technical Journal*, 61(8):1871–1894, 1982.

[2] E. de Souza e Silva and H.R. Gail. Calculating availability and performability measures of repairable computer systems using randomization. *Journal of the ACM*, 36(1):171–193, 1989.

[3] E. de Souza e Silva and H.R. Gail. An algorithm to calculate transient distributions of cumulative rate and impulse based reward. *Communications in Statistics - Stochastic Models*, 14(3), 1998 (to appear).

[4] E. de Souza e Silva, H.R. Gail, and R. Vallejos Campos. Calculating transient distributions of cumulative reward. In *Proc. Performance '95 and 1995 ACM SIGMETRICS Conf.*, pages 231–240, 1995.

[5] L. Donatiello and V. Grassi. On evaluating the cumulative performance distribution of fault-tolerant computer systems. *IEEE Trans. on Computers*, 40(11):1301–1307, 1991.

[6] H. Nabli and B. Sericola. Performability analysis: a new algorithm. *IEEE Trans. on Computers*, 45(4):491–494, 1996.

[7] M.A. Qureshi and W.H. Sanders. Reward model solution methods with impulse and rate rewards: An algorithm and numerical results. *Performance Evaluation*, 20(4):413–436, 1994.

[8] Q.R. Ren and H. Kobayashi. Transient solutions for the buffer behavior in statistical multiplexing. *Performance Evaluation*, 23:65–87, 1995.

[9] S.M. Ross. *Stochastic Processes*. John Wiley & Sons, 1983.

[10] H. Weisberg. The distribution of linear combinations of order statistics from the uniform distribution. *Annals Math. Stat.*, 42:704–709, 1971.

Edmundo de Souza e Silva
Federal University of Rio de Janeiro, NCE and CS Dept.
Cx.P. 2324, CEP 20001-970, Rio de Janeiro, Brazil

H. Richard Gail
IBM Thomas J. Watson Research Center
Yorktown Heights, NY 10598

João Carlos Guedes
Federal University of Rio de Janeiro, COPPE/Systems
Cx.P. 2324, CEP 20001-970, Rio de Janeiro, Brazil

IV. Reliability

P THOFT-CHRISTENSEN AND F M JENSEN
Reliability based optimization of passive fire protection on offshore topsides

1. Introduction

It is well known that fire is one of the major risks of serious damage or total loss of platforms/topsides used for oil and gas production. This paper presents a methodology and software for reliability-based optimization of the layout of passive fire protection (PFP) of firewalls and of structural members on offshore structures. The paper is partly based on research performed within the EU supported research project B/E-4359 *"Optimised Fire Safety of Offshore Structures"* and partly on research supported by the Danish Technical Research Council. Special emphasis is put on the optimization software developed within the project.

Optimization of a topside of an offshore structure involves optimization of the passive fire protection, the active fire protection system, the safety equipment, the primary and secondary structural elements, the Temporary Safe Refuge, and Escape, Evacuation and Rescue Systems. However, such a complex optimization is not realistic with the current knowledge in this field.

Since PFP is very important for the safety of offshore structures this paper focuses on the optimization of PFP. The overall optimization problem formulated is to minimise the cost of the PFP with constraints on the minimum acceptable safety. The design variables are the type and amount of PFP and to some extent whether PFP is to be applied to walls/structural elements or not. Uncertainties are related to the fire loading, the thermal properties of the structural steel, the insulation and to material and strength parameters.

Methodologies for optimization of PFP and corresponding computer programs have been developed. A program OPTIWALL[1] for optimization of the PFP on firewalls and a program OPTIBEAM[2] for optimization of PFP on structural members have been implemented.

The program OPTIWALL performs deterministic and reliability-based optimization of the PFP attached to firewalls. The program determines the optimal thickness and material for the PFP for one or more firewalls subjected to thermal loads while minimising the cost.

The program OPTIBEAM performs deterministic and reliability-based optimization of the PFP attached to structural members (beams or columns). The program determines the optimal thickness (and material) for the PFP for one or more scenarios while minimising the cost of PFP.

Additionally, the effect of other mitigation measures such as deluge/sprinkler systems can be taken into account. In both programs constraints are related to the reliability of the wall/structural members using limit states on the maximum temperature and on general buckling/yielding failure using API, AISC and ECCS models.

2. Architecture of the Otimization System

Only optimization of PFP is considered. The topside layout is assumed given (location of the firewalls is given and thee structural elements protected using PFP are identified). The design variables are the amounts and types of PFP. The optimization problem is then to minimize the cost of the PFP with requirements on the minimum acceptable safety.

The optimization methodology consists of number of steps. Not all steps are obligatory.

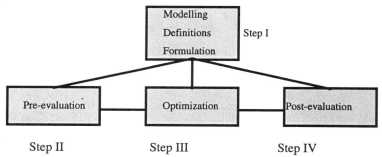

Figure 1. Architecture of the optimization system

3. Step I. Modelling, Definitions and Formulation

This first step consists of a number of actions such as: selection of the structural model, definition of a FEM model, grouping of structural elements, definition of fire scenarios, definition of failure modes and corresponding limit states, and the stochastic modelling.

4. Step II. Pre-evaluation

This pre-evaluation step is a very useful tool. In many cases the optimization of PFP can be performed using only the pre-evaluation modules. In the pre-evaluation the following actions are performed: a FEM analysis of the structure is performed and the potential failure modes are evaluated, the structure is modified if one or more limit states are violated, sensitivity analysis parameters are defined, a sensitivity analysis is performed to obtain a feasible design without reanalysis of the structure, design variables are added or removed based on the results of the sensitivity analysis, a corresponding, deterministic optimization problem is formulated (optional) and the deterministic optimization problem is solved, the reliability index and its derivatives are calculated so that limit states, stochastic variables etc. may be deleted/added.

5. Step III. Optimization

This step is the main step, but in some cases it is not used since it may be very time-consuming. At this step the following actions are performed: the reliability-based optimization problem is defined (design variables, objective function and constraints), the reliability-based optimization problem is solved.

6. Step IV. Post evaluation

At this step the following actions are performed: the optimization results may be modified, e.g. rounding up of some design variables to the nearest allowable value, the optimization results are evaluated to ensure that all assumptions are valid, a new grouping of elements or the use of new PFP material may be done and a new optimization performed, i.e. the optimization is repeated from the beginning.

7. Formulation of the Problem

The reliability based optimization problem can be formulated in the following way

$$\min_{b} \; C(\bar{b})$$
$$\bar{b}^T = (b_1,\ldots,b_n)$$
$$\text{s.t.} \quad \beta_j(\bar{b},\bar{x},T,s_i) \geq \beta_j^{\min} \qquad j = 1,\ldots,M \qquad (1)$$
$$\beta^{sys}(\bar{b},\bar{x},T,s_i) \geq \beta^{sys,\min}$$
$$b_i^{\min} \leq b_i \leq b_i^{\max} \qquad i = 1,\ldots,n$$

where C is the objective function (cost function) and $\bar{b}^T = (b_1,\ldots,b_n)$ are the design variables. s_i is fire scenario i and T is a reference time. The reference time could be the time where the fire is maximum or the time to evacuate all personnel. $\bar{x}$ is a vector of stochastic variables, M is the number of constraints and n is the number of design variables. The solution to this problem is $\bar{b}_{opt}^i$ where superscript "i" indicates scenario i. Problem (1) is solved for all N scenarios and as the final optimal solution the maximum value for each design variable is used.

Optimization of PFP on the topside is divided into two parts:
- optimization of PFP on non-structural parts (firewalls) using the software package OPTIWALL,
- optimization of PFP on structural members using the software package OPTIBEAM.

The programs OPTIWALL and OPTIBEAM are able to find optimal PFP for both firewalls and structural members subjected to pool and/or jet fires.

8. The Software Package OPTIWALL

The program OPTIWALL combines a fire analysis program, a reliability assessment program for reliability evaluation of firewalls subjected to fire (consisting of a program for calculation of heat transfer to firewalls and a program for reliability evaluation), and a nonlinear optimization program.

It is assumed that all firewalls have insulation material, that the geometry of the fire wall is constant and that only insulation on the hot side of the firewall is optimized. There are only two design variables for a firewall, namely the thermal conductivity of the PFP material and the thickness of the insulation material. The objective function is the cost of the PFP modelled as a function of the thickness and of the thermal conductivity and a constant term related to the installation. A constraint is in the deterministic case imposed on the temperature at the interior face of the insulation, which at the reference time T (60 minutes for A60 walls and 90 minutes for A90 walls) must be lower than some specified limit state temperature. In the reliability-based formulation the constraints are related to the probability that the temperature in the firewall exceeds a limit value.

In figures 2, 3, and 4 output screens from OPTIWALL, namely the reliability index as function of the PFP thickness, the history of the objective function, and the history of the design variables, are shown for illustration.

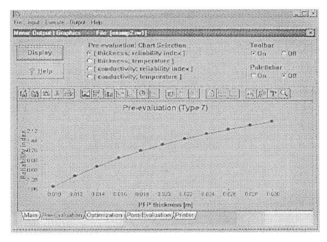

Figure 2. OPTIWALL. Pre-evaluation: Reliability index as a function of the PFP thickness.

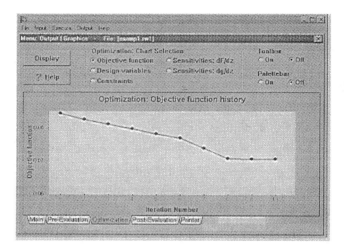

Figure 3. OPTIWALL. Optimization: History of the objective function..

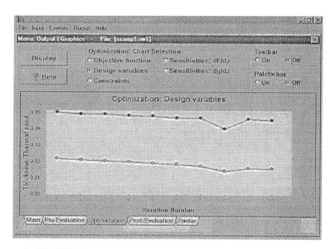

Figure 4. OPTIWALL. Post-evaluation: History of the design variables.

9. The Software Package OPTIBEAM

The OPTIBEAM program combines the modules for reliability assessment with the modules for optimization. OPTIBEAM performs deterministic and reliability based optimization of PFP attached to structural members. The design variables are the thickness of the PFP on topside beams/columns. Since the number of structural elements on a standard topside structure may be quite large, grouping of the design variables into a number of groups is implemented in OPTIBEAM in order to reduce the number of design variables.

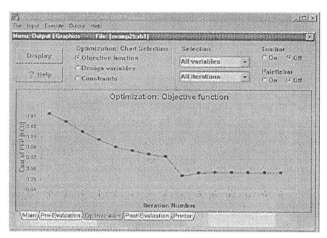

Figure 5. OPTIBEAM. Optimization: History of the objective function..

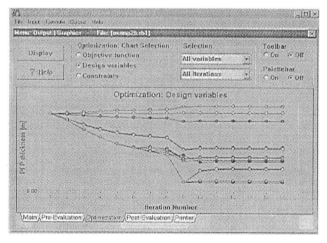

Figure 6. OPTIBEAM. Optimization: History of design variables.

In order to take into account the effect of other mitigation measures (AFP, improved lay-out, etc.) a third term may be included in the objective function. The objective function is the sum of the total cost of PFP and the expected failure costs. It is assumed that the expected failure costs are proportional to the initial cost of the structure without PFP. Constraints are related to a limiting temperature failure criterion or to member failure by buckling/yielding (using the API/AISC model or the ECCS model).

In figures 5, 6, and 7 output screens from OPTIBEAM, namely the history of the objective function, the history of the design variables, and a sensitivity analysis, are shown for illustration.

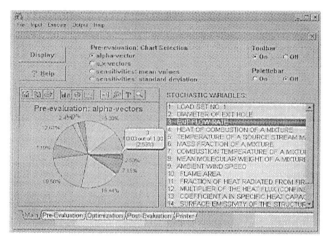

Figure 7. OPTIBEAM. Post-Evaluation. Sensitivity Analysis.

9. CONCLUSIONS

Major achievements in this paper with regard to reliability-based optimization can be summarized as:

- A formulation of reliability based optimization problems for both PFP on firewalls and structural members has been specified.

- A methodology and specifications for prototype software for PFP optimization including pre- and post-evaluation of firewalls and structural members have been developed.
- A DOS program OPTIWALL for optimization PFP (including pre- and post-evaluation) of PFP on firewalls has been implemented and tested.
- A DOS program OPTIBEAM for optimization PFP (including pre- and post-evaluation) of PFP on structural members has been implemented and tested.
- A Windows GUI for OPTIWALL and OPTIBEAM has been developed.

9. Acknowledgement

This paper presents work performed in the EU supported research project B/E-4359 *"Optimised Fire Safety of Offshore Structures"*. The partners in the project are: RINA, Snamprogetti and Tecnomare, Italy; Germanisher Lloyd, Germany; TU of Lisbon (IST), Portugal; APS and WS-Atkins (AST), UK; and CSR, Denmark. The research is also financially supported by the Danish Technical Research Council. We would like to thank CSRconsult, Aalborg, Denmark for permission to publish this paper.

10. References

[1] OPTIWALL., CSRsoftware, CSRconsult, P.O. Box 218, DK-9000 Aalborg, Denmark.
[2] OPTIBEAM, CSRsoftware, CSRconsult, P.O. Box 218, DK-9000 Aalborg, Denmark.

DIMITAR CHRISTOZOV
Evaluation of the quality of an option compared to its alternatives

1. Introduction

Comparing options, an essential problem in decision making, is addressed. Options can be goods, products, ideas, suppliers, services, technologies, even applicants for a given position. They are the alternative solutions to a problem. The common in all these cases is that a manager evaluates an option among the set of its alternatives, comparing it with the other members of the same set; and also that every option is evaluated according to a given list of criteria (**indicators**).

The goal is to assess the options in a way to support the decision making process. Two general approaches can be distinguished:

- **Designing an integral indicator** $I: R^n \to R$ to assign a single value to each of the options (**quality of the option**). The option with the highest grade is chosen.
- **Step-by-step decision procedure** to select the "best" option, without evaluating formally every one of the alternatives.

An approach to design of an integral indicator of the options' quality, which benefits from between indicators dependencies, is presented as well as an interactive selection procedure. A piece of software is developed to demonstrate these procedures.

1. Information model

To identify the problems and to explain the selected approaches and algorithms we shall use the terms as they are presented in Table 1:

All options are described according to a common list of **single indicators** I_i (further only indicators). An indicator measures the quality of every option according to a given property. $X = \{x_{ij}\}$, $x_{ij} \geq 0$, is called **information matrix**. The decision makers preferences are presented as indicators' **weights** W_i. Direction (**sign**) S_i of an indicator shows whether the quality of options grows with increasing the value of the indicator or if it goes down. "+1" indicates the higher value of the indicator the higher quality of the option and "-1" indicates the opposite direction. This notation represents the case when the infinite growth of an indicator's value leads to infinite growth of the option's quality. To preserve the same notation for the case when the quality, measured by an indicator I_r, depends on the distance of its value to a given finite optimal quantity y_r the transformation $\hat{x}_{jr} = abs(x_{jr} - y_r)$ has to be applied, and $s_r = -1$. We can consider every **option** O_j as a numerical vector, every element of which represents the value of the respective indicator.

indicator's name	dimension	weight	sign	Option 1 O_1	Option 2 $O2$	...	Option n On
I_1	D_1	W_1	S_1	x_{11}	x_{21}		x_{n1}
I_2	D_2	W_2	S_2	x_{12}	x_{22}		x_{n2}
...	...	...	...	...	...		...
I_j	D_j	W_j	S_j	x_{1j}	x_{2j}		x_{nj}
...	...	...	...	...	...		...
I_k	D_k	W_k	S_k	x_{1k}	x_{2k}		x_{nk}

Table 1: Information model

An **integral indicator** is a measure combining a set of single properties.

The two cases, according to the domain of an indicator's value are considered:

- quantitative indicators, measured by an interval scale – the value of each indicator for a given option is presented as a numeral;
- qualitative indicators, but the scale used for measuring it, allows ordering the options in "better-worse" direction, the indication about the option's level is given also as a number.

The third option -- qualitative indicators, measured by a scale, which does not allow ordering of the options are not used. Such does not provide any valuable for our purpose information.

Example: To illustrate the proposed approaches, let us consider the case of six options compared according to thirteen indicators. The source data, presented in the following table, is randomly generated.

Indicator	D	W	S	O 1	O 2	O 3	O 4	O 5	O 6
I 1	m	2	+	33.33	38.44	43.34	35.24	39.11	33.23
I 2	n	2	+	42.22	45.99	57.23	44.12	51.10	37.00
I 3	b	2	+	674.00	618.00	683.00	628.00	729.00	600.00
I 4	v	2	+	350.00	396.00	400.00	388.00	381.00	360.00
I 5	c	2	+	856.00	831.00	940.00	950.00	761.00	940.00
I 6	x	2	+	1400.00	1200.00	1800.00	1400.00	1200.00	1400.00
I 7	z	2	-	160.00	150.00	140.00	140.00	160.00	140.00
I 8	a	2	-	70.00	30.00	45.00	40.00	30.00	35.00
I 9	s	1	-	38.50	40.20	36.20	39.10	38.00	71.80
I 10	d	1	-	43.00	43.50	41.00	44.70	43.40	48.00
I 11	f	1	-	22.00	23.00	22.50	27.00	30.00	40.00
I 12	g	1	+	16.00	16.50	18.50	17.50	17.00	15.00
I 13	h	1	+	4.00	4.40	4.70	4.30	3.70	3.50

2. Integral Indicator

To illustrate the idea, the approach derived from, let us consider the case of comparing two engines A and B, measured by two indicators: power ("+") and weight ("-"); and let A = {100, 100}, and B = {90, 90}. One can say that the engine A is better, because it is more powerful; other can chose B, because it is lighter; a third can argue that the two are equal, because the ratio power-per-pound (an integral indicator) of the two are equal. According to the third, to preserve the level of quality of an engine, the shift of the value of one of the indicators (independent) requires an "expected" shift of the value of the other (dependent). Let us consider another option C = {85,80} and let "weight" is the independent indicator. The expected value for the "power" of C is 80, but the actual value is 85, the sign of power is "+", and we can argue that the quality of C is higher than the quality of the two A and B.

The idea is to choose one of the indicators (**main**), which measures the shift among the options, and to use it to define the "expectation" for the shift of the value of an other indicator, which remains the idea of regression. The regression between every ordinary indicator and the main indicator is established and the expected value of this indicator for every option is calculated. The residuals define the quality of the options. A residual decreases or increases the quality of the option according to the sign of the indicator.

The four linear regression models between indicators I_i and I_j are used:

1. $x_i = a_{ij} + b_{ij} x_j$
2. $x_i = a_{ij} + b_{ij} \ln x_j$
3. $x_i = \exp(a_{ij} + b_{ij} x_j)$
4. $x_i = \exp(a_{ij} + = b_{ij} \ln x_j)$

Evaluation the parameters *{a,b}* could be done with any standard technique (e.g. LSM). To avoid computational problems in calculating logarithms, all values in the matrix X have to be either different from zero or the first of the models can be applied only.

The procedure for constructing the integral indicator starts with establishing the relation between any pair of indicators I_i and I_j as the regression (the model from the above list) with the highest correlation coefficient C_{ij}^m. The following four matrices are calculated:

regressions' coefficients $A = \{a_{ij}\}$ and $B = \{b_{ij}\}$
correlation coefficients $C = \{c_{ij}\}$
number of the selected model $M = \{m_{ij}\}$

The second step is to select the main indicator (I_g) It can be chosen either by the user or as the one with the largest sum of correlation coefficients, when serves as the independent variable ($I_g = \max_i \sum_j c_{ij}$). In all of the models, one of the indicators plays the role of the argument (independent) and the other - the role of the function (dependent). Further, the main indicator serves only as the argument in all of the regressions.

The values q_{ij} (quality of the option O_j, according to the indicator I_i) are achieved, after normalization of x_{ij}, with the formula

$$q_{ij} = \frac{x_{ij}}{f(a_{ig}, b_{ig}, m_{ig}, x_{ig})^{s_i}}, \text{ where } i = 1, 2, ..., n; j = 1, 2, ..., g-1, g+1, ..., k; \text{ and}$$

$$f(a,b,m,x) = \begin{array}{ll} a+bx & m=1 \\ a+b\ln x & m=2 \\ \exp(a+bx) & m=3 \\ \exp(a+b\ln x) & m=4 \end{array}$$

q_{ij} is as higher as the respective value x_{ij} is greater than the expected value $f(a_{ig},b_{ig},m_{ig},x_{ig})$ in the case of positive sign and vice versa.

The integral quality of the option O_j is evaluated by the expression

$$Q_i = \frac{1}{\sum_{j=1,j\neq g}^{n} w_j c_{ij}} \sum_{j=1,j\neq g}^{n} w_j c_{ij} q_{ij}, \; i=1, 2, ..., n$$

The case:
Regressions between **I 4** and **I 6** are:
1. I_4 = 0.02I_6 + 356.17, c_{46}=0.236523   selected model
2. $\ln(I_4)$ = 0.00 I_6 + 5.88 c_{46}=0.195672
3. I_4 = 23.42 $\ln(I_6)$ + 213.64 c_{46}=0.225224
4. $\ln(I_4)$ = 0.06 $\ln(I_6)$ + 5.52 c_{46}=0.184812

Sum of the correlation coefficients for any of the indicators are:

Indicator	Sum of Correlation Coefficients
I 1	7.1188
I 2	7.3221
I 3	5.5104
I 4	6.5116
I 5	6.2304
I 6	7.7846 - selected indicator
I 7	6.8628
I 8	5.4765
I 9	6.6282
I 10	6.7135
I 11	4.7334
I 12	7.7635
I 13	7.3943

x_{14}=350, x_{16} = 1400, m_{46} = 1, s_4 = +1
f(0.02, 356.17,1,1400) = 384.17,
q_{14}=1.097
Q_1(O 1) = 1.0332 (close to, but higher then the average)

The main indicator represents the quantitative shift among the set of options and correlation coefficients show how a given indicator fits to this tendency. The above formula includes c_{ij} as a measure of these relations.

The reason to use simple regressions (regression with a single argument) is the amount of data. Every option can be considered as an observation. A quite large number of options is required to apply more complex models.

3. Missing Values

Missing values are an expected problem. We can either exclude the indicator or option where a missing value occur, or we have to apply a forecast procedure, with minimal (or limited) impact on the final results, to fill the gaps. Our approach is to choose the "expected" value as a forecast for the missing one. The problem is that the empty cells have to be filled up before establishing of the regressions and selecting the main indicator.

The simpler case -- known main indicator and known regression between the main indicator and the indicator containing the missing value -- is not considered here. In the other case we need to increase the probability that the forecast will use the relation with the candidate-main indicators.

The assumption that indicators are not independent, allows us to use the redundant information within the matrix X. Let us assume that the information matrix X is given and all cells, except the cell x_{ij}, are filled. The problem is to find a forecast $\overline{x_{ij}}$, which value is expected to be "close" to the "true" value of x_{ij}. The problems with the intuitive definitions of the terms "close" and "true value" will not be discussed here.

The method derives from the way the integral indicator was designed. It can be explained by the following algorithm:

1. Exposure of the cells of X, which contain missing values
$$x_m = \{x_{i_1 j_1}, x_{i_2 j_2}, \ldots, x_{i_m j_m}\}$$

2. Set up a matrix F by excluding the rows and columns of X, which cross on a cell containing a missing value.

3. Start a loop to calculate a forecast for the pth missing value, which will be placed in the cell $\{i_p, j_p\}$ (Let us assign $l = i_p$ and $q = j_p$.)

4. Calculation of the linear regression coefficients **a** and **b**, and correlation coefficient **c**. It is done in the same way as in Section 2. It represents a relation between $x_i = \{x_{i1}, \ldots x_{iq-1}, x_{iq+1}, \ldots, x_{in}\}$ and $x_l = \{x_{l1}, \ldots x_{lq-1}, x_{lq+1}, \ldots, x_{ln}\}$, where i = 1, 2,...l-1, l+1,...k.

5. Calculation of the values $\overline{x_{lq}^i} = f(a_i, b_i, m_i, x_{li})$, where i = 1,2, ..., l-1,l+1,...,k.

6. The five $\overline{x_{lq}^i}$, derived from regressions with the highest c_i, are used for calculation of the forecast $\overline{x_{lq}} = \dfrac{1}{\sum c_i} \sum_{i=1}^{5} c_i \overline{x_{lq}^i}$

7. Continue for the next missing value.

4. Step--by--step selection

The natural way, a human is choosing the "best" option among a given set of alternatives, is short-listing. On every step, a subset of the unsuitable options are excluded from further consideration. Such exclusion is made on the basis of a single criterion.

An interactive step-by-step procedure to simulate the process of extracting the best option is developed. On every step, the set of options are split into two subsets "good" and "bad", according to the values of one indicator. Splitting is done either by an user-defined threshold or by splitting the set into two clusters – cutting between the two most distant values.

Let us call that an option O_i dominates on option O_j, when each of the values x_{ip} "is better than" x_{jp} (considering signs). The option O_j, cannot become the "best option" according to the given selection procedure. Such options are excluded in the beginning. This procedure can be applied to define the order (ranks) of the options.

The case:

Step 1	I 8 (s = -1)	O 2	30.00	
		O 5	30.00	
		O 6	35.00	
		O 4	40.00	
		O 3	45.00	
		-----------	------	
		O 1	70.00	excluded
Step 2	I 1 (s = +1)	O 3	43.34	
	threshold = 37	O 5	39.11	
		O 2	38.44	
		-----------	------	
		O 4	35.24	excluded
		O 6	33.23	excluded
Step 3	I 11 (s = -1	O 3	22.50	
		O 2	23.00	
		-----------	------	
		O 5	30.00	excluded
Step 4	I 7 (s = -1)	O 3	140	**the best**
		-----------	------	
		O 2	150	excluded

Conclusion

1. The following properties of the proposed integral indicator are highly important, when apply to evaluate quality of options in a decision making process:

- Simplicity:
  - From user's point of view:
    - natural way of presenting data, preferences, and relations;
    - easy to interpret;
    - capable for simulations.
  - From developer's point of view: simple and clear algorithm with limited computational problems.
- Data requirements: The method does not require difficult for proving preconditions, e.g. independence between indicators. Because simple models are used for regressions,

only small number of observations are needed.

- Applicability:

The area of the most useful application of the proposed method is a set of options, providing alternative solutions to a given problem. The presented approach allows existence of small quantitative shifts among the options.

2.   The sequence of indicators applied in the selection procedure is critical. Indicators can be ordered according to heir weights, but in general the problem of defining a feasible sequence is open.

References

Christozov, D. G., Denchev S. G. and Ugarchinsky B.V. (1988) *Analysis and Evaluation of mechanical technologies*, ICSTI, Moscow (in Russian)

Dimitrov, B. N. (1997) *Some Approaches to the Unified Estimation of the Total Quality of a Product*, 18th IFIP TC7 Conference on System Modeling and Optimization, Detroit

VLADIMIR DRAGALIN
Control charts for monitoring the mean of a multivariate normal distribution

1. Statement of the Problem.
It is assumed that the stochastic behavior of the process of observations $\{X_i\}$ is described in terms of the vector of parameters θ. Under the desirable conditions, this vector belongs to the set Θ_0. If a control procedure gives a signal in this environment, it is classified as a false alarm. At some point in time ν, called the change point, θ can abruptly change to some value that belongs to a rejectable set Θ_1. The control chart is then supposed to detect this change as soon as possible. It is frequently assumed that the acceptable behavior of $\{X_i\}$ is associated with a specified in-control density $f_0(x)$ and that, at some unknown moment ν of time it may switch to an out-of-control density $f_\theta(x)$, $\theta \in \Theta_1$.

One of the most important cases in many practical situations involves monitoring the mean vector of a p-variate normal distribution $\mathcal{N}_p(\mu, \Sigma)$. This case, being at the same time the most simple for theoretical investigations, has received a great attention in the literature. A most part of the analysis, design and comparison of different multivariate control charts is concerned with the case of multivariate normal distribution (see e.g. Wierda (1994) for a good survey). But even for this simple case, the analysis of control charts is mainly based on Monte Carlo simulations. In order to elaborate on more delicate points related to comparison in performance between several detection procedures we propose an asymptotic approach. We restrict ourselves to multivariate control charts for the change in the mean vector of the multivariate normal distribution with fixed, known covariance matrix Σ. The in-control mean vector μ_0 is supposed to be known.

The X_n are observed sequentially with the goal of detecting the change point ν by a *stopping rule* N which minimizes the conditional average delay time (CADT)

$$\overline{E}_\mu(N) = \sup_{\nu \geq 1} E_{\mu,\nu}(N - \nu + 1 \mid N \geq \nu) \qquad (1)$$

for all $\mu \in \Theta_1$ subject to the constraint

$$E_\infty(N) \geq T \qquad (2)$$

for some specified large level T. Here $E_{\mu,\nu}$ denotes expectation under the hypothesis that the true change point is ν and the out-of-control mean vector is μ while E_∞ denotes the expectation under the hypothesis of no change whatever.

Depending on the specified out-of-control situation Θ_1, one can distinguish between different changes in the mean vector, which may be classified as:

case 1: $\boldsymbol{\mu}_1$ is known, i.e. $\boldsymbol{\Theta}_1 = \{\boldsymbol{\mu}_1\}$
case 2: known magnitude but unknown direction of change, i.e.
$\boldsymbol{\Theta}_1 = \{\boldsymbol{\mu} : (\boldsymbol{\mu} - \boldsymbol{\mu}_0)' \boldsymbol{\Sigma}^{-1} (\boldsymbol{\mu} - \boldsymbol{\mu}_0) = \lambda^2\}$
case 3: known direction but unknown magnitude of change, i.e.
$\boldsymbol{\Theta}_1 = \{\boldsymbol{\mu} : \boldsymbol{\mu} = \boldsymbol{\mu}_0 + r\boldsymbol{R}\}$, where $\boldsymbol{R}$ is the unit vector of the change direction
case 4: known lower bound for the magnitude but unknown direction of change, i.e.
$\boldsymbol{\Theta}_1 = \{\boldsymbol{\mu} : (\boldsymbol{\mu} - \boldsymbol{\mu}_0)' \boldsymbol{\Sigma}^{-1} (\boldsymbol{\mu} - \boldsymbol{\mu}_0) \geq \lambda^2\}$
case 5: completely unknown $\boldsymbol{\mu}_1$, i.e. $\boldsymbol{\Theta}_1 = \{\boldsymbol{\mu} : \boldsymbol{\mu} \neq \boldsymbol{\mu}_0\}$

The control charts can and, may be, should be defined and used differently according to the various levels of the available *a priori* information about $\boldsymbol{\Theta}_1$.

2. Definition of the Control Charts.

The likelihood ratio principle for testing two simple hypotheses $H^{(n)}$ (of no change point in the sequence $\boldsymbol{X}_1, \ldots, \boldsymbol{X}_n$) against $H_k^{(n)}$ (that at k ($1 \leq k \leq n$) there is a change point) states that H is to be rejected if and only if

$$\ell_{n,k}(\boldsymbol{X}_1^n; \boldsymbol{\mu}_0, \boldsymbol{\mu}_1, \boldsymbol{\Sigma}) = \log\left[\frac{L(\boldsymbol{X}_1^n \mid H_k)}{L(\boldsymbol{X}_1^n \mid H)}\right] \geq h,$$

where $L(\boldsymbol{X}_1^n \mid \cdot) = L(\boldsymbol{X}_1, \ldots, \boldsymbol{X}_k, \boldsymbol{X}_{k+1}, \ldots, \boldsymbol{X}_n \mid \cdot)$ is the likelihood function when the respective hypothesis holds and h is some fixed constant ($h \geq 0$).

Motivated by minimax arguments, we propose the following stopping rule

$$\tau = \inf\{n : \max_{1 \leq k \leq n} \ell_{n,k}(\boldsymbol{X}_1^n; \boldsymbol{\mu}_0, \boldsymbol{\mu}_1, \boldsymbol{\Sigma}) \geq h\}.$$

In the special case 1 that both in-control and out-of-control mean vectors are completely specified, we obtain a well-known (univariate) CUSUM procedure

$$\tau_1 = \tau_1(\boldsymbol{\mu}_1) = \inf\{n : \max_{1 \leq k \leq n} (n - k + 1)[(\boldsymbol{\mu}_1 - \boldsymbol{\mu}_0)' \boldsymbol{\Sigma}^{-1} (\overline{\boldsymbol{X}}_{n,k}^* - \boldsymbol{\mu}_0) - \frac{1}{2}\lambda^2(\boldsymbol{\mu}_1)] \geq h\},$$

where $\overline{\boldsymbol{X}}_{n,k}^* = \frac{1}{n-k+1}\sum_{i=k}^{n} \boldsymbol{X}_i$ and $\lambda^2(\boldsymbol{\mu}_1) = (\boldsymbol{\mu}_1 - \boldsymbol{\mu}_0)' \boldsymbol{\Sigma}^{-1} (\boldsymbol{\mu}_1 - \boldsymbol{\mu}_0)$ (see also Healy (1987)). Note that $(\boldsymbol{\mu}_1 - \boldsymbol{\mu}_0)' \boldsymbol{\Sigma}^{-1} (\boldsymbol{X}_i - \boldsymbol{\mu}_0)/\lambda(\boldsymbol{\mu}_1)$ has a standard univariate normal distribution when $\boldsymbol{X}_i$ has mean equal to $\boldsymbol{\mu}_0$ and a univariate normal distribution with mean $\lambda(\boldsymbol{\mu}_1)$ and variance 1 when $\boldsymbol{X}_i$ has mean equal to $\boldsymbol{\mu}_1$. Therefore, all of the theory available for calculating ARL's and the control limit h for univariate normal CUSUM's can also be used for the multivariate normal CUSUM's in the case 1.

But in other cases enumerated above, the parameter $\boldsymbol{\mu}_1$ is assumed to be unknown. We have to consider methods by which this unknown parameter can be eliminated from the log likelihood ratio $\ell_{n,k}(\boldsymbol{X}_1^n; \boldsymbol{\mu}_0, \boldsymbol{\mu}_1, \boldsymbol{\Sigma})$. One of such methods is the generalized likelihood ratio (GLR) approach, in which the log likelihood ratio $\ell_{n,k}(\boldsymbol{X}_1^n; \boldsymbol{\mu}_0, \boldsymbol{\mu}_1, \boldsymbol{\Sigma})$ is replaced by $\sup_{\boldsymbol{\mu} \in \boldsymbol{\Theta}_1} \ell_{n,k}(\boldsymbol{X}_1^n; \boldsymbol{\mu}_0, \boldsymbol{\mu}, \boldsymbol{\Sigma})$.

Using Lagrange's method, we get the generalized log likelihood ratio statistic

$$\hat{\ell}_{n,k} = \sup_{\boldsymbol{\mu} \in \boldsymbol{\Theta}_1} \ell_{n,k}(\boldsymbol{X}_1^n; \boldsymbol{\mu}_0, \boldsymbol{\mu}, \boldsymbol{\Sigma})$$

in each case. Let $\|\boldsymbol{\mu}\| = (\boldsymbol{\mu}' \boldsymbol{\Sigma}^{-1} \boldsymbol{\mu})^{\frac{1}{2}}$.

Case 2
$$\hat{\ell}_{n,k} = (n-k+1)\left(\lambda\|\overline{\boldsymbol{X}}^*_{n,k} - \boldsymbol{\mu}_0\| - \frac{\lambda^2}{2}\right);$$

Case 3
$$\hat{\ell}_{n,k} = \frac{n-k+1}{2}\boldsymbol{R}'\boldsymbol{\Sigma}^{-1}\boldsymbol{R}\ r^2_{n,k},$$

where
$$r_{n,k} = \frac{\boldsymbol{R}'\boldsymbol{\Sigma}^{-1}(\overline{\boldsymbol{X}}^*_{n,k} - \boldsymbol{\mu}_0)}{\boldsymbol{R}'\boldsymbol{\Sigma}^{-1}\boldsymbol{R}};$$

Case 4
$$\hat{\ell}_{n,k} = \frac{n-k+1}{2}\left[\|\overline{\boldsymbol{X}}^*_{n,k} - \boldsymbol{\mu}_0\|^2 - \left(\lambda - \|\overline{\boldsymbol{X}}^*_{n,k} - \boldsymbol{\mu}_0\|\right)^2 I_{\{\|\overline{\boldsymbol{X}}^*_{n,k}-\boldsymbol{\mu}_0\|\leq\lambda\}}\right],$$

where $I_{\{\cdot\}}$ is the indicator function;

Case 5
$$\hat{\ell}_{n,k} = \frac{n-k+1}{2}\|\overline{\boldsymbol{X}}^*_{n,k} - \boldsymbol{\mu}_0\|^2.$$

The GLR control chart for each case is based on the respective statistic $\hat{\ell}^{(i)}_{n,k}$ and its stopping time will be denoted by τ_i, i.e.
$$\tau_i = \inf\{n : \max_{1\leq k\leq n} \hat{\ell}^{(i)}_{n,k} \geq h\}$$

for each case i, $i = 2,\ldots,5$.

Note that in the univariate case, i.e. $p = 1$, the GLR control chart τ_2 becomes the two-sided CUSUM chart, while τ_3, τ_4 and τ_5 coincide with the generalized CUSUM procedures suggested and studied by many authors (e.g. Barnard (1959), Lorden (1971), Dragalin (1993), Siegmund & Venkatraman (1995)).

The GLR control chart τ_2 was considered also by Nikiforov (1980) and Pignatiello & Runger (1990) (although in a slightly different form and motivated by other arguments). For the case 2, another control chart, which gives relatively similar ARL's performance, was proposed by Crosier (1988). For other cases and still other procedures see Basseville & Nikiforov (1993).

3. Asymptotic Comparative Performance.

The principal results of this paper are the following two theorems.

Theorem 1. *As* $h \to \infty$
$$E_\infty(\tau_2) \sim K_2 h^{-\frac{p-1}{2}} \exp(h)$$
$$E_\infty(\tau_3) \sim K_3 h^{-\frac{1}{2}} \exp(h)$$
$$E_\infty(\tau_4) \sim K_4 h^{-\frac{p}{2}} \exp(h)$$
$$E_\infty(\tau_5) \sim K_5 h^{-\frac{p}{2}} \exp(h)$$

where K_i, $i = 2,\ldots,5$ *are some (rather complicated) positive constants depending on p, but not depending on h.*

These asymptotic results give us approximations to the control level h of the respective control chart.

Theorem 2. Let $h_i(T)$ be the control level of the respective control chart such that
$$E_\infty(\tau_i) \sim T \quad as \ T \to \infty.$$
Then as $T \to \infty$

$$\overline{E}_\mu(\tau_1) = \frac{1}{(\boldsymbol{\mu}_1 - \boldsymbol{\mu}_0)' \boldsymbol{\Sigma}^{-1}(\boldsymbol{\mu} - \boldsymbol{\mu}_0) - \frac{1}{2}\|\boldsymbol{\mu}_1 - \boldsymbol{\mu}_0\|^2} \left[\log T\right] + O(1) \tag{3}$$

for all $\boldsymbol{\mu}$ such that the denominator is positive ($\boldsymbol{\mu}_1$ is the mean used in the definition of τ_1);

$$\overline{E}_\mu(\tau_2) = \frac{1}{\lambda(\|\boldsymbol{\mu} - \boldsymbol{\mu}_0\| - \frac{\lambda}{2})} \left[\log T + \frac{p-1}{2} \log\log T\right] + O(1) \tag{4}$$

for all $\boldsymbol{\mu}$ such that the denominator is positive (λ is the magnitude of the change in the mean, used in the definition of τ_2);

$$\overline{E}_\mu(\tau_3) = \frac{2\|\boldsymbol{R}\|^2}{[\boldsymbol{R}' \boldsymbol{\Sigma}^{-1}(\boldsymbol{\mu} - \boldsymbol{\mu}_0)]^2} \left[\log T + \frac{1}{2} \log\log T\right] + O(1) \tag{5}$$

for all $\boldsymbol{\mu}$ such that the denominator is positive ($\boldsymbol{R}$ is the unit vector used in the definition of τ_3);

$$\overline{E}_\mu(\tau_4) = \frac{1}{\rho(\boldsymbol{\mu})} \left[\log T + \frac{p}{2} \log\log T\right] + O(1) \tag{6}$$

for all $\boldsymbol{\mu}$ such that

$$\rho(\boldsymbol{\mu}) = \begin{cases} \|\boldsymbol{\mu} - \boldsymbol{\mu}_0\|^2/2 & if \ \|\boldsymbol{\mu} - \boldsymbol{\mu}_0\| \geq \lambda, \\ \lambda(\|\boldsymbol{\mu} - \boldsymbol{\mu}_0\| - \lambda/2) & otherwise \end{cases}$$

is positive (λ is the lower bound for the magnitude of the change in the mean, used in the definition of τ_4);

$$\overline{E}_\mu(\tau_5) = \frac{2}{\|\boldsymbol{\mu} - \boldsymbol{\mu}_0\|^2} \left[\log T + \frac{p}{2} \log\log T\right] + O(1) \tag{7}$$

for all $\boldsymbol{\mu} \neq \boldsymbol{\mu}_0$.

The complete proof of these Theorems (based on results of nonlinear renewal theory, Woodroofe (1982), and repeated likelihood ratio tests, Lalley (1983)) as well as the accuracy of these approximations will be discussed in a future paper.

Based on the optimality result for the univariate CUSUM procedure (Moustakides (1986)) and the result announced at the beginning of section 2 that the specification of both in-control and out-of-control means of a p-variate normal distribution yields a univariate CUSUM procedure, we conclude that the control chart τ_1 can serve as a benchmark in a comparative analysis of different multivariate control charts.

Note that for $\boldsymbol{\mu} = \boldsymbol{\mu}_1$ we have by (3)

$$\overline{E}_{\boldsymbol{\mu}_1}(\tau_1) = \frac{2}{\|\boldsymbol{\mu}_1 - \boldsymbol{\mu}_0\|^2}\left[\log T\right] + O(1) \quad \text{as } T \to \infty \tag{8}$$

and therefore the right hand side of (8) characterizes the asymptotic optimal delay in detecting the change from $\boldsymbol{\mu}_0$ to $\boldsymbol{\mu}_1$.

Observe that the GLR control chart τ_2 is first-order asymptotically optimal for detecting the change in the mean from $\boldsymbol{\mu}_0$ to all $\boldsymbol{\mu}$ of the anticipated magnitude $\lambda = \|\boldsymbol{\mu} - \boldsymbol{\mu}_0\|$, i.e.

$$\overline{E}_{\boldsymbol{\mu}}(\tau_2) = \frac{2}{\|\boldsymbol{\mu} - \boldsymbol{\mu}_0\|^2}\left[\log T + \frac{p-1}{2}\log\log T\right] + O(1) \quad \text{as } T \to \infty.$$

The second term is the price for the unknown direction of the change. This procedure is more powerful in detecting the change of magnitude λ than the control charts τ_4 and τ_5. On the other hand, this procedure loses even the first-order asymptotic optimality in detecting the shifts of other than λ magnitude.

The GLR control chart τ_3 is first-order asymptotically optimal for detecting the change in the mean from $\boldsymbol{\mu}_0$ to all $\boldsymbol{\mu}$ in the direction of the unit vector $\boldsymbol{R}$, $\boldsymbol{\mu} = \boldsymbol{\mu}_0 + r\boldsymbol{R}$ for all real r, i.e.

$$\overline{E}_{\boldsymbol{\mu}}(\tau_3) = \frac{2}{\|\boldsymbol{\mu} - \boldsymbol{\mu}_0\|^2}\left[\log T + \frac{1}{2}\log\log T\right] + O(1) \quad \text{as } T \to \infty.$$

The second term is the price for the unknown magnitude r of the change (in the known direction $\boldsymbol{R}$). This expansion shows that the asymptotic expected delay of this procedure does not depend on p and this procedure is more powerful in detecting the change in the direction $\boldsymbol{R}$ than the control charts τ_2, τ_4 and τ_5. However, it loses even the first-order asymptotic optimality in detecting the shifts in other than $\boldsymbol{R}$ direction.

The GLR control chart τ_4 is first-order asymptotically optimal for detecting the change in the mean from $\boldsymbol{\mu}_0$ to all $\boldsymbol{\mu}$ with a magnitude $\|\boldsymbol{\mu} - \boldsymbol{\mu}_0\| \geq \lambda$ but it loses this optimality property for changes with a smaller than λ magnitude. The GLR control chart τ_5 is first-order asymptotically optimal for detecting any change in the mean (from the in-control $\boldsymbol{\mu}_0$). Comparing (4), (6) and (7) suggests that τ_4 is a compromise between τ_2 and τ_5.

Observe that (asymptotically) the price for the unknown magnitude of the change is an increase in the expected delay in detection of order $\frac{1}{\|\boldsymbol{\mu} - \boldsymbol{\mu}_0\|^2}\log\log T$, while for the unknown direction of the change, the increase is of order $\frac{p-1}{\|\boldsymbol{\mu} - \boldsymbol{\mu}_0\|^2}\log\log T$.

4. Comparative Performance Based on Simulations.

Restricted simulation studies (see, e.g. Crosier (1988), Pignatiello and Runger (1990), Lowry, Woodall, Champ and Rigdon (1992), Wierda (1994)) of comparison in performance of different control charts for detection of a shift in the mean vector of a p-variate normal distribution have shown that the multiple univariate CUSUM (Woodall and Ncube (1985)), the mean estimating CUSUM (Pignatiello and Runger (1990)) and shrinking CUSUM (Crosier (1988)) procedures give relatively similar ARL performance under a "slippage" situation, i.e. when only one of the p means shifts to an out-of-control while all of the remaining means remain in-control. However, the first performs less for shifts of equal magnitude in each component. This sensitivity to the direction of the shift was shown to be more pronounced as the dimensionality p of the process increases. On the other hand, these multivariate control charts have a superior ARL performance over the multivariate Shewhart chart and the CUSUM procedure based on Hotelling's T^2 statistic.

In this section we use a Monte Carlo simulation to compare the following control charts on the basis of their ARL performance:

1. GLR control charts $\tau_2^{(1)}$ and $\tau_2^{(2)}$, designed to detect any shift of magnitude 1 and 1.5 respectively
2. GLR control chart τ_5
3. GLR control charts $\tau_3^{(1)}$ and $\tau_3^{(2)}$, designed to detect any shift in the direction of the first component and along the bisectrix respectively
4. multivariate Shewhart χ^2 control chart

In Table 1, we compare the out-of-control ARLs of these procedures with the optimal CUSUM *envelope* for detecting the shifts $\boldsymbol{\mu}$ in the mean of a p-variate normal distribution with a large range of magnitude $\lambda(\boldsymbol{\mu})$ for $p = 2, 3, 5$ and 10. The CUSUM envelope is defined for a fixed level of false alarm T as a function $V(\mu; T)$ of the shift in the mean and is equal to the out-of-control ARL of an optimal CUSUM procedure τ_1 (oriented to the shift from $\boldsymbol{\mu}_0$ to $\boldsymbol{\mu}$) with the control level h chosen to achieve the in-control ARL $E_\infty(\tau_1) = T$. The univariate CUSUM envelope obtained by Dragalin (1994) is used for this purpose.

The ARL comparison is simplified because the ARL's of the control charts τ_2, τ_5 and χ^2 depend on the values of the shift $\boldsymbol{\mu}$ and the covariance matrix $\boldsymbol{\Sigma}$ only through the magnitude $\lambda(\boldsymbol{\mu})$ of the shift. The same is true also for the ARL's of the control chart τ_3 if the shift is in the anticipated direction. Hence, without loss of generality, the in-control mean $\boldsymbol{\mu}_0$ is assumed to be equal to 0 and the covariance $\boldsymbol{\Sigma}$ to be a unit matrix.

Note that $\boldsymbol{R}' \boldsymbol{\Sigma}^{-1}(\boldsymbol{X}_i - \boldsymbol{\mu}_0)/\|\boldsymbol{R}\|$ has a standard univariate normal distribution when $\boldsymbol{X}_i$ has mean equal to $\boldsymbol{\mu}_0$ and a univariate normal distribution with mean $\lambda(\boldsymbol{\mu}) = \boldsymbol{R}'\boldsymbol{\Sigma}^{-1}(\boldsymbol{\mu} - \boldsymbol{\mu}_0)/\|\boldsymbol{R}\|$ and variance 1 when $\boldsymbol{X}_i$ has mean equal to $\boldsymbol{\mu}$. If $\boldsymbol{\mu} = \boldsymbol{\mu}_0 + r \boldsymbol{R}$ then $\lambda(\boldsymbol{\mu}) = r \|\boldsymbol{R}\|$. It is clear that both the in-control ARL and the out-of-control ARL for shifts in the anticipated direction do not depend on p. However the out-of-control ARL for shifts in other directions follows more complicated patterns. For τ_3, we confine ourselves considering only two cases of the shift (of the same magnitude λ) in the mean: one in the anticipated direction and another along the bisectrix, i.e. each component is shifted by $\delta = \lambda/\sqrt{p}$. The respective ARL's are in the columns with the headings $\tau_3^{(1)}$ and $\tau_3^{(2)}$. It is clear that both the in-control ARL and the out-of-control ARL for shifts in the anticipated direction do not depend on p. The ARLs of $\tau_3^{(1)}$ were independently simulated for the out-of-control conditions for all cases to verify the invariance of τ_3 by p.

λ	$V(\lambda,T)$	$\tau_2^{(1)}$	$\tau_2^{(2)}$	τ_5	$\tau_3^{(1)}$	$\tau_3^{(2)}$	χ^2
$p=2$	h	5.477	5.788	6.655	5.065	5.065	10.6
0.0	200.0	200.8	199.5	200.5	200.5	200.5	200.0
0.5	19.34	33.16 (.27)	44.02 (.40)	33.56 (.22)	28.12 (.19)	47.26 (.35)	115.7
1.0	7.40	10.26 (.05)	10.98 (.07)	11.06 (.06)	9.31 (.05)	16.30 (.10)	42.0
1.5	4.04	5.86 (.02)	5.42 (.03)	5.65 (.03)	4.85 (.03)	8.40 (.05)	15.8
2.0	2.61	4.16 (.01)	3.61 (.01)	3.59 (.02)	3.11 (.02)	5.31 (.03)	6.9
2.5	1.87	3.24 (.01)	2.75 (.01)	2.54 (.01)	2.23 (.01)	3.75 (.02)	3.5
3.0	1.47	2.71 (.01)	2.24 (.01)	1.95 (.01)	1.73 (.01)	2.85 (.01)	2.2
$p=3$	h	6.502	6.903	7.938	5.065	5.065	12.85
0.0	200.0	199.67	199.51	199.58	200.5	200.5	201.0
0.5	19.34	35.90 (.28)	48.71 (.44)	37.83 (.25)	27.90 (.19)	62.63 (.48)	130.0
1.0	7.40	11.33 (.06)	12.15 (.08)	12.24 (.07)	9.31 (.06)	22.72 (.15)	52.6
1.5	4.04	6.58 (.02)	6.03 (.03)	6.33 (.03)	4.83 (.03)	11.74 (.07)	20.5
2.0	2.61	4.67 (.01)	4.00 (.01)	3.95 (.02)	3.13 (.02)	7.36 (.04)	8.8
2.5	1.87	3.66 (.01)	3.05 (.01)	2.77 (.01)	2.22 (.01)	5.17 (.03)	4.4
3.0	1.47	3.06 (.01)	2.49 (.01)	2.10 (.01)	1.72 (.01)	3.88 (.02)	2.6
$p=5$	h	8.289	8.825	10.123	5.065	5.065	16.75
0.0	200.0	199.25	200.12	200.02	200.5	200.5	200.0
0.5	19.34	40.10 (.30)	52.18 (.46)	42.94 (.28)	28.25 (.20)	84.69 (.70)	145.0
1.0	7.40	13.13 (.06)	13.53 (.08)	13.99 (.08)	9.29 (.06)	33.67 (.24)	68.1
1.5	4.04	7.73 (.03)	6.85 (.03)	7.21 (.03)	4.84 (.03)	17.85 (.11)	28.5
2.0	2.61	5.55 (.02)	4.62 (.02)	4.48 (.02)	3.09 (.02)	11.19 (.07)	12.4
2.5	1.87	4.37 (.01)	3.56 (.01)	3.14 (.01)	2.24 (.01)	7.75 (.04)	5.99
3.0	1.47	3.63 (.01)	2.92 (.01)	2.37 (.01)	1.72 (.01)	5.81 (.03)	3.31
$p=10$	h	12.064	12.841	14.716	5.065	5.065	25.2
0.0	200.0	199.99	199.15	199.2	200.5	200.5	200.8
0.5	19.34	46.03 (.31)	58.65 (.51)	51.54 (.33)	28.56 (.19)	120.01 (1.05)	162.0
1.0	7.40	16.50 (.07)	16.00 (.09)	17.15 (.09)	9.36 (.05)	55.26 (.42)	92.8
1.5	4.04	10.01 (.03)	8.50 (.03)	8.68 (.04)	4.87 (.03)	30.71 (.21)	44.7
2.0	2.61	7.34 (.02)	5.89 (.02)	5.41 (.02)	3.11 (.02)	19.48 (.13)	20.6
2.5	1.87	5.80 (.01)	4.57 (.01)	3.74 (.02)	2.23 (.01)	13.53 (.08)	9.9
3.0	1.47	4.83 (.01)	3.77 (.01)	2.82 (.01)	1.71 (.01)	10.10 (.06)	5.2

TABLE 1. *Estimated Average Run Lengths of the GLR Control Charts*

The values of the control level h were chosen so that the in-control ARLs of the control charts are (approximately) $T = 200$. The Monte Carlo estimates of the out-of-control ARLs of the GLR control charts based on 10000 replications are shown in a single line with their standard errors in parentheses.

Simulation results confirm our conclusions based on the asymptotic performance comparisons. The out-of-control ARLs of the chart τ_3 are almost the same for different values of p, while that of the other charts are increasing in p. The control chart τ_3 is the best in detecting shifts in the anticipated direction but may have a poor performance in other directions. The control chart τ_2 has a superior performance for the shifts of the magnitude close to one for which it was designed. The control chart τ_5 is the most robust to both the magnitude and direction of the change. All GLR control charts offer a considerably superior performance over the multivariate Shewhart χ^2 chart.

Acknowledgments. This research was made during a Research Fellowship at the *Institute for Applied Mathematics and Statistics, University of Würzburg* supported by the *Alexander von Humboldt Foundation*. I thank Professors William Woodall and Joe Sullivan for helpful discussions concerning this work.

References

1. Barnard, G.A. (1959). "Control Charts and Stochastic Processes". *Journal of Royal Statistical Society, Ser. B*, 21, pp. 239–257.
2. Basseville, M. and Nikiforov, I.V. (1993). *Detection of Abrupt Changes: Theory and Applications*. PTR Prentice-Hall, New Jersey.
3. Crosier, R.B. (1988). "Multivariate Generalizations of Cumulative Sum Quality-Control Schemes". *Technometrics*, 30, pp. 291–303.
4. Dragalin, V. (1993). "Optimality of Generalized CUSUM Procedure in Quickest Detection Problem". In *Proceedings of Steklov Math. Institute: "Statistics and Control of Stochastic Processes"*, 1993, v.202, pp.107–119.
5. Dragalin, V. (1994). "Optimal CUSUM Envelope for Monitoring the Mean of Normal Distribution". *Economic Quality Control*, 9, pp. 185-202.
6. Healy, J.D. (1987). "A Note on Multivariate CUSUM Procedures". *Technometrics* 29, pp. 409–412.
7. Lalley, S.P. (1983). "Repeated Likelihood Ratio Tests for Curved Exponential Families". *Z. Wahrsch. verw. Gebiete*, 62, 293–321.
8. Lorden, G. (1971). "Procedures for Reacting to a Change in Distribution". *Annals of Mathematical Statistics*, 42, pp. 1897–1908.
9. Lowry, C.A., Woodall, W.H., Champ, C.W. and Rigdon, S.E. (1992). "A Multivariate Exponentially Weighted Moving Average Control Chart". *Technometrics*, 34, pp. 46–53.
10. Moustakides, G.V. (1986). "Optimal Stopping Times for Detecting Changes in Distributions". *Annals of Statistics* 14, pp. 1379–1387.
11. Nikiforov, I.V. (1980). "The Modification and Investigation of the Cumulative Sum Procedure". *Avtomatika i Telemekhanika*, 8, pp. 74–80 (in Russian).
12. Pignatiello, Jr.J.J. and Runger G.C. (1990). "Comparisons of Multivariate CUSUM charts". *Journal of Quality Technology* 22, pp. 173–186.

13. Siegmund, D. and Venkatraman, E.S. (1995). "Using the Generalized Likelihood Ratio Statistic for Sequential Detection of a Change-Point". *Annals of Statistics*, 23, pp. 255–271.
14. Wierda, S.J. (1994). "Multivariate Statistical Process Control – Recent Results and Directions for Future Research". *Statistica Neerlandica* 48, pp. 147–168.
15. Woodall, W.H. and Ncube, M.M. (1985). "Multivariate CUSUM Quality-Control Procedures". *Technometrics* 27, pp. 285–292.
16. Woodroofe, M. (1982). *Nonlinear Renewal Theory in Sequential Analysis*. SIAM, Philadelphia.

DEPARTMENT OF BIOSTATISTICS
UNIVERSITY OF ROCHESTER
ROCHESTER, NY, 14642
e-mail: dragalin@bst.rochester.edu

A general and simple approach to economic process control

1. Introduction

Quality of a process with a well-defined purpose should be defined as a measure how well the system meets its objective. If *profit* constitutes the objective, then *profitability* reflects its quality best. Lack of profitability, i.e. quality may be caused by gradually increasing wear-out phenomena, which are characterized by the fact that their future development depends on the past and the present states of the process and therefore may be predicted to a certain extent. Thus, by observing the actual state of the process and removing the wear-out in due time, quality deterioration may be avoided. Besides process wear, sudden disturbances effect process quality. The occurrence of sudden disturbances doesn't depend on the process and therefore cannot be avoided by preventive actions.

Process wear generally generates dependency of the output and shifts the process parameters from the target values, causing an increase of nonconformities on the one hand, and an increase of the frequency of random disturbances on the other hand. Thus, according to common understanding we define the aim of *process maintenance* as to remove any sign of wear before it leads to a decrease of quality i.e. profitability. In contrast to maintenance, the aim of *process control* is defined as to detect and remove random disturbances before the resulting losses become substantially.

2. Process evolution

Any manufacturing process can be described by a stochastic process representing the output, and another stochastic process representing the probabilities of the occurrence of a disturbance. The output process is often given by the process of the quality characteristics X: $\{X_t(\theta(t))\}_{t \in \mathbb{R}^+}$ with $\theta(t) \in \Theta$, where Θ is the set of possible distributional parameters of X. The process of disturbance probabilities is given by: $\{p_t(\theta(t), \lambda(t))\}$ with $\lambda(t) \in \Lambda$, where Λ is the set of admissible distributional parameters. Thus process evolution is determined by the process

$$\{\theta(t), \lambda(t)\}_{t \in \mathbb{R}^+} \quad \text{with} \quad \theta(t) \in \Theta \text{ and } \lambda(t) \in \Lambda$$

It is assumed here that so-called target values denoted by $\theta_0 \in \Theta$ and $\lambda_0 \in \Lambda$ are known. These target values should be determined as to guarantee in a certain sense highest profitability of the process. It follows that an ideal process is characterized by a process of independent and identically distributed random variables with distributional parameter $\theta_t = \theta_0$ and constant disturbance probability $p_t(\theta(t), \lambda(t)) = p_0$. Any departure from this ideal process leads to a decrease in process profitability. As to process evolution, we assume that

- for $t = 0$ the process is on target: $\theta(0) = \theta_0$ and $\lambda(0) = \lambda_0$
- for $t > 0$ the evolution of $\theta(t)$ and $\lambda(t)$ is given by: $\theta(t) = \theta^{(c)}(t) + \theta^{(p)}(t)$ and $\lambda(t) = \lambda^{(c)}(t) + \lambda^{(p)}(t)$.

where $\theta^{(c)}(t)$ and $\lambda^{(c)}(t)$ are continuous processes and $\theta^{(p)}(t)$ and $\lambda^{(p)}(t)$ are point processes.

3. Process maintenance

As mentioned earlier, process wear has different effects which have to be modeled differently. The gradually developing dependency of the output X_t and its gradually developing deviation from target is modeled by $\theta^{(c)}(t)$ the continuous part of $\theta(t)$. The gradually increasing disturbance intensity is modeled by $\lambda^{(c)}(t)$ the continuous part of $\lambda(t)$.

Examples: *In the theory of automatic process control the continuous part of θ is sometimes modeled by a linear process $\theta^{(c)}(t) = \theta_1 t + \theta_2 + \epsilon(t)$ where $\theta_i \in \mathbb{R}$ and $\epsilon(t)$ is a white noise process. In maintenance theory $\lambda^{(c)}(t)$ is often assumed to be the Weibull failure intensity $\lambda^{(c)}(t) = (\lambda \cdot \beta)(\lambda t)^{\beta-1}$ with $\lambda > 0$ and $\beta > 1$.*

The objective of process maintenance is to prevent process evolution due to $\theta^{(c)}(t)$ and $\lambda^{(c)}(t)$. The two standard methods of process maintenance are on the one hand automatic process controllers, which continuously monitor the relevant process parameters and automatically adjust them in order to keep $\theta^{(c)}(t) \approx \theta_0$ for all t and remove the dependency of the output, and on the other hand preventive replacements/renewals by means of which a too large deviation of $\lambda^{(c)}(t)$ from the target λ_0 is made impossible, i.e. $\lambda^{(c)}(t) \approx \lambda_0$ for all t. Assuming that a successful process maintenance has been implemented implies:

- $\theta(t) \approx \theta_0 + \theta^{(p)}(t)$ and $\lambda(t) \approx \lambda_0 + \lambda^{(p)}(t)$

It remains to solve the problem of detecting and removing the effects of random disturbances, which in general cannot be compensated by automatic process controllers. Not to repair them and to wait until the next preventive maintenance action could last too long and therefore may cause a considerable decrease in profit.

4. Process control

The main problem with the random disturbances is to detect their occurrence. Once they are detected, we assume that they can be removed completely, i.e. the process starts after a removal of a disturbance anew. For detecting a disturbance we need a process *monitoring policy*, i.e. a directive fixing the actions which have to be performed under the possible circumstances. In order to compare different monitoring policies we need an indicator which should reflect their contributions to overall process profitability.

A monitoring policy for detecting the occurrence of random disturbances is given by $\{(t_i, n_i, \gamma_i)\}_{i \in \mathbb{N}}$, where t_i is the ith monitoring timepoint, n_i describes the way of monitoring at time t_i, and γ_i is the decision rule. Here, we restrict ourselves to periodic monitoring policies, i.e.:

$$t_{i+1} - t_i = h \quad \text{for} \quad i = 1, 2, \cdots \text{ with } \quad t_0 = 0$$

with fixed sample size $n_i = n$. Aim of sampling at time t_i is to decide whether or not the process parameter $\theta(t)$ is still on target or has been shifted by the occurrence of a disturbance. By selecting a suitable shift θ_p, the decision problem can be solved by testing the nullhypothesis $H_0 : \theta(t) = \theta_0$ against the alternative hypothesis $H_1 : \theta(t) = \theta_0 + \theta_p$. Because of the assumptions, it makes sense to use only one fixed decision function $\gamma_i = \gamma \in \Gamma$ in all decision points t_i, $i = 1, 2, ...$, where Γ denotes the set of admissible decision functions. The error probabilities when using γ are denoted by $\alpha = \mathbf{P}(\gamma = 1 | \theta_0)$ and $\beta = \mathbf{P}(\gamma = 0 | \theta_0 + \theta_p)$. Now we are in a position to define the monitoring policy investigated here:

Definition *A simple periodic monotoring policy for detecting the occurrence of a random disturbance is given by three quantities (h, n, γ), and the following rule: Take every $h \in \mathbb{R}^+$ hours of operation a random sample of size $n \in \mathbb{N}_0$ with $0 \leq n \leq hv$ (where v denotes the production speed) from the process output, and decide by means of the decision function $\gamma \in \Gamma$ on the actual state of the system. In case the decision is in favor of State II (i.e. $\theta(t) = \theta_0 + \theta_p$) an alarm is released and the disturbance is removed. Otherwise the process is left alone.*

5. Process profitability

In order to compare different monitoring policies we need an indicator which should reflect the policy's contribution to process profitability. Here, we use the long run profit per item produced as a measure for profitability. Let $G(m)$ be the gain derived from the first m items produced, then

$$\Pi^* = \lim_{m \to \infty} \frac{G(m)}{m} \quad (1)$$

defines the long run profit per item. Π^* is a function of process design, process maintenance and the monitoring policy. As design and maintenance are assume to be given, Π^* is considered here only as function of the monitoring policy, i.e. $\Pi^* = \Pi^*(h, n, \gamma)$. The problem is to determine an optimal monitoring policy defined by:

Definition *A monitoring policy (h^*, n^*, γ^*) is called optimal with respect to Π^* and the set of admissible monitoring policies $\{(h, n, \gamma)\}$, if*

$$\Pi^*(h^*, n^*, \gamma^*) \geq \Pi^*(h, n, \gamma) \quad \text{for any} \quad (h, n, \gamma)$$

In order to find an explicit expression for Π^* the economic situation has to be described. The economic situation with respect to the output is given by:

g_I = expected profit derived from one item produced on target
g_{II} = expected profit derived from one item produced after occurrence of a disturbance disturbance

The monitoring policy includes three actions: sampling, intervention after false alarm (= inspection), intervention after true alarm (= renewal):

a^*n = cost for taking and evaluating a sample of size n
e^* = expected cost of an inspection after a false alarm
b^* = expected benefit per renewal

After a disturbance is detected, the process is renewed. Consecutive renewals divide the process in stochastically equivalent parts which are called renewal cycles. The following random variables refer to one renewal cycle:

A_I = number of monitoring actions when operating on target
A_F = number of false alarm
A_{II} = number of monitoring actions after the occurence of a disturbance
N = number of produced items
G = profit derived from the N produced items

then

$$E[G] = g_{II}E[A_I + A_{II}]hv + b^* - e^*E[A_F] - a^*nE[A_I + A_{II}]$$
$$E[N] = E[A_I + A_{II}]hv$$

Hence

$$\Pi^*(h, n, \gamma) = \frac{E[G]}{E[N]} = \frac{1}{hv}\left\{\frac{b^* - e^*E[A_F]}{E[A_I + A_{II}]} - a^*n\right\} + g_{II}$$

6. Optimization

In [2] a rather general methodology is developed to determine approximately optimal monitoring policies. It is shown that the optimization problem can be solved approximately in two separate steps. First an approximately optimal sampling plan $(\hat{n}^*, \hat{\gamma}^*)$ is determined. Once $(\hat{n}^*, \hat{\gamma}^*)$ is given the approximately optimal monitoring interval $\hat{h}^*$ can be calculated by the following explicit formula:

$$\hat{h}^* = E[\tau]\frac{\alpha^* + \frac{a^*}{e^*}\hat{n}^* + \sqrt{(\alpha^* + \frac{a^*}{e^*}\hat{n}^*)\frac{\frac{b^*}{e^*}(1-\beta^*)+\alpha^*}{1-0.5(1-\beta^*)}}}{\frac{b^*}{e^*} + 0.5\alpha^* - \frac{a^*}{e^*}\hat{n}^*(\frac{1}{1-\beta^*} - 0.5)}$$

where α^* and β^* denote the error probabilities for $(\hat{n}^*, \hat{\gamma}^*)$. It remains to determine the approximately optimal sampling plan $(\hat{n}^*, \hat{\gamma}^*)$.

7. Approximately optimal sampling plan

In the following the solution for some of the most frequently arising situations are briefly displayed. For details we refer to [2]. As to the economic input parameters, the following three parameters (called *economic key parameters*) must be necessarily known in order to obtain the approximately optimal economic monitoring policy.

sampling cost	inspection cost	benefit per renewal
a^*	e^*	b^*

Normal, one-sided case, fixed variance

Process characteristics:

Qual. Characteristic	Distribution	Target Value	Shifted Value
$X_t(\mu(t), \sigma^2)$	$N(\mu(t), \sigma^2)$	μ_0	$\mathbf{P}\left(\mu(t) = \mu_0 + \delta\sigma\right) = 1$

Set of process input parameters to be specified:

target value	variance	shift parameter
μ_0	σ^2	δ

Decision function:

$$\gamma(X_1, \cdots, X_n) = \begin{cases} 0 & \text{for} \quad \frac{\bar{X} - \mu_0}{\sigma}\sqrt{n} < c \\ 1 & \text{for} \quad \frac{\bar{X} - \mu_0}{\sigma}\sqrt{n} \geq c \end{cases}$$

Determination of $(\hat{n}^*, \hat{c}^*)$: Calculate $a_0 = \frac{a^*}{e^*} \cdot \frac{1}{\delta^2}$ and take for given a_0 the quantities y and z from Table 1. Then: $\hat{n}^* =$ closest positive integer to $\left(\frac{y}{\delta}\right)^2$ and $\hat{c}^* = z$.

a_0	0.0001	0.0005	0.0010	0.0050	0.01
y	4.70283	4.16945	3.91402	3.23682	2.89647
z	3.47853	3.02028	2.80396	2.214127	1.96448

Table 1: Normal, one-sided case, fixed variance

Normal, two-sided symmetric case, fixed variance

Process characteristics:

Qual. Characteristics	Distribution	Target Value	Shifted Values
$X_t(\mu(t), \sigma^2)$	$N(\mu(t), \sigma^2)$	μ_0	$\mathbf{P}\left(\mu(t) = \mu_0 - \delta\sigma\right) = 0.5$
			$\mathbf{P}\left(\mu(t) = \mu_0 + \delta\sigma\right) = 0.5$

Set of process input parameters to be specified:

target value	variance	shift parameter
μ_0	σ^2	δ

Decision function:

$$\gamma(X_1, \cdots, X_n) = \begin{cases} 0 & \text{for} & \left|\frac{\bar{X}-\mu_0}{\sigma}\sqrt{n}\right| < c \\ 1 & \text{for} & \left|\frac{\bar{X}-\mu_0}{\sigma}\sqrt{n}\right| \geq c \end{cases}$$

Determination of $(\hat{n}^*, \hat{c}^*)$: Calculate $a_0 = \frac{a^*}{e^*} \cdot \frac{1}{\delta^2}$ and take for given a_0 the quantities y and z from Table 2. Then: $\hat{n}^* =$ closest positive integer to $\left(\frac{y}{\delta}\right)^2$ and $\hat{c}^* = z$.

a_0	0.0001	0.0005	0.0010	0.0050	0.01
y	4.91206	4.40830	4.16945	3.54520	3.23682
z	3.66052	3.22443	3.02027	2.49551	2.24127

Table 2: Normal, Two-Sided Symmetric Case, Fixed Variance

Normal, two-sided unsymmetric case, fixed Variance

Process characteristics:

Qual. Characteristics	Distribution	Target Value	Shifted Values
$X_t(\mu(t), \sigma^2)$	$N(\mu(t), \sigma^2)$	μ_0	$\mathbf{P}\left(\mu(t) = \mu_0 - \delta_1\sigma\right) = w$
			$\mathbf{P}\left(\mu(t) = \mu_0 + \delta_2\sigma\right) = 1 - w$

Set of process input parameters to be specified:

target value	variance	shift parameters	δ_1-shift probability
μ_0	σ^2	δ_1 and δ_2	w

Decision function:

$$\gamma(X_1, \cdots, X_n) = \begin{cases} 0 & \text{for} & -c_1 < \frac{\bar{X}-\mu_0}{\sigma}\sqrt{n} < c_2 \\ 1 & \text{for} & \frac{\bar{X}-\mu_0}{\sigma}\sqrt{n} \leq c_1 \text{ or } \frac{\bar{X}-\mu_0}{\sigma}\sqrt{n} \geq c_2 \end{cases}$$

Determination of $(\hat{n}^*, \hat{c}^*)$: Calculate $a_0 = \frac{a^*}{e^*} \cdot \frac{1}{\delta^2}$ and $\Delta = \frac{\delta_1}{\delta_2}$ and take for given a_0, Δ and w the quantities y, z_1 and z_2 from Table 3. Then: $\hat{n}^* =$ closest positive integer to $\left(\frac{y}{\delta}\right)^2$ and $\hat{c}_1^* = z_1$, $\hat{c}_2^* = z_2$.

	$\Delta = 0.25$	,	$w = 0.2$		
a_0	0.0001	0.0005	0.0010	0.0050	0.01
y	9.97406	7.22355	6.10928	4.25170	3.62780
z_1	2.79716	2.29700	2.07814	1.57406	1.35386
z_2	5.42927	4.09091	3.55880	2.59313	2.21450

Table 3: Normal, two-sided unsymmetric case, fixed variance

Normal, two-sided symmetric case

Process characteristics:

Qual Characteristics	Distribution	Target Values	Shifted Values
$X_t(\mu(t), \sigma^2(t))$	$N(\mu(t), \sigma^2(t))$	μ_0, σ_0^2	$\mathbf{P}\left(\mu(t) = \mu_0 - \delta_1\sigma_0\right) = 0.5$
			$\mathbf{P}\left(\mu(t) = \mu_0 + \delta_2\sigma_0\right) = 0.5$
			$\mathbf{P}\left(\sigma^2(t) = (\delta_2\sigma_0)^2\right) = 1.0$

Set of process input parameters to be specified:

target values	shift parameter (mean)	shift parameter (variance)
μ_0, σ_0^2	$\delta_1 \geq 0$	$\delta_2 > 1$

Decision function:

$$\gamma(X_1, \cdots, X_n) = \begin{cases} 0 & \text{for } \frac{1}{n}\sum_{i=1}^{n}\left(\frac{X_i-\mu_0}{\sigma_0}\right)^2 < c \\ 1 & \text{for } \frac{1}{n}\sum_{i=1}^{n}\left(\frac{X_i-\mu_0}{\sigma_0}\right)^2 \geq c \end{cases}$$

Determination of $(\hat{n}^*, \hat{c}^*)$: Calculate $a = \frac{a^*}{e^*}$ and take for given a, δ_1 and δ_2 the approximately optimal sampling plan $\hat{n}^*$ and $\hat{c}^*$ from Table 4.

a \ δ_1	0.0	0.5	$\delta_2 = 2.25$ 1.0	1.5	2.0	2.5
0.0001	12 3.25	11 3.40	9 3.80	7 4.44	5 5.50	4 6.48
0.0005	9 3.28	9 3.31	7 3.80	6 4.23	4 5.46	3 6.72
0.0010	8 3.25	8 3.28	6 3.84	5 4.35	4 5.09	3 6.24
0.0050	6 3.09	5 3.39	4 3.90	4 4.05	3 4.93	2 6.56
0.0100	4 3.37	4 3.40	4 3.51	3 4.24	2 5.64	2 5.87

Table 4: Normal, two-sided and symmetric case

Poisson case

Process characteristics:

Qual. Characteristics	Distribution	Target Value	Shifted Value
$X_t(\mu(t))$	$Po(\mu(t))$	μ_0	$\mathbf{P}\left(\mu(t) = \delta\mu_0\right) = 1.0$

Set of process input parameters to be specified:

target value	shift parameter
μ_0	$\delta > 1$

Decision function:
$$\gamma(X_1, \cdots, X_n) = \begin{cases} 0 & \text{for} \quad \sum_{i=1}^{n} X_i < c \\ 1 & \text{for} \quad \sum_{i=1}^{n} X_i \geq c \end{cases}$$

Determination of $(\hat{n}^*, \hat{c}^*)$: Calculate $a = \frac{a^*}{e^*} \cdot \frac{1}{\mu_0}$ and $\Delta = \frac{\mu_1}{\mu_0}$ and take for given a and Δ the quantities y and z from Table 5. Then: $\hat{n}^*$ = closest positive integer to $\frac{y}{\mu_0}$ and $\hat{c}^* = z$.

a_0 \ Δ	1.50	2.00	3.00	4.00	5.00
0.0005	59.237 79	21.646 36	7.832 18	4.493 13	2.832 10
0.0010	49.792 66	18.757 31	7.031 16	4.223 12	2.580 9
0.0050	26.953 35	12.402 20	4.954 11	2.848 8	2.107 7
0.0100	18.057 23	9.433 15	4.127 9	2.566 7	1.846 6
0.0500	0.000 0	3.455 5	2.497 5	1.580 4	0.964 3

Table 5: Poisson case

Binomial case

Process characteristics:

Qual. Characteristic	Distribution	Target Value	Shifted Value
$X_t(p(t))$	$Bi(1, p(t))$	p_0	$\mathbf{P}(p(t) = p_1) = 1.0$

Set of process input parameters to be specified:

target value	shifted value
p_0	$p_1 > p_0$

Decision function:
$$\gamma(X_1, \cdots, X_n) = \begin{cases} 0 & \text{for} \quad \sum_{i=1}^{n} X_i < c \\ 1 & \text{for} \quad \sum_{i=1}^{n} X_i \geq c \end{cases}$$

Determination of $(\hat{n}^*, \hat{c}^*)$: Calculate $a = \frac{a^*}{e^*}$ and $\Delta = p_1 - p_0$ and take for given a, Δ and p_0 the approximately optimal sampling plan $(\hat{n}^*, \hat{c}^*)$ from Table 6.

a \ p_0	$\Delta = 0.05$				
	0.01	0.02	0.04	0.06	0.08
0.0005	(85,3)	(127,6)	(158,11)	(201,18)	(227,25)
0.0010	(59,2)	(87,4)	(119,8)	(149,13)	(157,17)
0.0050	(5,0)	(29,1)	(37,2)	(42,3)	(35,3)
0.0100	(4,0)	(8,0)	(11,0)	(0,0)	(0,0)

Table 6: Binomial case

References

[1] E. v. Collani, A Simple Procedure to Determine the Economic Design of an $\bar{X}$ Control Chart. Journal of Quality Technology 18(1986) 145.

[2] E. v. Collani, Determination of the Economic Design of Control Charts. In: *Optimization in Quality Control*, ed. K. S. Al-Sultan and M. A. Rahim. (Kluwer Academic Publisher, Boston 1997).

[3] E. v. Collani and J. Treml, Control of a Two-Dimensional Process-Quality-Indicator by means of a Screening Procedure. Economic Quality Control 8(1993)167.

[4] E. v. Collani & V. Dragalin, A Simplified Economic Design of CUSUM Charts for Monitoring a Normally Distributed Process Mean. Research Report of the WRQC, No. 69, 1996.

[5] E. v. Collani & K. Dräger, A Simplified Economic Design of Control Charts for Monitoring the Nonconforming Probability. Economic Quality Control 10(1996)231.

[6] E. v. Collani & K. Dräger, The Economic Design of Control Charts for Monitoring the Number of Nonconformities. Economic Quality Control 12(1997)15.

[7] A.J. Duncan, The Economic Design of $\bar{X}$ Control Charts Used to Maintain Current Control of a Process. Journal of the American Statistical Association 51(1956)228.

[8] C. Ho and K.E. Case, Economic Design of Control Charts: A Literature Review for 1981 - 1991. Journal of Quality Technology 26(1994)39.

Elart von Collani
University of Würzburg
Sanderring 2
D-97070 Würzburg
Germany

YANHONG WU
Sequential change point detection and estimation for multiple alternative hypothesis[1]

1. Introduction

Let $X_1, ..., X_\theta, X_{\theta+1}, ...$ be independent random variable sequence where X_k for $k \leq \theta$ follow the density $f_0(x)$ and X_k for $k > \theta$ follow one of K alternative densities $f_i(x)$ for $i = 1, 2, ..., K$. θ is the so-called change point and unknown. We assume that the total $k+1$ density functions are distinguishable in the sense that the Kullback-Leibler distances between $f_j(x)$ and $f_i(x)$ $D_{ji} = E_j \ln \frac{f_j(X_1)}{f_i(X_1)} > 0$ for all $i, j = 0, 1, ..., K$.

In this paper, the isolation and identification of the change alternative and change point are considered after sequential detection. Two methods are evaluated and compared: the Multi-dimensional CUSUM Procedure and the Multi-dimensional Shiryayev-Roberts Procedure. First, certain bounds and approximations for the average run length and false isolation probabilities are obtained. Second, numerical comparison is conducted between the two procedures particularly in terms of the estimation of the change point. It is found that the Shiryayev-Roberts procedure is slightly better, Third, an application to sequential segmentation for a randomly changing mean at two states is used for demonstration by using the two methods. One advantage is that both methods can be implemented by using parallel computing techniques. It is shown that the Shiryayev-Roberts procedure can be seen as the approximation for the Bayesian method under the Hidden Markov Chain model.

2. Method based on CUSUM Procedure:

By generalizing the one-dimensional CUSUM process(Page[2]), we define

$$Y_n^{(j)} = \max(0, Y_{n-1}^{(j)} + \ln \frac{f_j(x_n)}{f_0(x_n)}) = S_n^{(j)} - \min_{1 \leq k \leq n} S_k^{(j)},$$

where $Y_0^{(j)} = 0$ and $S_n^{(j)} = \sum_{i=1}^n \ln \frac{f_j(x_n)}{f_0(x_n)}$ is the log-likelihood ratio with alternative $f_j(x)$. Let

$$\tau_j = \min\{n > 0 : Y_n^{(j)} > c_j\}, \quad for \quad j = 1, 2, ..., K.$$

An alarm will be made at

$$\tau = \min\{\tau_1, ..., \tau_K\}$$

---
[1]This research is supported by NSERC of Canada

and the change alternative is estimated as

$$j^* = argmin_j \tau_j,$$

and the change time is estimated as

$$\theta^* = \max\{k \leq \tau_{j^*} : Y^{(j^*)}_{\tau_{j^*}} = 0\}.$$

Note that θ^* is actually the maximum likelihood estimator conditioning on that the change alternative is j^* and the change is detected.

Denote $ARL_j = E_j(\tau)$ as the average run length for $j = 0, 1, 2, ..., K$, and $e_{ij} = P_i(j^* = j)$ as the false isolation probability when the true alternative is f_i for $i \neq j, i, j = 1, 2, ..., K$.

Our first result gives a lower bound for the average in-control run length ARL_0.

Proposition 1:

$$ARL_0 \geq [\sum_{j=1}^{K}(E_0(\tau_j))^{-1}]^{-1}.$$

Proof: By using the renewal property of $\{Y_n^{(j)}\}$, we can write

$$E_0[\tau] = E_0[\tau_j] - E_0[\tau_j - \tau]$$
$$= E_0[\tau_j] - E_0[\tau_j - \tau; \tau_j \neq \tau]$$
$$= E_0[\tau_j] - E_0[\tau_j - \tau | \tau_j \neq \tau] P_0(\tau_j \neq \tau)$$
$$\geq E_0[\tau_j] - E_0[\tau_j - \tau | \tau_j \neq \tau; Y_\tau^{(j)} = 0] P_0(\tau_j \neq \tau)$$
$$= E_0[\tau_j] - E_0[\tau_j] P_0(\tau_j \neq \tau)$$
$$= E_0[\tau_j] P_0(\tau_j = \tau).$$

Thus,

$$E_0[\tau]/E_0[\tau_j] \geq P_0(\tau_j = \tau).$$

Summing up the above equation in j, we get the result.

A convenient choice for the control limits c_j is such that all $E_0[\tau_j]$'s are the same, in which case

$$ARL_0 \geq E_0[\tau_j]/K.$$

Accurate approximations for $E_0[\tau_j]$ have been given by Siegmund[7].

Example 1: (2-sided change detection for the normal mean) Suppose before the change $X_1, ..., X_\theta$ are iid N(0, 1) and after the change the mean can change to either μ_0 or μ_1 such that $\mu_0 < 0 < \mu_1$.

Let
$$\tau_{i+1} = \inf\{n > 0 : Y_i = \mu_i \sum_{i=1}^{n} x_i - \frac{n}{2}\mu_i^2 - \min_{1 \leq k \leq n}(\mu_i \sum_{i=1}^{k} x_i - \frac{k}{2}\mu_i^2) \geq c_{i+1}\}$$

for $i = 1, 2$, and $\tau = \min(\tau_1, \tau_2)$. Then from Siegmund[7, Lemma 2.62 and 2.64], it follows that if
$$|c_1/|\mu_0| - c_2/\mu_1| \leq \frac{1}{2}(|\mu_0| + \mu_1),$$
then, we have
$$1/E_0[\tau] = 1/E_0[\tau_1] + 1/E_0[\tau_2].$$

Before we give the approximation for ARL_j, we first estimate the false alarm probabilities $e_{ij} = P_i(\tau = \tau_j)$ whis is less than $P_i(\tau_j < \tau_i)$.

By using Wald's Likelihood Ratio Identity, we have
$$P_i(\tau_j < \tau_i) = E_j[e^{-\sum_{k=1}^{\tau_j} \ln \frac{f_j(X_k)}{f_i(X_k)}}; \tau_j < \tau_i]$$
$$\leq E_j[e^{-\sum_{k=1}^{\tau_j} \ln \frac{f_j(X_k)}{f_i(X_k)}}]$$
$$= O(e^{-E_j(\tau_j)E_j \ln \frac{f_j(X_1)}{f_i(X_1)}})$$
$$= O(e^{-c_j \frac{D_{ji}}{D_{j0}}}),$$
where $D_{ji} = E_j \ln \frac{f_j(X_1)}{f_i(X_1)} > 0$.

Now, we consider the estimate of ARL_j. Writing $ARL_0 = T$, under the design with same $E_0\tau_j$'s, we know that $c_i = O(\ln T)$. We thus have $E_i\tau_i = \frac{\ln T}{D_{i0}}(1 + o(1))$.

On the other hand, for detecting the change with a single alternative $f_i(x)$ under the same ARL_0, we denote the alarming time as τ_i'. It is obvious that
$$E_i\tau_i' = \frac{\ln T}{D_{i0}}(1 + o(1)).$$
However,
$$E_i\tau_i' = E_i[\tau_i'; \tau = \tau_i] + \sum_{j \neq i} E_i[\tau_i'; \tau = \tau_j]$$
$$\leq E_i[\tau_i; \tau = \tau_i] + \sum_{j \neq i} E_i[\tau_j; \tau = \tau_j]$$
$$+ \sum_{j \neq i} E_i[\tau_i']P_i(\tau = \tau_j)$$
$$= E_i[\tau] + E_i[\tau_i']P_i(\tau \neq \tau_i)$$

$$= E_i[\tau] + E_i[\tau_i']O(e^{-\min_{j \neq i}(c_j \frac{D_{jt}}{D_{j0}})}).$$

By using the obvious fact that $E_i[\tau] \leq E_i[\tau_i]$, we have

$$ARL_i = \frac{\ln T}{D_{i0}}(1 + o(1)).$$

3. Method based on Shiryayev-Roberts Procedure:

Similarly, by generalizing the one-dimensional Shiryayev-Roberts procedure(Shiryayev[6], Roberts[5]), we define the following multi-dimensional Shiryayev-Roberts process

$$R_n^{(j)} = (1 + R_{n-1}^{(j)})\frac{f_j(x_n)}{f_0(x_n)},$$

with $R_0^{(j)} = 0$ for $j = 1, 2, ..., K$.

Define

$$\tilde{\tau}_j = \inf\{n > 0 : R_n^{(j)} \geq B_j\}, \quad for \quad R_0^{(j)} = 0,$$

and $\tilde{\tau} = \min\{\tilde{\tau}_1, ..., \tilde{\tau}_K\}$, and

$$\tilde{j} = argmin(\tilde{\tau}_j).$$

Similarly, we denote the average run lengths and false isolation probabilities as $A\tilde{R}L_j$ and $\tilde{e}_{ij}$ respectively.

To estimate the change point, we define

$$Q_n^{(j)} = (n + Q_{n-1}^{(j)})\frac{f_j(x_n)}{f_0(x_n)},$$

and let

$$\tilde{\theta} = Q_{\tilde{\tau}}^{(\tilde{j})}/R_{\tilde{\tau}}^{(\tilde{j})}.$$

This can be derived by assuming that conditioning on the change being detected, the change point is uniformly distributed on $\{1, 2, ..., \tilde{\tau}\}$.

A convenient way is to let $B_j = B$, free of j. Under this choice we have the following lower bound for $A\tilde{R}L_0$.

Proposition 2: If $B_j = B$, then $E_0[\tilde{\tau}] \geq B/K$.

Proof: Using that fact that under H_0, $\{R_n^{(j)} - n\}$ are martingales for all j, we have

$$E_0[\tilde{\tau}] = E_0[R_{\tilde{\tau}}^{(j)}]$$

$$\geq E_0[R_{\tilde{\tau}i}^{(j)}; \tilde{\tau} = \tilde{\tau}_j]$$

$$\geq BP_0(\tilde{\tau} = \tilde{\tau}_j).$$

Summing it together in j, we get the expected result.

Similarly, we can obtain the same estimates for the false isolation probabilities and average out-of-control run lengths. Note that the approximation for ARL_1 in the single alternative case is obtained by Pollak[3].

Example 2: We consider the detection of change in either mean or variance. Let X_i follow N(0,1) for $i \leq \theta$ and X_i follow either $N(\delta, 1)$ or $N(0, 1+\Delta)$ for $i > \theta$ where δ and Δ are both positive. Define

$$R_n^{(1)} = (1 + R_{n-1}^{(1)})exp(\delta x_n - \frac{\delta^2}{2});$$

and

$$R_n^{(2)} = (1 + R_{n-1}^{(2)})exp(\frac{1}{2}(1 - \frac{1}{1+\Delta})x_n^2 - \frac{1}{2}\ln(1+\Delta)),$$

for $R_0^{(1)} = R_0^{(2)} = 0$. The alarm time will be $\tau = \min(\tau_1, \tau_2)$ where

$$\tau_i = \inf\{n > 0 : R_n^{(i)} > B\},$$

and B is designed such that $ARL_0 = E_0 \tau = T$.

4. Numerical Comparison:

The comparison between the two methods in terms of the average delay detection time or ARL_1 has been extensively studied; see for example Roberts[5], Pollak and Siegmund[4], Srivastava and Wu[8]. The general conclusion is that the procedures are equally competitive. In this section, we evaluate the two methods in terms of the estimation of the change point conditioning that the change is detected correctly. Thus, we can assume that there is only one alternative hypothesis without loss of generality. Suppose X_i follows N(0,1) for $i \leq \theta$ and $N(\delta, 1)$ for $i > \theta$ where $\delta > 0$ is known.

For the CUSUM procedure, we let

$$Y_n = \max\{0, Y_{n-1} + \delta X_n - \frac{\delta^2}{2}), \quad with \ Y_0 = 0.$$

The design for the control limit c is decided by the approximation given by Siegmund[7]:

$$ARL_0 \approx \frac{e^{c+1.166\delta} - 1 - (c + 1.166\delta)}{\delta^2/2}.$$

While for the Shiryayev-Roberts procedure defined as

$$R_n = (1 + R_{n-1})exp(\delta X_n - \frac{\delta^2}{2}), \quad with \ R_0 = 0,$$

Table 1: Comparison of bias and MSE for the change point estimator

			bias	$\sqrt{MSE}$
CUSUM	$\delta = 0.5$	$\theta = 25$	2.25	15.09
	c=4.41	50	-0.016	16.59
	$\delta = 1.0$	25	0.076	5.42
	c=5.06	50	-1.37	8.01
Shiryayev-Roberts	$\delta = 0.5$	25	3.72	12.29
	B=747.14	50	0.375	14.30
	$\delta = 1.0$	25	0.862	4.54
	B=558.22	50	0.210	6.21

the control limit B is decided by Pollak[3]'s approximation

$$ARL_0 \approx e^{0.583\delta} B.$$

These approximations have been show to be very accurate(Wu[9]).

For 1000 simulations and $ARL_0 = 500$, Table I gives the simulated results for the bias and square root of mean squared error for δ =0.5 and 1.0, and $\theta = 25$, and 50 based on the CUSUM procedure and Shiryayev-Roberts procedure respectively. All the results are obtained based on 500 simulations. The table also gives the designed control limit c and B in the second column.

Comparing the results, we find that the Shiryayev-Roberts procedure gives slightly better estimators in terms of the mean squared error. However, the CUSUM procedure gives smaller biases for small shifts and the Shiryayev-Roberts procedure has the reverse behavior. Also, we note that when the change location is too close to zero, the bias increases dramatically.

In a yet submitted paper(Ding and Wu[1]), the authors investigated the bias of the change point estimation θ^* in the above normal observation case after CUSUM detection. It is shown that as both d and θ approach ∞, conditioning on $\tau > \theta$, the average bias of θ^* is equal to

$$E[\theta^* - \theta|\tau > \theta] = \frac{1}{2\delta^2} - \frac{0.167}{\delta} + O(1),$$

as $\delta \to 0$. This shows that even as the change occurs far away from zero, the bias of θ^* is not negligible, particularly for small δ.

5. Sequential Segmentation for Randomly Changing Mean at Finite States

In this section, we demonstrate how the two methods can be used for sequential classification and segmentation when the mean randomly changes at finite states. We first look at the Bayesian method by assuming that the density of consecutive observations are affected by a finite state Markov Chain J_t with state space $\{1, 2, ...N\}$ and transition probability matrix $Q = (q_{ij})$ for $i,j = 1, 2, ..., N$. Assume the initial distribution $\pi_0^{(i)} = P(J_0 = i)$ and at state j, the density function of X_n is $f_j(x)$ for $j = 1, 2, ..., N$.

Let $\pi_n^{(j)} = P(J_n = j | x_1, ..., x_n)$ be the posterior distribution of J_n based on the observation $x_1, ..., x_n$. Using the Bayes formula, we get

$$\pi_n^{(j)} = \frac{\sum_{i=1}^{N} \pi_{n-1}^{(i)} q_{ij} f_j(x_n)}{\sum_{k=1}^{N} \sum_{i=1}^{N} \pi_{n-1}^{(i)} q_{ik} f_k(x_n)}.$$

The Bayesian method is to classify time n is at state j, say $\hat{J}_n = j$ if $\pi_n^{(j)}$ is the maximum.

Now, we assume that $q_{ii} \to 1$ for all $i = 1, 2, ..., N$ at the same order. That means, the average sojourn time at each state $1/(1 - q_{ii})$ goes to ∞ at the same order. Let us see how the Bayesian method can be used to obtain a sequential classification procedure.

Suppose at time 0, we classify $\hat{J}_0 = j$, i.e. $\pi_0^{(j)}$ is the maximum. Then we monitor the N-1 dimensional posterior ratio process as $\{\frac{\pi_n^{(k)}}{\pi_n^{(j)}}\}$ for $k \neq j$. A change of state is declared when one of the process is sufficiently large, say k. Then we classifying the changed state as k, and the procedure continues.

We first write

$$\frac{\pi_n^{(k)}}{\pi_n^{(j)}} = \frac{\pi_{n-1}^{(k)} + \sum_{i \neq k}(\pi_{n-1}^{(i)} - \pi_{n-1}^{(k)})q_{ij}}{\pi_{n-1}^{(j)} + \sum_{i \neq j}(\pi_{n-1}^{(i)} - \pi_{n-1}^{(j)})q_{ij}} \frac{f_k(x_n)}{f_j(x_n)}$$

$$\approx \left(\frac{\pi_{n-1}^{(k)}}{\pi_{n-1}^{(j)}} + \sum_{i \neq k}\left(\frac{\pi_{n-1}^{(i)}}{\pi_{n-1}^{(j)}} - \frac{\pi_{n-1}^{(k)}}{\pi_{n-1}^{(j)}}\right)q_{ik} - \frac{\pi_{n-1}^{(k)}}{\pi_{n-1}^{(j)}}\sum_{i \neq j}\left(\frac{\pi_{n-1}^{(i)}}{\pi_{n-1}^{(j)}} - 1\right)q_{ij}\right)\frac{f_k(x_n)}{f_j(x_n)}.$$

Define $R_{jn}^{(k)} = \frac{\pi_n^{(k)}}{\pi_n^{(j)}} \frac{1}{q_{jk}}$ for $j \neq k$.

Suppose as $q_{jk} \to 0$, $\pi_0^{(k)}/\pi_0^{(j)} \to 0$ such that $R_{j0}^{(k)} \to r_{j0}^{(k)}$. A careful investigation on the right hand side of the above equation gives us

$$R_{jn}^{(k)} = (R_{j(n-1)}^{(k)} + 1 + O(1 - q_{jj}))\frac{f_k(x_n)}{f_k(x_n)}$$

$$\approx (R_{j(n-1)}^{(k)} + 1)\frac{f_k(x_n)}{f_k(x_n)}.$$

This is nothing but the Shiryayev-Roberts process. Therefore, the Shiryayev-Roberts procedure can be considered as an approximation to the Bayesian method when the average sojourn time at each state is sufficiently large.

In the following, we assume that the mean changes alternatively at two states and compare the CUSUM procedure and the Shiryayev-Roberts procedure by evaluating the average error rate. More specifically, we assume $\{X_n\}$ are normal with unit variance and mean μ_n, and μ_n changes at two states: 0 and δ alternatively with equally distant intervals with length T.

The classification can be done either on-line or off-line. In the on-line case, we classify the mean immediately after each observation; while in the off-line case, we can trace back and reduce the error rate by estimating the change point.

For example, for the CUSUM procedure with initial mean 0, we first stop the process at

$$\tau_1 = \inf\{n > 0 : Y_n = max(0, Y_{n-1} + \delta X_n - \delta^2/2) > d\},$$

and denote $\nu_1 = \max\{n < \tau_1 : Y_n = 0\}$. In the on-line case, we classify $\mu_n = 0$ for all $n \leq \tau_1$; while in the off-line case, we classify $\mu_t = 0$ for $n \leq \nu_1$, and δ for $\nu < n \leq \tau_1$. At τ_1, a new CUSUM procedure starts with $-\delta^2/2$ changed to $\delta^2/2$ and X_n to $-X_n$ and the similar procedure continues.

The same rule applies to the Shiyayev-Roberts procedure.

The choice of c for the CUSUM method and B for the Shiryayev-Roberts method should be selected adaptively when the segment lengths are changing.

The error rate are affected by two factors: in the on-line case, by the whole misclassified period after a false alarm and the delayed detection time for true alarms, and in the off-line case, by the misclassified periods before and after an false alarm and the bias of the estimator when an alarm is true.

To demonstrate and compare the two methods, we conduct two simulation studies. In the first study, we generate 1000 data with the mean changing from 0 to 0.5 alternatively at equally spaced interval with length 50. We apply the CUSUM and Shiryayev-Roberts procedure with c=4.0 and B=747 respectively as in the example of Table 1. In both cases, we give the off-line segmented points. Both methods gives no false alarms. However, the Shiryayev-Roberts procedure(with error rate 0.143) is better than the CUSUM procedure(with error rate 0.189).

In the second study, the mean changes from to 1.0 alternatively at same equally spaced intervals. Again we apply the CUSUM and Shiryayev-Roberts procedures with c=5.0 and B=558 respectively to give the off-line segmented points.

In this case, both methods have two false alarms. However, the CUSUM procedure is able to eliminate the false periods in the off-line case(with error rate 0.076); while the Shiryayev-Roberts procedure is not successful which causes larger error rate (0.093).

Therefore, it is difficult to draw conclusions to say which method is better. Of course, the selection of c and B are highly dependent on whether we want to minimize the on-line misclassification rate or the off-line segmentation error rate. A detailed theoretical study and comparison will be presented somewhere else.

REFERENCES:

1. Ding, K. and Wu, Y.(1997). On the biases of change point and change magnitude estimation after CUSUM test.(submitted to *Ann. Statist.*)

2. Page,E.S.(1954). Continuous inspection schemes. *Biometrika*, **41**, 100-114.

3. Pollak,M.(1987). Average run lengths of an optimal method of detecting a change in distribution. *Ann. Statist.*,**15**, 749-779.

4. Pollak,M. and Siegmund,D.(1985). A diffusion process and its applications to detecting a change in the drift of Brownian motion. *Biometrika*, **72**, 267-280.

5. Roberts,S.W.(1966). A comparison of some control chart procedures. *Technometrics*,**8**, 411-430.

6. Shiryayev,A.N.(1963). On optimum methods in quickest detection problems. *Theo. Probab. Appl.* , **13**, 22-46.

7. Siegmund,D.(1985).*Sequential Analysis: Tests and Confidence Intervals.* Springer, Berlin.

8. Srivastava,M.S. and Wu,Y.(1993). Comparison of EWMA, CUSUM and Shiryayev-Roberts procedures for detecting a shift in the mean. *Ann. Statist.*, **21**, 645-670.

9. Wu,Y.(1994). Design of control charts for detecting the change point. In *Change Point Problems*, IMS Lecture Notes, Monograph Series, **23**, 330-345.

Yanhong Wu, Department of Mathematical Sciences, University of Alberta, Edmonton, AB, T6G 2G1, Canada.

V. Modelling

STEFAN DRESBACH
Modeling methodologies for modeling environments

1. Introduction
Problem solving in the fields of MS/OR is essentially done by using models. They can play the role of a catalyst in the problem solving process if (and only if) they function as problem descriptions. Problem-centered modeling has to be supported in two ways: technically, by modern modeling environments which potentially increase the productivity, quality, and frequency of use of model-based work [8] and conceptually, by modeling methodologies which give the theoretical basis for the model itself as well as for the modeling process.

There is no general theory for the object 'model' in the application context of Operations Research. Neither attempts to describe the term 'model' more precisely nor propositions for modeling strategies have ever led to a general solving-paradigm-free methodology, which is needed to design an advanced modeling environment. A solving-paradigm-specific style can be seen if a model contains special elements that relate to the solving method and not inevitably directly to the problem. Even if they are based on the same problem, models for different kinds of methods are not very alike. Only a few approaches on the way to solving-paradigm-free modeling methodologies exist: Structured Modeling (SM) [7], Modeling by Example [2], WWS analysis [3], or Modeling by Construction (MbC) [5] are examples. In this paper we concentrate on SM and MbC since these two are the most general methodologies. For a discussion of other approaches see [4].

Since a modeling methodology should serve as a basis for practically used modeling environments it must cover a system of sentences and statements concerning the elements as well as rules and methods that describe how to structure the modeling process. Epistemological spoken a modeling methodology must contain, first, a theory of models (i.e. a system of theorems and tenets completed by statements of its scope) and, second, statements about the application and methods of this theory. A modeling methodology must have 3 characteristics:
- It has to be meta-theoretic.
- It has to be normative.
- It has to deal with methods in the sense of application rules of the theory.

The first and the second point are self-descriptive. The third point represents the 'rules of the modeling game'. This stresses the analogy to Popper's [17] position who talks about the 'rules of the game of science' in a pure epistemological context. A modeling methodology has to fulfill three main requirements for practical reasons. It has to be user-oriented, problem-oriented, and solution-oriented. These requirements have pendants in modeling environments. They can be supported technically by, first, an adequate editor and user-interface; second, a model base system which allows the handling of complex problem structures (e.g. with object-oriented databases); and third, the integration of methods (e.g. by the use of method-base concepts).

2. Structured Modeling

"At its core, Structured Modeling is a conceptual framework for representing models based on a set of interrelated definitions of all the elements comprising a 'model'." [15, p. 697] It is based on discrete mathematics and uses hierarchical organized, partitioned, attributed, and acyclic graphs as model representations. It is a definitional system with some special properties: definitions belong to one out of five types, make dependencies explicit, are grouped by 'definitional similarity', organized hierarchically by 'conceptual similarity', definitions are free of circularity, and instances' data are organized in tables [7].

SM is explicitly designed as a basis for modeling environments. It "aims to provide the foundation for a new generation of modeling systems with all of the features..." [7, p. 550] listed later on. This shows that SM makes high demands on itself. These are in fact the requirements of a modeling methodology serving as basis for modeling environments. They become clearer when taking the above mentioned 'features' into account. These are in short:

one single model representation language suitable for managerial communication, mathematical use, and direct computer execution; independence of model representation and model solution; encompassing most of the solving paradigms and methods; support of most life-cycle phases; independence of general model structure and data instances; ability to be implemented on a desktop with a modern user interface; integrated data management; immediate expression evaluation.

First, an example and some basics shall give a feeling of what SM is. To emphasize the original terms of SM these are typed in a different font. Structured models consist of three levels with different types of abstraction. The elemental structure captures all of the definitional detail of a specific model instance. The generic structure (i.e. elements are grouped into genera) for a simple transportation model with suppliers in DALlas and CHIcago and with demand located in PITTSburgh, ATLanta, and CLEVeland is shown in fig. 1. The modular structure shows the hierarchy among element groups. Five types of elements are distinguished: primitive elements that are "undefined mathematically" [9, p. 34] and that are not value-bearing, compound elements that are also not value-bearing, attribute elements that have a constant value, function elements that have a value depending on a definite rule, and test elements that are like such function elements that have only the value 'true' or 'false'. All type definitions depend directly or indirectly on the definition of primitive elements. The definitional dependencies of model elements are constructed by tuples, the calling sequences, which are grouped inside into calling

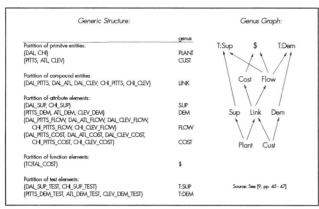

Fig. 1: generic structure of transportation model

sequence segments that are needed for the definition of the generic structure.

The grouping of elements for the generic structure is done in such a way that the genera fulfill generic similarity, i.e. all element types (with exception of primitive entities) must not contradict the three following conditions:

I. All elements in the genus have the same number of calling sequence segments.
II. All calls (i.e. definitional references) in a given segment are to elements in the same genus.
III. All elements call the same genera in corresponding calling sequence segments. [9]

There exists a formalized modeling language for SM called Structured Modeling Language (SML) [11]. There are different prototype implementations that use SM as the basis of a modeling environment [13].

The strength of SM is its mathematical basis which makes the formulation of models especially for Mathematical Programming (MP) very straightforward. One can match typical elements of MP models to SM element types (e.g. MP's variables correspond directly to SM's variable attribute elements). SM offers an explicit definition of all its elements. In addition to that the associated concepts and constructs are determined like the different types of graphs or the view concept of the modular structure as well as some proved propositions are given which make statements about the characteristics of the different SM elements. This formal specification makes it possible to build modeling environments with features like error-checking, partial consistence checking, model integration, model navigation, or independent treatment of models and solvers [12].

Because a completely specified structured model does not require the values (of attribute, function or test elements) the instantiating data can be organized separately. One only has to preserve that the elemental structure is evaluateable. Geoffrion suggests to organize data with the help of elemental detail tables that are part of SML but not of the formal SM definition. The skeletal structure of the elemental detail tables can be determined automatically from the schema. The benefits of this solution are not only dimensional flexibility, reusability, conciseness, stability, and error avoidance (these benefits result from the separation of general structure and instantiating data) but are also a high compatibility with relational data base theory and systems, the automatic generation of the relational structure (i.e. the database scheme), and standardization of data organization in all models (these benefits result from the elemental detail tables themselves) [11]. Summarizing Structured Modeling is said to be practically *and* theoretically strong.

But there are some pitfalls. In addition to some epistemological problems, one cannot deal adequately with time, cyclic dependencies, or soft information in SM. The grouping of elements into genera and modules is problematic. Last but not least, the modeling process is not strongly enough supported.

First, there are some details which are critizeable in a technic-epistemological point of view. A well designed theory (and therefore also a methodology) should be created with positive definitions. Therefore it is somehow unsatisfactory if the one element which all other definitions are based on (the primitive element) is defined solely as 'mathematically undefined'. Furthermore, the conventions about the right use of calling sequences are explicitly not part of the modeling framework. But such explanations of the application context should be part of the methodology. The last technic-epistemological point refers to the introduction of the concept of generic similarity. This concept is introduced to limit the number of possible generic structures. But there is neither a logic-based deduction nor a verifiable argumentation why generic similarity is defined the way it is. The only argument for this concept is Geoffrion's great modelling experience – this might be a little bit too weak for a methodology.

Second, there are some missing elements if SM shall fulfill the requirements of a paradigm-free modeling methodology. Modeling time concepts that are important for model building is not well supported because the elements in a genus (e.g. different points of time) have no order. The same is valid for indices. They are quoted as common and helpful means for grouping elements. But they do not become part of SM's definition. One can build genera but there is no defined concept to reference a special element (e.g. the first one) or relative one (e.g. the next period) in a genus. Sometimes there is the need for modeling simultaneous equations. E.g. if the commission depends upon profit and, of course, profit depends on the amount of commissions you have to build an equilibrium model with a cyclic dependency. But structured models have to be acyclic. And there are some difficulties if one has to handle soft information like fuzzy or uncertain data or relations.

A third group of pitfalls concerns the concept of generic similarity and the grouping of elements. It is remarkable that the definition of generic similarity is not explicitly stated as recursive. Checking conditions II and III, i.e. verifying whether elements relate to corresponding genera, requires to check the generic similarity of these genera first. That has practical impacts in modeling because the problem of grouping the elements is shifted on the level of primitive entities. Therefore the forming of genera depends on the construction of the primitive entities which itself is not methodological supported. To make the practical impacts clearer look at the transportation model again. Fig. 2 shows a subset of model elements, their calling sequences, and their genera. The question, which is to be answered, is whether the two genera T:SUP and T:DEM can be combined in a single genus and if this should be done. A methodology should support answering this question.

The first step is to check generic similarity of a hypothetically combined T:SUP-T:DEM-genus. Condition I holds because all elements have two calling sequence segments. Condition II also holds because there is only one element in each first calling sequence segment and all elements of each second calling sequence segment refer to the genus FLOW. To fulfill condition III all called genera of each first calling sequence segment would have to be the same (this condition holds for each second calling sequence segment as shown when checking condition II). So the new question is whether the genera SUP and DEM can be unified in one genus, i.e. is there generic similarity between all elements of SUP and DEM? Condition I holds because of only one calling sequence segment in each element. So condition II must hold, too. To check condition III successfully all elements of PLANT and CUST would have to be in the same genus. The new question is now: Is that possible and would that be good? But this question cannot be answered from the methodology because PLANT and CUST are primitive entities for which generic similarity is not valid. There is no methodological support if such a collection would make sense. In other words: A unified genus with all elements of T:SUP and T:DEM depends *only* on the definition of the primitive entities in PLANT and CUST. A single genus called LOC (for locations) would also be possible. The epistemological critique is that the definition of primitive entities has got effects and implications that cannot be seen and estimated in such an early phase in the modeling process since one would start modeling the primitive entities first. And additionally, there are no rules how to group elements into primitive entities genera.

genus	Element	calling sequence
Sup	Dal_Sup	(Dal)
Sup	Chi_Sup	(Chi)
Dem	Pitts_Dem	(Pitts)
...		
Link	Dal_Pitts	(Dal; Pitts)
...		
Flow	Dal_Pitts_Flow	(Dal_Pitts)
...		
Cost	Dal_Pitts_Cost	(Dal_Pitts)
...		
$	Total_Cost	(Dal_Pitts_Flow, Dal_Atl_Flow, Dal_Clev_Flow, Chi_Pitts_Flow, Chi_Clev_Flow; Dal_Pitts_Cost, Dal_Atl_Cost, Dal_Clev_Cost, Chi_Pitts_Cost, Chi_Clev_Cost)
T:Sup	Dal_Sup_Test	(Dal; Dal_Pitts_Flow, Dal_Atl_Flow, Dal_Clev_Flow)
T:Sup	Chi_Sup_Test	(Chi; Chi_Pitts_Flow, Chi_Clev_Flow)
T:Dem	Pitts_Dem_Tes	(Pitts; Dal_Pitts_Flow, Chi_Pitts_Flow)
T:Dem	Atl_Dem_Test	(Atl; Dal_Atl_Flow)
T:Dem	Clev_Dem_Test	(Clev; Dal_Clev_Flow, Chi_Clev_Flow)

Subset of model elements, their calling sequences and genera:

Fig. 2: subset of model elements

Since a methodology should have rules about the

application of theory, guidelines are needed for the modeling *process*. SM gives no hints for a starting point, nor does it make suggestions what strategy one should or even could use for model building. No remarks are given how to make refinements or which points should be of interest in different modeling phases. Even modeling phases are not defined. Summarizing these points there is no support for the *process* of modeling in SM.

The discussion of SM is closed with a general epistemological appreciation. SM "is intended as a *lingua franca* within which model classes from a wide variety of modeling paradigms can be expressed – much as English is so used, ..." [10, p. 11]. As the discussion above has shown SM has some gaps in covering different modeling paradigms. This is the result of the underlying interpretation pattern that models are definitional systems. Under epistemological considerations these limitations should have worked out more clearly so that there were statements about questions which kind of models are not covered by SM. But SM as a whole is quite near to what can be called a modeling methodology refering to the characteristics that a modeling methodology must have. SM is meta-theoretic as well as normative. There are some weak points in dealing with methods of its applications. It is fully solution-oriented, quite well problem-oriented, but has difficulties in fulfilling user orientation. The fact that SM problem statements and graphs are found quite abstract leads even to the recommendation to use SM as an *internal* representation scheme for computers [16]. This shows a real weak methodological point of SM. On the other hand most of the eight listed features of a new generation of modeling systems are covered by SM and there are many interesting ideas for future extensions [12].

3. Modeling by Construction

Modeling by Construction is a "framework for a step-by-step conception and formulation of general paradigm-free models for decision support in an easy-to-use and problem-driven manner" [5, p. 178]. It is based on two general core ideas. First, because a problem which occurs in reality is ex definitione not well-structured [1] a model that shall function as a problem description cannot be only a mapping of reality. So model building must be viewed as a creative act which needs constructive achievement (that is where the name comes from: modeling = construction of a model). Second, problems are based upon a situation that can be described as a conglomeration of facts, conditions and also developments. "Problems are conceptual entities or constructs" [18, p. 1491]. So one may dare to say problems consist of somehow distinguishable elements. Therefore it seems to be possible to handle the complexity of problem solving by dividing problems in parts that have less complexity.

This leads to the definition of six general types of problem elements: variables, relations, dimensions, time, grades of determinism, and aspects (i.e. abstract concept of a special part of the problem like e.g. interpretation patterns, basic conditions or even other aspects). A model that functions as a problem description omits some fragments of the problem and uses simplifications while, on the other hand, adding structure. The resulting imprecision and generalization are acceptable only if independences, which restrict negative side effects of abstraction, can be identified. Therefore MbC is based on a theoretical 'concept of independence' [6] which is the foundation of the definition of eight modeling levels likewise the OSI communication model: meta level, means-end level, entity level, detail level, vagueness / ambiguousness level, specification level, variable level, and instance level. These levels can function as components of the modeling process and can be grouped into a semantic scheme, a conceptual scheme, and a subtly differentiated scheme. [5] gives an overview of the MbC modeling levels and their corresponding independences.

MbC defines *model element types* for every modeling level that must not be confused with the six general types of *problem* elements mentioned above. The types (in *italic*) are the following: *Problem name*, *symptom*, *basic condition*, and *interpretation pattern* can be used actively (i.e. to be inserted) on the meta level. They play a declarative role and have to be operationalized later on. *Goal* and *means* are the active model element types of the means-end level, so the well-known means-end-conception can be integrated directly in the modeling process. On the entity level *decision* and *entity* but also *description rule*, *taxonomical rule*, and *conditional rule* are the actively usable types. One can construct complex relations adequately with help of the rule elements. *Simple index*, *derivative index*, *relational index*, and *hierarchical index* can be used actively on the detail level, while *accidentalness*, *risk*, and *fuzziness* are able to be used on the vagueness / ambiguousness level for modeling soft information. Model element types on the specification level, variable level, and instance level are the usual one that are used in OR. They are not defined in MbC to make the methodology more flexible for technical reasons. There are also precise definitions for the types of relation between the above enumerated model element types.

MbC defines a graphical modeling language as can be seen examplified in figure 4. There are also methodological rules as suggested by Popper. MbC contains elements for control of the modeling procedure, too. Both, top-down- and bottom-up-modeling is possible and even combineable. Additionally, MbC supplies elements for the graph-based modeling process called 'elementary construction primitives' that are defined by an identifier, a source graph, a sink graph, a restriction to its application, and an optional description. This allows a graph-grammar-based implementation which advantages are demonstrated in [14].

Since MbC has its foundations in the field of problem theory (with the help of the concept of independence) it can be seen as fully *problem-oriented*. Looking at *user-orientation* MbC comes quite near to it. E.g. one can find well-known concepts like goal definition, using ontological constructs or the means-end-conception. Subjectivity is an explicite element of MbC (e.g. the interpretation patterns). No deep mathematical skills are needed to build a model with the MbC methodology. The graphical modeling language make the communication very easy. And last but no least the different abstractional levels can support the inter-personal usage of models. But on the other hand MbC is a rather complex methodology. Ten methodological rules, 25 types of model elements, six elementary construction primitives, makes it possibly rather complicated. MbC requires relative strong efforts

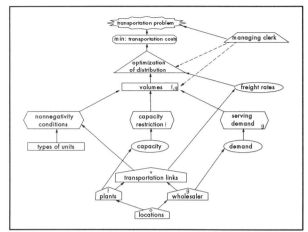

Fig. 4: MbC transportation model on the detail level

to learn it. This is a result of the broad applicability and the strong theoretic foundations.

There is some justified critique on the third main requirement. MbC needs to become more solution-oriented through some extensions. Until now there is no implementation for the whole methodology as a modeling environment. The most concrete levels (particularly the subtly differentiated scheme) must become more specified. Models on these levels cannot be paradigm-free. Modeling by Construction cannot give a guarantee that any specific model has got a calculateable solution. Questions of suitable data management must also be answered to evaluate the practical usability. But this third requirement should not be emphasized too much. Albert Einstein had answered the question what he would do to save world within only one hour: 'I would spend 55 minutes *defining* the problem and then only five minutes solving it.'

MbC has got the three main characteristics of a modeling methodology. It is meta-theoretic because of its methodological rules, it is normative because of the two core ideas it is based on, and it deals with methods of the application of the methodolgy. Using Geoffrion's features of a modeling system as a measure it is obvious from the discussion above that MbC can fully serve all of them but two. These are 'integrated data management' which probably can be integrated with the help of object-oriented database theory. And it is not clear how 'immediate expression evaluation' can be guaranteed. But this feature can only apply to an instantiated model, so it is not essential in the methodological point of view.

4. Outlook

The discussion has shown that there is much work left for improving the process of modeling. The best point to start is to search for weak points in the existing approaches. Based on this knowledge a step-wise refinement of methodologies will be possible. Since SM has its strength in solution orientation and MbC fulfills the requirements of problem and user orientation it seems to be worthy to try the integration of both methodological approaches. SM (with some extensions) can probably serve as a theoretical basis for MbC's subtly differentiated scheme. This research which is currently in progress has to be completed by implementation of prototypes of modeling environments.

5. References

[1]   Agre, G.: The concept of problem; in: Educational Studies 13 [1982] 2, pp. 121 - 142.
[2]   Angehrn, A. A. [1991]: Modeling by Example: A link between users, models and methods in DSS; in: European Journal of Operational Research 55 [1991], pp. 296 - 308.
[3]   Basadur, M. / Ellspermann, S. J. / Evans, G. W.: A new methodology for formulating ill-structured problems; in: OMEGA 22 [1994] 6, pp. 627 - 645.
[4]   Dresbach, S.: Modeling by Construction – Entwurf einer Allgmeinen Modellierungs-methodologie für betriebswirtschaftliche Entscheidungen; Aachen 1996.
[5]   Dresbach, S.: Modeling by Construction: A new methodology for constructing models for decision support; in: Proc. of the 29th Hawaii Int. Conf. Sys. Science, Vol. II, Washington - Brussels - Tokyo, 1996, pp. 178 - 187.
[6]   Dresbach, S.: The concept of independence – A step towards a general modeling methodology; Working Paper 4/95, University of Cologne (Germany), Cologne 1996.
[7]   Geoffrion, A. M.: An Introduction to Structured Modeling; in: Management Science 33 [1987], pp. 547 - 588.
[8]   Geoffrion, A. M.: Computer-based modeling environments; in: European Journal of Operational Research 41 [1989], pp. 33 - 43.
[9]   Geoffrion, A. M.: The formal aspects of Structured Modeling; in: Operations Research 37 [1989] 1, pp. 30 - 51.
[10] Geoffrion, A. M.: Integrated Modeling Systems; in: Computer Science in Economics and Management 2 [1989] 1, pp. 3 - 15.
[11] Geoffrion, A. M.: The SML language for Structured Modeling; in: Operations Research 40 [1992] 1, pp. 38 - 75.
[12] Geoffrion, A. M.: Structured Modeling: Survey and Future Research Directions; in: ORSA CSTS Newsletter 15 [Spring 1994] 1, pp. 1, 11 - 20.
[13] Geoffrion, A. M.: Structured Modeling; in: Gass, S.I. / Harris, C.M. (eds.): Encyclopedia of Operations Research and Management Science; Boston et al. 1996, pp. 652 - 655.
[14] Jones, C. V.: An introduction to graph-based modeling systems, in: ORSA Journal on Computing; part 1: 2 [1990] 2, pp. 136 - 151, and part 2: 3 [1991] 3, pp. 180 - 206.
[15] Lenard, M. L.: Fundamentals of Structured Modeling; in: Mitra, G. (ed.): Mathematical models for decision support; Berlin - Heidelberg 1988, pp. 695 - 713.
[16] Murphy, F. H. / Stohr, E. A. / Asthana, A.: Representation schemes for linear programming models; in: Management Science 38 [1992] 7, pp. 964 - 991.
[17] Popper, K. R.: The logic of scientific discovery; London 1959.
[18] Smith, G. F.: Towards a heuristic theory of problem structuring; in: Mangement Science 34 [1988] 12, pp 1489 - 1506.

BIRGER FUNKE AND HANS-JÜRGEN SEBASTIAN
An advanced modeling environment based on a hybrid AI-OR approach

1. Introduction

In this paper we want to introduce knowledge-based model building. In particular we describe the KONWERK tool-box[3] and show the modeling of optimization tasks with KONWERK. We use the model of the Nitra River Case, which is described in [6] as an example of the field of environmental research problems. Solving a real world optimization task generally consists of three main parts:

- modeling the task in a form which can be processed by a machine
- solving a problem assigned to the resulting model
- interpreting and explaining the results.

In many cases the modeling task is the most time-consuming part. The modeling process can be considered as the "transformation" of a given problem from a human formulation into a form which can be processed and solved by a machine (computer program). Why is this transformation so difficult and time-consuming? First of all, modeling an optimization problem generally includes at least three steps of transformation:

- from the real world problem into a human formulation (how experts think)
- from the human formulation into an analytic model which is used by methods of the Operations Research (OR model)
- from the OR model into a computer language or into a standard format file for a chosen solver program

Unfortunately, human experts' descriptions are often difficult to extract and sometimes incomplete. Experts often cannot describe and explain their knowledge in a proper way (even if they really want to do it[1]). Furthermore, the structure of the real world problem often does not match directly with the structure of an OR model. Hence, in many cases the

---

1. In the field of expert system developement the knowledge acquisition process plays an important role in the whole process of developing a decision support system. There are several techniques developed for this task like heuristic or hierarchic classification methods, different interview techniques and indirect acquisition techniques (see [5], [9]).

transformation includes reformulation of the model in a different way and sometimes even relaxation of restrictions and so on. Modeling is in general not just a straight forward transformation but often some kind of art.

People who want to develope the model

- have to know the specific application domain very well or
- need good experiences in knowledge acquisition methods
- need to be familiar with the properties of different methods of OR
- must know the available OR-tools and other software and
- usually also need to be experienced in writing computer programs in at least one computer language.

Summarizing, people who want to build a model for a real world optimization problem face many different tasks at the same time. The combination of all these subtasks makes the modeling process as a whole very complex and therefore, tough. A model builder, who wants to develop an OR model accurately representing all important facts of the real world, almost needs to be an allround genie. At the end of a complex model building process the expert himself will often not be able to read and understand the final model, which is for instance a computer program or a (low level) standard format like an **MPSX** file. Who can check whether this model really expresses the same one the expert has in mind? Can we trust the model outcomes?

This paper aims to show how the use of knowledge-based modeling techniques can significantly ease and speed up the model building process of real world optimization tasks. The resulting model can be easier read, checked and modified.

2. Brief Description of the Nitra River Case

In this section we give a very brief introduction into the Nitra River Case ([6], [7]). This application will then be used as a reference example in order to explain modeling with KONWERK and to show examples for representation of different types of knowledge.

In the Nitra River Case we consider a river basin consisting of a main river (Nitra) and its tributaries. The tributaries themselves may also have tributaries. Along the rivers there are different types of nodes. These are monitoring points, emission points and other types of nodes which are not important for the less detailed description of the model in this paper (see Figure 1). In order to reduce the pollution of the river and to improve the water quality of the river system there are treatments to be implemented at the emission points. For each emission point a set of different treatment technologies is considered. Only one treatment technology can be applied at an emission point.

There are 6 objectives considered:

- Maximize the minimum concentration of oxygen at the monitoring points of the river system (DOmin)
- Minimize the maximum concentration of carbonaceous oxygen and nitrogenous oxygen at the monitoring points (BODmax)
- Minimize the maximum concentration of ammonia at the monitoring points (NH4max)
- Minimize total investment costs of all applied treatments
- Minimize total operating and managment costs of all applied treatments
- Minimize total annual costs of all applied treatments.

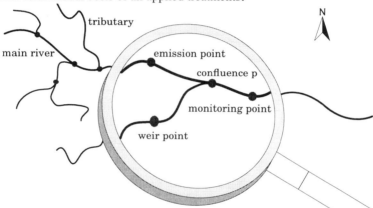

Figure 1: River System

The decision variables are the treatment technologies which are to be selected for the considered emission points. The Nitra River Case can be represented as a linear mixed integer problem (MIP). A more detailed description of the Nitra River Case is in [6].

3. Knowledge Base of KONWERK

The knowledge base of KONWERK is divided into distinct types of knowledge about the domain, one or several objectives and the intended solution procedure. In the following subsections we will introduce the most important parts of the knowledge in detail using simplified examples of the Nitra Case.

3.1. Hierarchy of Concepts:

The first step of building the knowledge base of a specific domain is always the definiton of concepts. All concepts the modeler wants to work with must become part of the taxonomic hierarchy. The root concept of the taxonomic hierarchy is predefined and named object. Any further concept of the domain is more specific than object and consequently at a deeper level of the taxonomic hierarchy. Figure 2 shows a (simplified) taxonomic hierarchy of the Nitra River Case.

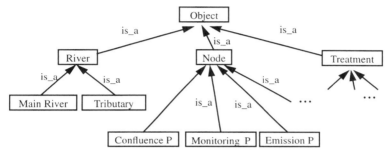

Figure 2: Hierarchy of Concepts

The object descriptions of a KONWERK knowledge base are represented by frames. The syntax used for the knowledge base of KONWERK is close to the syntax used in CLOS[2]. In this paper definitions of the knowledge base will be shown in their original form. However, the definition of concepts can be supported by a graphical user interface. The user (the model builder) does not need to learn the syntax shown in this paper. The following example is the definiton of the concept river:

```
(def-concept
    :name           river
    :super          object
    :parameter      ((length[0 100km])
                    (node_num [0 100])...) )
```

This definition can be read as follows: the concept river *is a*n object (the super concept is object) which has two parameters length and node_num. The length of any river of this domain is restricted to be between 0 and 100km. The parameter node_num can hold any number between 0 and 100.

The taxonomic hierarchy has to be constructed by defining all new concepts (in our example the concepts river, main_river, tributary, node ,...). The definition of a concept has

---

2. CLOS is the object oriented extension of the computer language Lisp.

to consist of a name and the super concept. Furthermore, it is possible to assign attributes to the objects (e.g. the length of the river or the number of nodes along the river). The super concept defines the position of the new concept in the taxonomic hierarchy.

3.2. Compositional Hierarchies:

In the taxonomic hierarchy only the concepts themselves are defined but not the relations between these concepts. A compositional hierarchy expresses the compositional structure of a concept. In the Nitra River Case we have to declare that rivers may have tributaries (see Figure 3), each node of our domain is belonging to a river (see Figure 4) and each

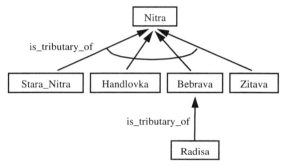

Figure 3: Tributaries of River Nitra

treatment object is a treatment of exactly one emission point. The implementation of compositional hierarchies has to be done in two steps. The first step is the definition of the compositional relations:

```
(def-relation                          (def-relation
    :name     has_nodes                    :name     has_tributaries
    :inverse  is_node_of                   :inverse  is_tributary_of
    :type     has_parts_relation)          :type     has_parts_relation)
```

Now, the definition of any object can include a declaration of a has_nodes or a is_node_of (has_tributaries or a is_tributary_of) relation to another concept. The second step of constructing compositional hierarchies is to add this information to the definition of the river as follows:

```
(def-concept
    :name       river
    :super      object
    :parameter  ((length[0 100km]) (node_num [0 100]))
    :relations  ((has_nodes [(a node) 0 100]})
                 (has_tributaries [(a tributary) 0 10])))
```

Now we have stated that a river may have from zero to 100 nodes. At the same time it may have between zero and ten tributaries which belong to the river.

The parameter node_num is intended to hold the number of nodes which are at a river. However, node_num is just a name and the knowledge that the parameter node_num has to be the number of nodes of the river is not explicitly declared. In order to express such relations between parameters of an object or between parameters of different objects we can use conceptual constraints, which will be introduced now.

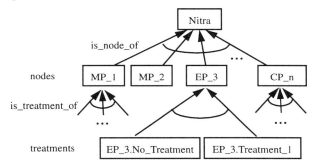

Figure 4: Compositional Hierachies of River Nitra

3.3. Conceptual Constraints

The conceptual constraints of KONWERK can express relations between different parameters of one, two or more objects of the domain. The constraints are called conceptual because they are defined at the level of concepts. During the computations these constraints will be applied and taken into account for all objects of concepts which are mentioned in the definiton of the conceptual constraint[3]. In general, the definition of conceptual constraints consists of a name, a set of object patterns and a formula or a set of formulas. The set of patterns is used to identify and filter the objects which have to fulfill the defined relation. The formula describes the relation itself. A constraint, guaranteeing that the number of nodes of a river is always equal to its parameter node_num, could be formulated as follows:

---

3. Similar to the use of the notion "object" in object oriented computer languages (where the instances of classes are called objects) we use "object" for the instances of concepts.

```
(def-conceptual-constraint
    :name      set_node_number
    :patterns  ((?riv :name node))
    :formula   "?riv.node_num = card(?riv.has_nodes)")
```
In this definiton ?riv is a kind of variable which can hold any instance of the concept river. Then an equation for the number in the parameter slot node_num and the cardinal number of entries in the has_nodes slot is formulated. This equation has to be fulfilled for all instances of the concept river. More interesting examples can be found in [2].

3.4. Conceptual Objectives

The definition of objectives is very similar to the definition of constraints. The only difference is the need to specify the desired direction of optimization (minimize or maximize). As an example we consider the problem of maximizing the minimum of oxygen concentrations (DOmin) for the monitoring nodes at any river of the system. Let us consider an additional parameter slot DOmin of the concept main river. DOmin is intended to hold an oxygen concentration. It is restricted by conceptual constraints to be less or equal to the oxygen concentration of any monitoring point of the main river and its tributaries. The objective can now be formulated as follows:

```
(def-conceptual-objective
    :name       Maximise_DOmin
    :patterns   ((?mr :name main_river))
    :direction  :max
    :formula    "?mr.DOmin")
```
This definition can be read like the conceptual constraints. The formula of the definition consists of only one term. It is a single expression in this example: the parameter DOmin of the main river.

As we know, in the Nitra Case there are 6 objectives and we have to consider the multi objective task. The objective value of the single objective optimization (maximizing the DOmin parameter) can also be used as the utopia value of DOmin for the reference point method (see [8], [6]). Therefore, we introduce an additional parameter utopia_DOmin for main_river. We define a further constraint which restricts the parameter utopia_DOmin to be less than or equal the DOmin value of main_river

```
(def-conceptual-constraint
    :name      restriction_of_utopia_DOmin
    :patterns  ((?mr :name main_river))
    :formula   "?mr.utopia_DOmin <= ?mr.DOmin")
```
and we replace the previous objective by:

```
(def-conceptual-objective
    :name       Maximise_DOmin
    :patterns   ((?mr  :name main_river))
    :direction  :max
    :formula    "?mr.utopia_DOmin")
```

All six objectives of the Nitra Case (see section 2.) are defined in a similar manner. These objectives can be used for the optimization of single objectives as well as for multi-criteria approaches like the reference point method. However, in order to run an optimization with KONWERK, an optimization task has to be defined.

3.5. Optimization Tasks

The definition of an optimization task in KONWERK allows the user to assign several requests and informations to an intended optimization problem. We consider the computation of the utopia_DOmin value. At first we define a single objective optimization task, but we want to know the utopia value in order to use it as input value for a multi-criteria task later.

```
(def-param-problem
    :name           determine_utopia_DOmin
    :objectives     Maximise_DOmin
    :assign-values  ( :slots          (utopia_DOmin)
                      :upper_bounds   (nadir_DOmin)
                      :lower_bounds   (nadir_BODmax) (nadir_NH4max) ... )
                      (nadir_TAC) (nadir_inv_cost) (nadir_op_cost)) )
```

The optimization task determine_utopia_DOmin uses the previously defined objective Maximise_DOmin. We define the optimization task in a way that only the value of utopia_DOmin will be set to the value which was found by the MIP solver. All the other variables (computed by the solver) will not be stored. This is because we are just interested in the utopia value of DOmin. We do not need to know which policy is leading to this result. The values computed by the MIP solver for nadir values[4] can be used as new upper or lower bounds for the real nadir values. Further entries in the optimization task definition can include a suggestion for a solver to be used, specific parameters for the solver programs or a list of conceptual constraints which have to be taken into account for the optimization. It is possible to define several optimization tasks for one domain. This can be used to generate a sequence of optimizations. In our case, we can calculate the single objective tasks for all objectives and continue with the multi objective task taking into account the

---

4. The nadir parameters of main_river are defined similar to utopia values. For instance, nadir_DOmin is greater than or equal to DOmin.

previously computed results. In order to do so, we have to define in which order the optimization tasks have to be processed in the session. This has to be done by strategies and will be shown in the next subsection.

3.6. Strategies

The original idea of KONWERK is to provide a framework for defining and solving configuration tasks [3]. Configuration processes usually base on the idea of decomposition of complex configuration tasks into smaller subtasks, which are easier to work out. The subtasks may be further decomposed themselves. Strategies are used to control the decision making process. In KONWERK there are four base types of decision tasks distinguished:

- Decomposition of a (complex) object into its parts
- Integration of objects
- Specialization of objects
- Specification of a parameter of an object

The optimization modules of KONWERK add another type:

- Optimization task [1]

KONWERK puts all *single decisions* which have to be performed in the decision process on an agenda. The agenda entries are relating to one of the introduced types of decision tasks. They refer to a particular object, parameter or optimization task. An agenda entry could for instance represent the specialisation of a particular node object or the decomposition step of assigning a treatment object to a has_treatments slot of a nodes treatments of a emission_point.

Strategies specify the decisions for a specific part of the decision process. They can also be used to define an order of these decisions and to indicate a subset of methods, which may be applied to make a single decision [4]. An example and more details about strategies in KONWERK can be found in [2]. Using the strategies it is possible to express even complex methodologies like the Reference Point Method in KONWERK. For the Nitra River Case this method was implemented using existing modules only. Results can be found in [2].

4. Conclusion and Outlook

In this paper we reported on the developement of a knowledge-based model of the Nitra River Case using the KONWERK tool-box. Though KONWERK is not a professional tool and exists as a prototype, the complete model of the Nitra Case could be implemented in a rather short time and adjustment to model changes and the enlargement of the model was easily realized several times. The success of the experiment to use a knowledge-base of an AI tool-box for configuration tasks for modeling and implemention an optimization task can be seen as a hint to the potential of the knowledge-based modeling approach in general. The advantages of knowledge-based modeling can be summarized as follows. The knowledge-based model is

- easy to develope
- easy to read and understand
- easy to change (adjustment to new situations)

Therefore, overcoming some remaining weaknesses of KONWERK, like missing of a good graphical user interface and further development of specific multi-criteria optimization approaches (e.g. Reference Point Method) could lead to an even more effective tool for model building and evaluation. In addition, the integration of knowledge-based model building and developing a core model of a specific domain [6] is a base for further investigations.

5. Literature

[1] B. Funke: Using Optimization methods and Modeling of Optimization Problems (in german), in Wissensbasiertes Konfigurieren: Ergebnisse aus dem Projekt Prokon, A.Günter, Ed. Sankt-Augustin: Infix, 1995
[2] B. Funke, H.-J. Sebastian: Knowledge-based model building with KONWERK, WP-96-105, IIASA, Laxenburg, 1996
[3] A. Günter, H. Dörner, H. Gläser, B. Neumann, C. Posthoff, and H.-J. Sebastian: The PROKON project: Problem specific tools for knowledge-based configuration (in german), PROKON-BERICHT Nr. 1, BMFT Verbundprojekt PROKON, 1991
[4] A. Günter, R. Cunis: Flexible Control in Expert Systems for Construction Tasks, Journal of applied Intelligence 2,369-385, 1992
[5] W. Karbach, M. Linster: Knowledge Acquisition fpr Expert Systems: Techniques, Models, and Softwaretools (in german), Munich, Vienna: Hanser 1990
[6] M. Makowski: Methodology and Modular Tool for Multiple Criteria Analysis of LP Problems, WP-94-102, IIASA, Laxenburg, 1994
[7] M. Makowski, L. Somlyódy, and D. Watkins: Multiple Criteria Analysis for Regional Water Quality Management: the Nitra River Case, WP-95-022, IIASA, Laxenburg, 1995
[8] R.E. Steuer, L.R. Gardiner: Interactive Multiple Criteria Programming: Concepts, Current Status, and Future Directions, in Readings in Multiple Criteria Decision Aid, C.A. Bana E. Costa, Ed. Springer, 1990
[9] A. Wierzbicki: Multicriteria Aid of Intuition in Decision Making, 1994

YOSHITERU NAKAMORI, MINA RYOKE
AND KAZUTAKA UMAYAHARA
Interactive fuzzy modeling system

1. Introduction

Model building is an interactive process in which person-computer or interpersonal communication is needed to greater or lesser degrees. We have been developing an interactive modeling methodology which integrates structural and statistical modeling techniques systematically. The use of the highly interactive computer system based on this methodology is an interactive process in the course of which the subjective expert knowledge can be utilized to a great advantage, and interpersonal communication is greatly enhanced as well.

The already published version called the interactive modeling support system[8] consists of combined modeling techniques using statistical and graph-theoretical approaches, and multi-stage person-computer dialogues. The data analysis facility gives understandable graphical expressions of measurement data so that one can consider model structures before statistical modeling. The structural analysis facility is helpful for organizing one's thinking with respect to the system under study. Additionally, the statistical analysis facility includes stepwise linear modeling, hypothesis testing, residual analysis, partial and total tests, to evaluate the model from the statistical standpoint.

The system has been improved to carry out fuzzy modeling, based on a fuzzy modeling methodology developed in [9], where the main technical contribution is a clustering algorithm that detects fuzzy subsets based on our desire about their shapes, where interaction is very important in creating a balance between continuity and linearity of the data distribution within clusters. After developing a number of fuzzy subsets, linear substructures of the system under study are identified. The second technical contribution in [9] is related to the integration of rules, that is, selection of conditional variables, identification of membership functions, and evaluation of a fuzzy model.

The current system also contains a facility of using a variety of fuzzy clustering methods developed in Dunn[5], Gustafson and Kessel[6], Bezdek et al.[1][2][3], Dave[4], Hathaway and Bezdek[7] and ourselves[10]. In the last paper, an existing approach[7] to the simultaneous determination of a data partition and regression equations is modified in such a way that the shapes of clusters are changed dynamically and adaptively in the clustering process. Another proposal in [10] is related to the identification of premises of a fuzzy model. The membership functions are defined on the principal axes of clusters and optimized by an nonlinear optimization method. These techniques are integrated into the modeling support system which is now working under the environments of both WINDOWS and UNIX.

In all these works, one of the unsolved problems is the detection of clusters of different dimensionalities. This paper proposes a new objective function for detecting clusters of different dimensionalities, and then a new type of fuzzy model defined by elliptic membership functions. Before introducing this new approach, this paper presents a modified version of already proposed method in [10], which is related to the determination of adaptive clustering parameters in the process of simultaneous determination of clusters and regression models.

2. Interactive fuzzy modeling

The inputs to the modeling system are a set of variables, a binary relation between variables, and a set of data. The outputs from the system are structural models, statistical models, and fuzzy models. Main facilities are data screening and transformation, graphical expressions of data and relations, classical regression analysis, simulation, linear programming, clustering and fuzzy modeling.

The model structure should be considered first. A crisp relation or a fuzzy relation between variables should be defined before going into quantitative modeling. The system has facilities to develop crisp or fuzzy structural models based on pairwise relationships between variables.

The clustering stage has several important clustering algorithms including the hyperellipsoidal clustering method[9] which was developed by ourselves. The standard and extended fuzzy clustering algorithms are also available. This clustering stage together with the modeling stage help us to develop fuzzy models.

At the simulation stage, we can carry out fuzzy simulation to see the behavior of the developed fuzzy model. But, often, especially in future prediction, we have to develop fuzzy rules without any numerical data. This stage helps us to consider future relationships of variables.

The fuzzy prediction model is a nonlinear model which consists of a number of rules. Instead of its original form[11], this paper treats the following type of rules:

$$\text{Rule } R_i: \quad \text{if } \boldsymbol{z} \text{ is } F_i, \text{ then } \boldsymbol{y} = \boldsymbol{g}_i(\boldsymbol{x};\Omega) \equiv \boldsymbol{a}_{i0} + A_i\, \boldsymbol{x}. \tag{1}$$

Here, $\boldsymbol{x} = (x_1, x_2, \cdots, x_s)^\top$ is the vector of consequence variables, $\boldsymbol{z}=(z_1, z_2, \cdots, z_t)^\top$ is the vector of premise variables, and $\boldsymbol{y} = (y_1, y_2, \cdots, y_r)^\top$ is the vector of response variables. Often, there is an intersection between two variable sets $\{x_1, x_2, \cdots, x_s\}$ and $\{z_1, z_2, \cdots, z_t\}$. F_i denotes a fuzzy subset with the membership function $f_i(\boldsymbol{z})$ which has premise parameters to be identified. The regression parameters $\Omega = \{\boldsymbol{a}_{i0} \in \boldsymbol{R}^r,\ A_i \in \boldsymbol{R}^{r\times s};\ i = 1, 2, \cdots, c\}$ are called consequence parameters. The prediction of $\boldsymbol{y}$ is given by

$$\hat{\boldsymbol{y}} = \frac{\sum_{i=1}^{c} f_i(\boldsymbol{z}^*)\cdot \boldsymbol{g}_i(\boldsymbol{x}^*)}{\sum_{i=1}^{c} f_i(\boldsymbol{z}^*)}, \tag{2}$$

where x^* and z^* denote actual inputs, and c is the number of rules.

3. Classification and regression

One of the recent interests in the field of fuzzy clustering is the simultaneous determination of a fuzzy partition of a given data set and parameters of assumed models of different shapes.

Let $\{(x_1, y_1, z_1), \cdots, (x_n, y_n, z_n)\}$, $x_k \in R^s$, $y_k \in R^r$, $z_k \in R^t$, be the set of standardized data corresponding to consequence, response and premise variables, respectively. The clustering is carried out in the space defined by the union of all variables.

Consider the well-known fuzzy partition matrix U with u_{ik} for the (i,k)-entry, satisfying

$$0 \leq u_{ik} \leq 1, \quad i = 1, 2, \cdots, c, \quad k = 1, 2, \cdots, n \tag{3}$$

$$0 < \sum_{k=1}^{n} u_{ik} < n, \quad i = 1, 2, \cdots, c, \tag{4}$$

$$\sum_{i=1}^{c} u_{ik} = 1, \quad k = 1, 2, \cdots, n. \tag{5}$$

Define the degree of fitness of the k-th data to the i-th model by

$$E_{ik}(\Omega) = \|y_k - g_i(x_k; \Omega)\|^2, \quad i = 1, 2, \cdots, c, \quad k = 1, 2, \cdots, n. \tag{6}$$

The objective function of the fuzzy clustering is then defined by

$$J_{fcrm}(U, \Omega) = \sum_{k=1}^{n} \sum_{i=1}^{c} (u_{ik})^q E_{ik}(\Omega), \tag{7}$$

where $q(>1)$ is the smoothing parameter indicating the degree of fuzziness. This formulation is given in Hathaway and Bezdek[7], and the method is called the fuzzy c-regression models (FCRM).

This approach provides a fuzzy partition of the given data set, and at the same time, a set of regression models corresponding to the data partition. But, since this method does not take into account the data distribution, it is not necessarily appropriate for fuzzy modeling.

In this paper, the FCRM is modified based on the idea in Dave[4]. The modified version can be called the adaptive fuzzy c-regression models (AFCR). Denote the set of centers of clusters in the space of premise variables by $V = \{\bar{z}_1, \cdots, \bar{z}_c\}$, which are also parameters to be determined in the clustering:

$$\bar{z}_i = \frac{\sum_{k=1}^{n}(u_{ik})^q z_k}{\sum_{k=1}^{n}(u_{ik})^q}, \quad i = 1, 2, \cdots, c. \tag{8}$$

Introduce an objective function that takes into account a balance between the minimization of regression errors and the minimization of variances within clusters:

$$J_{afcr}(U, \Omega, V, \alpha_1, \cdots, \alpha_c, \eta_1, \cdots, \eta_c) = \sum_{k=1}^{n} \sum_{i=1}^{c} (u_{ik})^q L_{ik}(\Omega, V, \alpha_i, \eta_i). \tag{9}$$

Here, the function $L_{ik}(\Omega, V, \alpha_i, \eta_i)$ is defined by

$$L_{ik}(\Omega, V, \alpha_i, \eta_i) = \alpha_i \, E_{ik}(\Omega) + (1 - \alpha_i) \, \eta_i \, D_{ik}(V), \tag{10}$$

and $D_{ik}(V)$ is the square distance between $\bar{z}_i$ and the k-th data point z_k in the space of premise variables:

$$D_{ik}(V) = \|z_k - \bar{z}_i\|^2, \quad i = 1, 2, \cdots, c, \quad k = 1, 2, \cdots, n. \tag{11}$$

The parameters α_i ($0 \leq \alpha_i \leq 1$) are changed in the clustering process adaptively as in Dave[4]. The parameters η_i play a role of making a balance between the absolute values of the first and second terms in the objective function. Unlike in Dave's adaptive fuzzy c-elliptotypes clustering algorithm, here E_{ik} and D_{ik} are distance measures defined over different spaces, hence this factor is needed. The appropriate values of η_i depend on a given data set. One idea is that they will be determined by the ratio of the data spread in two spaces.

Let $\lambda_{i1p}, \lambda_{i2p}, \cdots, \lambda_{i(s+1)p}$ be the eigenvalues of the fuzzy scatter matrix S_{ip} calculated by using the data in the space of a response variable y_p and all consequence variables:

$$S_{ip} = \sum_{k=1}^{n} (u_{ik})^q \, (w_{kp} - \bar{w}_{ip})^{\top}(w_{kp} - \bar{w}_{ip}), \quad \bar{w}_{ip} = \frac{\sum_{k=1}^{n} (u_{ik})^q \, w_{kp}}{\sum_{k=1}^{n} (u_{ik})^q}. \tag{12}$$

Let $\mu_{i1}, \mu_{i2}, \cdots, \mu_{it}$ be the eigenvalues of the scatter matrix S'_i calculating by using the data in the space of premise variables:

$$S'_i = \sum_{k=1}^{n} (u_{ik})^q \, (z_k - \bar{z}_i)^{\top}(z_k - \bar{z}_i), \quad \bar{z}_i = \frac{\sum_{k=1}^{n} (u_{ik})^q \, z_k}{\sum_{k=1}^{n} (u_{ik})^q}. \tag{13}$$

The parameter α_{ip} is defined by using the variances of eigenvalues:

$$\alpha_{ip} = \frac{Var\{\lambda_{ijp}\}}{Var\{\lambda_{ijp}\} + Var\{\mu_{ij}\}}, \quad i = 1, 2, \cdots, c, \quad p = 1, 2, \cdots, r. \tag{14}$$

If the variance of eigenvalues in the space where we are building a linear model is large, there is a chance to develop a good linear model. In this case we give added weight to the first term of the function L_{ik}. In order to obtain better models for all objective variables, the minimum one is used for α_i:

$$\alpha_i = \min\{\alpha_{i1}, \alpha_{i2}, \cdots, \alpha_{ir}\}, \quad i = 1, 2, \cdots, c. \tag{15}$$

On the other hand, η_{ip} is defined by

$$\eta_{ip} = \frac{\sum_{j=1}^{s+1} \lambda_{ijp}}{\sum_{j=1}^{t} \mu_{ij}}, \quad i = 1, 2, \cdots, c, \quad p = 1, 2, \cdots, r. \tag{16}$$

The sum of eigenvalues is one of the indicators to show how the data are scattered. Since we want to develop each linear model in a smaller region, the maximum one is used for η_i:

$$\eta_i = \max\{\eta_{i1}, \eta_{i2}, \cdots, \eta_{ir}\}, \quad i = 1, 2, \cdots, c. \tag{17}$$

The clustering algorithm is omitted in this paper because it is quite similar to the fuzzy c-means method.

4. Different dimensional clusters

One of the unsolved problems is the detection of clusters of different dimensionalities. In this section, a new objective function for detecting clusters of different dimensionalities is proposed.

Let $\{x_1, x_2, \cdots, x_m\}$ be the set of variables, $\{x_{1i}, x_{2i}, \cdots, x_{ni}\}$ be the standardized data set of x_i, and $\boldsymbol{w}_k$ be the k-th data vector of all variables:

$$\boldsymbol{w}_k = (x_{k1}, x_{k2}, \cdots, x_{km})^T, \quad k = 1, 2, \cdots, n. \tag{18}$$

The problem is to obtain c fuzzy clusters $C_1, \cdots, C_c$ by partitioning the data set $\{\boldsymbol{w}_1, \cdots, \boldsymbol{w}_n\}$. Denote the cluster centers by $\boldsymbol{v}_1, \cdots, \boldsymbol{v}_c$.

We propose the following objective function:

$$J_{fvd}(U,V) = \sum_{k=1}^{n} \sum_{i=1}^{c} (u_{ik})^q E_{ik}(\boldsymbol{v}_i) \tag{19}$$

where

(a) $E_{ik}(\boldsymbol{v}_i) = \beta_i^0 \dfrac{G_{ik}^0(\boldsymbol{v}_i)}{m} + \sum_{r=1}^{m-1} \beta_i^r \dfrac{G_{ik}^r(\boldsymbol{v}_i)}{m-r},$

(b) $G_{ik}^r(\boldsymbol{v}_i) = \begin{cases} \|\boldsymbol{w}_k - \boldsymbol{v}_i\|^2, & r = 0, \\ \|\boldsymbol{w}_k - \boldsymbol{v}_i\|^2 - \sum_{j=1}^{r} |<\boldsymbol{w}_k - \boldsymbol{v}_i, \boldsymbol{e}_{ij}>|^2, & r = 1, 2, \cdots, m-1, \end{cases}$

(c) $\beta_i^r = \dfrac{\gamma_i^r}{\sum_{r=0}^{m-1} \gamma_i^r}$,

(d) $\gamma_i^r = \begin{cases} (\lambda_{im})^l, & l \geq 0, \quad r = 0 \\ (\lambda_{ir} - \lambda_{i,r+1})^l, & l \geq 0, \quad r = 1, 2, \cdots, m-1. \end{cases}$

The ideas of using the function (19) are summarized as follows:

- The square distances from the data point w_k to the linear varieties with all dimensionalities V_i^r ($r = 1, 2, \cdots, m-1$) are considered as in (b) to detect clusters with different dimensionalities.

- In order to compare distances from a point to linear varieties with different dimensionalities, the square distance G_{ik}^r is divided by $m - r$ which is the dimensionality of the orthogonal complement of the linear variety V_i^r as shown in (a).

- The weight β_i^r in (c) is defined by using the difference between λ_{ir} and $\lambda_{i,r+1}$ as shown in (d). If β_i^r is relatively large, the possibility of the dimensionality of the cluster C_i being r is relatively high.

- The exponent l in (d) is called the degree of linearity. If $l = 0$, the shapes of all clusters become vague. On the other hand, if $l = 1$, the shapes of clusters are expected to reflect the data distribution. A real number greater than one can be used for l to stress the linearity in the data distribution.

The clustering is a process to determine

$$\{u_{ik}, \; \boldsymbol{v}_i; \; i = 1, 2, \cdots, c, \; k = 1, 2, \cdots, n\}$$

that minimize the objective function with assumed parameters c, q, and l. The algorithm is similar to other fuzzy clustering algorithms.

5. Elliptic fuzzy models

The above fuzzy clustering provides the information about locations and shapes of clusters in terms of the cluster centers, and the eigenvalues and eigenvectors of the scatter matrices. Here, we consider a type of fuzzy model consisting of obtained clusters as rules.

First, the cluster C_i is identified with the rule R_i of the fuzzy model, and the rule R_i is considered as a fuzzy subset F_i with the membership function defined by

$$f_i(\boldsymbol{z}) = \exp\left\{-\sum_{j=1}^{m} \dfrac{1}{2\lambda_{ij}} \left[(\boldsymbol{z} - \boldsymbol{v}_i)^\top \boldsymbol{p}_{ij}\right]^2\right\}, \quad \boldsymbol{z} \in \boldsymbol{R}^m. \tag{20}$$

This model does not have the concept of input and output *a priori*. But, if the values of some variables are given, the model predicts the possible distributions of the rest of variables by respective membership functions. For instance, the possible distribution of a variable x_j is defined by the membership function $f_i(z)$ of the fuzzy subset F_{ij} on the x_j axis.

To put it concretely, let X be the set of variables whose values are given, and x^* be the input vector. Let Y be the set of variables whose values are not given, and $|Y|$ be the number of elements in Y. The membership function of the fuzzy subset $F_{ij}(x^*)$ of the variable $y_j \in Y$ is defined by

$$f_{ij}(y_j \mid x^*) = \sup_{a \in R^{|Y|-1}} f_i(y_j, a, x^*). \tag{21}$$

Then, $F_{ij}(x^*)$ can be regarded as a fuzzy output from the rule R_i corresponding to the input x^*. Here, a is the vector consisting of the variables in Y except y_j. The above function maps the maximum values of $f_i(z)$ onto the respective axes of variables in Y except y_j, that is, for any value of y_j, it calculates the maximum membership value corresponding to the input x^*.

While the output of a rule is fuzzy, the output of the fuzzy model $\hat{y}_j(x^*)$ can be given by the weighted average of the centers of fuzzy output $\hat{y}_{ij}(x^*)$ by the membership values of the centers $w_i(x^*)$. That is, using

$$\hat{y}_{ij}(x^*) = \arg\sup_{y_j} f_{ij}(y_j \mid x^*), \qquad w_i(x^*) = \sup_{y_j} f_{ij}(y_j \mid x^*), \tag{22}$$

we define

$$\hat{y}_j(x^*) = \frac{\sum\limits_{i=1}^{c} w_i(x^*) \cdot \hat{y}_{ij}(x^*)}{\sum\limits_{i=1}^{c} w_i(x^*)}. \tag{23}$$

Since this model does not distinguish between the premise and the consequence parts, in addition, this model does not use regression models, it is not necessary to carry out selection of premise and consequence variables. Because of this characteristic, the distinguish between simulation and optimization vanishes.

6. Concluding remarks

In this paper, we presented the detection of linear substructures in the context of fuzzy modeling. The techniques used are fuzzy clustering, principal component analysis and regression analysis. We introduced two methods: the first is the adaptive fuzzy clustering and regression, which determines clusters and linear models in a fixed dimensional space simultaneously and adaptively, the second is the detection of clusters of different dimensionalities. Then, we introduced an elliptic-type fuzzy model and a simulation technique based on this model. This model is not the final model, but we can see the relation of variables in a high-dimensional space.

References

[1] Bezdek, J C (1981) *Pattern Recognition with Fuzzy Objective Function Algorithms*, Plenum Press, New York.

[2] Bezdek, J C et al. (1981) Detection and characterization of cluster substructure I. linear structure: fuzzy c-lines, *SIAM J. Appl. Math.*, Vol. 40, No. 2, pp. 339-357.

[3] Bezdek, J C et al. (1981) Detection and characterization of cluster substructure II. fuzzy c-varieties and convex combinations thereof, *SIAM J. Appl. Math.*, Vol. 40, No. 2, pp. 358-372.

[4] Dave, R N (1990) An adaptive fuzzy c-elliptotype clustering algorithm, Proc. NAFIPS 90: Quater Century of Fuzziness, Vol. I, pp. 9-12.

[5] Dunn, J (1974) A fuzzy relative of the ISODATA process and its use in detecting compact well-separated clusters, *J. Cybernetics*, Vol. 3, pp. 32-57.

[6] Gustafson, D E and Kessel, W C (1979) Fuzzy clustering with a fuzzy covariance matrix, Proc. IEEE CDC, pp. 761-766, San Diago, CA.

[7] Hathaway, R J and Bezdek, J C (1993) Switching regression models and fuzzy clustering, *IEEE Trans. on Fuzzy Systems,* Vol. 1, No. 3, pp. 195-204.

[8] Nakamori, Y (1989) Development and application of an interactive modeling support system, *Automatica*, Vol. 25, No. 2, pp. 185-206.

[9] Nakamori, Y and Ryoke, M (1994) Identification of fuzzy prediction models through hyperellipsoidal clustering, *IEEE Trans. on Systems, Man and Cybernetics*, Vol. 24, No. 8, pp. 1153-1173.

[10] Nakamori, Y and Ryoke, M (1995) Adaptive fuzzy clustering for fuzzy modeling, Proc. of 6th International Fuzzy Systems Association World Congress, Vol. II, pp. 65-68, Sao Paulo, Brazil, July 22-28.

[11] Takagi, T and Sugeno, M (1985) Fuzzy identification of systems and its applications to modeling and control, *IEEE Trans. on Systems, Man and Cybernetics*, Vol. 15, No. 1, pp. 116-132.

Y. Nakamori and M. Ryoke are with School of Knowledge Science, Japan Advanced Institute of Science and Technology, 1-1 Asahidai, Tatsunokuchi, Ishikawa 923-1211, JAPAN, and K. Umayahara is with Center for Tsukuba Advanced Research Alliance, University of Tsukuba, 1-1-1 Tennodai, Tsukuba, Ibaraki 305-0006, Japan.

DIETER FRANKE
Rule-based design and 2D-analysis of a hybrid system

1. Introduction

The three tank switched server arrival system discussed by *Chase, Serrano* and *Ramadge* [1] has been cited by many other authors as an interesting example of a hybrid dynamical system. In this process refilling of a tank is started as soon as the tank gets empty. This rule basis, however, will cause chattering phenomena whenever two or three tanks get empty at nearly the same time. Therefore the question should be posed if the undesirable chattering phenomena can be avoided by using a different switching strategy which nevertheless guarantees a task oriented operation of the plant. Such a strategy will be proposed in this paper. On this occasion, the restrictive assumption that the net flow out of the three tanks be constant and equals the net flow into the tanks will be dropped.

2. Problem formulation

In the three tank system outlined in Fig. 1 the server position is described by vector $\mathbf{u} = [u_1 \ u_2 \ u_3]^T$ as in [1]. Hence, $u_i(k) = 1$ means that tank i is served at step k. It is assumed that the flow out of the server is available as an additional control input taking values 0 and q_0. Therefore the following Boolean coding is near at hand:

$$u_4 = \begin{cases} 0, \text{ flow 0 out of server} \\ 1, \text{ flow } q_0 \text{ out of server} \end{cases} \tag{1}$$

Hence, the over-all Boolean control vektor in step k takes the form

$$\hat{\mathbf{u}}(k) = \begin{bmatrix} \mathbf{u}(k) \\ u_4(k) \end{bmatrix}. \tag{2}$$

The flows out of the tanks, q_1, q_2 and q_3, are no longer assumed to be constant but may be arbitrary functions of time t in some known interval $[0, q_{max}]$, where

$$3 \cdot q_{max} < q_0 \tag{3}$$

with regard to a well-posed problem.

Let $x_i(t)$, $i = 1, 2, 3$, be the levels of liquid. Then the normed state equations of the continuous system take the form

$$\frac{d}{dt}\begin{bmatrix} x_1(t) \\ x_2(t) \\ x_3(t) \end{bmatrix} = -\begin{bmatrix} q_1(t) \\ q_2(t) \\ q_3(t) \end{bmatrix} + q_o u_4(k) \cdot \left\{ \begin{bmatrix} 1 \\ 0 \\ 0 \end{bmatrix} u_1(k) + \begin{bmatrix} 0 \\ 1 \\ 0 \end{bmatrix} u_2(k) + \begin{bmatrix} 0 \\ 0 \\ 1 \end{bmatrix} u_3(k) \right\} \qquad (4)$$

As can be seen from Fig. 1 threshold values x_{max} (tank full) and $x_{min} > 0$ have been introduced for each of the tanks. Whenever the level in a tank passes one of these thresholds this event will be communicated to a controller device such that the control input $\hat{u}(k)$ can be changed immediately if necessary. It should be noted that the value of x_{min} is not specified in advance, it should rather be determined as part of the problem (see below for details). In (4), integer k is the counter for discrete events.

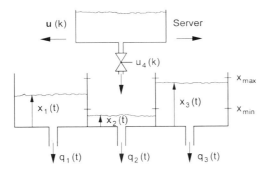

Fig.1. Three tank switched server arrival system

Next the levels of liquid will be coded as follows via Boolean variables v_i and w_i, $i = 1, 2, 3$:

$$v_i = \begin{cases} 0 \text{ if } x_i(t) < x_{min}, \\ 1 \text{ if } x_i(t) \geq x_{min}. \end{cases} \qquad (5)$$

$$w_i = \begin{cases} 0 \text{ if } x_i(t) < x_{max}, \\ 1 \text{ if } x_i(t) \geq x_{max}. \end{cases} \quad (6)$$

Hence

(a) $v_i = 0$ implies $w_i = 0$.
(b) $w_i = 1$ implies $v_i = 1$.
(c) The combination $v_i = 0, w_i = 1$ cannot occur. This reduces the over-all number of possible Boolean states (v, w) from $2^6 = 64$ to $3^3 = 27$.

The initial state of the continuous process, $x(0)$, is allowed to take any value in the box $x_{min} \leq x_i(0) \leq x_{max}$, $i = 1, 2, 3$.

The problem to be posed consists in determining a switching strategy (rule basis) which maps the Boolean process state $(v(k), w(k))$ to the Boolean control $\hat{u}(k)$ such that

1. the continuous state $x(t)$ does never exceed the box $0 < x_i(t) \leq x_{max}$, $i = 1,2,3$, for all $t \geq 0$ (which requires to consider the sensor location x_{min} as well !) and
2. chattering phenomena should strictly be avoided in favour of an as calm as possible switching behaviour.

As a result the rule basis will turn out to be a deterministic Boolean *automaton* rather than just a Boolean function $\hat{u}(k) = f(v(k), w(k))$.

3. Establishing the rule basis

The following basic elements are underlying the rule basis to be established:
I) A refilling procedure for a tank i having started will always be continued until the event $w_i = 1$ occurs (tank i full).
II) This is why not each event will cause a change of control $\hat{u}(k)$. Control $\hat{u}(k)$ will rather depend on $u_4(k-1)$, too.
III) No refilling will be started at step k if $v_i(k) = 1, i = 1, 2, 3$.
IV) Refilling will be started at step k if at least <u>one</u> $v_i(k) = 0$ <u>and</u> no other tank is just being refilled.

By means of these strategic elements chattering phenomena will strictly be avoided. On the other hand $o < x_i(t) < x_{min}$ is explicitly allowed in up to two tanks.

Table 1 outlines the 27 Boolean states (v, w) introduced in Chapter 2. Redundant values acc. to (a), (b) in Chapter 2 are in brackets. Moreover it can be observed that some rows can be obtained by other rows via permutation of index i. This results in the following groups of rows:

Rows 2, 4, 10.   Rows 3, 7, 19.   Rows 5, 11, 13.
Rows 9, 21, 25.   Rows 15, 17, 23.   Rows 18, 24, 26.
Rows 6, 8, 12, 16, 20, 22.

Rows 1, 14 and 27 have to be considered separately. The above mentioned symmetries are reflected by corresponding symmetries in the control strategy. Therefore only rows 1, 2, 3, 5, 6, 9, 14, 15, 18 and 27 of Table 1 need to be studied in detail. Table 2 outlines the Boolean rule basis for these 10 rows following the basic elements I) to IV) given above.

Table1. Boolean states of the plant

	v_1	w_1	v_2	w_2	v_3	w_3
1	0	0	0	0	0	(0)
2	0	0	0	0	1	0
3	0	0	0	0	(1)	1
4	0	0	1	0	0	(0)
5	0	0	1	0	1	0
6	0	0	1	0	(1)	1
7	0	0	1	1	0	(0)
8	0	0	1	1	1	0
9	0	0	1	1	(1)	1
10	1	0	0	0	0	(0)
11	1	0	0	0	1	0
12	1	0	0	0	(1)	1
13	1	0	1	0	0	(0)
14	1	0	1	0	1	0
15	1	0	1	0	(1)	1
16	1	0	1	1	0	(0)
17	1	0	1	1	1	0
18	1	0	1	1	(1)	1
19	1	1	0	0	0	(0)
20	1	1	0	0	1	0
21	1	1	0	0	(1)	1
22	1	1	1	0	0	(0)
23	1	1	1	0	1	0
24	1	1	1	0	(1)	1
25	1	1	1	1	0	(0)
26	1	1	1	1	1	0
27	1	1	1	1	(1)	1

The rule basis constructed in this way is complete and nonconflicting. It can be transferred from table notation to an analytical Boolean expression using known standard methods. The over-all Boolean rule basis obviously takes the form

$$\hat{u}(k) = f(v(k), w(k), v(k-1), \hat{u}(k-1)) \tag{7}$$

and therefore represents a deterministic Boolean automaton.

Table 2. Boolean rule basis for control

	$u_4(k-1) = 0$	$u_4(k-1) = 1$
$(v(k), w(k))^{(1)}$	$\hat{u}(k) = \begin{bmatrix} u(k-1) \\ 1 \end{bmatrix}$	$\hat{u}(k) = \hat{u}(k-1)$
$(v(k), w(k))^{(2)}$	$\hat{u}(k) = [1,0,0,1]^T$	$\hat{u}(k) = \hat{u}(k-1)$
$(v(k), w(k))^{(3)}$	$\hat{u}(k) = [1,0,0,1]^T$	$v_1(k-1)\ v_2(k-1)\quad \hat{u}(k)$ $0\quad 0\quad [1,0,0,1]^T$ $0\quad 1\quad [1,0,0,1]^T$ $1\quad 0\quad [0,1,0,1]^T$ $1\quad 1\quad [1,0,0,1]^T$
$(v(k), w(k))^{(5)}$	$\hat{u}(k) = [1,0,0,1]^T$	$\hat{u}(k) = \hat{u}(k-1)$
$(v(k), w(k))^{(6)}$	$\hat{u}(k) = [1,0,0,1]^T$	$u_3(k-1)\quad \hat{u}(k)$ $0\quad \hat{u}(k-1)$ $1\quad [1,0,0,1]^T$
$(v(k), w(k))^{(9)}$	$\hat{u}(k) = [1,0,0,1]^T$	
$(v(k), w(k))^{(14)}$	$\hat{u}(k) = \hat{u}(k-1)$	
$(v(k), w(k))^{(15)}$	$\hat{u}(k) = \hat{u}(k-1)$	$u_3(k-1)\quad \hat{u}(k)$ $0\quad \hat{u}(k-1)$ $1\quad \begin{bmatrix} u(k-1) \\ 0 \end{bmatrix}$
$(v(k), w(k))^{(18)}$	$\hat{u}(k) = \hat{u}(k-1)$	$u_1(k-1)\quad \hat{u}(k)$ $0\quad \begin{bmatrix} u(k-1) \\ 0 \end{bmatrix}$ $1\quad \hat{u}(k-1)$
$(v(k), w(k))^{(27)}$	$\hat{u}(k) = \begin{bmatrix} u(k-1) \\ 0 \end{bmatrix}$	

4. Worst-case analysis for sensor location

After having established the rule basis for Boolean control the question of sensor location x_{min} has to be tackled. A value x_{min} is said to be admissible if in combination with the above rule basis no tank will be emptied any time. A lower bound for x_{min} will be derived by a dynamical worst-case analysis of the over-all hybrid system.
Obviously, the worst-case is characterized by

$$q_i(t) \equiv q_{max}, \quad i = 1, 2, 3, \quad t \geq 0, \tag{8}$$

and by the initial state

$$x_i(t=0) = x_{min}, \quad i = 1, 2, 3. \tag{9}$$

The corresponding Boolean state can be seen in row 14 of Table 1.

In order to determine the evolution of states with respect to time t and counter k, the 2D-analysis proposed by *Franke* [2] will be chosen. This requires to solve at each discrete step the continuous state equations (4) subject to the Boolean rule basis established above. At the end $(t = t_k)$ of each step k the levels $x_i(t_k, k)$, $i = 1, 2, 3$, have to be determined. From the requirement that $x_i(t_k, k) > 0$, $i = 1, 2, 3$, lower bounds for x_{min} / x_{max} are obtained analytically as functions of q_{max} / q_0. Figure 2 outlines these lower bounds for steps $k = 0$ to $k = 5$.

Since in addition $q_{max} / q_0 < 1/3$ has to be required w. r. t. (3), curve c),

$$\frac{q_{max}}{q_0} \cdot \left(2 - \frac{q_{max}}{q_0}\right) < \frac{x_{min}}{x_{max}} \quad (<1), \tag{10}$$

turns out to be significant. Therefore the hatched domain in Fig. 2 is the admissible one for sensor location x_{min} / x_{max}. Obviously for growing counter k the lower bounds for

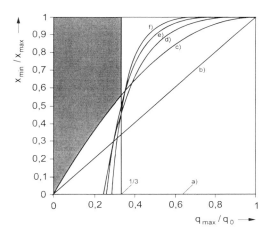

Fig. 2. Admissible domain (hatched) for x_{min} / x_{max}. Curves a) to f) represent lower bounds for x_{min} / x_{max} corresponding to the discrete steps $k = 0$ to $k = 5$, respectively.

x_{min}/x_{max} are approaching the vertical bound $q_{max}/q_0 = 1/3$ which is clearly plausible from the physical nature of the process.

5. Conclusion

A rule basis redesign has been proposed for a three tank switched server arrival system with some more realistic assumptions than in previous work on this subject. Since the flows out of the tanks are arbitrary within given bounds, so are the levels of liquid. Therefore questions concerning e. g. chaos or periodicity are irrelevant.

6. References

[1] C. Chase, J. Serrano and P. Ramadge, Periodicity and Chaos from Switched Flow Systems: Contrasting Examples of Discretely Controlled Continuous Systems, *IEEE Trans. Aut. Control*, 38, 1993, 70-83.

[2] D. Franke, 2-D Modelling of Hybrid Continuous - Boolean Dynamical Systems, in M. H. Hamza (Ed.): *Proc. IASTED / ISMM Internat. Conference "Modelling and Simulation"*, ACTA Press, Calgary, 1996, 297-300.

XUEZHANG HOU AND SZE-KAI TSUI
A mathematical model for control of flexible robot arms

1. Introduction

Structural flexibility in robotic systems has been emerging as an issue of increasing concern, for it is only realistic to include the vibration of such a system in the design of control to secure a certain degree of accuracy. The demands for high speed and low cost are driving the research for control of lightweight flexible robots. In this paper, we first formulate a mathematical model for flexible robot arms. This model describes a one-dimensional vibrating robot arm with a moving base. In general, a Cartesian robot consists of components which are flexible robot arms. There have been many investigations of the subject. Amongst them we list a few, such as works of Cannon and Schmitz [1] in 1984 and more recent work of Z.H. Luo, etc. [6,7,8]. Many of these works approach the subject from the design points of view. They have specific "goal items" to be controlled and designed controls accordingly. Our approach is more theoretical and general. First, we take a fourth order partial differential equation, the beam equation, to model the dynamics of the Cartesian flexible robot arm with several boundary conditions. Then, we consider a corresponding state-space control system in which the parameter matrix has its entries differential operators. In this setting we are able to determine the spectrum of the parameter matrix (see Section 2), and subsequently show that the system is both controllable and observable (see Section 3). In this infinite dimensional control analysis, one needs a heavy dose of functional analysis and operator theory in order to investigate the controllability and observability. This work has laid down a foundation for the design of a real-time closed loop feedback control for a flexible Cartesian robot. It is becoming more urgent that the traditional design of robot arms dependent on only the kinematics needs a makeover to include the dynamics of the system into the control. Our work fits nicely in this thrust of research which is becoming the focus of the research of dynamical robotics. The results of this article are taken from [4]. Further work is presently being pursued.

2. The model and spectral structure of a flexible robot system

2.1 The model and the evolution equation of the system Consider a Cartesian robot with a long tip arm illustrated in Fig. 1. Since any motion in the x-y plan can be decomposed into its x and y components, the vibrations in the x-direction and y-direction can be considered independently. Here is the picture of a Cartesian robot.

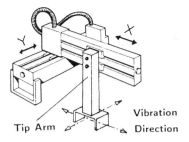

Fig. 1. A Cartesian robot with a long tip arm.

Fig. 2 shows motion of an x-y robot in the x-direction. m represents the mass of a moving body driven by a control motor. The one end with a payload M of the flexible arm is attached to this moving body.

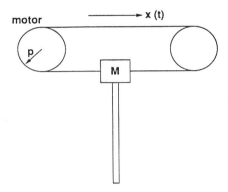

Fig. 2. Motion of an x-y robot in the x-direction.

Let the amplitude of vibration of the flexible arm at time t and position r be $w(t,r)$. Then the dynamic model for vibration of this flexible arm in x-direction can be written as follows [6]:

$$\begin{cases} \rho \ddot{w}(t,r) + EIw''''(t,r) = -\rho \ddot{x}(t), & 0 < r < l, t > 0 \\ w(t,0) = w'(t,0) = 0 \\ M[\ddot{w}(t,l) + \ddot{x}(t)] - EIw'''(t,l) = 0 \\ J\ddot{w}'(t,l) + EIw''(t,l) = 0 \\ w(0,r) = w_0(r) \ \dot{w}(0,r) = w_1(r), \end{cases} \quad (2.1)$$

where $\ddot{x}(t)$ denotes the acceleration of the moving body, "·" denotes the time derivative and "'" denotes the spatial derivative, ρ denotes the line density of mass for the arm, EI denotes the bending rigid degree of the flexible arm, l denotes the length of the arm, $w_0(r)$, $w_1(r)$ denote the initial displacement and initial velocity of the arm, respectively, and J denotes the turning inertia.

For the motor system, we shall establish the following control equation:

$$m\ddot{x}(t) = u(t) - EIw'''(t,0), \tag{2.2}$$

where the sliding friction was neglected and $u(t)$ is a control.

Let $y(t,r)$ be the total displacement in x-direction of the flexible arm. Thus, we have

$$y(t,r) = w(t,r) + x(t). \tag{2.3}$$

Substituting (2.3) into (2.1) yields the following controlled closed-loop system equation about state $y(t,r)$:

$$\begin{cases} \rho\ddot{y}(t,r) + EIy''''(t,r) = 0, & 0 < r < l, \quad t > 0 \\ y'(t,0) = 0 \\ m\ddot{y}(t,0) + EIy'''(t,0) = u(t) \\ M\ddot{y}(t,l) - EIy'''(t,l) = 0 \\ J\ddot{y}'(t,l) + EIy''(t,l) = 0 \end{cases} \tag{2.4}$$

In order to investigate the system (2.4) under the abstract frame, we now consider a real Hilbert space $H = R^3 \times L_\rho^2(0,l)$ equipped with the inner product as

$$(\Phi_1, \Phi_2)_H = m\xi_1\xi_2 + M\eta_1\eta_2 + J\zeta_1\zeta_2 + \langle \varphi_1, \varphi_2 \rangle_\rho,$$

where $\Phi_i = [\xi_i, \eta_i, \zeta_i, \varphi_i]^\tau \in H$, $i = 1,2$, $\langle \varphi_1, \varphi_2 \rangle_\rho = \int_0^l \rho\varphi_1(x)\overline{\varphi}_2(x)\,dx$, and τ means the transpose. We define a linear operator A with domain $D(A)$ in H as follows:

$$A\tilde{\varphi} = \begin{bmatrix} \dfrac{EI}{m}\varphi'''(0) \\ -\dfrac{EI}{M}\varphi'''(l) \\ \dfrac{EI}{J}\varphi''(l) \\ \dfrac{EI}{\rho}\varphi''''(\cdot) \end{bmatrix}, \text{ for } \tilde{\varphi} = \begin{bmatrix} \varphi(0) \\ \varphi(l) \\ \varphi'(l) \\ \varphi(\cdot) \end{bmatrix} \in D(A),$$

where $D(A) = \{\tilde{\varphi} \in H : \varphi, \varphi', \varphi'', \varphi''', \varphi'''' \in L_\rho^2(0,l), \varphi'(0) = 0\}$.

Using the operator A, (2.4) becomes the following second-order abstract evolution equation in H:

$$\frac{d^2\tilde{y}(t)}{dt^2} + A\tilde{y}(t) = bu(t), \qquad (2.5)$$

where $\tilde{y}(t) = [y(t,0), y(t,l), y'(t,l), y(t,\cdot)]^\tau$, $b = [\frac{1}{m}, 0, 0, 0]^\tau$.

2.2 The spectral structure of the operator A In this section we shall investigate the spectrum of the operator A in (2.5).

Theorem 2.1 $A: D(A) \to H$ *is a nonnegative self-adjoint operator.*

Proof. It is clear that $\overline{D(A)} = H$. For any $\tilde{\varphi}, \tilde{\psi} \in D(A)$, using integration by parts, we have

$$(A\tilde{\varphi}, \tilde{\psi})_H = EI \int_0^l \varphi''(x)\psi''(x)\, dx = (\tilde{\varphi}, A\tilde{\psi})_H,$$

and A is the symmetric operator; moreover,

$$(A\tilde{\varphi}, \tilde{\varphi})_H = EI \int_0^l |\varphi''(x)|^2 \, dx \geq 0, \qquad \tilde{\varphi} \in D(A).$$

Thus, A is nonnegative. It can be checked that A is a closed linear operator. Consider the restriction of A on the orthogonal complement K of the kernel of A. $A\mid_K$ is densely defined and closed, and by the symmetry of $A\mid_K$ we know that $A(K)$ is dense in K. Hence by the open mapping theorem and the fact that $A(K)$ is of second category in K, we have that the range of $A\mid_K$ is open and the range of A is K [10]. Therefore, $A\mid_K$ and A are self-adjoint.

Let λ be an eigenvalue of A, and suppose that $\tilde{\varphi}$ is the eigenvector corresponding to λ. Then, $A\tilde{\varphi} = \lambda\tilde{\varphi}$, namely,

$$\begin{cases} EI\varphi''''(x) = \lambda\rho\varphi(x), & 0 < x < l \\ \varphi'(0) = 0 \\ EI\varphi'''(0) - \lambda m\varphi(0) = 0 \\ EI\varphi'''(l) + \lambda M\varphi(l) = 0 \\ EI\varphi''(l) - \lambda J\varphi'(l) = 0. \end{cases} \qquad (2.6)$$

It is obvious that $\lambda_0 = 0$ is an eigenvalue of A with its eigenvectors of the form $\beta[1,1,0,1]^\tau$ for some scalar β. We denote $[1,1,0,1]^\tau$ by $\tilde{\varphi}_0$. Let $\nu^4 = \lambda\rho/EI (\lambda > 0)$. Then it follows from (2.6) that

$$\begin{cases} \varphi''''(x) = \nu^4\varphi(x) \\ \varphi'(0) = 0 \\ \varphi'''(0) - \nu^4 \frac{m}{\rho}\varphi(0) = 0 \\ \varphi'''(l) + \nu^4 \frac{M}{\rho}\varphi(l) = 0 \\ \varphi''(l) - \nu^4 \frac{J}{\rho}\varphi'(l) = 0. \end{cases} \qquad (2.7)$$

The general solution of (2.7) can be obtained as follows:

$$\varphi(x) = \frac{\tilde{a}}{2}(\cosh \nu x + \cos \nu x) + \frac{\tilde{b}}{2\nu^2}(\cosh \nu x - \cos \nu x) + \frac{\tilde{c}}{2\nu^3}(\sinh \nu x - \sin \nu x), \quad (2.8)$$

where $\tilde{a} = \varphi(0)$, $\tilde{b} = \varphi''(0)$, $\tilde{c} = \varphi'''(0)$. Let $y = \nu l$. Then it follows that

$$\varphi(l) = \frac{\tilde{a}}{2}(\cosh z + \cos z) + \frac{\tilde{b}}{2\nu^2}(\cosh z - \cos z) + \frac{\tilde{c}}{2\nu^3}(\sinh z - \sin z)$$

$$\varphi'(l) = \frac{\tilde{a}\nu}{2}(\sinh z - \sin z) + \frac{\tilde{b}}{2\nu}(\sinh z + \sin z) + \frac{\tilde{c}}{2\nu^2}(\cosh z - \cos z)$$

$$\varphi''(l) = \frac{\tilde{a}\nu^2}{2}(\cosh z - \cos z) + \frac{\tilde{b}}{2}(\cosh z + \cos z) + \frac{\tilde{c}}{2\nu}(\sinh z + \sin z)$$

$$\varphi'''(l) = \frac{\tilde{a}\nu^3}{2}(\sinh z + \sin z) + \frac{\tilde{b}\nu}{2}(\sinh z - \sin z) + \frac{\tilde{c}}{2}(\cosh z + \cos z).$$

The proofs of the next three theorems can be found in [4].

Theorem 2.2 (i) Let $\tilde{\varphi} = [\varphi(0), \varphi(l), \varphi'(l), \varphi(\cdot)]^\tau$ be any nonzero eigenvector corresponding to an eigenvalue $\lambda > 0$ of A, then $\varphi(0) \neq 0$, and $\varphi'(l) \neq 0$.

(ii) Every eigenvalue of A is of multiplicity one, that is, every eigenspace corresponding to an eigenvalue of A is one dimensional.

Theorem 2.3 *The resolvent of A is compact operator.*

Theorem 2.4 *The spectrum of A consists of only nonnegative eigenvalues with single multiplicity.*

3. The Wellposedness, Controllability and Observability of the System

3.1 The existence and uniqueness of the solution to the system For any $r > 0$, let $A_r \triangleq A + rI$ (I denotes the identity operator on H). Denote $\mathcal{H} = H \times H$ equipped with the inner product

$$\langle \Phi, \Psi \rangle = (\xi_1, \eta_1)_H + (\xi_2, \eta_2)_H, \text{ for every } \Phi = [\xi_1, \xi_2]^\tau, \ \Psi = [\eta_1, \eta_2]^\tau \in \mathcal{H}.$$

It is easy to see that with the inner product defined above $\mathcal{H}$ is a Hilbert space.

We now define a linear operator $\mathcal{A}$ on $\mathcal{H}$ below. For r in $\rho(A)$, the resolvent set of A, we define

$$\mathcal{A} = \begin{bmatrix} 0 & A_r^{\frac{1}{2}} \\ -A_r^{\frac{1}{2}} + rA_r^{\frac{1}{2}} & 0 \end{bmatrix}$$

with $D(\mathcal{A}) = D(A_r^{\frac{1}{2}}) \times D(A_r^{\frac{1}{2}})$. We also denote $[0, b]^\tau$ by $\mathcal{B}$, where b is defined in (2.5). Consider a subspace $\mathcal{S}$ of $\mathcal{H}$ consisting of $z = [z_1, z_2]^\tau$, where $z_1 = A_r^{\frac{1}{2}}\tilde{y}$ and $z_2 = \dot{\tilde{y}}$ and $\tilde{y}$ is defined in (2.5). In this notation, (2.5) with initial conditions

$$\begin{cases} \tilde{y}(0) \triangleq \tilde{y}_0 \\ \dot{\tilde{y}}(0) \triangleq \tilde{y}_1 \end{cases} \tag{3.1}$$

becomes a first-order evolution equation in $\mathcal{S}$ with an initial condition as follows:

$$\begin{cases} \frac{dz}{dt} = \mathcal{A}z + \mathcal{B}u \\ z(0) = [A_r^{\frac{1}{2}}\tilde{y}_0, \tilde{y}_1]^\tau \end{cases} \tag{3.2}$$

We denote the spectrum of $\mathcal{A}$ by $\sigma(\mathcal{A})$, a resolvent of A by $R(\lambda, A)$, and the resolvent set of A by $\rho(A)$. Proofs for theorems in this section can be found in [4].

Lemma 3.1 (i) $\mathcal{A}^* = \begin{bmatrix} 0 & -A_r^{\frac{1}{2}}+rA_r^{\frac{1}{2}} \\ A_r^{\frac{1}{2}} & 0 \end{bmatrix}$

(ii) $D(\mathcal{A}) = D(\mathcal{A}^*)$.

(iii) $\sigma(\mathcal{A}) = \sigma(\mathcal{A}^*)$.

(iv) The resolvents of $\mathcal{A}$ and $\mathcal{A}^*$ are compact.

(v) $\mathcal{A}$ is densely defined closed linear operator and the spectrum of $\mathcal{A}$ denoted by $\sigma(\mathcal{A})$ is equal to $\{\mu_n : \mu_0 = 0, \mu_n = \pm i\sqrt{\lambda_n}, n = 1, 2, \cdots\}$, where $\lambda_n, n = 1, 2, \cdots$, are the eigenvalues of A.

Lemma 3.2 The operator $\mathcal{A}$ is the infinitesimal generator of a C_0 semigroup $T(t)$, $t \geq 0$ on $\mathcal{H}$ satisfying $\|T(t)\| \leq e^{\|\mathcal{D}\|t}$, where $\mathcal{D}$ is defined in the proof of Lemma 2.1.

Theorem 3.3 (i) If $\tilde{z}_0 \in D(A^{\frac{1}{2}})$, $\tilde{z}_1 \in H$, $u \in L^2(0, T)(0 < T \leq +\infty)$, then the abstract Cauchy problem (3.2) has a unique mild solution;

(ii) if $\tilde{z}_0 \in D(A)$, $\tilde{z}_1 \in D(A^{\frac{1}{2}})$, $u \in C^1(0, T)(0 < T \leq +\infty)$, the Cauchy problem (2.3) has a unique classical solution.

Remark The existence and uniqueness of problem (3.2) are equivalent to the existence and uniqueness of original flexible robot control system.

3.2 The Approximate Controllability and Observability of the System

The following two theorems can be found in [2], and their proofs are not included here.

Theorem 3.4 Let $T(t)$ be the C_0-semigroup generated by $\mathcal{A}$. The system (3.2) is approximately controllable, if and only if one of the following is satisfied.

(i) If $\mathcal{B}^*T(t)^*\Psi = 0$, then $\Psi = 0$ for Ψ in $\mathcal{H}$.

(ii) If $\mathcal{B}^*R(\lambda, \mathcal{A})^*\Psi = 0$ for all λ in $\rho(\mathcal{A})$, then $\Psi = 0$ for Ψ in $\mathcal{H}$.

Theorem 3.5 Let $W(t) = \langle C, z \rangle_{\mathcal{H}}$ be an observation equation. The system (3.2) is observable if and only if one of the following is satisfied

(i) If $C^*T(t)\Phi = 0$, then $\Phi = 0$ for Φ in $\mathcal{H}$.

(ii) If $C^*R(\lambda, \mathcal{A})\Phi = 0$ for all λ in $\rho(\mathcal{A})$, then $\Phi = 0$ for Φ in $\mathcal{H}$.

Now we are ready for the following theorem.

Theorem 3.6 Let $b \in H$, $\mathcal{B} = [0, b]^{\tau}$, then system (3.2) is approximately controllable if and only if
$$(b, \tilde{\varphi}_n)_H \neq 0, \qquad n \geq 0. \tag{3.4}$$

Theorem 3.7 The flexible robot system (3.2) is approximately controllable.

Next, we shall discuss the observability of the flexible robot system. Let
$$b_1 = b, \quad b_2 = [0, 1/M, 0, 0]^{\tau}, \quad b_3 = [0, 0, 1/J, 0]^{\tau}.$$

Now we consider the following six major point observations for system (3.4).
$$W_{i1}(t) = (b_i, \tilde{y}(t))_H, W_{i2}(t) = (b_i, \dot{\tilde{y}}(t))_H, \quad i = 1, 2, 3,$$

where $W_{i1}(t), W_{i2}(t), i = 1, 2, 3$ denote the displacement $x(t)$, the line velocity $\dot{x}(t)$ of the moving body on the end of arm, the total displacement $z(t, l)$, its velocity $\dot{z}(t, l)$ of the end of the arm, the turning angle $w'(t, l)$, and the angular velocity $\dot{w}'(t, l)$ of the end of the arm, respectively.

Similarly, we consider the following observation equations for system (3.2)
$$W_i(t) = \langle C_i, z \rangle, i = 1, 2, \cdots, 6.$$

Here
$$C_1 = \begin{bmatrix} A_\gamma^{-\frac{1}{2}} b_1 \\ 0 \end{bmatrix}, \quad C_2 = \begin{bmatrix} 0 \\ b_1 \end{bmatrix}, \quad C_3 = \begin{bmatrix} A_\gamma^{-\frac{1}{2}} b_2 \\ 0 \end{bmatrix}$$

$$C_4 = \begin{bmatrix} 0 \\ b_2 \end{bmatrix}, \quad C_5 = \begin{bmatrix} A_\gamma^{-\frac{1}{2}} b_3 \\ 0 \end{bmatrix}, \quad C_6 = \begin{bmatrix} 0 \\ b_3 \end{bmatrix}$$

Explicitly, $W_{2i-1}(t) = W_{i1}(t)$; $W_{2i}(t) = W_{i2}(t), i = 1, 2, 3$.

Theorem 3.8 *The system* (3.2) *is observable for the measured data* $W_1(t)$ *and* $W_5(t)$, *while it is not observable for the measured data* $W_2(t)$, $W_3(t)$, $W_4(t)$, $W_6(t)$. *In other words, the flexible robot system* (2.4) *is observable for the displacement* $x(t)$ *of the moving body on the end of the arm and the angular velocity* $W'(t,l)$ *of the end of the arm, but for other measured data the flexible robot system is not observable.*

References

[1] Cannon, R. H., Jr. and Schimitz, E. (1984) Initial experiments on end-point control of a flexible one-link robot, *Int. J. Robotics Res.*, **3** issue 3, 62–75.

[2] Curtain, R. F. and Prichard, A. J. (1978). *Infinite Dimensional Linear Systems Theory*, Lecture in control and Information Sciences, 8. Springer-Verlag.

[3] Dunford, N. and Schwartz, J. (1968). *Linear Operator III*, Interscience Publishers, Wiley and Sons.

[4] Hou, X.-Z. and Tsui, S.-K. A control theory for Cartesian flexible robot arms, to appear.

[5] Kato, T. (1980). *Perturbation Theory for Linear Operators*, 2nd edition, Springer-Verlag: New York.

[6] Luo, Z. H. (1993). Direct strain feedback control of flexible robot arms: new theoretical and experimental results, *IEEE Trans. Autom. Contr.*, **38** issue 11, 1610–1622.

[7] Luo, Z. H., Kitamura, N. and Guo, B. (1995). Shear force feedback control of flexible robot arms, *IEEE Trans. Robotics and Automat.*, **11** issue 5, 760–765.

[8] Morgül, Ö. (1991). Orientation and stabilization of a flexible beam attached to a rigid body planar motion, *IEEE Trans. Auto. Control*, **36** issue 8, 953–965.

[9] Pazy, A. (1983). *Semigroup of Linear Operators and Applications to Partial Differential Equations*, Springer-Verlag: New York.

[10] Saperstone, S. (1981). *Semidynamical Systems in Infinite Dimensional Space* Springer-Verlag: New York.

[11] Taylor, A. and Lay D. (1980). *Introduction to Functional Analysis*, John Wiley and Sons: New York, Chichester, Brisbane, Toronto.

[12] Vidyasgar, M. (1978). *Nonlinear Systems Analysis*, Prentice-Hall: Englewood Cliffs, New Jersey.

Department of Mathematical Sciences,
Oakland University
tsui@oakland.edu

VI. Automotive

SCOTT AMMAN, MANOHAR DAS, MIKE BLOMMER AND NORM OTTO
Identification of powertrain noise in a low SNR environment using synchronous time averaging

1. Introduction

In the automotive industry, sound recordings are made in the passenger compartment in order to characterize what customers hear under typical driving conditions. Many times the recordings are contaminated with sources other than that of the desired one. This is typically the case in the recording of a vehicle's powertrain sound. Contamination can come from random sources such as wind and road noise. Synchronous Time Averaging (STA) has been used in the past for such problems as fault monitoring and detection in gears [1]-[3], engine vibration monitoring [4], alternator noise analysis [5] and turbocharger fault monitoring [6]. STA has the inherent capability of "averaging" out random sources not correlated to a periodic synchronizing signal. If N is the number of synchronous data frames to average, reductions of the noise component can approach $1/\sqrt{N}$ for the white noise case. Since powertrain sound recordings are usually taken in conjunction with a tachometer pulse from the crankshaft of the engine, a synchronizing signal that is correlated to the engine sound is readily available. By triggering the averaging process with this tachometer pulse, the fundamental rotational frequency of the engine and its corresponding harmonics (commonly referred to as "orders") can be estimated. In addition, harmonics corresponding to multiples of one-half the engine rotational frequency may be estimated by synchronizing the averaging process with every-other tachometer pulse.

The next section provides theoretical background on the concept of STA and the implications of RPM jitter on the averaging process. The method used to explore the effectiveness of STA in the estimation of the powertrain noise is discussed in the third section. Experimental results will then follow in section four for a steady-state powertrain load and speed condition when wind noise is present in the signal.

2. Synchronous Time Averaging

The engine noise signal consists of a fundamental frequency component based on the rotational frequency of the engine. This component and its harmonics exist and comprise the "full" or "integer" engine orders of the sound. Additionally, half engine orders are also present. These half engine orders are based off events that occur once every two rotations of the engine. These can be caused by cylinder-to-cylinder and cycle-to-cycle combustion variability in the engine. The identification of full and half engine orders utilizes STA for the removal of noise that is uncorrelated with the rotation of the engine crankshaft. The time averaging is synchronized with the crankshaft rotation via an optical tachometer pickup that produces one pulse per engine revolution. If the time averaging is synchronized with every revolution of the engine crankshaft, random noise and all orders other than full orders will be suppressed. If the averaging is performed at every other crankshaft rotation, random noise will be suppressed and both full and half orders will be kept intact.

The STA process is fully described in [7] and is briefly outlined here. STA is described by equation (1).

$$y(iT) = \frac{1}{N} \sum_{r=0}^{N-1} x(iT - rMT), \qquad i = 0, 1, 2, ...M-1 \qquad (1)$$

where: N = the number of data segments
x(iT) = raw data
y(iT) = averaged data
M = number of points per period of averaging process
T = sampling period
MT = period of averaging process

Applying the z-transform to (1), the STA process is represented as:

$$H(z) = \frac{Y(z)}{X(z)} = \frac{1}{N} \frac{1-z^{-MN}}{1-z^{-M}} \qquad (2)$$

The corresponding magnitude and phase of the STA process are,

$$|H(f)| = \frac{1}{N} \frac{\sin(\pi N f / f_p)}{\sin(\pi f / f_p)} \qquad (3)$$

$$\phi_H(f) = -\pi(N-1)f / f_p \qquad (4)$$

where f_p is the frequency of the averaging process (i.e., 1/MT). The magnitude response of (3) is that of a comb filter with main lobes centered at integer multiples of f_p. As a result, frequency components at k × f_p, with k being an integer, are preserved and broadband components rejected. Figure 1 shows the effect of changing N. As N increases the comb filters main lobes become narrower.

A normalization of H(f) can be obtained by substituting the variable f/f_p for f. Since H(f) is periodic with period f_p, $H(f/f_p)$ is periodic with unity period. The equivalent noise bandwidth (ENB) of $H(f/f_p)$ is given as the integral of $|H(f/f_p)|^2$ over the range -0.5 to 0.5.

$$ENB = \int_{-0.5}^{0.5} \left(\frac{1}{N} \frac{\sin(\pi N x)}{\sin(\pi x)} \right)^2 dx = \frac{2}{N^2} \int_{0}^{0.5} \left(\frac{\sin(\pi N x)}{\sin(\pi x)} \right)^2 dx = \frac{1}{N}, \qquad (5)$$

where x = f/fp. This result is to be expected since when averaging uncorrelated noise, the variance decreases as 1/N.

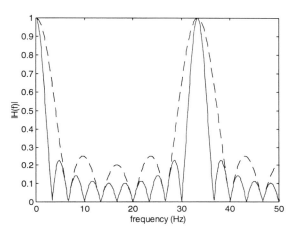

Figure 1. Effect of changing the number of frames used in the synchronous time average, N=10 (solid), N=5 (dashed).

At this point, a discussion of how variation or jitter in engine RPM can influence the STA result is needed. The main result of the engine speed not being completely stationary is similar to that of the trigger error effect discussed by Braun [7]. The result of this error is a decrease in the recovered signal due to the comb filters mistuning. The frequency difference error is proportional to the harmonic order of the signals components. Therefore, attenuation increases with harmonic order which results in the higher harmonics being "averaged" out. Figure 2 demonstrates this concept. Figure 2 (a) shows a fundamental tone at 50 Hz with 50.5 and 51 Hz tones from subsequent frames overlaid. This is indicative of a 1% frequency (or RPM) error from frame to frame. Figure 2 (b) shows the average of the three frames. Minor amplitude attenuation is noticed toward the end of the averaged data. Figure 2 (c) shows the 4th harmonic (200 Hz) overlaid with 202 and 204 Hz tones, also resulting in a 1% frequency error. The amplitude attenuation is much more dramatic as shown in Figure 2 (d).

It should be mentioned that methods which use an exponential synchronous time average (ESTA) are available for non-stationary data; however, the price paid is an increase in the main lobe bandwidth and deterioration in the noise reduction of the STA technique [7].

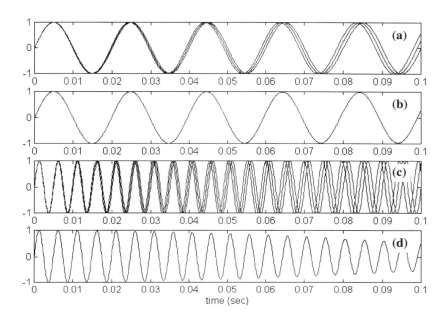

Figure 2. Effect of RPM jitter on STA. (a) Overlay of three frames with frequencies 50, 50.5 and 51 Hz. (b) STA of 50, 50.5 and 51.5 Hz components. (c) Overlay of three frames with frequencies 200, 202 and 204 Hz. (d) STA of 200, 202 and 204 Hz components.

3. Method

In order to study the effectiveness of synchronous time averaging on the estimation of powertrain noise mixed with wind noise, sound data was collected from both a chassis dynamometer and a wind tunnel for a mid-sized sedan. The chassis dynamometer sound data contains almost "pure" powertrain noise since the wind and road inputs have been minimized. The data was recorded at a sampling rate of 44.1 kHz at wide-open throttle and 4000 RPM. The wind tunnel sound data represents a situation where the sound data is virtually "pure" wind noise without powertrain or road noise influences. Wind noise data was acquired in a wind tunnel (50 MPH wind speed) on the same type of vehicle in order to get pure wind noise.

With virtually uncontaminated sources for the wind and powertrain in hand, the two sources can be mixed at various signal-to-noise ratios (SNR's) to investigate the limits of using STA in estimating the powertrain noise. The powertrain noise estimation can be validated by comparing an objective measure of signal energy, such as overall dB level, to the original powertrain sound. Since it is well known that pure signal energy indicators such as dB level do not adequately describe the subjective impression of acoustic data for psychoacoustic aspects such as the loudness of the noise, a comparison between the measured loudness of the original and estimated powertrain noise was made using the method described in [8]. This measure more closely describes the subjective impressions of loudness of the estimated powertrain sound, since auditory processing concepts such as variation of frequency weighting with level and masking effects are not taken into account.

The gain of the wind noise was adjusted to give a 10 dB, 0 dB and -10 dB SNR when combined with the powertrain noise. These SNR values were subjectively determined to span the range of audibility of both the powertrain and wind noise. A SNR of 10 dB was subjectively judged to be the point at which the added windnoise was barely audible. The -10 dB SNR signal was representative of a mixed sound in which the powertrain sound was almost entirely masked by the wind noise.

In order to identify the limitations of STA in proper powertrain noise estimation, STA was performed on the 10 dB, 0 dB and -10 dB SNR signals. Additionally, STA was performed on the original powertrain recording with no noise added. This was done since the original powertrain recording contains frequency components that are not integer multiples of one-half the rotational frequency of the engine. Some sources of such sounds include front end accessories (alternators, engine cooling fan, etc.), broad band components due to induction noise, chassis dynamometer induced noise, etc. Therefore, implementing STA on the powertrain sound recording with no noise reduces these effects and gives a basis for comparison to the estimated powertrain signal.

4. Experimental Results

Figure 3 shows the mixed (0 dB SNR) time domain data and the corresponding tachometer pulses which occur once every crankshaft revolution. The power spectral densities of the original powertrain signal, wind noise signal and mixed signal for the 0 dB SNR case are shown in Figure 4.

Figure 5 shows the results of using N=10 (20 engine revolutions) in the application of STA to the powertrain+wind noise (0db SNR) signal. The data frame size used in the averaging was 16k samples or approximately 0.36 seconds. We can see that the noise floor has been significantly reduced and the engine orders and half orders are the dominant component in the averaged signal. Another point to observe in Figure 5 is that it appears that there may be some evidence of RPM fluctuation. Some of the engine orders above 700 Hz appear to be underestimated. Upon further investigation of the tachometer data, it was found that the frequency deviation from frame to frame was as much as 0.06 %. At 10th engine order, the frequency difference would be $\approx$ 0.4 Hz. This translates into a 51° phase difference at the end of a 0.36 sec. data segment. Alternatively, the first order component would have a phase difference of only 5.1°. Therefore, it would be reasonable to assume that attenuation of the high order components due to RPM jitter is likely.

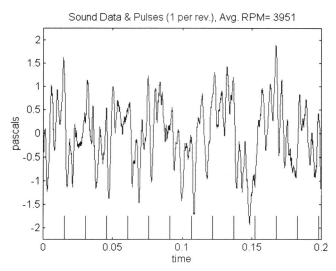

Figure 3. Powertrain + wind noise sound data with tachometer pulses from the engine crankshaft along the bottom.

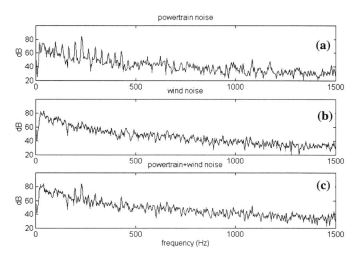

Figure 4. Power spectral densities of a) Powertrain sound, b) Wind noise, c) Powertrain + wind noise, SNR=0 dB.

406

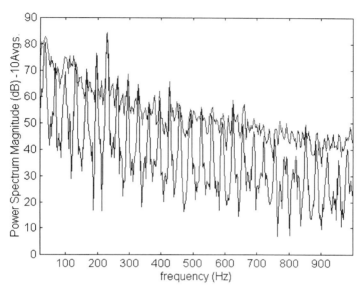

Figure 5. Power spectral densities of powertrain+wind noise (light line) and synchronous time average of powertrain+wind noise (dark line).

Table 1 shows the objective results of the synchronous time averaged signals. It can be observed that the STA performs quite well for the 10 dB and 0 dB SNR cases. The dB level differs by only 1 dB when comparing the STA of the original powertrain sound to the STA of the 0 dB SNR mixed signal. The loudness levels differ by approximately 2 sones or 6%. The STA begins to deteriorate significantly at the -10 dB SNR level. For this case, the differences are much more evident. The signal energy is off by 5 dB and the loudness is off by 12 sones or 35%.

Table 1. Comparison of Signal Energy (dB) and ISO 532B Loudness (sones) Metrics

	dB (re 20µPa)		ISO 532B Loudness (sones)	
	original	*after STA*	*original*	*after STA*
powertrain w/no noise added	90.3	88.6	43.1	35.1
powertrain+noise (10dB SNR)	90.9	88.7	43.8	35.3
powertrain+noise (0dB SNR)	93.8	89.6	49.9	37.3
powertrain+noise (-10dB SNR)	101.3	93.7	74.1	47.4

4. Conclusions

In the above case it was shown that STA can be a useful tool in increasing the SNR of an engine sound recording in situations in which the SNR is greater than 0 dB. This is due to the comb filtering effect of the STA process. However, particular attention must be paid to the RPM jitter that can reduce the levels of the higher frequency harmonics of the signal. Many times, these order components have a negligible effect on the perception of the signal due to their low level relative to the lower frequency orders. If these components are important, care must be taken to maintain a steady RPM or methods such as exponential averaging could be employed to reduce the effect.

References

1. P. McFadden, "Determining the Location of a Fatigue Crack in a Gear from the Phase of the Change in the Meshing Vibration," *Mechanical Systems and Signal Processing*, vol. 2, no. 4, pp. 403-409, 1988.
2. J. Mathew, A. Szczepanik, "Monitoring the Vibrations of Low and Variable Speed Gears," *ASME 12th Biennial Conference on Mechanical Vibration and Noise*, pp. 23-30, 1989.
3. D. Futter, "Techniques for Condition Monitoring of Large Gearboxes in the Power Industry," *Insight*, vol. 37, no.8, August 1995.
4. J. Sauw-Yoeg Tjong, D. Ki Chang, T. Lau, Z. Reif, "Monitor Engine Condition Via Vibration Signals," *11th International Modal Analysis Conference*, pp.1252-1258, Kissimmee, Florida, 1993.
5. B. Wilson, "The Analysis of Diesel Engine Turbocharger Vibration Using Synchronous Time Averaging," *SAE International Off-Highway & Powerplant Congress & Expo.*, Milwaukee, WI, SAE Paper 932500, 1993.
6. D. Frederick, G. Lauchle, "Aerodynamically-Induced Noise in an Automotive Alternator," *Noise Control Engineering Journal*, vol. 43, no. 2, pp. 29-37, Mar.-Apr. 1995.
7. S. Braun, "The Extraction of Periodic Waveforms by Time Domain Averaging," *Acustica*, vol. 32, pp. 69-77, 1975.
8. ISO 532: "Method for Calculating Loudness Level," 1975.

JOSEPH R ASIK[*]
Hot wire mass air flow sensor modeling and lag compensation

1. Introduction

Hot wire mass air flow sensors (MAFS) are a key sensor for the measurement and estimation of mass air flow into the engine intake system and cylinders (Arai (1990), Arai (1993), Loesing (1989), Nishimura (1983), Nishimura (1989), Sasayama (1982), and Sumai (1984)). Prior work on sensor frequency domain characterization was reported by Follmer (1988) and by Ziesmer (1993). In these two papers models of the MAFS were presented, although methods for compensating for the lag of the sensor were not discussed. In the work of Grizzle (1993) a simplified MAFS model was presented in which the first order time constant the MAF sensor was assumed constant. In addition, lead compensation of the sensor lag dynamics was presented and compared to experimental data.

2. Purpose and Experimental Data

The purpose of this work is to present recent measurements and a model for the MAFS. Both small signal (Bode) and large signal (step response) data are presented. The data are used to generate a model for the MAFS which forms the basis for lead compensation of the intrinsic lag of the MAFS. Fig. 1 shows the overall model for air flow, starting from the atmosphere, through the throttle body, through the intake manifold, and into the engine. Fig. 2 shows information flow for the transformation of throttle body air flow into an event based throttle air charge, a port charge (filtered by the manifold), and the final cylinder air charge Q_a. The latter is a two event extrapolation of the port air charge. The desired fuel charge Q_f, delivered by the fuel injectors, is calculated from $Q_f = Q_a*(F/A)$, where F/A is the desired fuel to air ratio, usually near stoichiometric. Figs. 3 and 4 illustrate the construction and circuit, respectively, for a the hot wire MAFS (Nishimura (1983)).

Fig. 5 shows the small signal experimental apparatus. The set point is adjusted with a blower while a loudspeaker modulates the air flow. Typical data are shown in Fig. 6. Using a first order data fit, the variation of the time constant with air flow is shown in Fig. 7. The large signal experimental apparatus is shown in Fig. 8, in which step air flow changes of 10 to 105 and 10 to 210 kg/hr, both increasing and decreasing, as well as static measurements, are conducted. The static calibration of the MAFS is shown in Fig. 9. A 0.4 power of the air flow provides an excellent fit. Typical step response data are shown in Figs. 10a, 10b, 11a, and 11b. It is clear that the step down time response is slower than the step up time response. Large signal step response data are shown in Figs. 12 and 13.

Fig. 14 summarizes the voltage response data analysis indicating that the step down time constant is approximately 2 to 4 times the step up time constant. For this analysis no correction for the nonlinearity of the air flow response was made. If this correction is made prior to the time constant analysis, the step down time constant is approximately 0.5 that of the step up (10.8 ms vs. 20.9 ms). Both of these analyses are somewhat in error, however, because of the highly nonlinear behavior of the MAFS.

3. Model

Eq. (1) states the proposed MAFS model.

$$MAFS(V) = \frac{C_1(Wa)^a}{1 + s\tau_0 \left(\frac{Wao}{Wa}\right)^b}, \quad \text{Eq.(1)}$$

where $a \approx 0.4$ from static analysis, $b \approx 0.8$ from large signal analysis, C_1 is a constant, and τ_0 is an effective time constant dependent on operating conditions. See Figs. 9 and 15.

For lead compensation, Eq. (1) is solved for Wa(k), given sampled data measurements of MAFS(k). One generally used approximation (#1) neglects τ_0 and applies only the static inversion:

$$Wa(k) = [(1/C_1)(MAFS(k)]^{2.5} \quad \text{Eq.(2)}$$

A second approximation (#2), used here, applies a sampled data lead compensation prior to the inversion:

$$Wa(k) = [(1/C_1)(MAFS(k) + \tau/\Delta t(MAFS(k) - MAFS(k-1)))]^{2.5} \quad \text{Eq.(3)}$$

Figs. 16 and 17 show the comparison of throttle position, uncompensated, and compensated throttle air charge, for step up and step down responses. Fig. 18 shows the comparison of lead compensated throttle air charge and final cylinder air charge.

4. Conclusion

A nonlinear model for the hot wire mass air flow sensor (MAFS) was proposed, based on extensive experimental data. The MAFS time constant is found to vary strongly with air flow both magnitude and sense of magnitude. This results in different effective time constants for small signal applications and large signal applications. It was shown that lead compensation of the MAFS can provide improved transient air flow and air charge accuracy

*Address: Powertrain Control Systems Dept., Ford Research Laboratory, Box 2053, Dearborn, MI 48121-2053. Email: jasik@pobox. srl.ford.com.

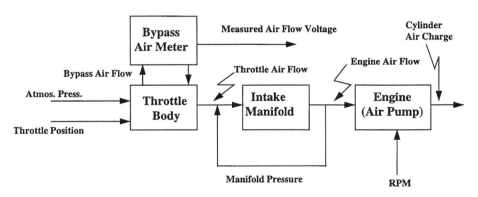

Fig. 1. Cylinder Air Flow Model

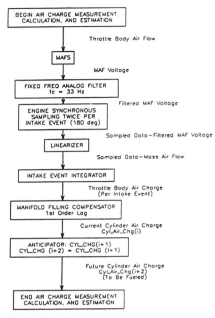

Fig. 2. Final Cylinder Aircharge Estimation Information Flow

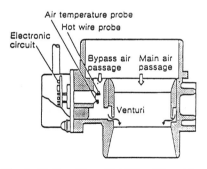

Fig. 3. ByPass Type Hot Wire Mass Air Flow Sensor

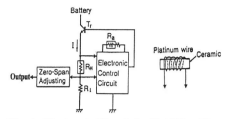

Fig. 4. Electronic Circuit for Hot Wire Mass Airflow Sensor

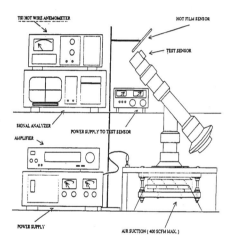

Fig. 5. Experimental Arrangement for Small Signal Measurement

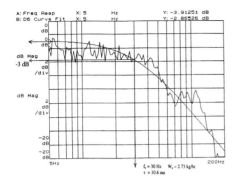

Fig. 6. Typical Small Signal Gain vs. Frequency Response

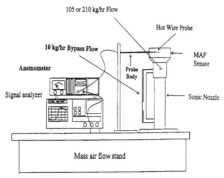

Fig. 8. Experimental Arrangement for Large Signal Measurement

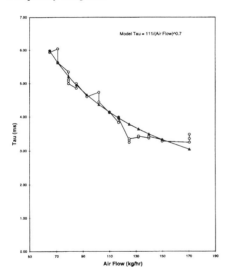

Fig. 7. Behavior of First Order Time Constant τ as a Function of Air Flow

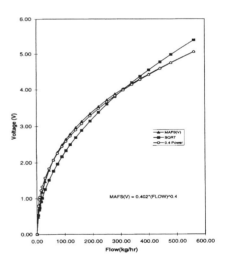

Fig. 9. Static Calibration of MAFS Output Voltage vs. Airflow, Comparing Data, Square Root, and 0.4 Power Fits

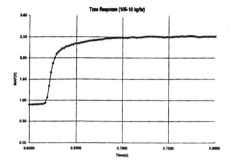

Fig. 10a. MAFS Time Response to A Step Increase in Airflow, 10 to 105 kg/hr

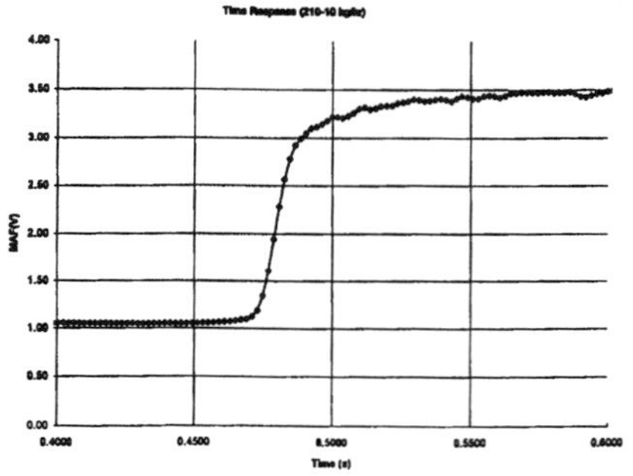

Fig. 11a. MAFS Time Response to A Step Increase in Airflow, 10 to 210 kg/hr

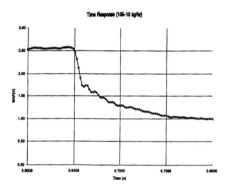 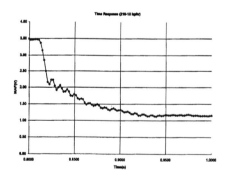

Fig. 10b. MAFS Time Response to A Step Decrease in Airflow, 105 to 10 kg/hr

Fig. 11b. MAFS Time Response to A Step Decrease in Airflow, 210 to 10 kg/hr

STEP	Time Constant ± STD DEV		
% Change Analyzed	Hot Wire Time Constant (ms)		
	(1) *	(2) **	
A. 10-105 (kg/hr)			
0-63%:	5.8 ± 1.2	5.8 ± 1.2	✓
10-90%:	4.4 ± 0.8	4.4 ± 0.8	✓
B. 105-10 (kg/hr)			
100-37%:	11.5 ± 0.7	11.5 ± 0.7	✓
90-10%:	8.2 ± 1.7	38.8 ± 13.8	✗
C. 10-210 (kg/hr)			
0-63%:	7.8 ± 0.7	7.8 ± 0.7	✓
10-90%	4.8 ± 0.6	4.8 ± 0.6	✓
D. 210-10 (kg/hr)			
100-37%	10.0 ± 1.2	10.0 ± 1.2	✓
90-10%:	7.1 ± 0.5	37.2 ± 2.4	✗

* (1) Oscillations in data.
** (2) Time constants corrected for oscillations.

Fig. 12. Voltage Response Analysis to Step Response in Airflow; Average Standard Deviation of Various Response Parameters

STEP	6 sample average		
% Change Analyzed	AVG (ms)	σ (ms)	% σ
A. 10-105 (kg/hr)			
0-63%:	13.2	0.8	6.1
10-90%:	38.4	2.1	5.5
B. 105-10 (kg/hr)			
100-37 %:	31.0	3.2	10.3
90-10%:	83.5	5.5	6.6
C. 10-210 (kg/hr)			
0-63%:	13.2	0.9	6.8
10-90%:	33.8	1.7	5.0
D. 210-10 (kg/hr)			
100-37%:	30.6	5.4	17.6
90-10%:	86.5	3.3	3.8

Fig. 13. Analysis of Voltage Response to Airflow Step Response; Time Constant and Standard Deviation

MAFS Voltage Response Analysis

Ratio of Fall to Rise Time Constant	Δ = 95 kg/hr Airflow 10 - 105 vs. 105 - 10 (kg/hr)	Δ = 200 kg/hr Airflow 10 - 210 vs. 210 - 10 (kg/hr)
100-37 % 0-63 %	2.6	3.8
90-10 % 10-90 %	1.3	1.7

Fig. 14. Comparison of Step Up vs. Step Down Time Constants Resulting from Voltage Response Analysis

Air Flow Step	Measured System Response Time	Corrected MAFS Response Time*
10-210 kg/hr	53.3 ± 3.5 ms	45.9 ± 3.4 ms
210-10 kg/hr	59.3 ± 2.6 ms	23.8 ± 4.8 ms
Air Flow Step	Calculated Time Constant	Calculated Time Constant
10-210 kg/hr	24.1 ms	20.9 ms
210-10 kg/hr	27.0 ms	10.8 ms

Fig. 15. Mass Airflow Analysis. Comparison of Step Up and Step Down Time Constants

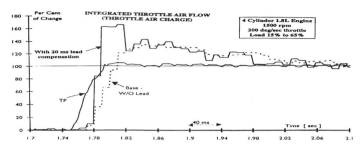

Fig. 16. Comparison of Throttle Position (TP), Uncompensated Throttle Air Charge, and Compensated Throttle Air Charge for Step Increase in Airflow vs. Time

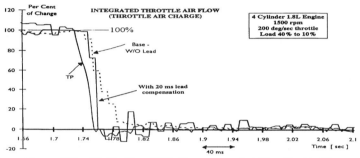

Fig. 17. Comparison of Throttle Position (TP), Uncompensated Throttle Air Charge, and Compensated Throttle Air Charge for Step Decrease in Airflow vs. Time

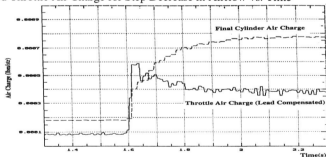

Fig. 18. Comparison of Final Cylinder Air Charge and Compensation Throttle Air Charge vs. Time

5. References

Arai, N., Sekine, Y., Osawa, T., and Tsutsui, M. (1990), *Advanced Design for Bypass Type of Hot-Wire Air Flow Meter*, SAE Paper **900259**, pp. 7-13.

Arai, N. and Igarashi, S. (1993), *Fluid Mechanics in Multi-Bypass Air Flow Sensor with Wide Dynamic Range*, SAE Paper **930231**, pp. 1-7.

Follmer, W. C. (1988), *Frequency Domain Characterization of Mass Air Flow Sensors*, SAE Paper **880561**, pp. 111-115.

Grizzle, J. W., Cook, J. A., and Milam, W. P. (1993), *Improved Cylinder Air Charge Estimation for Transient Air Fuel Ratio Control*, ACC Preprint.

Loesing, K. *et al.* (1989), *Mass Air Flow Meter-Design and Application*, SAE Paper **890779**, pp. 1-8.

Nishimura, Y., *et al.* (1983), *Hot Wire Air Flow Meter for Engine Control System*, SAE Paper **830615**, pp. 1-6.

Nishimura, Y. *et al.* (1989), *A Hot Wire Air Flow Meter for Intake Air Flow Measurement*, SAE Paper **890301**, pp. 23-26.

Sasayama, T. *et al.* (1982), *A New Electronic Engine Control System Using a Hot-Wire Air Flow Sensor*, SAE Paper **820323**, pp. 87-94.

Sumai, J. and Sauer, R. (1984), *Bosch Mass Air Flow Meter: Status and Further Aspects*, SAE Paper **840137**, pp. 19-28.

Ziesmer, D. A., Chuey, M. D., and Hazelton, L. (1993), *Frequency Domain Characterization of Mass Airflow Sensors*, SAE Paper **930325**, pp. 47-53.

KA C CHEOK, SHINICHI NISHIZAWA AND WILLIAM J YOUNG
Dynamic clustering technique with application to onboard traffic monitoring

1. Introduction

As city and highway traffic levels increase, so does the potential for automobile accidents. In an effort to make automobile travel safer, a number of researchers are investigating devices and methods to improve driver safety. One method, known as traffic monitoring, employs active sensors on the vehicle in concert with an onboard computer to warn the driver of potentially dangerous situations.

Several types of sensors have been experimented with, including sonar, millimeter-wave radar, computer vision, and lidar (laser radar) [1,2]. Of these, lidar appears to be very promising due to some distinct advantages. Commercial lidar are multiple-beam devices that can return several points of data per sensor. Lidar require no mechanical actuators to achieve this multiple-beam behavior, making them more robust for automotive applications [3]. Also, lidar response tends to be limited to automotive taillight reflectors, which reduces the potential number of false detection by the sensor.

Any traffic monitoring system employing lidar, as well as other sensors, must be able to correlate reported data points with vehicle motion. This job becomes more difficult as the number and type of sensors used increases. For the task of traffic monitoring with lidar, a clustering method must be employed to extract information about vehicle positions and velocities from the data points.

2. Problem Description

2.1 Moving Clusters

Figure 2.1 illustrates a situation where classifying clusters may become a problem when clusters of moving particles (data) cross paths and intersect. The figure shows three clusters moving with independent velocities. The particles in a cluster may move within the cluster and therefore the cluster shape may also change as it moves. Figure 2.1(a) shows an initial situation where the moving particles are clearly separated and so conventional static clustering methods can be used to classify each cluster. In contrast, Figure 2.1(b) depicts a situation when the three clusters have

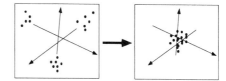

(a) Readily separable clusters (b) Intersecting clusters
Figure 2.1 Moving clusters

moved and now meet in an intersection. The static conventional method [4,5] will not be able to distinguish the individual clusters in this situation.

2.2 Problem Statement

The main problem in this paper is to develop a computer method for tracking and identifying the moving clusters. To distinguish from conventional static clustering methods, the developed technique will be called Dynamic Clustering Method (DCM). As an illustration, a traffic monitoring application using the DCM will be presented.

2.3 Assumptions

For proper scope of operation, the DCM will be subjected to the following practical nonrestrictive assumptions concerning the dynamic motions and measurements of the data.

(A.1) The position of each particle at time k can be correlated with its position at time $k+1$, so as to allow for the estimation of the particle velocity. The assumption can readily be satisfied if a sensor or an array of sensors covers measurements of particle locations over a wide area.

(A.2) All signal model and measurement noise are zero-mean Gaussian white noise with known covariance. This assumption approximately reflects most situations in the particle motion and measurements, and provides the basis for applying Kalman filtering.

(A.3) Each cluster moves in a continuous or non-abrupt manner. This requires that the particles do not appear and disappear in disparate regions of space, except when a cluster is newly spawned or when a old cluster that is no longer valid is deleted. In other words, it is desirable for the clusters and the particles in the cluster to move in a natural, continuous manner.

3. Dynamic Clustering Method (DCM)

3.1 Definitions

For ease of reference, variables, parameters and symbols in this paper are defined below.

Variables

n	Number of data points.	$D_{Threshold}$	Threshold for distance norm	τ	Sampling time
n_j	Number of data points belonging to j'th cluster	$V_{Threshold}$	Threshold for velocity norm	m	Number of clusters
p	Data vector	K	Kalman gain	v	Measurement noise
c	Center vector of cluster	w	Signal model noise	x,y	Position
D_{ij}	Norm of difference of distance between p_i and c_i	R	Covariance of measurement noise	$\dot{x}, \dot{y}$	Velocity
V_{ij}	Norm of difference of velocity between p_i and c_i	Q	Covariance of system noise		

Subscripts

i,j	Data ID, Cluster ID	k	Discretized time	x,y	x and y axis component

Symbols

C	Cluster set	KF	Kalman Filter	^	Estimated quantity
Z^{-1}	Delay operator	PD	One-Step Predictor	~	Predicted Quantity

3.2 Overview of the Dynamic Clustering Method

An overview of the proposed DCM is shown in . The DCM is a three-phase process involving tracking, clustering and prediction algorithms. The inputs to the DCM are the sensor measurements ($\mathbf{z}_i = [x_{pi} \; y_{pi}]'$, $i=1,...,n$, where n is the number of particles) for the location of each particle, and the outputs are estimated location and speed of the clusters ($\hat{\mathbf{c}}_j = [\hat{x}_{cj} \; \hat{\dot{x}}_{cj} \; \hat{y}_{cj} \; \hat{\dot{y}}_{cj}]'$, $j=1,...,m$, where m is the number of clusters found). Note that the initial value of m (the number of clusters) before the DCM is started is set to zero. Once it is running, the DCM will automatically estimate the value of m.

The *particle tracking algorithm* uses a Kalman filter on each sensor measurement ($\mathbf{z}_i$), to produce a smoothed estimate of the position and velocity state vector ($\hat{\mathbf{p}}_i = [\hat{x}_i \; \hat{\dot{x}}_i \; \hat{y}_i \; \hat{\dot{y}}_i]'$) for the i-th particle. The *cluster generation algorithm* then uses the estimated dynamic state information ($\hat{\mathbf{p}}_i$) to group the particles into clusters. It yields the number (m) of clusters found and position/velocity vectors ($\hat{\mathbf{c}}_j = [x_{cj} \; \dot{x}_{cj} \; y_{cj} \; \dot{y}_{cj}]'$) for the centers of these clusters. Finally, a *prediction algorithm* is applied to each cluster state $\hat{\mathbf{c}}_j(k)$ to predict its one-step ahead cluster motion state vector $\tilde{\mathbf{c}}_j$. The predicted vector is used in the clustering block at the next sampling iteration as an initial condition for the clustering algorithm.

3.3 Particle Tracking Algorithm

Each particle can be modeled as a second order stochastic dynamical process described by the state equation:

$$\frac{d\mathbf{p}_i}{dt} = \begin{bmatrix} 0 & 1 & 0 & 0 \\ 0 & 0 & 0 & 0 \\ 0 & 0 & 0 & 1 \\ 0 & 0 & 0 & 0 \end{bmatrix} \mathbf{p}_i + \begin{bmatrix} 0 \\ w_x \\ 0 \\ w_y \end{bmatrix} \quad (1)$$

where $\mathbf{p}_i = [x_{pi} \; \dot{x}_{pi} \; y_{pi} \; \dot{y}_{pi}]'$ are the longitudinal and lateral positions and velocities, and the w_x and w_y represents the accelerations of the particle. For the formulation, the acceleration of the vehicle motion can be regarded as zero-mean Gaussian white noise, i.e., $w_x \sim N(0, Q_x)$ and $w_y \sim N(0, Q_y)$. The measurement of the particle position can be modeled as the output equation:

$$z_i = \begin{bmatrix} 1 & 0 & 0 & 0 \\ 0 & 0 & 1 & 0 \end{bmatrix} \begin{bmatrix} x_{pi} \\ \dot{x}_{pi} \\ y_{pi} \\ \dot{y}_{pi} \end{bmatrix} + \begin{bmatrix} v_x \\ v_y \end{bmatrix} \quad (2)$$

where are $v_x \sim N(0, R_x)$ and $v_y \sim N(0, R_y)$ are the measurement noise. Note that Equations (1) and (2) satisfy Assumptions (A.1) and (A.2). Also note that the dynamical model (1) is extendible to a higher order process. Based on the measurement (2), an estimate of the state for the particle motion can be generated by a Kalman filter given by

$$\frac{d\hat{\mathbf{p}}_i}{dt} = \begin{bmatrix} 1-K_x & 0 & 0 & 0 \\ -K_{\dot{x}} & 1 & 0 & 0 \\ 0 & 0 & 1-K_y & 0 \\ 0 & 0 & -K_{\dot{y}} & 1 \end{bmatrix} \hat{\mathbf{p}}_i + \begin{bmatrix} K_x & 0 \\ K_{\dot{x}} & 0 \\ 0 & K_y \\ 0 & K_{\dot{y}} \end{bmatrix} z_i \quad (3)$$

where the closed form steady state filter gains, determined from the Riccati equation associated with the filter [6], can be shown to be

$$K_x = \sqrt{2\sqrt{Q_x/R_x}} \qquad K_{\dot{x}} = \sqrt{Q_x/R_x} \quad (4a)$$

$$K_y = \sqrt{2\sqrt{Q_y/R_y}} \qquad K_{\dot{y}} = \sqrt{Q_y/R_y} \qquad (4b)$$

The estimated motion for each particle is used in the cluster generation process.

3.4 Dynamic Clustering Algorithm (DCA)

As mentioned earlier, clustering techniques, such as the nearest neighborhood (NN) method, exist for grouping static particles or data points into clusters. The dynamic clustering algorithm developed in this paper is an extension of the static NN clustering technique that utilizes the information from the estimated position-velocity motion vector of the dynamic particles. The DCA is a simple and fast algorithm that automatically groups the dynamic particles and determines the number of dynamic clusters, as well as estimates the motion vector of each cluster. Another important difference in the DCA from the NN method is the use of the predicted motion vector $\tilde{\mathbf{c}}_j$ of the cluster centers as references for clustering decisions. This feature greatly improves the accuracy in clustering when cluster motion is smooth or non-abrupt, as stipulated by Assumption (A.3). A flow diagram for the DCA is shown in **Figure 3.1**. For each i-th particle, the distance (D_{ij}) from the particle to the predicted center of each j-th cluster, and the relative velocity (V_{ij}) between the same particle and cluster is determined. These calculations are shown in Equations (5a) and (5b):

$$D_{ij} = \left\| \hat{\mathbf{p}}_i - \tilde{\mathbf{c}}_j \right\|_{N_D}^{M_D} \text{ where } \mathbf{M}_D = \begin{bmatrix} 1 & 0 & 0 & 0 \\ 0 & 0 & 0 & 0 \\ 0 & 0 & 1 & 0 \\ 0 & 0 & 0 & 0 \end{bmatrix}, \; N_D = \infty \qquad (5a)$$

$$V_{ij} = \left\| \hat{\mathbf{p}}_i - \tilde{\mathbf{c}}_j \right\|_{N_V}^{M_V} \text{ where } \mathbf{M}_V = \begin{bmatrix} 0 & 0 & 0 & 0 \\ 0 & 1 & 0 & 0 \\ 0 & 0 & 0 & 0 \\ 0 & 0 & 0 & 1 \end{bmatrix}, \; N_V = \infty \qquad (5b)$$

where if $\mathbf{x} = [x_1 \; \ldots \; x_n]'$ and $\mathbf{M} = diag(m_1, \ldots, m_n)$, then the weighted N-norm of $\mathbf{x}$ is defined as $\left\| \mathbf{x} \right\|_N^M = \left(\sum_{k=1}^{n} m_k |x_k|^N \right)^{1/N}$. If the minimum D_{ij} is less than $D_{Threshold}$, and the minimum V_{ij} is less than $V_{Threshold}$, then the i-th particle is considered to be a member of the j-th cluster labeled as set C_j. If this condition is not true for $j = 1, \ldots, m$, then a new (*m+1*)-th cluster C_{m+1} is created, consisting of particle *i*. After classifying each particle, if there remain any clusters with no members, they are deleted.

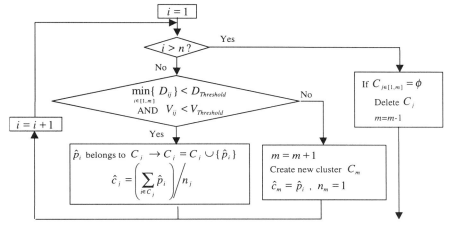

Figure 3.1 Dynamic clustering algorithm

3.5 Prediction Algorithm

The future position and velocity of each cluster ($\tilde{c}_j$) is predicted by using a simple one-step ahead predictor as shown in Equation (6). The predicted motion vector of each cluster is then used in the dynamic clustering algorithm, as described above. This step is important since it provides a more accurate reference for the clustering decisions.

$$\tilde{c}_j = \begin{bmatrix} 1 & \tau & 0 & 0 \\ 0 & 1 & 0 & 0 \\ 0 & 0 & 1 & \tau \\ 0 & 0 & 0 & 1 \end{bmatrix} \hat{c}_j \qquad (6)$$

4. Application to Traffic Monitoring Systems

4.1 System Description

The DCM has been verified in a real lidar-based traffic monitoring system installed on a military truck shown in **Figure 4.1**. The system consists of two identical lidar units mounted on the front of the truck. The two lidar are arranged on the truck such that their beam patterns intersect, as in **Figure 4.2**. This arrangement was chosen for two reasons: to widen the overall effective sensing area in front of the truck, and to increase the effective angular resolution by interlacing the lidar beams. The two lidars can together report up to six data points (one per beam) per sample. One cannot assume that the left sensor detects only vehicles on the left, since its beams spread equally in both directions. Therefore, a clustering method must be applied to associate the data points with actual vehicles in the path of the truck. For this experiment, both NN and DCM clustering techniques were applied and compared.

Figure 4.1 Laser radar installed a truck

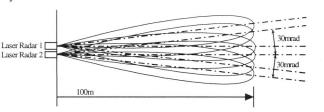

Figure 4.2 Lidar beam intersection

4.2 Traffic Monitoring Using DCM

Experiments have shown that DCM works well for tracking vehicles with lidar. **Figure 4.3** contains data generated by a typical highway driving situation. Each graph represents the time history of distance information collected by each lidar. The designations on the left indicate which beam of each lidar the graph represents (LL means Left lidar, Left beam, etc). The labels *a, b, c, d,* identify specific points in time that are of interest. This data set represents the motion of two vehicles of approximately equal width moving in highway traffic. One is on the left and the other on the right. At time *a*, the vehicles are well separated and have different velocities. At time *b*, the two vehicles have nearly equal position but opposite velocities. At time *c*, both vehicles are near and have similar velocities. At time *d*, the vehicles are very near each other and have a slight difference in velocity.

Figure 4.3(2) is the result of the NN method attempting to cluster the data correctly. At point *a*, NN correctly identifies each vehicle using distance information. At point *b* it correctly identifies two vehicles, but wrongly assigns all but one of the rightmost vehicle's particles to the leftmost vehicle. At time *c*, NN correctly identifies each vehicle. At time *d*, however, NN identifies only one vehicle when there are actually two.

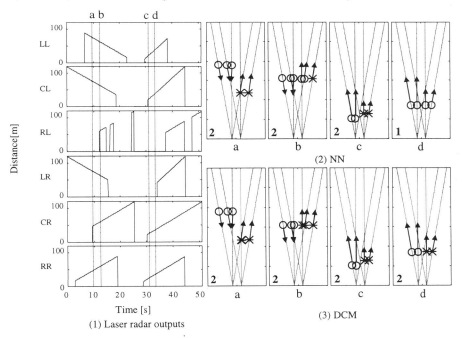

Figure 4.3 Comparison of NN and DCM

(1) graphs of lidar output over time. (2) Two dynamic clusters merge to become one when NN is applied. (3) Two dynamic clusters maintain their identity when DCM is employed.

Figure 4.3(3) shows the results of applying DCM to the same data set. Note how DCM correctly identifies both the number of clusters and the particle assignment to each cluster. DCM is able to perform so well because it accounts for not only particle position but also velocity. Thus, in situations such as *b* and *d*, cluster identity can be more reliably resolved.

5. Conclusion

The Dynamic Clustering Method (DCM) was developed for identifying and tracking moving clusters with constituent particles that move relative to their clusters. The proposed DCM is based on the conventional NN clustering method, but has a few key differences. First, DCM uses velocity information to classify moving particles into clusters. Also, DCM is intended to be used in real-time systems, continually tracking clusters and predicting their state for use at the next time step. DCM possesses distinct advantages over the NN clustering method. Most significantly, DCM can distinguish intersecting clusters by comparing their velocities. Also, DCM can predict initial cluster state vectors for each time step instead of "starting from scratch", which affords more reliable cluster tracking. The validity of DCM has been proven by means of simulation and actual application to a traffic monitoring system that utilizes laser radar sensors.

6. References

[1] Masakazu Iguchi, Takuro Miyazaki and Masayoshi Aoki "Promotion of the Program of Advanced Safety Vehicles for 21st Century," *Proc.IFAC 13th Triennial World Congress*, San Francisco,USA, pp.159-164
[2] Mark Hischke "Collision Warning Radar Interference" *Intelligent Transportation: Realizing the Benefits, Proc. of the 1996 Annual Meeting of ITS America*, pp.594-600
[3] Shinobu Kawashima, Kajiro Watanabe and Kazuyuki Kobayashi "Traffic Monitor by Kalman Filtering of the Laser Radar Signals," *Proc. IFAC 13th, Triennial World Congress*, San Francisco, USA, pp.165-169, 1996
[4] G.H.Pall and K.J.Hall "ISODATA: An Interactive Method of Multi-variate Data Analysis and Pattern Classification," *Proc. IEEE International Communication Conf.*, pp.116-117, 1966
[5] Donald E.Gustafson and William C.Kessel "Fuzzy Clustering with a Fuzzy Covariance Matrix," *Proc. IEEE CDC*, San Diego, CA, pp.761-766, 1979
[6] F. Lewis, *Optimal Filtering*, Wiley & Sons, 1990.

J HENRY, A VIEL AND J P YVON
Modeling of currents in a zirconium oxygen sensor

1 Introduction

Exhaust-gas sensors are commonly used in the automotive industry as a key element for the control of efficiency of the engine. The most commonly used oxygen sensor — the λ-sensor (see Fig. 1) — involves a measurement of the open circuit potential generated by the difference of oxygen partial pressure between exhaust gas and air. The underlying physical phenomenon is the migration of various species (electrons, holes, oxygen vacancies) throughout a crystal of Yttria stabilized Zirconia, this corresponds to a property of "semi-permeability" to oxygen. Such a system involves many chemical reactions due to the fuel mixture and the coating of the electrode (spinel). In what follows we will concentrate only on migration inside the zirconia, considering that external oxygen partial pressures are given (eventually function of time).

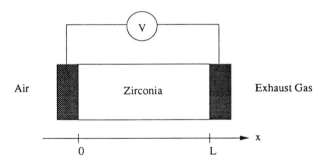

Figure 1: The Sensor

Classical literature on the subject utilizes an explicit expression of the voltage by means of the Nernst formula :

$$V = \frac{RT}{4F} \log \frac{P_{O_2,\text{air}}}{P_{O_2,\text{exhaust}}}.$$

Nevertheless, referring to the Nernst-Planck theory, the above relation is strictly verified only if the current carried by oxygen vacancies is null. In the sequel we will not set any *a priori* hypothesis on the currents or electroneutrality.

Among the great number of papers devoted to modeling and experimentation of oxygen sensors, few are devoted to precise analysis of migration of species in zirconia. In this paper we mainly analyze the models based upon a physical approach of the transport of charges in zirconia. In [9] a similar approach, but in a particular situation, does not take into account the variation of concentration of oxygen vacancies inside of the zirconia. This variation is emphasized in [10] where the presence of boundary layers for vacancy concentration is shown. This implies that there exist boundary layers for the potential and, clearly, this is closely related to the lack of electroneutrality near the boundaries of the zirconia.

2 Modeling : general equations

2.1 The chemical framework

An oxygen sensor (see Fig.1) is essentially made of a crystal of yttria-stabilized zirconia in which are created oxygen vacancies. Due to these vacancies, migration of oxygen through the sensor is permitted, this semi-permeability effect creates a current and a difference of potential which is used as the output of the sensor.

The general formulation of the problem is close to the one of semiconductors and a number of analogies will be invoked to settle the equations.

Without getting into the details of the chemical process, let say that yttria-stabilized zirconia is obtained by means of a chemical doping of the zirconia by Y_2O_3 following the reaction :

$$2Y'_{Zr} + V_o^{\cdot\cdot} \rightleftharpoons 0 \tag{1}$$

where $V_o^{\cdot\cdot}$ denotes the (doubly positive) charged oxygen vacancies (we have employed the notation of Kröger and Vink [3] : the positive charge is denoted by a dot in exponent and the negative charge by a quote) and Y'_{Zr} is the dopant species which is fully ionized and considered as being *immobile*. From now on we may consider the concentration of ions Y'_{Zr} as given and fixed (a great number of data can be found in the existing literature and mainly in [6]), it will be denoted by c_Y in the sequel.

At the boundary of the crystal (which usually is an interface between a porous electrode and the zirconia) there is fixation of oxygen in the vacancies giving the following reaction :

$$\frac{1}{2}O_2 + V_o^{\cdot\cdot} \rightleftharpoons O_o^x + 2h^{\cdot}. \tag{2}$$

This reaction indicates that occupancy of a vacancy by an oxygen creates simultaneously two *holes* which will contribute to the electric current.

The third species – electrons – is related to holes via the reaction

$$h^{\cdot} + e' \rightleftharpoons 0. \tag{3}$$

In order to formulate the dynamical equations we will denote the concentrations of the various species by c_i where i stands for e (electrons), h (holes), V (oxygen vacancies) or Y (dopant Y'_{Zr}).

2.2 Transport of species

In what follows we will consider concentrations as function of one variable x, $0 \leq x \leq L$, where L is the thickness of the zirconia. In classical situations both ends of the electrode are in contact with gas : air (reference electrode) at one end and exhaust gas (measurement electrode) at the other end ; in this case the output of the sensor is the voltage (λ-sensor) [1]. In the same way the *electric potential* will be denoted by $\phi(x)$. The classical Nernst-Planck equation can be used to describe the migration of each species :

$$\frac{\partial c_i}{\partial t} = -\nabla N_i + R_i \quad (4)$$

where R_i is a reaction term — which is discussed later on — and N_i, the flux of species i, is given by

$$N_i = -D_i \nabla c_i - z_i F u_i c_i \nabla \phi \quad (5)$$

where D_i refers to the ionic diffusion coefficient, z_i is the charge number, F is the Faraday constant and u_i is the mobility. The potential is governed by the classical Poisson's equation :

$$-\varepsilon \Delta \phi = F \sum_{i=e,h,V,Y} z_i c_i. \quad (6)$$

Remark 2.1 The classical electroneutrality condition would lead to

$$z_V c_V + z_e c_e + z_h c_h + z_Y c_Y = 0, \quad (7)$$

but, as we plan to study the possible boundary layers of potential, it is essential to start the study in a general framework, even if we know that the coefficient ε is small. □

By analogy with the semiconductor framework, the reaction term in (4) corresponds to a recombination-generation term – for holes and electrons – and can be written as

$$R_e = R_h = k_0(n_i^2 - c_h c_e) \quad (8)$$

where n_i is the intrinsic carrier concentration. The relation (8) can be viewed as the law of mass action for reaction (3). This term is irrelevant for the oxygen vacancies.

---

[1] Some other types of electrode can be used, in particular the blocking electrode (see [9] and section 4.2) where there is no gas exchange at the reference electrode. In this situation a difference of potential is fixed and the current intensity is the output of the sensor.

Remark 2.2 Introduction of the previous reaction term (8) indicates that electron and hole concentrations are somewhat independent and we do not assume thermodynamical equilibrium, expressed by the relation $c_e c_h = n_i^2$, *inside* the zirconia but, as we will see later on, we will assume it holds only at the *boundaries*. □

3 Boundary conditions

3.1 General analysis

A preliminary remark on notations : the analysis of boundary conditions will be made at *any* boundary of the zirconia, for this reason we will not indicate if the condition is written at $x = 0$ or at $x = L$.

For this type of system the main governing parameter is the partial pressure of oxygen at the boundaries (at least in the case of a λ-sensor). A classical approach consists in considering that reaction (2) is at equilibrium at each time ; by using the mass action law this leads to

$$c_h^2 [O_O^x] - k_e P_{O_2}^{1/2} c_V = 0 \qquad (9)$$

where k_e is a convenient equilibrium constant and $[O_O^x]$ is the concentration of fixed oxygen moles.

Remark 3.1 The equation (9) can be found in the classical literature (see, for instance, [1]). In most cases one considers both C_V and $[O_O^x]$ as constant and the hole concentration is directly given by

$$c_h = K P_{O_2}^{1/4} \qquad (10)$$

where K is a "true" constant, i.e. independent of the concentration of fixed oxygen and vacancies. This assumption is made by [9] and [6] where the corresponding relation is given by

$$\sigma_i = \alpha_i \exp\left(-\frac{\beta_i}{kT}\right) P_{O_2}^{1/4}, \qquad i = e, h$$

with

$$\sigma_i = |z_i|^2 F^2 c_i u_i, \qquad i = e, h.$$

In fact in many papers (e.g. in [1]) the relation (10) is considered as being valid *inside* the zirconia, but this will not be considered here. □

3.2 A first model : electroneutral boundary conditions

The classical boundary conditions for semiconductor–metal junction (see for instance the book by Markowich [4]) lead to assume that :

- the system is in thermodynamical equilibrium at the boundary :

$$c_e c_h = n_i^2, \tag{11}$$

- electroneutrality holds at the boundary (see (7)):

$$z_V c_V + z_e c_e + z_h c_h + z_Y c_Y = 0, \tag{12}$$

For oxygen sensors we need to introduce the reaction of fixation of oxygen, which is governed by equation (9), and one can write

$$c_h^2(\bar{c}_V - c_V) - k_e P_{O_2}^{1/2} c_V = 0 \tag{13}$$

where $\bar{c}_V$ is the initial concentration of vacancies (resulting from the doping).

Remark 3.2 It should be noted that, at a given boundary, the system of equations (11) (12) (13) can be solved for any given oxygen pressure. This implies that for λ-sensors the system of partial differential equations (4)(6) to be solved is associated to Dirichlet boundary conditions for the species. The boundary conditions for the potential remain to be analyzed. □

The usual conditions for metal-semiconductor contacts imply equality of so called "Fermi levels" on both sides of the junction. For electrons the Fermi level ψ_e is defined by the relation

$$C_e = n_i \exp\left(-\frac{\psi_e - \phi}{U_T}\right) \tag{14}$$

with $U_T = \frac{RT}{F}$.

Then, if V_a is the applied potential at the metal in contact with the electrode, we must have $\psi_e = V_a$ at the boundary. This leads to the classical relation :

$$\phi = V_{bi} + V_a \tag{15}$$

where the so-called "built in" potential V_{bi} is given by

$$V_{bi} = U_T \log \frac{C_e}{n_i}. \tag{16}$$

Remark 3.3 Actually, as there is an affine relationship between the Fermi level and the *electrochemical* potential, this analysis simply shows that, at the boundary, the applied potential and electrochemical potential of the electrons are equal. The discontinuity of the electric potential is compensated by the pure chemical potential. The conditions (15)(16) give appropriate boundary conditions to solve the Poisson's equation (6). □

3.3 Boundary conditions : a more general case

3.3.1 Natural conditions

Let us consider one end of zirconia, for instance $x = 0$, and let us denote by $\tilde{c}_i$, $i = h, e, V$, the concentrations and $\tilde{\phi}$ the given potential at this point.

It is clear that analysis of contacts between semiconductors and metals is a quite difficult problem, (see, for instance, [8]) and the situation is probably worse in this context. We now propose "natural" boundary conditions. In the framework of semiconductor analysis, the boundary conditions which are compatible with the limit of electroneutrality are

$$c_h - c_e = 0$$

which means continuity of total concentration of mobile charges through the interface. By analogy we propose the following condition :

$$2\tilde{c}_V + \tilde{c}_h - \tilde{c}_e = 0. \tag{17}$$

If we assume that there is no limitation of the electron mobility and of oxygen diffusion at the interface and thermodynamical equilibrium we get, once again, the relations (11)(13) :

$$\tilde{c}_e \tilde{c}_h = n_i^2, \tag{18}$$
$$\tilde{c}_h^2 (\bar{c}_V - \tilde{c}_V) = k_e P_{O_2}^{1/2} \tilde{c}_V. \tag{19}$$

The boundary conditions on the potential are, in this case, simply $\phi = \tilde{\phi}$.

3.3.2 The reduced problem

If we take the limit under electroneutrality, it is possible to follow the reasoning of [2]. If we introduce the real $d > 0$ solution of equation :

$$2d^2 \tilde{c}_V + d\tilde{c}_h - d^{-1}\tilde{c}_e - c_Y = 0$$

the boundary conditions for the reduced problem will be

$$c_e(0) = d^{-1} \tilde{c}_e, \quad c_h(0) = d\tilde{c}_h, \quad c_V(0) = d^2 \tilde{c}_V,$$

and, for the potential

$$\phi(0) = \tilde{\phi} - U_T \log d.$$

Boundary conditions at $x = L$ can be obtained in a similar way. It must be noticed that the potential ϕ is determined by the continuity equations (4) and the electroneutrality condition (12) which holds true at every point x, $0 \leq x \leq L$.

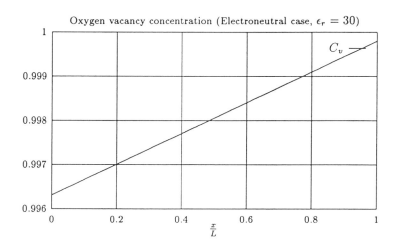

Figure 2: Oxygen vacancy concentration under electroneutrality at BC

4 Numerical analysis and results

4.1 Numerical approximation

In what follows we will restrict ourselves to the *stationary case* (the evolution case can be treated by the same approach). It is clear that, whatever the boundary conditions, the main difficulty is due to the small parameter ε and the system to solve is similar to the one coming from semiconductor analysis. The numerical scheme used here is due to [5] and is similar to the classical Scharfetter-Gummel [7] scheme for the approximation of fluxes.

4.2 An example : the blocking-electrode

The first example is the case where one end, say at $x = 0$, of the zirconia is linked to an electrode which allow electrons and holes to pass while blocking the flow of oxygen vacancies (see [9]). At $x = 0$ the chemical reaction balance (13) is replaced by the simple condition
$$N_V(0) = 0, \tag{20}$$
where N_V is given by (5) with $i = V$.

Under electroneutrality condition at both ends (this corresponds to the situation described in section 3.2) the numerical results are very similar to those obtained

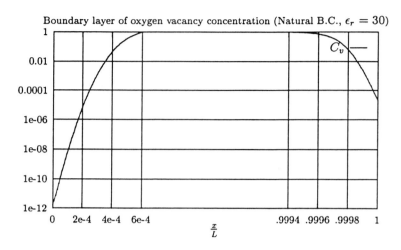

Figure 3: Boundary layer for the vacancy concentration

by [9], *although they are obtained under different assumptions on the model*. Actually the condition $N_V = 0$ holds true everywhere and the potential ϕ is linear and approximately constant. The corresponding results are shown on Fig.2 for oxygen-vacancy concentration, the current $N_h - N_e$ being constant along the zirconia.

On Fig.3 the boundary layers for the vacancy concentrations are shown in the case of a blocking electrode. In particular one can notice the strong differences between the concentrations at the two ends : in $x = 0$ where is the blocking electrode and at $x = L$ where is the gas to analyse.

References

[1] Desportes Ch. et al., *Electrochimie des solides*, Presses Universitaires de Grenoble, 1994.

[2] Henry J., Louro B., *Asymptotic analysis of reaction-diffusion-electromigration systems*, Asymptotic Analysis, No 10, pp 1-24, North-Holland, 1995.

[3] Kröger F.A., Vink H. J., in *Solid State Physics – Advances in Research and Applications*, Vol 3, p 307, Academic Press, New-York, 1956.

[4] Markowich P.A., *The Stationnary Semiconductor Device Equations*, Springer-Verlag, Wien, 1986.

[5] Montarnal Ph., Perthame B., *Asymptotic analysis of the drift-diffusion equations and Hamilton-Jacobi equations*, Mathematical models and methods in Applied Sciences (to appear) and INRIA Report No 2885 (1996).

[6] Park J.H., Blumenthal R.N., *Electronic Transport in 8 Mole Percent Y_2O_3-ZrO_2*, J. Electrochem. Soc., Vol. 136, No 10, pp 2867-2876, 1989.

[7] Scharfetter D.L., Gummel H. K., *Large-signal analysis of a silicon read diode oscillator*, IEEE Trans. on Electron Devices, ED-16 No 1, pp 64-77, 1969

[8] Schroeder D., *Modelling of Interface Carrier Transport or Device Simulation*, Springer-Verlag, 1994.

[9] Verbrugge M. W., Dees D.W., *Theoretical Analysis of a Blocking-Electrode Oxygen Sensor for Combustion-Gas Streams*, J. Electrochem. Soc., Vol 140, No 7, pp 2001-2010, 1993.

[10] Wang T., Soltis R.E., Logothetis E.M. et al., *Static Characteristics of ZrO_2 Exhaust Gas Oxygen Sensors*, SAE Techn. Paper No 930352, SAE International Congress and Exposition, Detroit Michigan, 1993.

Authors adresses

Jacques Henry
INRIA, B.P. 105, 78153 LE CHESNAY Cedex - France
e-mail : Jacques.Henry@inria.fr

Antoine Viel
UTC, Dept. GI, B.P. 649, 60206 COMPIEGNE Cedex - France
e-mail : Antoine.Viel@utc.fr

Jean-Pierre Yvon
INSA Rennes, 35043 RENNES Cedex - France
e-mail : Jean-Pierre.Yvon@insa-rennes.fr

ILYA KOLMANOVSKY, PAUL MORALL, MICHIEL VAN NIEUWSTADT AND ANNA STEFANOPOULOU

Issues in modelling and control of intake flow in variable geometry turbocharged engines

1 Introduction

Advanced hardware components are increasingly being considered for production of passenger car internal combustion engines to meet stricter emission regulations and customer demands of improved fuel economy and drivability. When integrated in a single engine configuration, these advanced hardware components may result in significant nonlinearities and interactions thereby requiring advanced control methods. In this paper we consider one such situation when demands for increased engine power, improved fuel economy and drivability provide a rationale for utilizing a variable geometry turbocharger while emission regulations necessitate an external exhaust gas recirculation system. However, careful modelling and analysis of the integrated system (Figure 1) uncovers difficult issues that a control designer must face in order to realize the benefits of advanced hardware components.

An automotive turbocharger consists of a compressor and a turbine coupled by a common shaft. The engine exhaust gas drives the turbine which drives the compressor which, in turn, compresses ambient air and directs it into intake manifold [5]. Since the increased quantity of air can be delivered to engine cylinders, a larger quantity of fuel can be burnt thereby providing larger torque output as compared to non-turbocharged engines. The turbocharging also improves fuel economy due to improved efficiency of engine operation at lean air-to-fuel ratios.

Turbocharging also affects regulated engine emissions such as particulates (PM), oxides of nitrogen (NOx), hydrocarbons (HC) and carbon monoxide (CO). Specifically, increased air-to-fuel ratio, air charge density and temperature resulting from turbocharging tend to reduce particulate emissions. However, increased charge temperature and oxygen availability tend to increase NOx formation. An intercooler is used to reduce air charge temperature and partially offset NOx increase caused by turbocharging. The intercooler also increases air charge density and thus reduces particulate emissions.

However, the use of intercooling alone is not sufficient to reduce NOx emissions to regulated levels. The exhaust gas recirculation (EGR) system is used to divert a portion of the exhaust gas back to the engine intake manifold to dilute the air supplied by the compressor. In the cylinders the recirculated exhaust gas acts as an inert gas and it increases the specific heat capacity of the charge. This reduces the burn rate, lowering peak flame temperatures, and, hence, decreases formation of NOx. The exhaust gas recirculation is, typically, accomplished with an exhaust gas recirculation valve that connects the exhaust manifold with the intake manifold. Some diesel engine configurations also include an EGR throttle between the compressor and the intake manifold to lower the manifold pressure and thus create a sufficient pressure drop across the EGR valve needed for high EGR rates. A judicious selection of EGR rates is important because excessive EGR rates lead to excessive particulate emissions. The EGR cooler is used to decrease EGR temperature and thus it contributes to further NOx reduction.

A turbocharger is typically sized to a particular engine so that it provides fast airflow response when the driver demands acceleration at low engine speeds. Fast increase in fresh air charge delivered to engine cylinders allows to rapidly increase fueling rate and, hence, engine torque. To achieve this effect the turbine flow area has to be relatively small to produce fast exhaust manifold pressure rise and, hence, fast increase in power generated by the turbine in response to a fueling rate increase. At high engine speeds and loads such a turbine may result in an excessive difference between exhaust and intake manifold pressures and thus impair fuel economy, and, in extreme cases, the resulting very high value of intake manifold pressure may damage the engine. One approach that avoids negative consequences of a small turbine area at high engine speeds and loads is to use a wastegate. The wastegate is opened at high engine speeds and loads to allow some of the exhaust gas to bypass the turbine. An alternative approach which avoids using the wastegate and grants more flexibility in shaping engine torque response, improving fuel economy and reducing emissions, is to use a variable geometry turbocharger (VGT). The turbine stator of a variable geometry turbocharger is equipped with a system of pivoted guide vanes. By operating the guide vanes the turbine flow area and the angle at which the exhaust gas is directed at the turbine rotor blades can be changed. By moving the guide vanes the turbine can be optimally sized for each engine operating condition to meet the requirements of fast torque response, fuel economy, low emissions and engine safety. An excellent description of various aspects of VGT operation can be found in [3].

Hereafter, we focus on the diesel engine configuration shown in Figure 1 without the EGR throttle and heat exchangers (intercooler and EGR cooler) and with nominal injection timing. The omissions do not change the fundamental aspects of the engine behavior.

2 Model

The investigation relies on a mean-value engine model developed by techniques described in [1, 2, 4]. The model has seven states. The six states, ρ_1, F_1, p_1, ρ_2, F_2, p_2 represent the gas dynamics in intake manifold and exhaust manifold. Specifically, ρ stands for gas mass (kg/m^3), F for burnt gas fraction, and p for pressure (kPa). The subscript 1 identifies the intake manifold and the subscript 2 identifies the exhaust manifold. The burnt gas fractions, F_1 and F_2, are defined as the density fraction of the inert combustion products in their mixture with air for the intake manifold and for the exhaust manifold, respectively. They are used to account for the amount of fresh air and burnt gas recirculated back to the engine. Because of lean combustion, as much as half of the flow through the EGR valve may be fresh air that can participate in combustion and, if properly accounted for, can be used to burn additional fuel. The seventh state is the turbocharger speed, N_{tc} (rpm). Very high turbocharger speeds (up to 200 krpm) are not unusual for medium size diesel engines. The control inputs are the EGR valve position χ_{egr} (between 0 and 1, where 1 is fully open) and the VGT actuator position χ_{vgt} (between 0 and 1, where 1 is fully open). Other external inputs include the fueling rate, W_f (kg/hr), and the engine speed, N (rpm).

The plant model is represented by the following equations which follow from the fundamental laws of mass and energy conservation for intake and exhaust manifolds and from the torque balance on the turbocharger shaft:

$$\begin{aligned}
\dot{\rho}_1 &= \frac{1}{V_1}\left(W_{c1} + W_{21} - W_{1e}\right), \\
\dot{F}_1 &= \frac{W_{21}(F_2 - F_1) - W_{c1}F_1}{\rho_1 V_1}, \\
\dot{p}_1 &= \frac{\gamma R}{V_1}\left(W_{c1}T_{c1} + W_{21}T_2 - W_{1e}T_1 - W_{12}T_1 - \frac{\dot{Q}_1}{c_p}\right), \\
\dot{\rho}_2 &= \frac{1}{V_2}\left(W_{e2} - W_{2t} - W_{21} + W_{12}\right), \\
\dot{F}_2 &= \frac{W_{e2}(F_{e2} - F_2)}{\rho_2 V_2}, \\
\dot{p}_2 &= \frac{\gamma R}{V_2}\left(W_{e2}T_{e2} - W_{2t}T_2 - W_{21}T_2 + W_{12}T_1 - \frac{\dot{Q}_2}{c_p}\right), \\
\dot{N}_{tc} &= \frac{30^2 c_p}{\pi^2 I_{tc}}\left(\frac{\eta_{tm}W_{2t}(T_2 - T_{tout}) - W_{c1}(T_{c1} - T_{amb})}{N_{tc}}\right).
\end{aligned} \qquad (1)$$

The variables in the right-hand side of these equations are either constant parameters or can be expressed as nonlinear functions of the seven states and inputs.

Specifically, W stands for a mass flow rate (kg/sec) where the first subscript identifies the flow upstream location while the second subscript identifies the flow downstream location. The subscript c stands for compressor, t for turbine, e for engine, 1 for intake manifold and 2 for exhaust manifold. The backflow through the EGR valve is represented by W_{12} and either $W_{21} = 0$ or $W_{12} = 0$. We do not model the backflow through either the compressor or the turbine since during normal operation these backflows do not occur.

The temperatures (K) of the flows are denoted by T with two subscripts and the same convention as for the flows. The temperature, T_{tout}, is the turbine outlet temperature. The temperatures inside intake and exhaust manifolds are denoted by T_1 and T_2, respectively. Due to low gas velocities the differences between static and dynamic pressures and temperatures are neglected.

The heat transfer rate to the surroundings for the intake manifold is denoted by $\dot{Q}_1$ and for the exhaust manifold it is denoted by $\dot{Q}_2$. The heat transfer effects in the intake manifold can be neglected, $\dot{Q}_1 = 0$. Because of high temperatures the heat transfer effects in the exhaust manifold can be quite significant. However, we still may use $\dot{Q}_2 = 0$ in Equations (1) because the temperature of the flow out of the engine, T_{e2}, is specified by a static map that already accounts for heat transfer effects in steady-state. Indeed, this map is developed from the experimental data for the gas already in the exhaust manifold. A better approximation is to use a transient heat correction term that is zero in steady-state, introduced e.g. by modelling exhaust manifold wall temperature dynamics.

The constant parameters include the volumes (m^3) of intake manifold, V_1, and of exhaust manifold, V_2, the specific heats at constant pressure and constant volume (kJ/kg/K), c_p and c_v, and their ratio and difference, $\gamma = \frac{c_p}{c_v}$, $R = c_p - c_v$, the turbocharger inertia (kg m^2), I_{tc}, and the turbocharger mechanical efficiency (between 0 and 1), η_{tm}, which accounts for turbine power losses due to friction. Strictly speaking, both temperature and composition affect c_p, c_v, R and γ. However, this dependence is relatively weak and may be treated as an uncertainty. The ambient pressure and temperature are denoted by p_{amb} and T_{amb}.

We summarize the dependencies of the intermediate variables in the right hand side of Equations (1) on state variables and inputs variables in Table 1. Some of these dependencies are obtained by fitting steady-state experimental engine mapping data with appropriate nonlinearities while others follow from physics. Similar static maps are employed to represent engine brake torque and emissions (smoke, PM, HC, CO and NOx). Engine speed is treated as a time-varying parameter.

We now make several observations pertinent to variable geometry turbine modelling. The lack of space prevents us from going further into details. The experimental data indicated that the variable geometry turbine mass flow rate, W_{2t}, is not a strong function of the turbocharger speed, N_{tc}. Therefore, this dependence was omitted (see Table 1) and a modification of the orifice flow equation with effective flow area, modelled a polynomial function of χ_{vgt}, was used to calculate W_{2t}. The turbine isentropic efficiency map, used to calculate T_{tout} however, does depend on N_{tc}.

The engine model also needs to account for engine cycle delays. For example, the change in fueling rate does not instantaneously affect the temperature of the exhaust gas leaving the engine since it takes one engine event until the exhaust valve opens. The engine cycle delays are engine speed (and, hence, time) dependent and they can be important for low engine speeds (e.g. at idle) and are less important for medium and high engine speeds. The delays can be included by using finite-dimensional approximations or their effect can be approximately captured by slightly increased exhaust manifold volume.

3 Control Objectives

A control system for a diesel engine must meet driver torque demand while satisfying constraints on emissions. There are two types of emission constraints. The first type are pointwise-in-time constraints on smoke emissions arising from customer requirements that no visible smoke emissions can be tolerated. The second type are cumulative over time constraints on NOx, HC and particulate emissions that need to be met during the government emission testing cycles (e.g. FTP cycle in the US and the Euro cycle in Europe). Visible smoke can be avoided by keeping the air-to-fuel ratio sufficiently lean. The treatment of constraints on NOx, HC and particulates is more complex because they are integral constraints only on a certain class of trajectories of the closed loop system. Neither emissions nor torque can be measured in production vehicles due to the cost and reliability issues associated with sensors.

It is important to understand the main steps in the diesel engine controller development. First, a static map is developed that provides the demanded steady-state fueling rate, W_f, as a function of N and the driver's pedal position. The analysis of the emission cycles yields optimal steady-state set-points for χ_{egr} and χ_{vgt} for each N and W_f. These set-points may also be uniquely defined by specifying the values of two internal process variables instead of χ_{egr}, χ_{vgt}. If sensors or estimators for these internal variables are available, a feedback controller can be designed that generates commands for χ_{egr} and χ_{vgt} to force these two internal variables to follow the set-points. The use of internal variables for feedback often assures better robustness properties of the engine operation. Typically, diesel engines are equipped with sensors for intake manifold pressure, p_1, and compressor mass flow rate, W_{c1}, and the feedback on these two variables is used. During transients caused, for example, by the driver's pedal tip-in the main objective of the control system is to provide sufficient fresh air charge to the engine as soon as possible so that the fuel demanded by the driver can be burnt without causing visible smoke. Lack of fresh air causes the control system to limit the fueling rate and is responsible for undesirable "turbo-lag" or sluggish diesel engine torque response. The secondary objective during transients is to provide the engine, whenever possible, with a sufficient amount of recirculated burnt gas to reduce NOx emissions. Large tracking errors may be admissible and, in fact, may be required during transients to meet these objectives.

4 Plant Properties

We use the diesel engine model to exhibit important properties of VGT/EGR diesel engine dynamics. These observations may be useful early on in the control design process when selecting an appropriate control system configuration.

First, note that for each quadruple $(N, W_f, \chi_{egr}, \chi_{vgt})$ there exists a unique equilibrium of the diesel engine model which is asymptotically stable. Figure 2 shows the equilibrium values of p_1, W_{c1}, F_1, in-cylinder air-to-fuel ratio $AFR = W_{1ea}/(W_f)$, $W_{1ea} = (1 - F_1)W_{1e}$, and p_2 as functions of χ_{egr} and χ_{vgt} for $N = 2000$ rpm, $W_f = 5$ kg/hr. Both F_1 and AFR are important because they affect engine emissions. Increasing the value of F_1 tends to reduce NOx emissions while small values of AFR lead to excessive smoke and high levels of particulate emissions. Opening the EGR valve increases F_1 but decreases AFR in a monotonic way. On the other hand, the dependence of AFR and F_1 on χ_{vgt} is nonmonotonic: For a fixed χ_{egr}, smallest values of F_1 result when χ_{vgt} is approximately 0.5. Typically, AFR decreases when F_1 increases. Since a feedback controller relies on sensor measurements of p_1 and W_{c1} it is of interest to examine equilibrium values of these variables. The effect of χ_{egr} on steady-state values of p_1 and W_{c1} is monotonic throughout the operating region: increase in χ_{egr} causes p_1 and W_{c1} to decrease. An increase in χ_{vgt} causes reduction in p_1. Hence, VGT can act as a wastegate and prevent overboosting the engine at high fueling rates. Note that p_2 is well-correlated with p_1 and the difference $p_2 - p_1$ increases when p_1 increases. Hence, opening the VGT for high fueling rates may improve fuel economy by reducing pumping losses associated with large values of the difference $p_2 - p_1$. Closing the VGT for low fueling rates helps maintain a larger pressure ratio across the EGR valve and thus increase the exhaust gas recirculation and reduce NOx emissions. The effect of χ_{vgt} on steady-state values of W_{c1} is nonmonotonic when the EGR valve is large open: when VGT is almost closed, an increase in χ_{vgt} causes W_{c1} to increase (point "b") while when VGT is almost completely open an increase in χ_{vgt} causes W_{c1} to decrease (point "c"). If the EGR valve is almost closed but χ_{vgt} is the same as for point "b", an increase in χ_{vgt} causes W_{c1} to decrease (point "a"). Since the regions where this dc-gain reversal takes place are uncertain, the analysis suggests that to avoid possible loss of stability it is best not to use the VGT to track setpoints in W_{c1} but use the EGR valve for this purpose.

To understand the dynamic properties of the plant, consider the Bode magnitude plots in Figure 3 for a plant linearization at four operating points specified by the quadruples $(N, W_f, \chi_{egr}, \chi_{vgt})$: (a) $[1000, 2, 0.8, 0.2]$; (b) $[3500, 7, 0.8, 0.2]$; (c) $[2500, 5, 0.8, 0.2]$; (d) $[2500, 5, 0, 0.2]$. The outputs have been scaled as $\bar{W}_{1ea} = 60W_{1ea}$, $\bar{W}_{c1} = 60W_{c1}$ and $\bar{p}_1 = p_1/100$. In all cases the same amount of fuel is injected per single engine stroke. Comparing cases (a), (b) and (c) we verify that the dynamics become faster as the engine speed increases. In case (a), the slowest eigenvalue is at -0.26, in case (b) it is at -2.34, and in the case (c) it is at -1.44. The dynamics slow down when the EGR valve opens up. In case (d) the slowest eigenvalue is at -2.28 as compared to -1.44 for case (c). From Figure 3 we see that at low engine speeds (point (a)) the EGR valve can increase fresh air flow to the engine, W_{1ea}, at a much slower rate than at higher engine speeds (point (b)), and that at low engine speeds the VGT can increase W_{1ea} more rapidly than the EGR valve. Consequently, a coordinated action of EGR valve and VGT is beneficial in the low speed region to rapidly increase fresh airflow in response to a driver's tip-in.

From Figure 3 it is obvious that we are dealing with a highly nonlinear plant that is also strongly coupled. For example, the dc-gains of the plant with inputs (χ_{egr}, χ_{vgt}) and outputs $(\bar{W}_{c1}, \bar{p}_1)$ at the four operating points are,

$$H_a = \begin{bmatrix} -0.56 & 0.68 \\ -0.11 & -0.05 \end{bmatrix}, H_b = \begin{bmatrix} -1.41 & 2.53 \\ -0.30 & -1.07 \end{bmatrix}, H_c = \begin{bmatrix} -0.80 & 1.27 \\ -0.11 & -0.68 \end{bmatrix},$$

$$H_d = \begin{bmatrix} -2.27 & -1.26 \\ -0.50 & -1.25 \end{bmatrix}.$$

Note that the dc-gain from VGT to W_{c1} changes its sign from point (c) to point (d), consistent with the previous discussion. The relative gain array analysis at operating points (b), (c), (d) suggests that if a decentralized control architecture is desired, EGR valve should be used to control W_{c1} and VGT should be used to control p_1. However, at the operating point (a) the relative gain array analysis suggests that the preferred pairing is to use EGR valve to control p_1 and VGT to control W_{c1}.

Because the plant is strongly coupled the existence of right-half plane zeros should not be surprising. Specifically, the transfer function from χ_{egr} to p_1 has a right-half plane zero while the transfer function from χ_{vgt} to W_{c1} has a right-half plane zero when EGR valve is closed or nearly closed. These nonminimum phase properties can be explained from physics, by demonstrating that the step response exhibits an undershoot. When the EGR valve opens, first the flow through the EGR valve increases thereby increasing the intake manifold pressure, p_1. However, since a smaller portion of the exhaust gas is supplied to the turbocharger and the exhaust manifold is emptied at a higher rate when the EGR valve opens, the intake manifold pressure, eventually, decreases. The latter effect is, however, delayed because of the turbocharger dynamics. Similarly, when the VGT opens instantaneously, the flow through the VGT increases instantaneously thereby increasing the power transferred to the compressor and the compressor mass flow rate, W_{c1}. Since the exhaust manifold is emptied at a higher rate, eventually, the exhaust manifold pressure will decrease causing the power transferred to the compressor and the compressor mass flow rate to drop. If the EGR valve opening is sufficiently small, the compressor mass flow rate drops to a steady-state value that is lower than the initial one. However, when the EGR valve is large open and we open the VGT, the final value of the compressor mass flow rate might be larger than the initial one and the response will no longer exhibit an undershoot but it will exhibit an overshoot. The MIMO system with inputs (χ_{egr}, χ_{vgt}) and outputs $(\bar{W}_{c1}, \bar{p}_1)$ can also be shown to be nonminimum phase. For the four operating points the system has a single right-half plane zero at (a) 68.97, (b) 62.0, (c) 73.19, (d) 64.7. A physics-based explanation for this nonminimum phase property is also available, suggested by M. Jankovic. Suppose a controller has been designed that holds p_1 and W_{c1} fixed, the system is initially at an equilibrium and there is an instantaneous increase in the exhaust manifold pressure p_2. To instantaneously compensate for possible increase in W_{c1} and p_1 the controller is forced to close EGR valve and VGT. But this leads to a further increase in p_2. Therefore the dynamics consistent with fixed p_1 and W_{c1} are unstable and the system is nonmimimum phase.

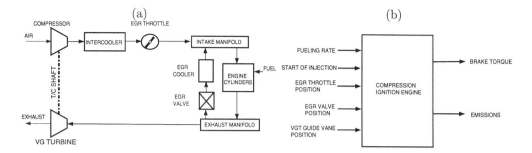

Figure 1: VGT diesel engine: (a) Main subsystems. (b) Inputs and outputs.

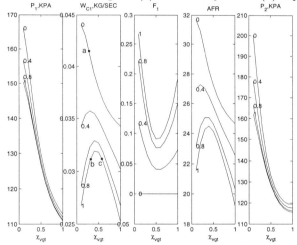

Figure 2: Equilibria for $N = 2000$ rpm, $W_f = 5$ kg/hr. Each line corresponds to constant χ_{egr} (shown on the plot) and varying χ_{vgt}.

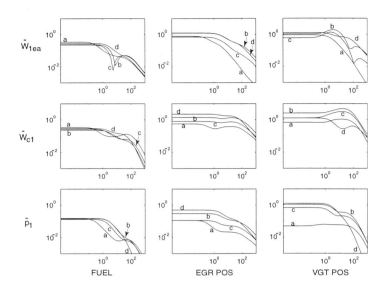

Figure 3: Bode magnitude plots at four operating points.

Variable	Depends as/on	Source
T_1	$T_1 = p_1/(\rho_1 R)$	ideal gas law
T_2	$T_2 = p_2/(\rho_2 R)$	ideal gas law
T_{c1}	p_1, N_{tc}	compressor isentropic efficiency map
T_{tout}	$p_2, T_2, N_{tc}, \chi_{vgt}$	turbine isentropic efficiency map
W_{c1}	p_1, N_{tc}	compressor flow map
W_{21}	$p_1, p_2, \rho_2, \chi_{egr}$	orifice equation
W_{12}	$p_1, p_2, \rho_1, \chi_{egr}$	orifice equation
W_{2t}	p_2, ρ_2, χ_{vgt}	turbine flow map
W_{1e}	ρ_1, N, T_1, p_2	engine volumetric efficiency map
W_{e2}	$W_{e2} = W_f + W_{1e}$	engine mass conservation
T_{e2}	T_1, F_1, W_f, W_{1e}	engine temperature rise map
F_{e2}	F_1, W_f, W_{1e}	stoichiometric combustion balance
$\dot{Q}_1$	$\dot{Q}_1 = 0$	neglected
$\dot{Q}_2$	$\dot{Q}_2 = 0$	heat transfer is partially accounted for in the map that specifies T_{e2}

Table 1: Static maps used in diesel engine model.

References

[1] Amstutz, A., and Del Re, L.R., "EGO sensor based robust output control of EGR in diesel engines," *IEEE Transactions on Control System Technology,* vol. 3, no. 1, 1995.

[2] Kao, M., and Moskwa, J.J., "Turbocharged diesel engine modeling for nonlinear engine control and estimation," *ASME Journal of Dynamic Systems, Measurement and Control*, Vol. 117, 1995.

[3] Moody, J.F., "Variable geometry turbocharging with electronic control," *SAE paper No. 860107*, 1986.

[4] Porter, B., Ross-Martin, T.J., and Truscott, A.J., "Control technology for future low emissions diesel passenger cars," *Proceedings of the Institution of Mechanical Engineers*, paper C517/035, 1996.

[5] Watson, N., and Janota, M.S., *Turbocharging the Internal Combustion Engine*, Wiley Interscience, New York, 1982.

ANDRZEJ W OLBROT, JOSEPH R ASIK AND MOHAMAD H BERRI
Robust controller based on Smith predictor for an electric EGR valve control system

Abstract

We provide an experimental proof that the Smith Predictor outperforms the PID under hard robustness constraints. It is well-known that the Smith predictor delivers excellent performance for the nominal system but its robustness is much poorer than of a corresponding PID controller. For this reason, the Smith controller never became popular in typical industrial applications. In this paper we change dramatically this viewpoint; we show that the Smith predictor can be desensitized (robustified) to achieve robust performance and stability over a wide range of parameter variations and can visibly beat the PID controller in terms of performance vs. robustness trade-offs. The basic idea for the robustification is to detune the nominal design to allow for less than optimal nominal performance. In particular, we require smaller than optimal rise time. As a result, the Nyquist plot of the nominal system is modified in such a way that large delay variations can be tolerated with moderate only loss of performance in the nominal case. The theoretical considerations are supported by a successful application to an automotive problem: An Exhaust Gas Recirculation (EGR) control system. The EGR system is briefly presented. Robustness is one of the main issues in this system due to nonlinearities, time delays, and the effects of complex exhaust gases dynamics. It was found after extensive measurements, identification and simulation experiments that a good approximation of the system dynamics consists of a first order transfer function with a time delay where both transfer function coefficients and the time delay strongly depend on the operating conditions (load, r.p.m., and the reference signal: desired EGR rate). The Smith predictor, well tuned to the nominal case, was found to have an excellent performance for the nominal system while the robustness was rather poor. The trade-off between the performance and robustness was achieved by relaxing the specifications for the rise time. Simulation results for both nominal and robust designs are presented as well as comparable results for an optimally tuned robust PID controller.

1. Introduction

Smith predictor or "Dead-Time Compensator" (DTC) was still a notable control research topic in the late seventies and early eighties though Smith's original paper was published as early as in 1957, [1]. Consequently, a large number of works have appeared (see [2-14] and refs. therein). Initially, it was very difficult to implement DTC algorithms because of the unavailability of digital process controllers. In the early eighties, some microprocessor-based industrial process controllers offered the DTC as a standard algorithm like the PID. However, since the DTC is composed of a primary controller and a model of the open loop process its tuning is not trivial and required the determination of a relatively large number of parameters even for simple process models. The tuning of DTC controllers is directly related to the stability and robustness properties of this model-based algorithm. The difficulty in tuning the Smith controller caused that it never became popular in typical industrial applications. It is well-known that the Smith predictor can deliver excellent performance for the nominal system but its robustness is much poorer as compared with a PID controller. In this paper we change dramatically this viewpoint; we show that the Smith predictor can be desensitized (robustified) to achieve robust performance and stability over a wide range of parameter variations and can visibly beat the PID controller in terms of performance vs. robustness trade-offs. The basic idea for the robustification is to detune the nominal design to allow for less than optimal nominal performance. In particular, we require smaller than optimal rise time. As a result, the Nyquist plot of the nominal system is modified in such a way that large delay variations can be tolerated with moderate only loss of performance in the nominal case. The theoretical considerations are supported by a successful application to an automotive problem: An Exhaust Gas Recirculation (EGR) control system, [15].

In this paper we will apply a modified (robustified) version of the Smith predictor to an automotive problem: An Exhaust Gas Recirculation (EGR) control system. The EGR control system is a subsystem of the electronic engine control system which controls the amount of exhaust gas recirculation (Fig. A). Engine cylinder temperature can reach more that $3000°$ F under normal operating conditions. The higher the temperature, the higher the exhaust NOx emissions. A small amount of exhaust gas (5 to 20%) is introduced into the cylinders to replace normal intake air. This results in lower NOx emissions. The control algorithm determines when the EGR valve is turned OFF or ON (Fig. B). The EGR valve is turned OFF during cranking, cold engine temperature (engine warm-up), idling, acceleration, or other conditions demanding high torque or when EGR is not desirable. The EGR control signal is determined by using input from engine speed (RPM), coolant temperature and engine load (mass air flow sensor). The EGR signal can control a valve opening which is detected by a valve position sensor or a differential pressure sensor. To perform the open-loop air/fuel ratio calculations the computer must know how much exhaust gas is being fed into the air intake. This is determined by using a sensor similar to the throttle position sensor or differential pressure sensor that gives an electrical signal which is related to the amount of opening of the EGR valve. The electronic engine control (EEC) calculates the desired EGR flow as a function of engine operating conditions and then uses the difference between this number and the actual EGR flow, as inferred from the measured pressure drop across an orifice in the EGR system, as input to a digital filter the output of which determines the duty cycles of the signal applied to the EGR valve drive.

2. Robustness issues

Although the electric EGR valve control problem can be approached by classical feedback control methods [2], there are a number of factors which render it difficult to achieve maximum possible performance by use of classical feedback design techniques. First, the plant gain is a very strong function of manifold vacuum. This means that, with fixed controller gains, those gains must be set fairly low in order to insure stability when the plant gain is high. Thus, under most operating conditions overall system response is not satisfactory. Another factor is that system non-linearities also require that controller be detuned. Examples have been recorded where the system is stable for a step input of 0 to 10% EGR, but goes into oscillation when a step to 20% EGR is required. Still another obstacles are large disturbances due to engine vibrations, pressure fluctuations and high order unmodeled dynamics. These uncertainty factors make the problem of achieving good control system performance very difficult. These difficulties are well illustrated by the results of real plant measurements and plant model identification.

3. Plant Model Identification

Since measurements of system inputs and outputs for all relevant operating conditions are practically very elaborate we chose to deal with only four representative operating conditions and then use interpolation to describe plant models and controllers for operating conditions other than the chosen four.

For identification we had one hundred twenty real experiment data files divided into three groups, in terms of the reference signal (EGR%) square wave: (5-10% EGR), (5-15% EGR), and (10-20% EGR). Out of each group, we selected good files with small disturbances, and decided to cut out windows to further minimize the impact of disturbances on the identification. Then we applied the standard Matlab identification procedures followed by manual fine tuning. As a result, we found that the first order models with time delay in the form of the following z-transfer functions

$$G_i(z) = \left(\frac{k_i}{z - p_i}\right) z^{-d_i} \qquad i = 1,2,3,4 \qquad (2.4.9)$$

are almost as accurate as higher dimensional models. Concluding the identification of the four first order models corresponding to four operating conditions $(EGR\%)_i$ $i = 1,2,3,4$. we obtained the model parameters as shown in Table 1 (sampling time is 10 msec).

Model #i: EGR%	Gain k_i	Delay d_i [msec]	Pole p_i
#1: 5%	88	50	0.9
#2: 10%	110	150	0.9
#3: 15%	180	130	0.9
#4: 20%	220	60	0.9

Table 1

As seen from the table the uncertainties in the model are very large: gain variation ratio is 220 : 88 and delay variations are 15 : 5. The sampling time is 10 msec. Hence, the largest time delay (150 msec in model #2) is larger than the plants time constant corresponding to the pole $p = 0.9$. This fact shows that the time delay is not negligible and, moreover, will have a substantial effect on the control performance. Therefore it is an interesting problem whether some controller structures, specially tailored for time delays, will provide sufficient performance under robustness constraints. We will show below that the Smith predictor actually outperforms the PID in our application.

3. Smith Predictor Control

The general structure of the Smith (predictor) controller follows from the (Smith) principle that if a controller $C_0(z)$ provides a given step response in the closed loop system with plant $G_0(z)$ then the Smith controller $C(z)$ for the plant

$$G(z) = z^{-d} G_0(z) \qquad (1)$$

provides an identical step response except that it is delayed by d sampling intervals. The formula resulting from this principle is as follows:

$$C(z) = \frac{C_0(z)}{[1+C_0(z)G_0(z)(1-z^{-d})]} \quad (2)$$

Of course, a similar formula is valid in continuous time but we will use the discrete time version only. By the very design principle, the Smith predictor compensates the impact of the time delay on the closed loop system performance in the case of perfect matching of the process by the model. On the other hand, Smith predictor is known to be very sensitive with respect to process/model mismatch [1], [2], [3], and [4]. We also support this claim in our case.

4. Sensitivity of the Original Smith Predictor

Consider the plant model #3, that is,

$$G_0(z) = \frac{180}{z - 0.9}, \qquad d = 13 \quad (3)$$

An original Smith predictor controller can be designed for this model based on, say, a PID type controller $C_0(z)$ optimized for fast step response and small overshoot. For instance, the controller

$$C_0(z) = 0.003\left(\frac{z - 0.87}{z - 1}\right) \quad (4)$$

provides an excellent performance in terms of the rise time and the overshoot. The rise time is 25msec and the overshoot is less than 3% (see Fig. 5). The corresponding Nyquist plot (see Fig. 6) for the nominal system shows classical "robust stability" with excellent gain (> 6db) and phase (> 45°) margins. In spite of having such good classical margins, this Smith predictor is very sensitive to the delay changes. When we increase the time delay by two units (sampling periods) the closed loop system becomes unstable and has a complicated oscillating time response which shows several critical frequencies as opposed to just one critical frequency of oscillations for typical systems without delays, (see Fig. 7). In the corresponding Nyquist plot (Fig. 8), the number of crossings of the real axis in the vicinity of the critical point -1 is fairly large. Therefore the number of frequency ranges over which the loop transfer function needs to be corrected is also large. Similar effects are obtained when we decrease the delay by two sampling intervals. It seems very difficult to modify the frequency response of this control loop by adding some extra lead-lag compensation in order to achieve robustness with respect to delay perturbations, that is, to get required delay margins. We will therefore abandon this excellent nominal controller and try a slightly detuned $C_0(z)$.

5. Robustified Smith Predictor vs. Robust PID

Experimental results on both real system and simulations convinced the authors that it is possible to obtain better performance / robustness trade-off as compared to the classical PID control if the controller $C_0(z)$ is mild in the sense that it does not try to optimize the transient response for the plant without delay $G_0(z)$ but rather achieves a suboptimal transient. This observation leads to a simple but important rule yielding a generation of control algorithms essentially better for practical applications in delay systems than traditional PID's. We will prove this point by successfully designing a Smith controller which beats any PID controller in performance under given robustness constraints for our application example (EGR valve).

Let us first determine the achievable performance and robustness for classical PID controllers. To set standards for robustness, let us assume that the overshoot will not exceed 25% for all parameter variations (However, this will be verified for our four models only since we do not have real data relative to operating conditions other than the chosen four). The following PID controller

$$C_0(z) = 0.000075\left(\frac{z-0.7}{z-1}\right) \quad (5)$$

achieves the shortest rise time (250 msec) at the maximum admissible overshoot (25%) for the nominal model #3 (15% EGR), see Fig. 1. The stability robustness is guaranteed as seen from Figs. 2, 3, and 4. A very poor performance of 800 msec rise time can be observed for model #1 (5% EGR) but this cannot be improved within the PID structure without violating the limits for overshoot for model #3.

Consider now the desensitized Smith predictor with $C_0(z)$ as follows

$$C_0(z) = 0.00035\left(\frac{z-0.7}{z-1}\right) \quad (6)$$

Note that its gain is more than four times larger than the gain of the robust PID controller (5) but also about nine times less then that of controller (4) which is used by the "optimal" Smith controller tuned to model #3. The simulation results for all four models are presented on Figs. 1a - 4a. For model #3 (15% EGR), which has the largest delay, the Smith predictor delivers 100 msec rise time (vs. 250 msec of PID) at overshoot less than 15%, see Figs. 1&1a. For model #1 (5% EGR), both Smith and PID have slow time responses (Smith has somewhat smaller rise time) but the integrated square error is definitely smaller for Smith controller since its step response is initially much steeper, see Figs. 2&2a. For model #2 (10% EGR), Smith is not only evidently faster but also does not have an overshoot as opposed to PID, see Figs. 3&3a.. Finally, for model #4 (20% EGR), the Smith predictor shows some nervousness symptoms; it delivers an oscillatory time response (undershoot before crossing the steady-state value) due to largely overestimated value of the delay but still its speed is excellent, almost no overshoot and much smaller integrated square error. Thus, we have proved that if both the PID and the Smith predictor are required to satisfy robust stability constraints (< 25% overshoot) then the Smith predictor clearly outperforms the PID controller.

6. Conclusions

We have examined the performance and robustness of the Smith predictor and the traditional PID control algorithm for one particular automotive application (electric EGR valve). Due to nonlinear dependence of system dynamics on the operating point and uncertainty in the time delay, the problem of designing a high performance robust controller for the EGR valve is very difficult. We have demonstrated a possibility for effectively solving the trade-off problem between robustness and performance for the Smith predictor by assuming milder specifications for the rise time and consequently achieving excellent robustness properties. As a result, we have obtained a controller which outperforms the PID in both robustness and performance.

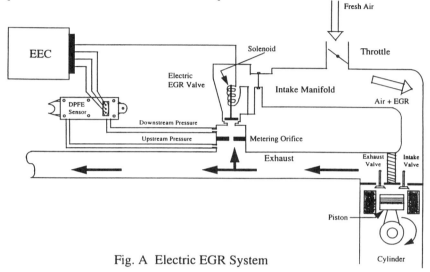

Fig. A Electric EGR System

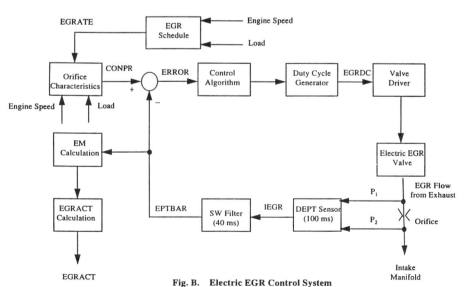

Fig. B. Electric EGR Control System

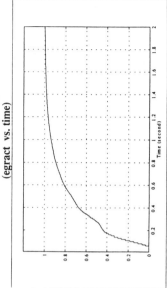

Figure 2
Step response for the robust PID controller
tuned for model #3 (15% EGR) applied to model #1
(egract vs. time)

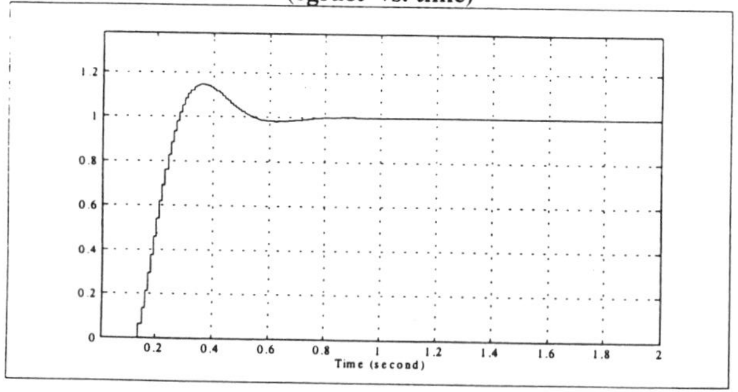

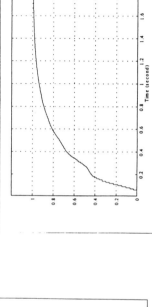

Figure 2a
Step response for the robust Smith predictor
tuned for model #3 (15% EGR) applied to model #1
(egract vs. time)

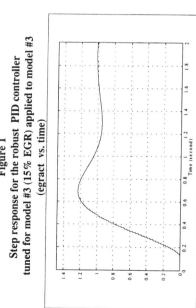

Figure 1
Step response for the robust PID controller
tuned for model #3 (15% EGR) applied to model #3
(egract vs. time)

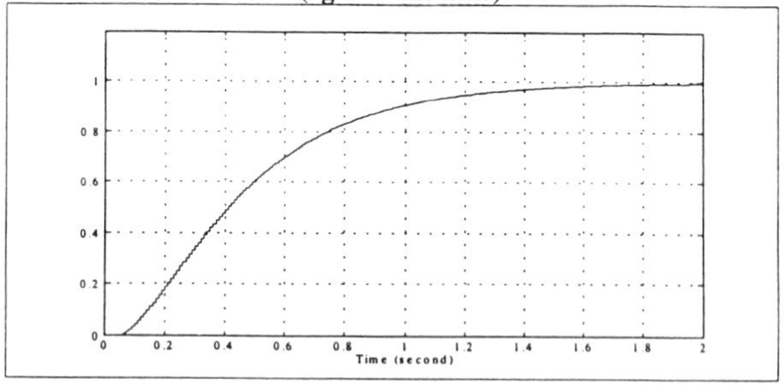

Figure 1a
Step response for the robust Smith predictor
tuned for model #3 (15% EGR) applied to model #3
(egract vs. time)

Figure 3
Step response for the robust PID controller
tuned for model #3 (15% EGR) applied to model #2
(egract vs. time)

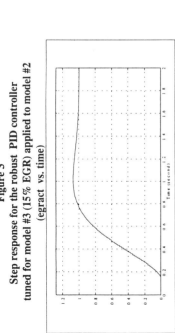

Figure 4
Step response for the robust PID controller
tuned for model #3 (15% EGR) applied to model #4
(egract vs. time)

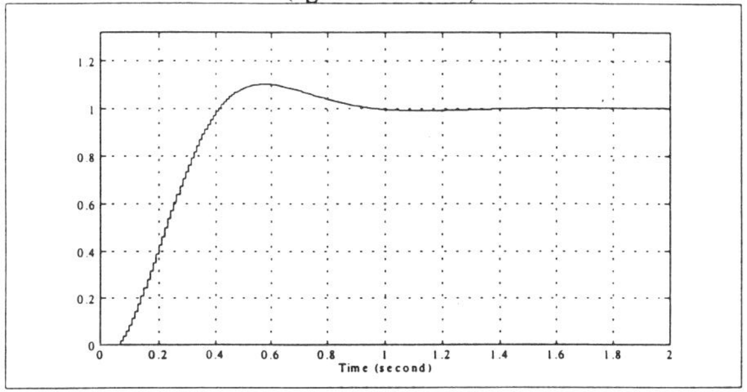

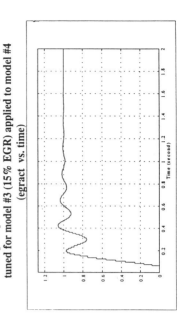

Figure 3a
Step response for the robust Smith predictor
tuned for model #3 (15% EGR) applied to model #2
(egract vs. time)

Figure 4a
Step response for the robust Smith predictor
tuned for model #3 (15% EGR) applied to model #4
(egract vs. time)

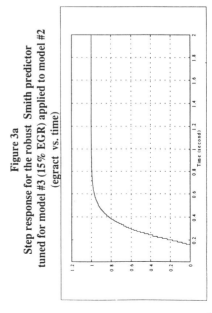

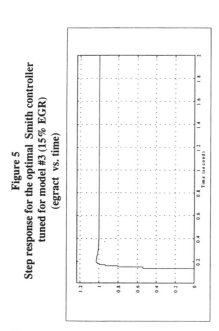

Figure 5
Step response for the optimal Smith controller tuned for model #3 (15% EGR)
(egract vs. time)

Figure 7
Step response for the optimal Smith controller tuned for model #3 (15% EGR, d = 13) applied to model #3 with perturbed delay (d = 15)
(egract vs. time)

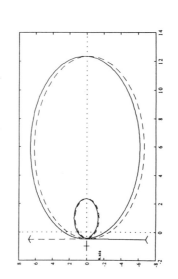

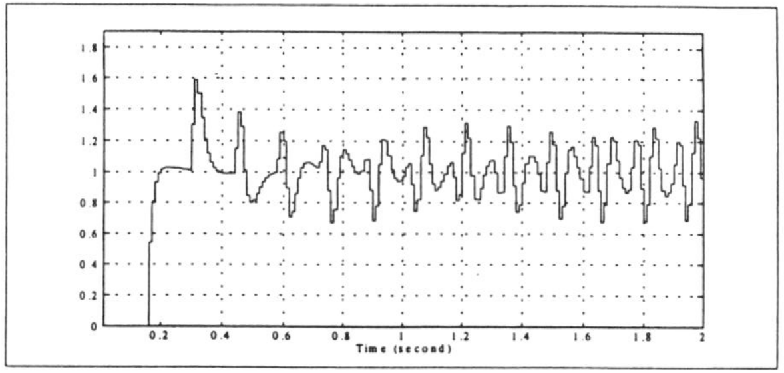

Figure 6
Nyquist plot for the optimal Smith controller tuned for model #3 (15% EGR)

Figure 8
Nyquist plot for the optimal Smith controller tuned for model #3 (15% EGR, d = 13) applied to model #3 with perturbed delay (d = 15)

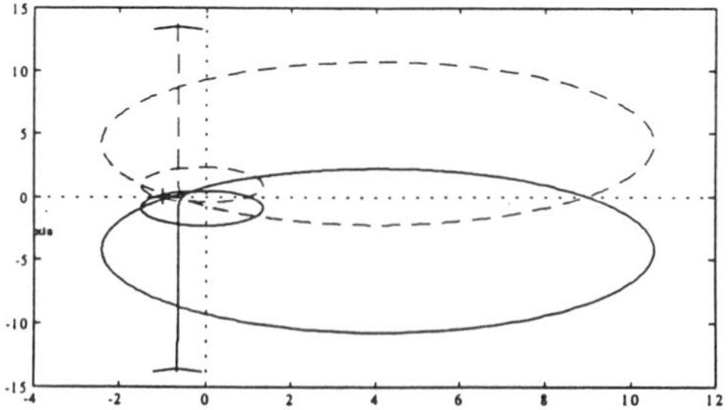

References

[1] O. J. Smith, "A controller to overcome dead time" ISA J. vol. 6, n0.2, Feb. 1959, pp.28-33.
[2] J. E. Marshall, (1979), Control of time delay systems , Stevenage, U.K., Peter Peregrinus.
[3] K.I Astrom, C. C Hang , and B. C. Lim, "A new Smith predictor for controlling a process with an Integrator and Long Dead-Time,". IEEE Trans. Automat. Contr. vol. 39, No.2, Feb. 1994.
[4] K. J. Astrom, Frequency domain properties of Otto Smith regulators, International Journal of Control, 26(2), 1977, pp. 307-314.
[5] Horowitz, I., 1983, Some properties of delayed controls (Smith regulator) International Journal of Control, 38(5), 977-990.
[6] D. L. Laughlin, H. Rivera, and M. Morari, (1987), "Smith predictor design for robust performance", International Journal of Control, 46 (2) ,pp. 477-505
[7] Ioannides, A C., G. J. Rogers and Latham (1979). Stability limits of a Smith controller in simple systems containing a time delay. Int. J. Contr., 29, pp. 577-563.
[8] Palmor, Z. (1980). Stability properties of Smith dead-time compensator controllers. Int. J. Control, 32, 937-949.
[9] Z. J. Palmor, and M Blau, "An auto-tuner for Smith dead time compensator", International Journal of Control, vol. 60 No.1, pp. 117-135.
[10] Z. J. Palmar, and Halevi, Y. (1994), "On the design and properties of multivariable dead-time copensators", pp. 255-264, Aut. 1994.
[11] Z. J. Palmar, (1982), "Properties of optimal stochastic control systems with dead-time", Automatica, 18 (1), pp. 107-116
[12] Parrish, J. R. and C. B. Brosilow (1985). Inferential control applications. Automatica, 21, pp.527-538.
[13] K. Watanabe and M. Ito, (1981), "A process model control for linear systems with delay", IEEE Trans. Automat. Contr. Vol. AC-26, No.6, DEC. 1981.
[14] K. Yamanake, K., and E. Shimemura, "An I-O property of control systems with mismatched Smith controller", IFAC 9th World Congress, Budapest, (1984),vol.III, pp. 140-145.
[15] A. W. Olbrot, M. H. Berri, and J. R. Asik, "Robust Parametrized Controller Design with an Application to Exhaust Gas Recirculation (EGR) System", Ford Research Tech. Report No. SR-96-011
[16] A. W. Olbrot, M. H. Berri, and J. R. Asik, "Robust Parametrized Controller Design with an Application to Exhaust Gas Recirculation (EGR) System", Proc. IEEE Conference on Decision and Control, 1995, pp. , 3557 - 3560 (Abbreviated version of [15]).
[17] M. H. Berri, "Robust Parametrized Controller Design with Applications to Engine Control", Ph.D. Dissertation, Wayne State Univ., Dept. ECE, March 1996, Advisor: A. W. Olbrot.
[18] A. W. Olbrot and M. H. Berri, "Perfectly Robust Deadbeat Controller for Systems with Unknown Delays", 1996 IEEE Conference on Decision and Control, Kobe, Japan, Dec. 1996.
[29] A. W. Olbrot, J. R. Asik, and M. H. Berri, "Delay Identification Experiments, Identification Method, And Results for the Electric EGR Valve Control System", Ford Research Technical Report No. SR-96-122.
[20] A. W. Olbrot, J. R. Asik, and M. H. Berri, " Parameter Scheduling Controller for Exhaust Gas Recirculation (EGR) System", 1997 SAE Congress, Detroit, Feb. 1997

VII. Applications

YEON-WOOK CHOE AND GI-SIG BYUN
Robust controller design for a VTOL plane

1. Introduction
A plane with the ability of VTOL (vertical take-off and landing) and hovering such as a helicopter is of great use in various fields by taking advantage of its features. The dynamic characteristics of the flying object are, however, unstable. Moreover, as the parameters or sensitivity of the plane are apt to be changeable in accordance with the flight circumstances, it is quite difficult to precisely control its attitude while it is hovering.
In this paper, we adopt a small sauce-type plane with four fans as a plant[1], which has a distinctive feature being able to hover over one place. Although it is a small-sized model, its flight performance is almost similar to that of a real one. Therefore, a robust controller is needed to cope with the existing internal or external uncertainties of the plane.

A lot of papers related to the control of a flying object (especially on a helicopter) have been published since the1990s [1]~[3], and the H∞ algorithm was adopted to design a robust controller in most of the papers. It seems, however, that the greatest attention has not been given to the modeling process of a plant when compared to that of the control system design. So, we carefully describe the modeling process of the unstable plant in the first half of the paper and in the second half of the paper, the design process for $H_2/H\infty$ controllers is outlined in detail.

2. Modeling
2.1 Plant
The sauce-type plane shown in Fig.1 was composed of four propellers and four motors. It was originally designed so that the propellers in the front and rear rotate counterclockwise and the ones on the right and left rotate clockwise. Consequently, the plane not only has symmetry with respect to its center of gravity, but there is little interference between the rotatory motion of each propeller. Therefore, it is possible to simplify the modeling process of the plane by taking advantage of this distinctive quality [1].

---
[1] As a basic model, we made use of the product made by The KEYENCE Corporation, JAPAN

In order to make a free flight by properly adjusting the speed of each fan, the plane has two built-in gyros. One called the free-gyro. It is used to adjust the voltage applied to the motors in proportion to the angular velocity and the slope of the plane while it is in rolling pitching motion. This enables us to freely move the plane in every direction. The other is called rate gyro. It is supposed to regulate the voltage applied to the motors proportional to the angular velocity of the plane while it is in yawing motion. That is, the direction of the plane facing the front can be altered owing to the role of the rate gyro.

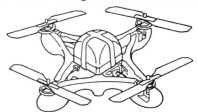

Fig.1 A Sauce-Type VTOL Plane

2.2 Nominal Model for Controller Design

Since there is little interference between each motion as mentioned above, it is possible to build up a model with three sub-systems on the assumption that each motion should be operating independently. However, because there is almost no knowledge of the feedback coefficients of the built-in gyros, the complete model cannot be obtained through only physical analysis. Therefore, we have to use observations from the plant in order to fit the model's properties to those of the real ones. For this purpose, we manufactured the experimental apparatus by which each motion could be operated independently. It is assumed that the transfer function of the pitching motion is identical with that of the rolling motion because of their symmetrical characteristics.

It has already been known that there are two poles at z=1 (s=0) by the physical analysis of the plant. Generally, it is quite difficult to identify the parameters corresponding to the unstable mode. We can, however, cope with this difficulty provided that there is little measurement noise. If the information having two poles at z=1 is utilized, the following relation (n_c=0)

$$F(z^{-1})|_{z=1} = 0 \qquad \frac{d}{dz}F(z^{-1})|_{z=1} = 0 \tag{1}$$

can be obtained, where $F(z^{-1})$ is the denominator of the identified model through some identification method such the LS method.

The final reduced-order transfer functions are:

$$G_{roll}(s) = 10^5 \times \frac{-3.736s^2 + 252.2s + 3837.1}{s^2(s^3 + 41.7s^2 + 7640.4s + 89320)}$$

$$G_{yaw}(s) = \frac{30s^3 + 1560s^2 - 25660s - 130480}{s(s^3 + 16.3s^2 + 180s + 520)}$$

3. Feedback Controller Design
3.1 Determination of Weighting Functions

Here, we adopt the multiplicative uncertainty as follows:

$$\hat{G}(s) = G_{nom}(s)\{1 + \Delta(s)W_m(s)\} \tag{2}$$

where $\Delta(s)$ is a variable stable transfer function satisfying $\| \Delta(j\omega) \|_\infty \leq 1$ ($\forall \omega$) and $G_{nom}(s)$ means a nominal model (rolling and yawing) which is to be used for designing a controller. $W_m(s)$ is a weighting function which represents the variation of uncertainty with respect to frequency.

We generally use the LS method to obtain the frequency response of the plant from the measured input-output data. However, there are several problems to be solved when one applies the LS method to the actual system. We adopt the Multi-decimation (MD) method [6] to acquire the frequency response of the plant as exactly as possible.

By carrying out the MD method ten times with ten different measured input-output data, we get ten-frequency responses of the plant.

In order to apply the H∞ algorithm, the uncertainty has to be described as a frequency function. At first, we need to determine the model $\hat{G}(s)$ which represents the transfer function of the real plant. This is however, impossible to get. The models that have already been obtained through the MD method will suffice for the $\hat{G}(s)$ because of its comparative accuracy depending on the frequency regions. That is, we can select $\hat{G}(s)$ as follows:

$$\hat{G}_r(s) = \frac{156.3s^4 + 4083s^3 + 47510s^2 + 36490s + 1518}{s^5 + 11.93s^4 + 142.5s^3 + 224.2s^2 + 52.8s + 0.4925}$$

$$\hat{G}_y(s) = \frac{139.2s^4 + 1602s^3 + 61350s^2 + 106000s + 3208}{s^5 + 11.77s^4 + 141.6s^3 + 178.3s^2 + 5.914s + 0.005736}$$

From Eq. (2),

$$\Delta(s)W_m(s) = \frac{\hat{G}(s)}{G_{nom}(s)} - 1 \tag{3}$$

is easily derived. As a nominal plant $G_{nom}(s)$, we adopt ten models that can be identified through the method of the previous section 2. In Fig.3, there are ten lines representing gain characteristic of $\varDelta(s)W_m(s)$. The weighting function $W_m(s)$ can be obtained by choosing a function that has to cover all the gain characteristics of $\varDelta(s)W_m(s)$.

$$W_{m\_r}(s) = \frac{1100s^4 + 5662s^3 + 25390s^2 + 18110s + 364.7}{s^4 + 58.82s^3 + 759.9s^2 + 16570s + 280.7}$$

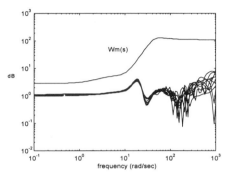

Fig.2 Identified Gain Response and $W_m(s)$

3.2 Generalized Plant

There are three specifications that should be considered in the controller design:
① The stability of the system should be guaranteed in spite of parameter variations and modeling errors in the high frequency region (Robust Stable).
② It should be possible to cope with the disturbance added to the system (Disturbance Suppression).
③ While the plane is hovering, the deviation owing to the rolling or pitching phenomenon of the plane itself should be less than ±5 [cm], and rotation angle of the fuselage be less than ±15 [Degree].

It is well known that the H∞ algorithm is currently the best way to satisfy these three specifications, To apply this algorithm, the given specifications should be reformulated as a generalized plant (Fig.3). As a result of several trial and errors, we selected the weighting function $W_s(s)$ and $\varepsilon(s)$ as follows:

$$W_{s\_r}(s) = \frac{9.58s + 4.48}{s(s + 2.198)} \quad W_{s\_y}(s) = \frac{1.791s + 22.98}{s} \quad \varepsilon_s(s) = \frac{s}{s + 628.8} \times 10^{-2}$$

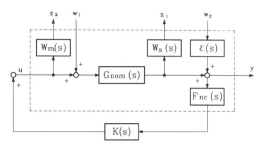

Fig.3 Generalized Plant for H∞

4. Feedforward Controller Design

As mentioned earlier, the servo system structure with two-degree-of-freedom was adopted (Fig.4) because the transient response could be improved by the action of a feedforward path. Therefore, it is necessary to design the feedforward controller to be consistent with this purpose. To do this, we make use of the H_2 control algorithm because of its effectiveness to time-domain responses.

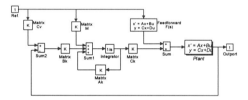

Fig.4 A Servo System with T.D.O.F

In Fig.4, $v(t)$ represents the internal state of the reference signal generator, that is,

$$\frac{d\mathbf{v}(t)}{dt} = \Theta \mathbf{v}(t), \quad \mathbf{r}(t) = \mathbf{C}_V \mathbf{v}(t) \tag{4}$$

From this figure, a state-space equation of the entire system is easily obtained:

$$\frac{d\mathbf{w}(t)}{dt} = \mathbf{A}_K \mathbf{w}(t) + \mathbf{B}_K (\mathbf{r}(t) - \mathbf{y}(t)) + \mathbf{M}\mathbf{v}(t) \tag{5}$$

$$\mathbf{y}(t) = \mathbf{C}_Z \mathbf{z}(t), \quad \mathbf{C}_Z = [\mathbf{C} \ \ 0], \quad \mathbf{u}(t) = \mathbf{C}_K \mathbf{w}(t) + \mathcal{L}^{-1}[\mathbf{F}(s)]\mathbf{v}(t) \tag{6}$$

$$\mathbf{z}(t) = [\mathbf{x}^T(t) \ \ \mathbf{w}^T(t)]^T, \mathbf{K}(s) = \mathbf{C}_k (s\mathbf{I} - \mathbf{A}_k)\mathbf{B}_k$$

where $w(t)$ represents the state of the designed controller.

4.2 Design of Feedforward Controller F(s)
(A) <u>Derivation of State Space Equation</u>

We adopt the performance criterion as follows:

$$J_{FF} = \left\| \begin{array}{c} R(s) - Y(s) \\ \rho G_{ur}(s) \end{array} \right\|_2 \tag{7}$$

where $G_{ur}(s)$ represents the transfer function from r(t) to u(t), and R(s) and Y(s) are Laplace transforms of r(t) and y(t), respectively. The criterion (15) can be interpreted as follows: the first term indicates the tracking error ($\int_0^\infty (r(t)-y(t))^2 dt$), and the second term means the control input energy ($\int_0^\infty u(t)^2 dt$) with regard to impulse input. Accordingly, if we make the first term less, the tracking performance may be improved. On the contrary, the magnitude of the control input can be diminished by making the second term smaller. In this case, an integer ρ is used as a trade-off.

To design a controller minimizing Eq. (7), we must transform this into the standard H_2 problem. At first, the first term of J_{FF} becomes

$$\text{First term} = \{C_V - [C\ 0]\Phi(s)\begin{bmatrix} BF(s) \\ (*) \end{bmatrix}\}V(s) \tag{8}$$

where $\Phi(s) = (sI - A_z - B_z K_{fb})^{-1}$, $K_{fb} = [0\ C_K]$

$$A_Z = \begin{bmatrix} A & 0 \\ -B_K C & A_K \end{bmatrix}, B_Z = \begin{bmatrix} B \\ 0 \end{bmatrix}$$

In this process, it was assumed that the reference r(t) was a step function with the intention of improving step response. And on the second term, by assuming v(t) as an impulse function,

$$u(s) = K_{fb}\Phi(s)\begin{bmatrix} BF(s) \\ (*) \end{bmatrix} + F(s) \tag{9}$$

is obtained. In Eq. (8) and (9), (*) stands for $B_K C_V + M$. We regarded it as 0 in the process of the H_2-norm computation since this part was not affected by the feedforward controller F(s) [9].

It is required to minimize the magnitude of Eq. (8) and (9) simultaneously by making use of F(s). This is a standard H_2 optimal control problem. In order to do this, we introduce the generalized plant as follows:

$$\begin{bmatrix} z_1 \\ z_2 \\ y \end{bmatrix} = G(s)\begin{bmatrix} w \\ u \end{bmatrix} = \begin{bmatrix} s^{-1}C_V & s^{-1}[C\ 0]\Phi(s)B_Z \\ 0 & K_{fb}\Phi(s)B_Z + I \\ I & 0 \end{bmatrix}\begin{bmatrix} w \\ u \end{bmatrix} \tag{10}$$

$$u(s) = F(s)y(s)$$

where $z_1(s)$ and $z_2(s)$ correspond to the first and second term of Eq. (8), respectively. Therefore, by properly designing $F(s)$, the H_2-norm of the transfer function T_{zw} from $w(t)$ to $[z_1^T(s)\ z_2^T(s)]^T$ can be minimized

(B) Solution of H_2- Problem

The H_2 controller that stabilizes $G(s)$ internally and minimizes T_{zw} is easily obtainable by using software on the market. In this case, the general solution procedure is not directly applicable because the generalized plant Eq. (10) does not meet the standard assumptions required of the output feedback control H_2 problem. This is due to the integrator of the feedback controller $K(s)$. However, it is possible to cope with this difficulty through some modifications.

5. SIMULATION

5.1 Performance of the Controller

The design process leads to a controller with as many states as the interconnection structure of the generalized plant of Fig.4. By using a controller order reduction method, we could obtain the reduced-order controllers that guarantee the closed-loop stability and the closed-loop performance as well. The frequency response of sensitivity function $S(j\omega)$ and complementary sensitivity function $T(j\omega)$ are shown in Fig.6 to confirm the robustness of the design.

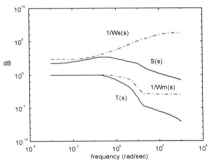

Fig.5 Frequency Response of $S(s)$, $T(s)$, $1/W_s(s)$ & $1/W_m(s)$ (Rolling & Pitching)

5.2 Time Simulation

As we adopted the servo system structure with two-degree-of-freedom, it is expected that the transient response of the system should be improved.

The step response of the closed-loop system is shown in Fig.8. In order to confirm the robustness of the controller, we have also shown the result when the parameters of the nominal model were modified within the degree of ±50% (there must be no difference between the number of unstable poles of the nominal and modified model).

6. Experiment

Our final feedback controllers are

$$K_{roll}(s) = \frac{53.16s^3 + 16.59s^2 + 40.27s + 0.6386}{s(s^5 + 6.138s^4 + 27.8s^3 + 34.24s^2 + 13.9s + 0.6559)}$$

$$K_{yaw}(s) = \frac{7.5693s^4 + 16.635s^3 + 699.71s^2 + 58.836s + 0.082}{s(s^4 + 132.25s^3 + 1728.8s^2 + 42158s + 65.921)}$$

The discrete controller $K(z)$ obtained through the bilinear transformation from $K(s)$ are realized by making use of DSP equipment, and apply to the real plant. At this step, we adopted 20 [ms] as a sampling time.

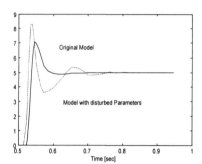

Fig.8 Step Response of System

The experimental apparatus is composed of two PSD cameras, an AD-DA converter, a Filter and a computer.

The experiment on the plane's attitude control was carried out. We made the plane come to a standstill in the air and checked the variations of rolling and yawing motion. Fig.7 shows the displacement in regard to the horizontal axis (rolling and yawing). We know that the region of error is ±6 [cm] at rolling and ±15[Deg] at yawing motion, which seem to be both within an allowable error.

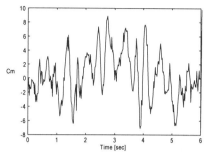

Fig.6 (a) Hovering data of Rolling Motion   (b) Hovering data of Yawing Motion

6. Conclusion

In this paper, a robust controller was designed to control the attitude of a VTOL plane while it was in a state of hovering.

In the process of the experiment, there were two problems: First, we had difficulty obtaining current attitude of the VTOL in real-time because of the poor quality of PSD cameras. Second, since the plane was controlled by a remote transmitter, it caused some transmission problem resulting from noise.

Although there were problems in the experiment, we have confirmed that the H∞/H_2 control algorithm and the servo system with the feedforward path were quite useful to control a flying object despite the existing external disturbance and the modeling error

7. References

[1] Apkarian, P C et al. (1990), *Proc. of 29th CDC, IEEE*, pp 2600/2606
[2] Mammar, S & Duc, G (1992*), Proc. IEEE*, ACC 820/ 828
[3] Yue, A & Postlethwaite, I (1990), *Proc. Vol.137, Pt.D*, No.3, 115/129
[4] Isermann, R (1981), *Digital Control Systems*, Springer-Velag
[5] Ljung, L (1987), *System Identification, Theory for the User* Prentice Hall
[6] Adachi, S (1994), *System Identification Theory for the User* SICE Publishing, Japan
[7] Hirata, M et el (1993), *SICE, Trans. Vol.29*, No.1, 71/77
[8] Hagiwara, T et al (1992), *SICE Trans., Vol.28*, No.1, 77/86
[9] Choe, Y W & Lee, K W(1995) *'95 KACC*, 780/783
[10] Sugie, T & Yoshikawa, T (1986), *IEEE Trans., AC-31*, 551/554
[11] Balas, G J et al (1993), *Matlab Toolbox, μ-Analysis and Synthesis*
* Department of Control and Instrumentation Engineering
 Pukyong National University, Pusan, Korea 608-739

J P KENNE AND E K BOUKAS
Optimal control of failure prone manufacturing systems with corrective maintenance

1 Introduction

This paper deals with the control problem of a stochastic manufacturing system consisting of different machines in flowshop configuration. The capacity of the system is improved by controlling the machine repair rates. Then, decision variables are input rates to the machines and their repair rates which influence the number of parts in the buffers between machines and the surplus. Since the WIP inventories and the surplus are considered as state variables, we are facing to a state constrained control problem which is to choose admissible input rates of machines and their repair rates to minimize the inventory/backlog and repair costs over an infinite horizon. Many authors contributed in the sphere of the production planning problem of flexible manufacturing systems (FMS).

Based on the Rishel formulation [10], Older and Suri [8] presented a model for FMS with unreliable machines whose failures and repairs are described by certain homogeneous Markov process. The main difficulty with this approach is the lack of efficient methods for solving the optimization problem characterized by stochastic Hamilton-Jacobi-Bellman equations (HJB). Akella and Kumar [1] solved analytically a one-machine one-part type problem and Sharifnia [11] studied the steady state probability distribution of a multiple-machine one-part type problem. Investigation in the same direction gave rise to the extension presented by Boukas and Haurie [4] where the authors considered a machine age dependent matrix generator for the Markov process which describes the machines states. However, with such a model, it remains difficult to obtain the optimal control of large scale FMS.

Based on the presence of state constraints and observing that explicit optimal solution as in [1] does not exist, Lou et al. [7] extended the problem in [1] to a two-machine flowshop problem and conducted a rigorous study of the dynamic properties of the system and the related boundary conditions. An analysis of the m-machine flowshop ($m \geq 2$) in the context of obtaining piecewise deterministic optimal control problem is discussed in [9]. Such a system has been studied in the work of Bai and Gershwin [2], [3] where the authors constructed an hierarchical controller to regulate the production. It is the purpose of this paper to extend the problem in [9] by controlling both input rates of machines and their repair rates.

2 Problem statement

In this paper, we consider the flow control problem for a tandem production system with m ($m \geq 2$) unreliable machines. Each machine has two states (up and down denoted by 1 and 0 respectively), resulting in a system with a p-state Markov chain $k(t) = (k_1^1(t), \cdots, k_m^1(t), k_1^p(t), \cdots, k_m^p(t))$ on the probability space $(\Omega, \mathcal{F}, P)$ with values in an finite set $\mathcal{M}$; $k_i^j(t)$, $i = 1, \cdots, m$; $j = 1, \cdots, p$ is the capacity of the machine M_i in mode j at time t. We use $u_i(t)$ to denote the input rate to M_i and $x_i(t)$ to denote the number of parts in the buffer between M_i and M_{i+1} ($i = 1, \cdots, m-1$). Finally the difference between the cumulative production and the cumulative demand, called surplus, is denoted by $x_m(t)$.

The dynamics of the system can be written as follows:

$$\dot{x}_i(t) = u_i(t) - u_{i+1}(t), \quad x_i(0) = x_i, \quad i = 1, \cdots, m-1 \quad (1)$$
$$\dot{x}_m(t) = u_m(t) - u_{m+1}, \quad x_m(0) = x_m \quad (2)$$

where $u_{m+1} := d$ is a given constant demand rate. In matrix notation, the system of equations (1)-(2) becomes:

$$\dot{x}(t) = \boldsymbol{A}\boldsymbol{u}(t) \qquad \boldsymbol{x}(0) = \boldsymbol{x} \quad (3)$$

where $\boldsymbol{A}$ is an $(m+1) \times m$ matrix, $\boldsymbol{u}(t) = (u_1(t), \cdots, u_m(t))$, $\boldsymbol{x}(t) = (x_1(t), \cdots, x_m(t))$ and $\boldsymbol{x} = (x_1, \cdots, x_m)$.

Let $\boldsymbol{u}_r = (u_{r1}(t), \cdots, u_{rm}(t))$ denote the vector of machines repair rates, considered here as another control variables. For $\boldsymbol{k} = (k_1^j, \cdots, k_m^j)$, $k_i^j \geq 0$, $i = 1, \cdots, m$, $j = 1, \cdots, p$, let

$$U(\boldsymbol{k}) = \Big\{(\boldsymbol{u}, \boldsymbol{u}_r) = (u_1, \cdots, u_m, u_{r1}, \cdots u_{rm}) : 0 \leq u_i \leq k_i^j U_{max}^i; \text{ and } \underline{u}_r^i \leq u_{ri} \leq \bar{u}_r^i\Big\}$$

and for $\boldsymbol{x} \in S = [0, \infty)^{m-1} \times \mathbb{R}$, let

$$\mathcal{A}(\boldsymbol{x}, \boldsymbol{k}) = \Big\{(\boldsymbol{u}, \boldsymbol{u}_r) : (\boldsymbol{u}, \boldsymbol{u}_r) \in U(\boldsymbol{k}); \ x_i = 0 \Rightarrow u_i - u_{i+1} \geq 0, \ i = 1, \cdots, m-1\Big\}$$

$\mathcal{A}(\boldsymbol{x}, \boldsymbol{k})$ denote the set of all admissible controls with respect to $\boldsymbol{x} \in S$ and $\boldsymbol{k}(0) = \boldsymbol{k}$. The control problem consists of finding an admissible control law $(\boldsymbol{u}(.), \boldsymbol{u}_r(.))$ that minimize the cost function $J(.)$ given by:

$$J(\boldsymbol{x}, \boldsymbol{k}, \cdot) = \mathrm{E}\Big\{ \int_0^\infty e^{-\rho t} \big[g(\boldsymbol{x}(t)) + c(\boldsymbol{u}(t)) + h(\boldsymbol{u}_r(t))\big] dt \Big| \boldsymbol{x}(t) = \boldsymbol{x}, \ \boldsymbol{k}(t) = \boldsymbol{k}\Big\} \quad (4)$$

where $g(\boldsymbol{x}(t))$, $c(\boldsymbol{u}(t))$ and $h(\boldsymbol{u}_r(t))$ denote the cost of inventory/shortage, the cost of production and the cost of repair respectively and ρ is the discount rate. The value function of the planning problem is given by:

$$v(\boldsymbol{x}, \boldsymbol{k}) = \inf_{(\boldsymbol{u}, \boldsymbol{u}_r) \in \mathcal{A}(\boldsymbol{x}, \boldsymbol{k})} J(\boldsymbol{x}, \boldsymbol{k}, \boldsymbol{u}(t), \boldsymbol{u}_r(t)) \quad (5)$$

3 Optimality conditions

In this section, we shall give some properties of the value function and show that it satisfies a set of coupled partial derivatives equations (HJB) derived from the application of the dynamic programming approach. We let $G(x(t), w(t)) = g(x(t)) + c(u(t)) + h(u_r(t))$ where $w(t) = (u(t), u_r(t))$.

3.1 Properties of the value function

We show that the value function given by (5) is strictly convex in x provided that the inventory, production and corrective maintenance cost function $G(.)$ is so. Moreover it is shown to be continuously differentiable. For this purpose, we need the following lemma who gives a constructive proof for the Lipschitz property of the value function.

Lemma 3.1 *Given $x = (x_1, \cdots, x_m)$, $x' = (x'_1, \cdots, x'_m)$ and $k \in \mathcal{M}$, let $u \in \mathcal{A}(x, k)$, there exists $u' \in \mathcal{A}(x', k)$ such that:*

(i) $u'_1(t) = u_1(t)$ for all $t \geq 0$;

(ii) $0 \leq u'_i(t) \leq u_i(t)$ for all $t \geq 0$ and for all $i = 2, \cdots, m$;

(iii) $\int_0^t [u_i(s) - u'_i(s)]ds \leq (x_{i-1} - x'_{i-1})^+ + \int_0^t [u_{i-1}(s) - u'_{i-1}(s)]ds$, for all $i = 2, \cdots, m$.

with the notation $a^+ = \max(a, 0)$ and $a^- = \max(-a, 0)$ for any real number a.

Proof: The proof of this lemma can be found in [9]. ∎

Theorem 3.1 *The value function given by (5) is convex, continuous on S and satisfies the condition:*

$$|v(x, k) - v(x', k)| \leq C(1 + |x|^{k_h} + |x'|^{k_h})|x - x'| \quad (6)$$

for some positive constant C and for all $x, x' \in S$.

Proof: Given $x, x' \in S$ and $k \in \mathcal{M}$, it follows from lemma 3.1 that, for any $u = (u_1, \cdots, u_m) \in \mathcal{A}(x, k)$, there is a $u' = (u'_1, \cdots, u'_m) \in \mathcal{A}(x', k)$ such that (i)-(iii) in lemma 3.1 hold. By virtue of Lipschitz assumptions on $g(.)$, the Lipschitz property (given by the inequality (6)) of the value function can be obtained as in [9].

To prove the convexity of the $v(., k)$, let $x^1, x^2 \in S$, $k \in \mathcal{M}$ and $\epsilon > 0$. Let also $w^1(.) \in \mathcal{A}(x^1, k)$ and $w^2(.) \in \mathcal{A}(x^2, k)$ denote ϵ-optimal controls. For any $0 \leq \lambda \leq 1$,

$$\lambda w^1(.) + (1 - \lambda)w^2(.) \in \mathcal{A}(\lambda x^1 + (1 - \lambda)x^2, k)$$

Let $x^1(.)$ and $x^2(.)$ denote the trajectories corresponding to $(x^1, w^1(.))$ and $(x^2, w^2(.))$ respectively. Let also $w(t) := \lambda w^1(t) + (1 - \lambda)w^2(t)$ and $x(.)$ the trajectory with the initial value $\lambda x^1 + (1 - \lambda)x^2$ and control $w(t)$. For any $\lambda \in [0, 1]$

$$\lambda J(x^1, k, w^1) + (1 - \lambda)J(x^2, k, w^2) = \mathbb{E} \int_0^\infty e^{-\rho t} \Big[\lambda G(x^1(t), w^1(t)) + (1 - \lambda)G(x^2(t), w^2(t))\Big] dt$$

Recall that
$$\mathrm{E}\int_0^\infty e^{-\rho t}G(\boldsymbol{x}(t),\boldsymbol{w}(t))dt = J(\lambda\boldsymbol{x}^1+(1-\lambda)\boldsymbol{x}^2,\boldsymbol{k},\lambda\boldsymbol{w}^1(.)+(1-\lambda)\boldsymbol{w}^2(.))$$
Then
$$\lambda J(\boldsymbol{x}^1,\boldsymbol{k},\boldsymbol{w}^1)+(1-\lambda)J(\boldsymbol{x}^2,\boldsymbol{k},\boldsymbol{w}^2) \geq J(\lambda\boldsymbol{x}^1+(1-\lambda)\boldsymbol{x}^2,\boldsymbol{k},\lambda\boldsymbol{w}^1(.)+(1-\lambda)\boldsymbol{w}^2(.))$$
This means that $J(.,\boldsymbol{k},.)$ is jointly convex. Therefore one have:
$$\begin{aligned}v(\lambda\boldsymbol{x}^1+(1-\lambda)\boldsymbol{x}^2,\boldsymbol{k}) &\leq J(\boldsymbol{x},\boldsymbol{k},\boldsymbol{w}(.))\\&\leq \lambda J(\boldsymbol{x}^1,\boldsymbol{k},\boldsymbol{w}^1)+(1-\lambda)J(\boldsymbol{x}^2,\boldsymbol{k},\boldsymbol{w}^2)\end{aligned} \quad (7)$$
For each $\boldsymbol{x}^i$, $\boldsymbol{k}$, $\boldsymbol{w}^i \in \mathcal{A}(\boldsymbol{x}^i,\boldsymbol{k})$, $i=1,2$, is an ϵ-optimal control, i.e.,
$$J(\boldsymbol{x}^i,\boldsymbol{k},\boldsymbol{w}^i) \leq v(\boldsymbol{x}^i,\boldsymbol{k})+\epsilon \quad (8)$$
Combining (7) and (8), one have:
$$v(\lambda\boldsymbol{x}^1+(1-\lambda)\boldsymbol{x}^2,\boldsymbol{k}) \leq \lambda v(\boldsymbol{x}^1,\boldsymbol{k})+(1-\lambda)v(\boldsymbol{x}^2,\boldsymbol{k})+\epsilon+\epsilon$$
Letting $\epsilon \to 0$, we have
$$v(\lambda\boldsymbol{x}^1+(1-\lambda)\boldsymbol{x}^2,\boldsymbol{k}) \leq \lambda v(\boldsymbol{x}^1,\boldsymbol{k})+(1-\lambda)v(\boldsymbol{x}^2,\boldsymbol{k})$$
which prove that $v(\boldsymbol{x},\boldsymbol{k})$ is convex. ∎

3.2 The dynamic programming equation

Due to state constraints, the state process may move along the boundary of the state space. Because we need an appropriate boundary condition, we shall apply the concept of directional derivatives to the value function who was proved previously to be a convex function.

Definition 3.1 *A function $f(\boldsymbol{x})$, $\boldsymbol{x} \in \mathbb{R}^m$, is said to have a directional derivative $f'_{\boldsymbol{p}}(\boldsymbol{x})$ along the direction $\boldsymbol{p} \in \mathbb{R}^m$ if there exists*
$$\lim_{\epsilon \to 0}\frac{f(\boldsymbol{x}+\epsilon\boldsymbol{p})-f(\boldsymbol{x})}{\epsilon} = f'_{\boldsymbol{p}}(\boldsymbol{x})$$

Note that if a function $f(\boldsymbol{x})$ is differentiable at $\boldsymbol{x}$, then $f'_{\boldsymbol{p}}(\boldsymbol{x})$ exists for every $\boldsymbol{p}$ and
$$f'_{\boldsymbol{p}}(\boldsymbol{x}) = \langle \nabla f(\boldsymbol{x}),\boldsymbol{p}\rangle$$
where $\nabla f(\boldsymbol{x})$ is the gradient of $f(\boldsymbol{x})$ and $\langle .,.\rangle$ is the scalar product.

With the notation of directional derivatives (DD), the optimality conditions, called HJBDD (Hamilton-Jacobi-Bellman in terms of DD), can be given by:

$$\rho v(\boldsymbol{x}, \boldsymbol{k}) = \inf_{\boldsymbol{w} \in \mathcal{A}(\boldsymbol{x}, \boldsymbol{k})} \{v'_{A\boldsymbol{u}}(\boldsymbol{x}, \boldsymbol{k}) + G(\boldsymbol{x}, \boldsymbol{w}) + Q(\boldsymbol{u}_r)v(\boldsymbol{x}, .)\} \tag{9}$$

Let ∂S denote the boundary of S. If there exits $x_i = 0$, $i = 1, \cdots, m-1$, then $\boldsymbol{x} \in \partial S$. Let the restriction of $v(\boldsymbol{x}, \boldsymbol{k})$ on some l-dimensioanl face, $0 < l < m$, of ∂S be differentiable at an inner point $\boldsymbol{x}_0$ of this face. Then there is a vector $\tilde{\nabla} v(\boldsymbol{x}_0, \boldsymbol{k})$ such that $v'_{\boldsymbol{p}}(\boldsymbol{x}_0, \boldsymbol{k}) = \langle \tilde{\nabla} v(\boldsymbol{x}_0, \boldsymbol{k}), \boldsymbol{p} \rangle$ for any admissible direction at $\boldsymbol{x}_0$. It follows from the continuity of the value function that in this case, the boundary condition on $v(., .)$ is given by:

$$\min_{\boldsymbol{w} \in U(\boldsymbol{x}_0, \boldsymbol{k})} \{\langle \tilde{\nabla} v(\boldsymbol{x}_0, \boldsymbol{k}), A\boldsymbol{u} \rangle + G(\boldsymbol{x}_0, \boldsymbol{w}) + Q(\boldsymbol{u}_r)v(\boldsymbol{x}_0, \boldsymbol{k})\}$$
$$= \min_{\boldsymbol{w} \in U(\boldsymbol{k})} \{\langle \tilde{\nabla} v(\boldsymbol{x}_0, \boldsymbol{k}), A\boldsymbol{u} \rangle + G(\boldsymbol{x}_0, \boldsymbol{w}) + Q(\boldsymbol{u}_r)v(\boldsymbol{x}_0, \boldsymbol{k})\} \tag{10}$$

For the interpretation of the condition (10), we refer the reader to [7] where the case of a two-machine system is studied.

4 The numerical approach

In this section, we develop the numerical method for solving the optimality conditions presented in the previous section. This method is based on the Kushner approach [4] and [6]. The basic idea behind this approach consists of using an approximation scheme for the directional derivative of the value function $v(\boldsymbol{x}, \boldsymbol{k})$. Let h_j, $j = 1, \cdots, m$, denote the length of the finite difference interval of the variable x_j. Using the finite difference interval, the system (9) becomes:

$$v^h(\boldsymbol{x}, \alpha) = \min_{(\boldsymbol{u}, \boldsymbol{u}_r) \in \mathcal{A}^h(\boldsymbol{x}, \alpha)} \Big\{ \frac{G(\boldsymbol{x}, \boldsymbol{w})}{Q_h^\alpha \{1 + \rho/Q_h^\alpha\}} + \frac{1}{\{1 + \rho/Q_h^\alpha\}} \Big(\sum_{j=1}^m p_j^\alpha(1)(v^h(x_1, \cdots,$$
$$x_j + h_j, \cdots, x_m, \alpha) + \sum_{j=1}^m p_j^\alpha(2) v^h(x_1, \cdots, x_j - h_j, \cdots, x_m, \alpha)$$
$$+ \sum_{\beta \neq \alpha} p^\alpha(\beta) v^h(\boldsymbol{x}, \beta) \Big) \Big\} \tag{11}$$

where $v(\boldsymbol{x}, \boldsymbol{k})$ is approximated by $v^h(\boldsymbol{x}, \boldsymbol{k})$, $\mathcal{A}^h(\boldsymbol{x}, \alpha)$ is the discrete control space or control grid and

$$Q_h^\alpha = |q_{\alpha\alpha}| + \sum_{j=1}^m \frac{|u_j - u_{j+1}|}{h_j} \qquad p_j^\alpha(1) = \begin{cases} \frac{u_j - u_{j+1}}{h_j Q_h^\alpha} & \text{if } u_j - u_{j+1} \geq 0 \\ 0 & \text{otherwise} \end{cases}$$

$$p_j^\alpha(2) = \begin{cases} \frac{d_j - u_{j+1}}{h_j Q_h^\alpha} & \text{if } u_j - u_{j+1} < 0 \\ 0 & \text{otherwise} \end{cases} \qquad p^\alpha(\beta) = \frac{q_{\alpha\beta}}{Q_h^\alpha}$$

Note that for each α, $\sum_{j=1}^{m}(p_j^\alpha(1)+p_j^\alpha(2))+\sum_{\beta\neq\alpha}p^\alpha(\beta)=1$. In this paper, we use the policy improvement technique to derive a solution of the approximating optimization problem. The algorithm of this technique can be found in [6].

As illustrative example, we consider a numerical example for a two-machine system (i.e., $m = 2$). The system capacity is described by a four-state Markov process with states $\alpha \in \bar{\mathcal{M}} = \{1,2,3,4\}$. This is equivalent to $\boldsymbol{k} \in \mathcal{M} = \{\boldsymbol{k}^1,\cdots,\boldsymbol{k}^4\}$ such that $\boldsymbol{k}^1 = (k_1^1, k_2^1) = (1,1)$, $\boldsymbol{k}^2 = (k_1^2, k_2^2) = (1,0)$, $\boldsymbol{k}^3 = (k_1^3, k_2^3) = (0,1)$, $\boldsymbol{k}^4 = (k_1^4, k_2^4) = (0,0)$.

c_1	c_2^+	c_2^-	$q_{10}^{1,2}$	$\underline{u}_{r1}$	$\underline{u}_{r2}$	$\bar{u}_{r1}$	$\bar{u}_{r2}$	U_{max}^1	U_{max}^2	c_{r1}	c_{r2}	ρ
1	1	10	0.02	0.04	0.04	0.08	0.08	0.2	0.19	0.25	0.3	0.2

Table 1: Parameters of the numerical example

We consider the following instantaneous cost:

$$G(\boldsymbol{x},\boldsymbol{w}) = c_1 x_1 + c_2^+ x_2^+ + c_2^- x_2^- + c_{r1} u_{r1} + c_{r2} u_{r2}$$

where c_1, c_2^+, c_2^- and c_{ri}, $i = 1,2$, are given constants. We use the computational domain $\mathcal{D}$ given by:

$$\mathcal{D} = \{(x_1, x_2) : 0 \leq x_1 \leq 6; \ -5 \leq x_2 \leq 5\} \quad (12)$$

with $h_1 = h_2 = 0.2$. The demand rate for the product is $u_3 := d = 0.18$. The table 1 summarizes other parameters used in this example. The policy improvement technique is used to solve the system of equations obtained from (11). The obtained results are presented in figures 1 to 3.

Results analysis: The production policy (production rate) for both machines M_1 and M_2 are illustrated in figure 1 at mode 1 where the machines are operational. This figure shows that there is no need to produce the part with suffisent stock levels both in the work-in-process (WIP), described by x_1 and in the downstream buffer described by x_2. For small stock levels, the obtained policy defines well the region in the domain (x_1, x_2) where a maximal production rate is optimal. A similar trend in the optimal production rate in mode 2 (for M_1 while M_2 is under repair) and in mode 3 (for M_2 while M_1 is under repair) is observed but with a different switching policy from U_{max}^j to d, from U_{max}^j to 0 or from d to 0 ($j = 1, 2$).

The corrective maintenance policy is illustrated in figures 2 to 3 for mode 2 (repair rate of M_2 while M_1 is up), mode 3 (repair rate of M_1 while M_2 is up) and mode 4 (repair rates of M_2 and M_1 while both are down). It is interesting to note that, in these modes, the repair rate is maximal when both x_1 and x_2 are negative. The obtained policy also recommends to repair each machine at minimal repair rate for significant stock levels with a switching point (from $\underline{u}_r$ to $\bar{u}_r$) being different for figures 2 to 3.

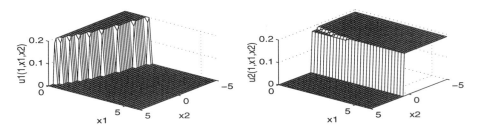

Figure 1: Production rate of M_1 ($u_1(1, x_1, x_2)$) and production rate of M_2 ($u_2(1, x_1, x_2)$) at mode 1

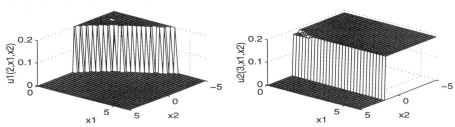

Figure 2: Repair rate of M_2 ($u_{r2}(2, x_1, x_2)$) and repair rate of M_1 ($u_{r1}(3, x_1, x_2)$) at modes 2 and 3 respectively

Comparative study

In [3], the authors explicitly introduce buffer sizes, average buffer levels, starvation and blockage fractions as control parameters. A desirable system state, called the hedging point, is then determined which is used to regulate the production. The starvation and blockage fractions, the buffer hedging level and space can first be determined as follows:

$$\begin{aligned} f_1^s &= f_2^s = 0; \quad f_1^b = \min\{1 - d/D_1, u_{r2}/(u_{r2} + q_{10}^2)\} \\ f_2^s &= \min\{1 - d/D_2, u_{r1}/(u_{r1} + q_{10}^1)\} \\ z^b &= (d/u_{r1})[1 - (u_{r1}/q_{10}^1)(f_2^s/(1 - f_2^s))] \\ z^s &= (d/u_{r2})[1 - (u_{r2}/q_{10}^2)(f_1^b/(1 - f_1^b))] \end{aligned}$$

where $D_i = u_{r1}(u_{ri} + q_{10}^i)^{-1} U_{max}^i$ is the isolated capacity of machine M_i, $i = 1, 2$. Next, with this notation, the hedging point is given recursively by:

$$z_2 = \triangle_2; \quad z_1 = z^b + z_2;$$

where

$$\triangle_i = \frac{u_{ri} q_{10}^i}{u_{ri} + q_{10}^i} \frac{d}{2} \Big(\frac{U_{max}^i}{U_{max}^i - d} \Big) \Big\{ \Big(\frac{1}{u_{ri}}\Big)^2 + \Big(\frac{f_i^s}{q_{10}^i}\Big)^2 + \Big(\frac{f_i^b}{q_{10}^i}\Big)^2 \Big\} \quad i = 1, 2$$

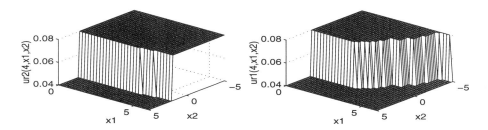

Figure 3: Repair rate of M_1 ($u_{r1}(4, x_1, x_2)$) and repair rate of M_2 ($u_{r2}(4, x_1, x_2)$) at mode 4

	$u_{ri}(x_1, x_2) = \frac{1}{2}(\underline{u}_{ri} + \overline{u}_{ri})$	$\underline{u}_{ri} \leq u_{ri} \leq \overline{u}_{ri}$
Bai and Gershwin [3]	$(z_1, z_2) = (16.44, 11.56)$	-
A numerical approach	$(z_1, z_2) = (1.5, 2.5)$	-
The proposed approach	-	$(z_1, z_2) = (0, 2)$

Table 2: Hedging points comparison

Table 2 summarizes the hedging point values $z = (z_1, z_2)$ in three different cases corresponding to the Bai and Gershwin approach [3], to a numerical solution of the dynamic programming equations for fixed repair rates and to the corrective maintenance situation. One can observe that, with data presented in table 1, $z = (16.44, 11.56)$ with the repair rates u_{ri}, $i = 1, 2$, set to $\frac{1}{2}(\underline{u}_{ri} + \overline{u}_{ri})$. Solving the discrete dynamic programming set of equations obtained form (11) gives $z = (1.5, 2.5)$ with $u_{ri} = \frac{1}{2}(\underline{u}_{ri} + \overline{u}_{ri})$. By controlling both the production and machines repair rates, with repair costs c_{r1} and c_{r2}, we obtain numerically $z = (0, 2)$ in mode 1.

The minimal value function with corrective maintenance is less than the one corresponding to the numerical approach (without any control on machines repair rates) which in turn is obviously less than the Bai and Gershwin [3] minimal value function. Thus, there is a significant difference between the system performances in the above situations (the proposal, the direct numerical solution of dynamic programming equations and the Bai and Gershwin [3] approaches). This observation shows the contribution of our proposal which is based on the corrective maintenance planning problem.

5 Conclusion

The production planning problem of a manufacturing system with machines in tandem in the presence of corrective maintenance has been proposed. We describe a procedure to construct a control for an state constrained control problem by extending the concept of directional derivatives to the system under study. By controlling both production and repair rates, we obtained a near optimal control policy of the system through numerical techniques. We illustrated and validated the proposed approach by a numerical example of a two-machine system. The contribution of the corrective maintenance has been illustrated by comparing our results to those presented in [3].

References

[1] Akella R. and Kumar P. R., *Optimal Control of Production Rate in a Failure Prone Manufacturing System*, IEEE Transactions on Automatic Control, Vol. AC-31, No. 2, pp. 116-126, February 1986.

[2] Bai S. X. and Gershwin S. B., *Scheduling Manufacturing Systems With Work-In-Process Inventory Control: Multiple-Part-Type Systems* , International Journal of Production Research, Vol. 32 No. 2, pp. 365-385, 1994.

[3] Bai S. X. and Gershwin S. B., *Scheduling Manufacturing Systems With Work-In-Process Inventory Control: Single-Part-Type Systems* , IIE Transactions, Vol. 27 pp. 599-617, 1995.

[4] Boukas, E. K. and Haurie A., *Manufacturing Flow Control and Preventive Maintenance : A Stochastic Control Approach* IEEE Transactions on Automatic Control, Vol. 33, No. 9, pp. 1024-1031, February 1990.

[5] Glassey C. R. and Hong Y. *Analysis of Behaviour of an Unreliable n-Stage Transfer Line With (n-1) Inter-Stage Storage Buffers*, International Journal of Production Research, Vol. 31 No. 3, pp. 519-530, 1993.

[6] Kushner H. J. and Dupuis P. G., *Numerical Methods for Stochastic Control Problems in Continuous Time*, Springer-Verlag, New York, 1992.

[7] Lou S. P. and Zhang Q. *Optimal Feedback Production Planning in a Stochastic Two-Machine Flowshop* European Journal of Operational research, No. 73, pp. 331-345.

[8] Older G. J. and Suri R. *Time optimal of part-routing in a manufacturing system with failure prone machines* in Proc. 19*th* IEEE Conf. Decis. Contr., Albuquerque, NM, 1980, pp. 722-727.

[9] Presman E., Sethi S. and Zhang Q., *Optimal Feedback Production Planning in a Stochastic N-machine Flowshop*, Automatica, Vol. 31, No. 9, 1995.

[10] Rishel R., *Control of Systems With Jump Markov Disturbances*, IEEE Transactions on Automatic Control, pp. 241-244, April 1975.

[11] Sharifnia A., *Production Control of Manufacturing System With Multiple Machine State*, IEEE Transactions on Automatic Control, Vol. 33, No. 7, July 1988.

Address: Mechanical Engineering Department, École Polytechnique de Montréal,
C.P. 6079, Succ. "Centre-ville", Montréal (Québec), Canada, H3C 3A7.

MAREK MAKOWSKI

Generation and analysis of a non-linear optimization problem: European Ozone model case study

Abstract

The paper presents an outline of the Ozone model being developed at IIASA and used for analysis of various policy measures aimed at reduction of air pollution in order to improve the corresponding air quality in Europe.

Generation of the Ozone model requires statistical analysis of a large amount of data, and analysis of the model requires solution of large scale non-linear optimization problems. A model generator has been developed not only for providing various optimization solvers with the model specification but also for performing data consistency check and in order to allow for implementation two techniques (soft constraints and regularization) necessary for a more complete model analysis.

Keywords: Generation of non-linear models, preprocessing of non-linear optimization problems, non-linear optimization, Ozone model

1 Outline of the problem

Major concerns about the environmental impacts of air pollution exist in many parts of the world. In some cases, such concerns have led to the introduction of measures to reduce the emissions of air pollutants in order to limit their negative effects.

Within Europe interest in ground-level ozone has intensified in recent years, with increasing experimental evidence that ozone can have adverse effects on crops, trees, materials and human health. Studies of the impacts of ozone have resulted in the establishment of critical levels for ozone in order to protect agricultural crops and forests, using a long-term exposure measure, the 'accumulated excess ozone' concept. A threshold concentration of 40 ppb has been set for both crops and trees. This exposure index is referred to as AOT40, the accumulated exposure over a threshold of 40 ppb. In many parts of Europe the critical levels are exceeded and measures to reduce ozone concentrations in these areas will be needed to protect the relevant ecosystems.

An essential requirement of an integrated assessment model for ozone is a simplified but reliable description of the ozone formation process in order to represent the source-receptor relationships involved. It is possible to envisage several ways of condensing the results of more complex models of ozone formation to achieve this. One approach is to use statistical techniques to summarize the results obtained from a complex mathematical model for a large number of emission reduction scenarios. The formation of ozone involves chemical reactions between NO_x (i.e., nitric oxide (NO) and nitrogen dioxide (NO_2) taken together) and VOCs (Volatile Organic Compounds) driven by solar radiation, and occurs on a regional scale in many parts of the world. Hence the reduction of ozone can be achieved by a reduction of NO_x and VOC emissions.

Cost effective measures aimed at the reduction of ground level ozone concentrations at several hundreds of receptors over Europe can be calculated by minimization of a cost function that corresponds to the costs related to reductions of NO_x and VOC emissions. The Ozone model (cf e.g. (Heyes, Schöpp, Amann, Bertok, Cofala, Gyarfas, Klimont, Makowski and Shibayev, 1997)) has been developed in order to analyze various policy options that lead to improvement of the air quality by implementing reductions of such emissions. However, the emissions of NO_x have also to conform to standards set at each receptors for acidification and for eutrophication. The latter problem is handled by the RAINS model, cf. e.g. (Alcamo, Shaw and Hordijk, 1990), but analysis of two separate models is cumbersome therefore recently the RAINS model has been included into the Ozone model[1]. This in turn requires joint consideration of not only emissions of NO_x and VOC but also of NH_x (ammonia) and SO_x (sulphur oxides).

2 Model definition

There is a set of sources of various types of air pollution, and a set of points (areas), where concentrations of various air quality indicators are measured. Conventionally, the names *emitter* and *receptor* are used for elements of these two sets.

Indices $i \in I$, $j \in J$ correspond to emitters and receptors, respectively. The numbers of elements in I and J correspond to numbers of countries (about 50) and number of grids (about 600).

2.1 Decision variables

n_i - annual emission of NO_x (nitrogen oxides),
v_i - annual emission of VOCs (Volatile Organic Compounds),
a_i - annual emission of NH_x (ammonia),
s_i - annual emission of SO_x (sulphur oxides),
y_j - violation of ozone concentration targets.

Each of decision variables is defined for elements of the emitters set ($i \in I$) and it is implicitly bounded by a corresponding definition of the domain of the corresponding cost function that define costs associated with reduction of emission (see Section 2.4). Additionally, the definitions of cost functions of type given by eq. (11) may contain linear constraints for n_i and v_i, cf eq. (12).

Variables y_j are implied decision variables (they correspond to soft constraints, see eq (4)) and are generated only for receptors at which for a given starting point (that defines all emissions) the resulting ozone concentration violates given environmental standards.

---

[1] Information on a current version of the RAINS model is available at URL:http://www.iiasa.ac.at/Research/TAP/. The currently developed Ozone model (which includes the RAINS model for Europe) will be documented later in 1997.

2.2 State variables

en_j - effective emissions of NO_x experienced at j-th receptor
nlv_j - a representation of another non-linear term at j-th receptor.
o_j - resulting ozone concentrations

State variables are defined for each receptor ($j \in J$). The o_j variables are defined only in the model description (they do not appear in the model actually generated for optimization). However, a function for computing o_j is implemented.

2.3 Constraints

The mean effective concentration of NO_x experienced at j-th receptor is given by:

$$en_j = \sum_{i \in I} e_{ij} n_i + enn_j \qquad (1)$$

where enn_j are given effective natural emissions of NOx.

The representation of another non-linear term defining ozone concentration at j-th receptor is defined by:

$$nlv_j = \sum_{i \in I} d_{ij} v_i \qquad (2)$$

The ozone concentration is defined by:

$$o_j = k_j + \sum_{i \in I}(a_{ij} v_i + b_{ij} n_i + \gamma_{ij} n_i^2) + \alpha_j en_j^2 + \beta_j en_j nlv_j \qquad (3)$$

and it is constrained at each receptor by:

$$o_j - y_j \leq o_j^{max} \qquad (4)$$

where o_j^{max} is a given maximum ozone concentration at j-th receptor.

The variables y_j are bounded by:

$$0 \leq y_j \leq y_j^{max} \qquad (5)$$

where

$$y_j^{max} = max(0, o_j(\cdot) + feas) \qquad (6)$$

and $o_j(\cdot)$ is the value defined by eq (3) for a given set of emissions (that defines starting point for optimization) and $feas$ is a given feasibility tolerance.

The sum of depositions of NO_x, NH_x and SO_x is constrained at each receptor by a given maximum acidification:

$$\sum_{i \in I} tn_{ij} n_i + \sum_{i \in I} ta_{ij} a_i + \sum_{i \in I} ts_{ij} s_i \leq da_j^{max} \qquad (7)$$

where tn, ta and ts are transport coefficients for nitrogen, ammonia and sulphur oxygen, respectively.

The sum of depositions of NO_x and NH_x is constrained at each receptor by a given maximum eutrophication:

$$\sum_{i \in I} tn_{ij} n_i + \sum_{i \in I} ta_{ij} a_i \leq de_j^{max} \tag{8}$$

A sulphur deposition is also constrained at each receptor by:

$$\sum_{i \in I} ts_{ij} s_i \leq ds_j^{max} \tag{9}$$

2.4 Cost functions

For each type of emission and for each emitter the annual costs related to reduction of a corresponding emission to a certain level is given by a non-linear function. All cost functions are strictly convex and smooth. Additionally, cost functions of type (10) are strictly decreasing.

The following two types of functions are used (for the sake of clarity the emitter index was omitted in the following two definitions):

$$cx(x) = \frac{a_0 + a_1 x}{1 + a_2 x + a_3 x^2} + a_4 \tag{10}$$

$$c(x, y) = \frac{a_0 + a_1 x}{1 + a_2 x + a_3 x^2} + \frac{a_4 + a_5 y}{1 + a_6 y + a_7 y^2} + \frac{a_8 + a_9 x + a_{10} y}{1 + a_{11} x + a_{12} y + a_{13} x^2 + a_{14} y^2 + a_{15} xy} + a_{16} \tag{11}$$

where x in definition (10) corresponds to either a (for ammonia) or s (for sulphur); x and y in definition (11) correspond to NO_x and VOC, respectively.

The following functions define the annual cost related to reducing the level of emission to a level given by argument(s) of the function: $ca_i(a_i)$ for NH_x, $cs_i(s_i)$ for SO_x, and joint cost functions $c_i(n_i, v_i)$ for NO_x and VOC.

Each cost function defines its domain by specifying lower and upper bounds for its argument(s). This implicitly defines lower and upper bounds for all emissions. Additionally, the cost functions (11) may further restrict the domain (of admissible NO_x and VOC) by defining linear constraints in the form of:

$$p0_{ki} n_i + p1_{ki} v_i \leq p2_{ki} \tag{12}$$

where $p0_{ki}, p1_{ki}, p2_{ki}$ are given parameters, $k \in \{1, ..., K_i\}$, and K_i is a number of such constraints for i-th emitter.

2.5 Goal function

The following goal function is minimized:

$$\sum_{i \in I} (ca_i(a_i) + cs_i(s_i) + c(n_i, v_i)) + \rho \sum_{j \in J} y_j^2 + \epsilon \|z - \bar{z}\| \tag{13}$$

where ϵ is a given small positive number and ρ is a large positive penalty coefficient.

A justification for selection of such a goal function is presented in Section 4.

3 Generation and solution of the model

3.1 Structure of the optimization problem

Table 1 illustrates in an informal way the structure of the optimization problem outlined in Section 2. The model has about 2000 variables and about 3700 constraints.

id	en	nlv	nox	voc	am	so	y	type	rhs	eq_no
o3	α_j, β_j, nlv_j	β_j, en_j	b_{ij}, γ_{ij}	a_{ij}			-1	$\leq$	$o_j^{max} - k_j$	4
acid			tn_{ij}		ta_{ij}	ts_{ij}		$\leq$	da^{max}	7
eutr			tn_{ij}		ta_{ij}			$\leq$	de^{max}	8
sulp						ts_{ij}		$\leq$	ds^{max}	9
dc??			$p0_{ki}$	pl_{ki}				$\leq$	$p2_{ki}$	12
endef	-1		e_{ij}					=	$-enn_j$	1
nldef		-1		d_{ij}				=	0	2

Table 1: Structure of the optimization problem.

The following notation is adopted for this table:
- Indexed variables (nlv_j and en_j) denote existence a non-linear term to which a corresponding variable is contributing.
- Single indices with a parameter denote linear terms defined by such a parameter and a corresponding variable.
- Parameters with indices $_{ij}$ denote linear terms defined by product a j-th transposed column of the corresponding parameter and a corresponding variable.
- Parameters $p0_{ki}$ and pl_{ki} denote linear terms defined by the such parameter and a corresponding variable, cf eq. (12).
- **-1** denotes a diagonal matrix composed of elements equal to -1.

3.2 Generating and solving the model

The Ozone model (for all European countries) is a large non-linear model. A commonly accepted rule of thumb is to try various solvers for optimization of a non-trivial nonlinear model. Therefore three solvers, namely CFSQP (Lawrence, Zhou and Tits, 1996), Conopt (Drud, 1996) and Minos (Murtagh and Saunders, 1987) are being used for solving a resulting optimization problem.

There are basically two approaches to generation and analysis of a mathematical programming problem. Namely, to develop a problem specific generator or to use a modeling system (such as GAMS, AMPL, AIMMS). The following issues should be considered when selecting one of these approaches:
- A modeling system greatly simplifies the task of a model specification, especially if compared with the amount of resources needed for a development of a model generator using traditional procedural programming languages like Fortran or C. However, use of C++ (especially with recently included Standard Template Library and with a class library for mathematical programming substantially reduces this difference.

- A model generator is more efficient in processing (including more sophisticated check of consistency) input data that is needed for a model specification.
- A modeling system has limited possibilities for an efficient interaction with a solver. This is not a problem for linear models (because preprocessing is a standard feature of any good LP solver) but preprocessing of non-linear models is much more difficult (see e.g. (Drud, 1997)). A model generator can generate a non-linear problem that is much easier (in comparison with a model generated by a modeling system) to be solved. Therefore it might be a better choice for large problems, for which solution time becomes an important issue.
- A modeling system typically does not exploit possibilities of an optimization algorithm used by a solver. This is not an important issue as long as the optimization problem is of a moderate size but for large problems exploitation of the problem structure may be necessary for reducing optimization time. In such a case a modeling system has to be augmented by a software tool, e.g. by Fragniere, Gondzio, Sarkissian and Vial (1997), for exploiting structure of optimization problems generated by a modeling language. A model generator can easily be adapted for generating a model in a form that allows the solver to exploit the model structure. Additionally, a modeling system provides interface to a small number of solvers, which makes it practically impossible to use other solvers (including simulation tools) for the model analysis.
- A modeling system releases a modeler from a complex task of providing code for computation of values of non-linear constraints and of non-linear elements of Jacobian. Especially the latter used to require substantial development time. However, recently tools that substantially ease this task are available, see e.g. (Bischof, Carle, Corliss and Griewank, 1992). A typical non-linear problem has only few formulas for non-linear part. Therefore one can use e.g. Mathematica (Wolfram, 1996) for generating C language code for formulas for the Jacobian and for values of constraints, and then include this code into a class that provides values of particular elements of the Jacobian and of constraints. One should note, that using the *FullSimplify* operator in the Mathematica may greatly simplify the formulas, hence it may also reduce the computation time.

The task of using several solvers is interesting from the software-engineering point of view. Each solver has a different interface (the way of formulating an optimization problem). However, majority the software components are common to all the solvers. Therefore, object-oriented programming approach was a natural choice because it greatly simplifies software development task by handling common parts in base classes and by providing solver-specific interfaces through inherited classes.

Hence, the model generator and postprocessing for the Ozone model have been developed in the C++ programming language. The solvers (Conopt and Minos are written in Fortran, CFSQP in C language) are available in forms of libraries that are linked with the generator and with postprocessing modules. Therefore, from a user point of view, generation of the model, solution of the resulting optimization problem and postprocessing (pre-analysis of the results) are done by one application.

3.3 Preprocessing and data handling

It is a commonly know fact, that preprocessing of an optimization problem can not result in a substantial reduction of computation time but may be necessary to solve the problem.

The current version of the Ozone model contains many dominated or empty constraints and many variables (e.g. y_j variables for receptors for which $y_j^{max} = 0$) that can be reduced from the resulting optimization problem. Therefore preprocessing techniques commonly used for linear programming problems have been adapted to the Ozone model generator and additional techniques are being implemented now. Preprocessing of the Ozone model dramatically reduces the size of the resulting optimization problem.

Another important element of the model preprocessing is its scaling. Absolute values of Jacobian elements (including the goal function) are in range of $[10^{-6}, 10^7]$, which would create numerical problems, if no problem specific scaling would have been applied.

The data used for the model specification comes from different sources, large part of data preprocessing is done at several institutions using computers with various operating systems. Therefore it is rational to use an efficient tool for handling data used for the model specification in a form of binary portable files. For this purpose the HDF (Hierarchical Data Format) public domain software has been adapted for the Ozone model by Haagsma (1996).

4 Model analysis

Application of a classical OR (optimization driven) approach to a model analysis for decision support may result in a failure. An overview of the related problems and techniques which can enhance applicability of optimization-based model analysis for decision support are discussed by Makowski (1994). Two of such techniques have been applied to the Ozone model, namely:

- The optimization problem has practically non-unique solution (i.e. there are many very different solutions having almost the same value of the goal function). Therefore a regularization technique has been used to provide a unique solution having some additional properties, cf eq. (13).
- Environmental targets may not be feasible at some receptors for some scenarios of policy constraints for emissions. Identification of the corresponding constraints in the optimization model is easy for linear constraints but may be difficult for non-linear constraints. All constraints that may cause infeasibility of the optimization problem are identified during the preprocessing of the optimization model and are converted into so-called *soft constraints*, cf eq. (4).

Therefore the goal function (13) is composed of three terms. The first term corresponds to the cost of emission reduction and it is the original goal function. The second term is the penalty term introduced to deal with the soft constraints defined by introduction of variables y_j into constraints (4). The third term[2] is a regularizing term introduced for avoiding very different solutions (with almost the same value of the original goal function) for

---

[2] $\epsilon \|z - \bar{z}\|$ where z denotes a vector composed of all decision variables.

problems that differ very little. This term assures that the optimal solution (for a problem that does not have a unique local optimal solution) will be the optimal solution closest to the point defined by given reference vector $\bar{z}$.

5 Conclusions

The paper illustrates optimization-based model analysis applied to analysis of various policy options aimed at improving the air quality in Europe. Extensions of a traditional OR methods that enhance usefulness of optimization for policy analysis have been presented. Software engineering issues pertinent to generation and analysis of complex and large non-linear models were discussed.

At the time of writing this paper a new version of the Ozone model is under development. It will include soft constraints for acidification, eutrophication and sulphur deposition. Therefore the readers of this chapter may want to check the Web page of the author for information about the new version of the Ozone model.

Acknowledgments

The RAINS and Ozone models have been developed since several years by the TAP (Transboundary Air Pollution) Project[3] at IIASA in collaboration with several European institutions. The author has been collaborating with the TAP Project since several years by taking part in various activities related to the development of methodology and of software for generation and optimization based analysis of RAINS and Ozone models. This long term collaboration has provided the author with a good stimulation for research on extensions of standard OR techniques by methods and corresponding software tools necessary for applications of optimization methods to policy making problems. Author thanks all members of the TAP Project led by Dr. Markus Amann for this collaboration, but especially Dr. Chris Heyes and Dr. Wolfgang Schöpp for many fruitful discussions.

The author acknowledges and appreciates the contribution of three participants of the IIASA's Young Summer Scientists Program, namely of Mr. Piotr Zawicki and of Mr Paweł Białoń, both of the Warsaw University of Technology, who have developed the prototypes (cf Zawicki (1995) and Białoń (1996), respectively) of model generators for the previous versions of the Ozone model, and Mr. Ijsbrand Haagsma of the Delft University, who has implemented a prototype for data handling through HDF (a platform independent software for data storage), see Haagsma (1996).

Finally, the authors thanks Dr Arne Drud of ARKI Consulting and Development A/S, Denmark, for continuously providing the latest versions of the Conopt libraries (Drud, 1996), and for his consultations on advanced topics related to formulation and solving non-linear optimization problems.

---

[3] http://www.iiasa.ac.at/Research/TAP/.

References

Alcamo, J., Shaw, R. and Hordijk, L. (eds): 1990, *The RAINS Model of Acidification*, Kluver Academic Publishers, Dordrecht, Boston, London.

Białoń, P.: 1996, Optimization-basded analysis of a simplified ozone model, *Working Paper WP-96-134*, International Institute for Applied Systems Analysis, Laxenburg, Austria.

Bischof, C., Carle, A., Corliss, G. and Griewank, A.: 1992, ADIFOR – generating derivative codes for fortran programs, *Scientific Computing* **1**(1), 1–29.

Drud, A.: 1997, Interactions between nonlinear programming and modeling systems, *Technical Report 1994.3*, Arki Consulting and Development, Bagsvaerd, Denmark. (preprint submitted to Elsview Science).

Drud, A. S.: 1996, CONOPT: A system for large scale nonlinear optimization, *Reference manual for CONOPT subroutine library*, ARKI Consulting and Development A/S, Bagsvaerdvej, Denmark.

Fragniere, E., Gondzio, J., Sarkissian, R. and Vial, J.: 1997, Structure exploting tool in algebraic modeling languages, *Technical Report 1997.2*, Department of Management Studies, University of Geneva, Geneva, Switzerland. URL:ecolu-info.unige.ch/~logilab.

Haagsma, I.: 1996, Platform independent storage of data: an application of HDF for simplified ozone model, *Working Paper WP-96-133*, International Institute for Applied Systems Analysis, Laxenburg, Austria.

Heyes, C., Schöpp, W., Amann, M., Bertok, I., Cofała, J., Gyarfas, F., Klimont, Z., Makowski, M. and Shibayev, S.: 1997, A model for optimizing strategies for controlling ground-level ozone in Europe, *Interim Report IR-97-002*, International Institute for Applied Systems Analysis, Laxenburg, Austria. Available on-line from http://www.iiasa.ac.at/cgi-bin/pubsrch?IR97002.

Lawrence, C., Zhou, J. and Tits, A.: 1996, User's guide for CFSQP version 2.4: A C code for solving (large scale) constrained nonlinear (minimax) optimization problems, generating iterates satisfying all inequality constraints, *Technical report*, Institute for Systems Research, University of Maryland, College Park, Maryland.

Makowski, M.: 1994, Design and implementation of model-based decision support systems, *Working Paper WP-94-86*, International Institute for Applied Systems Analysis, Laxenburg, Austria. Available on-line from http://www.iiasa.ac.at/~marek/pubs/.

Murtagh, B. A. and Saunders, M. A.: 1987, MINOS 5.1 user's guide, *Technical Report SOL 83-20R*, Stanford University, Stanford, CA 94305-4022, USA.

Wolfram, S.: 1996, *The Mathematica Book, Third Edition, Mathematica Version 3*, Cambridge University Press, Cambridge.

Zawicki, P.: 1995, Software tools for generation, simulation and optimization of the simplified ozone model, *Working Paper WP-95-107*, International Institute for Applied Systems Analysis, Laxenburg, Austria.

Marek Makowski
International Institute for Applied Systems Analysis,
A-2361 Laxenburg, Austria
    marek@iiasa.ac.at   http://www.iiasa.ac.at/~marek

KURT MARTI AND SHIHONG QU[*)]
Adaptive stochastic path planning for robots - real-time optimization by means of neural networks

1 Stochastic trajectory planning for robots

1.1 The trajectory planning problem for robots

The problem of trajectory planning for robots can be described [1-4] mathematically by the following variational problem:

$$\min_{\beta(.), q(.)} \int_{s_0}^{s_e} f_0(s, q, q', q'', \beta, \beta', p) ds \tag{1a}$$

s.t.

$$q(s_0) = q_0, \ q(s_e) = q_e, \ \beta(s_0) = \beta(s_e) = 0 \tag{1b}$$

$$q_{min} \leq q \leq q_{max}, \ \dot{q}_{min} \leq q'\sqrt{\beta} \leq \dot{q}_{max}, \tag{1c}$$

$$\tau_{min,i} \leq a_i(q, q', p)\beta' + b_i(q, q', q'', p)\beta + c_i(q, p) \leq \tau_{max,i},$$

$$i = 1, 2, ... n \tag{1d}$$

$$\beta(s) \geq 0, \ s_0 \leq s \leq s_f \tag{1e}$$

The optimal geometric path in configuration space $q = q(s)$ and the optimal velocity profile $\beta = \beta(s)$ must be determined such that a certain objective function, e.g., the total run time, the total energy consumption or a combination of the both, will be minimized and some restrictions are fullfilled. For a given geometric path $q = q_e(s)$ in the configuration space, only the velocity profile $\beta = \beta(s)$ has to be determined optimally.

1.2 The trajectory planning problem under stochastic uncertainty

In practice, the vector of model parameters

$$p = p(w), \ w \in (\Omega, \Theta, P), \tag{2}$$

is not a given fixed quantity, but due to stochastic variations of the material, manufacturing and adjustment errors, stochastic variations of the payload, uncertain obstacles, modelling uncertainties, etc., $p = p(w)$ must be considered as a random vector with a certain probability distribution $P_{p(.)}$.

[*)] Supported by the DFG-Priority Program "Real-Time Optimization of Large Scale Systems"

To get optimal velocity profiles $\beta^*(.)$ being **robust** with respect to random parameter variations, the methods of **stochastic optimization** are used, which presupposes an appropriate substitute problem. Depending on the decision theoretical viewpoint, one gets different deterministic substitute problems for the stochastic trajectory planning problem (1a-e). Two basic cases will be introduced in the following.

1.2.1 Chance constrained problem

We consider the expected value of the objective function and demand that the stochastic conditions (1d) are fulfilled with given minimum probabilities α_i. This yields then the Chance Constrained Problem:

$$\min_{q(.),\ \beta(.)} \int_{s_0(=0)}^{s_e} E_p\, f_0(s, q, q', q'', \beta, \beta', p)\, ds \qquad (3a)$$

s.t.

$$q(s_0) = q_0,\ q(s_e) = q_e,\ \beta(0) = \beta(s_e) = 0 \qquad (3b)$$

$$q_{min} \leq q \leq q_{max},\ \dot{q}_{min} \leq q'\sqrt{\beta} \leq \dot{q}_{max} \qquad (3c)$$

$$\beta(s) \geq 0,\ s_0 \leq s \leq s_f \qquad (3d)$$

$$P\{\tau_{min,i} \leq a_i\beta' + b_i\beta + c_i \leq \tau_{max,i}\} \geq \alpha_i,\ i = 1, 2, ...n; \qquad (3e)$$

1.2.2 Cost constrained problem

$$\min_{q(.),\ \beta(.)} \int_{s_0(=0)}^{s_e} E_p\, f_0(s, q, q', q'', \beta, \beta', p)\, ds \qquad (4a)$$

s.t.

$$q(s_0) = q_0,\ q(s_e) = q_e,\ \beta(0) = \beta(s_e) = 0 \qquad (4b)$$

$$q_{min} \leq q \leq q_{max},\ \dot{q}_{min} \leq q'\sqrt{\beta} \leq \dot{q}_{max} \qquad (4c)$$

$$\beta(s) \geq 0,\ s_0 \leq s \leq s_f \qquad (4d)$$

$$E_p \gamma_i(a_i\beta' + b_i\beta + c_i; \tau_{min,i}, \tau_{max,i}) \leq \Gamma_{max}^{(i)},\ i = 1, 2, ...n. \qquad (4e)$$

Here γ_i are given cost functions evaluating the violations of the constraints (1d). In this substitute problem we consider the expected value of the objectiv function and demand that the given upper bounds $\Gamma_{max}^{(i)}$ for the expected value of the cost functions should not be exceeded.

2 Adaptive stochastic trajectory planning

By means of measurement and estimation algorithms the uncertainty about the model parameters can be reduced. Hence, this way we may get new information about the robotic system and its enviroment. We suppose that the informations about the parameter vector p available at the initial and correction time points t_0, t_1, t_2, ... can be represented by the information set (σ-algebra of events up to t_j) $\mathcal{A}_j$. In order to utilize the new information $\mathcal{A}_j$, j=1,2,.., for improving the control of the robot, the velocity profile $\beta = \beta(s)$ (in general the geometric path $q = q(s)$, too) must be adjusted, if the new information $\mathcal{A}_j$ is available at the time point t_j. Consequently, we have an adaptive stochastic trajectory planning problem.

2.1 Description of the problem

For the sake of simplicity we consider in the following the problem with a **given** geometric path $q = q_e(s)$ in the configuration space. According to Section 1, at the time point t_j we have then one of the following substitute problems:

$$\min_{\beta(.)} \int_{s_j}^{s_f} E_p\left(f_0(s, q_e, q_e', q_e'', \beta, \beta', p)|\mathcal{A}_j\right) ds \tag{5a}$$

s.t.

$$\beta(s_j) = \beta_j, \ \beta(s_f) = 0 \tag{5b}$$

$$\dot{q}_{min} \leq q_e' \sqrt{\beta} \leq \dot{q}_{max} \tag{5c}$$

$$\beta(s) \geq 0, \ s_j \leq s \leq s_f \tag{5d}$$

$$P\{\tau_{min,i} \leq a_i\beta' + b_i\beta + c_i \leq \tau_{max,i}|\mathcal{A}_j\} \geq \alpha_i, \ i = 1, 2, ...n, \tag{5e}$$

having the "initial values" $(s_j, \beta_j, \mathcal{A}_j)$, or

$$\min_{\beta(.)} \int_{s_j}^{s_f} E_p\left(f_0(s, q_e, q_e', q_e'', \beta, \beta', p)|\mathcal{A}_j\right) ds \tag{6a}$$

s.t.

$$\beta(s_j) = \beta_j, \ \beta(s_f) = 0 \tag{6b}$$

$$\dot{q}_{min} \leq q_e' \sqrt{\beta} \leq \dot{q}_{max} \tag{6c}$$

$$\beta(s) \geq 0, \ s_j \leq s \leq s_f \tag{6d}$$

$$E_p\left(\gamma_i(a_i\beta' + b_i\beta + c_i; \tau_{min,i}, \tau_{max,i})|\mathcal{A}_j\right) \leq \Gamma_{max}^{(i)}, \ i = 1, 2, ...n. \tag{6e}$$

In the j-th step, j=0,1,...,

$$\beta^{(j)} = \beta^{(j)}(s), \ s_j \leq s \leq s_f, \ j = 0, 1, ..., \tag{7}$$

denotes the optimal velocity profile according to (5a-e) or (6a-e). The positions s_{j+1}, j=0,1,..., are defined here by
$$s_{j+1} := s^{(j)}(t_{j+1}), \tag{8}$$
where the transformation $s^{(j)} : [t_j, t_f^{(j)}] \to [s_j, s_f]$ is defined by the ordinary differential equation:
$$\dot{s}(t) := \sqrt{\beta^{(j)}(s)}, \quad s(t_j) = s_j. \tag{9}$$

Fig. 1 t-s-transformation at different stages

Remark 2.1 For the point-to-point problem or a problem with obstacles we have similar variational problems.

2.2 Solving the problem by means of finite element methods

Dividing the interval $[s_0, s_f]$ into disjoints sets S_l, l=0,1,...L,
$$[s_0, s_f] = S_0 \cup S_1 \cup ... S_L,$$
we describe the optimal solution of (5a-e), (6a-e), resp.,
$$\beta^{(j)} = \beta^{(j)}(s) = \tilde{\beta}(s; s_j, \mathcal{B}_j, \mathcal{A}_j), \quad s_j \leq s \leq s_f,$$
as linear combination of known basis functions (quadratic splines)
$$\beta^{(j)}(s) = \sum_{t=1}^{T_l} \gamma_t^{(l)} B_t(s; s_j), \quad s \geq s_j, \text{ for } s_j \in S_l, l = 0, 1, ...L. \tag{10}$$

Putting (10) into the variational problem (5a-e) or (6a-e), we get a parameter optimization problem for the unknown coefficients
$$\gamma_t^{(l)} = \gamma_t^{(l)}(\zeta_j), \quad \zeta_j := (s_j, \mathcal{B}_j, \mathcal{A}_j), \quad l = 0, 1, ...L, \tag{11}$$
which can be solved by standard optimization procedures, e.g. SQP-methods.

Remark 2.2 The same method can be used for more general trajectory planning problems (point-to-point problem or problems with obstacles).

Remark 2.3 Computation in real-time Because industrial robots, except the increasingly important class of **service robots**, move very quickly and the computing

given geometric path and about 30 minutes for the point-to-point path planning problem, the methods mentioned above can not be used on-line.

3 Adaptive trajectory planning in real-time by means of neural networks

Due to the heigh demand for computing time, the adaptive stochastic trajectory planning problem can not be solved on-line by means of standard parameter optimization methods. To approximate off-line the optimal control law $\tilde{\beta} = \tilde{\beta}(s; s_j, \beta_j, \mathcal{A}_j)$, $s_0 \leq s \leq s_f$, we utilise **Neural Networks (NN)**. At the correction time point t_j the optimal velocity profile $\beta^{(j)}$ for j-th stage can be determined then very fast.

3.1 Multilayer neural networks

We use a multilayer neural network:

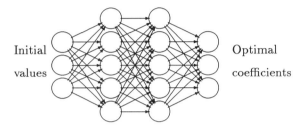

Fig. 2 Neural Networks

A multilayer neural network consists of one input layer, one or several hidden layers and one output layer. The structure of multilayer neural networks is very simple: The informations are transformed from one layer to the next layer. For the training of the network there is a so-called Back-Propagation-Algorithm [6].

For multilayer neural networks one can calculate the activation of every neuron as follows:

$$\xi_i^{(k)} = \phi(\sum_{j=1}^{n_{k-1}} w_{ij}^{(k)} \xi_j^{(k-1)}) = \phi(v_i^{(k)}), \quad k = 1, 2, 3, ...K, i = 1, ..n_k, \qquad (12)$$

where the following notations are used:

$\xi_i^{(0)}$: input to the i-th neuron of the input layer, $i = 1, 2, ..., n_0$,

$\xi_i^{(k)}$: activation of the i-th neuron on the k-th layer, $i = 1, 2, ..., n_k$,

$w_{ij}^{(k)}$: weight factor for the transformation from the i-th neuron on the (k-1)-th layer to j-th neuron on the k-th layer,

$\phi(.)$: activation function.

For given inputs $\xi_i^{(0)}$, $i = 1, 2, ..., n_0$, in the input layer the NN yields outputs $\xi_i^{(K)}$, $i = 1, 2, ..., n_K$, in the output layer.

If a NN is used to approximate a given vector valued function $\gamma = F(\zeta)$, the vector of weight factors w has to be determined such that the error

$$e = \sum_{\kappa=1}^{N} ||F(\xi_\kappa^{(0)}) - \xi_\kappa^{(K)}||, \qquad (13)$$

$$\xi_\kappa^{(0)} = (\xi_{\kappa 1}^{(0)}, ..., \xi_{\kappa n_0}^{(0)})' = (s_{(.)}, \beta_{(.)}, \mathcal{A}_{(.)})'_{(\kappa)} = \zeta_{(.)(\kappa)}, \qquad (14)$$

$$\xi_\kappa^{(K)} = (\xi_{\kappa 1}^{(K)}, ..., \xi_{\kappa n_K}^{(K)})' = (\gamma_1^{(l)}, ... \gamma_{T_l}^{(l)})'_{(\kappa)}, \qquad (15)$$

will be minimized. By means of gradient methods one can calculate $w_{ij}^{(k)}$ iteratively as follows:

$$w_{ij}^{(k),new} = w_{ij}^{(k),old} - \eta \frac{\partial e}{\partial w_{ij}^{(k)}}, \qquad (16)$$

where $\eta > 0$ denotes the learning rate (step size). For the quadratic error

$$e = \sum_{\kappa=1}^{N} ||F(\xi_\kappa^{(0)}) - \xi_\kappa^{(K)}||_E^2, \qquad (17)$$

the derivatives $\frac{\partial e}{\partial w_{ij}^{(k)}}$, can be computed by the Back-Propagation-algorithm [6]. At the output layer we have

$$\frac{\partial e}{\partial w_{ij}^{(K)}} = \sum_{\kappa=1}^{T} \delta_{\kappa i}^{(K)} \xi_{\kappa j}^{(K)}, \qquad (18)$$

where $\delta_{\kappa i}^{(K)} := (\xi_{\kappa i}^{(K)} - F_i(\xi_\kappa^{(0)}))\phi'(v_{\kappa i}^{(K)})$, and at the other layers we have

$$\frac{\partial e}{\partial w_{ij}^{(k)}} = \sum_{\kappa=1}^{T} \delta_{\kappa i}^{(k)} \zeta_{\kappa i}^{(k)}, \qquad (19)$$

where

$$\delta_{\kappa i}^{(k)} := (\sum_{j=1}^{n_{k+1}} w_{ij}^{(k+1)} \delta_{\kappa j}^{(k+1)}) \phi'(v_{\kappa i}^{(k)}). \qquad (20)$$

3.2 Adaptive trajectory planning by means of neural networks

If the initial values for the j-th stage $\zeta_j = (s_j, \beta_j \mathcal{A}_j)$ are given, problem (5a-e) or (6a-e) can be solved **off-line** by means of representation (10), and one obtains the coefficients $\gamma_t^{(l)*}$. Using different initial values $\zeta_j = (s_j, \beta_j, \mathcal{A}_j) = z^{(\kappa)}$, $\kappa = 1, 2, ..., N$, we can construct in this way the following table:

$z^{(1)}$	$\gamma_t^{(l)}(z^{(1)*})$
$z^{(2)}$	$\gamma_t^{(l)}(z^{(2)*})$
....	...
$z^{(\kappa)}$	$\gamma_t^{(l)}(z^{(\kappa)*})$
....	...
$z^{(N)}$	$\gamma_t^{(l)}(z^{(N)*})$

If the NN is trained off-line with these data, for the actual initial values it will yield on-line very fast the related optimal coefficients $\gamma_t^{(l)*}$ and then the optimal velocity profile $\beta^{(j)*}$.

4 Numerical results

4.1 Robotic model and task description

We consider the robot Manutec r3 with the fixed end effector; the dynamic equation and model parameters are taken from the dissertation of Türk [5]. The adaptive stochastic trajectory planning problem is solved then for a given geometric path in the configuration space. The accuracy of the NN-approximation and the computing time of the suggested method are considered. We suppose that the payload m_l is stochastic, hence, the information set $\mathcal{A}_j$ can be described by the conditional moments

$$\{E(m_l(\omega)|\mathcal{A}_j), E(m_l(\omega)^2|\mathcal{A}_j)\} = \{\bar{m}_l^{(j)}, (\bar{m}_l^{(j)})^2 + (\sigma_l^{(j)})^2\}. \qquad (21)$$

4.2 Numerical results I: Standard parameter optimization

We consider at first the solution of the adaptive stochastic trajectory planning problem by standard parameter optimization. At the starting time point $t_0 = 0$ ($s_0 = 0$), the information about the payload is given by $\bar{m}_l^{(0)} = 7.5$ and $\sigma_l^{(0)} = 4.33$. Of course, we have $\beta_0 = 0$. Solving problem (5a-e) with these data, we get $\beta = \beta^{(0)}(s), s_0 \leq s \leq s_f$ and $t = t^{(0)}(s), s_0 \leq s \leq s_f$. At the first correction time point $t_1 = 0.2733$ ($s_1 = 0.2$) one has new information about the payload m_l: $\bar{m}_l^{(1)} = 6.0$, $\sigma_l^{(1)} = 1.732$. With $\beta_1 := \beta^{(0)}(s_1)$ we can solve the stochastic planning problem (5a-e), and we get $\beta = \beta^{(1)}(s)$, $s_1 \leq s \leq s_f$ and $t = t^{(1)}(s)$, $s_1 \leq s \leq s_f$. Finally, at the correction time point $t_2 = 0.4248$ ($s_2 = 0.5$) we know the payload exactly: $\bar{m}_l^{(2)} = 6.0$, $\sigma_l^{(2)} = 0$. In the same way, using $\beta_2 := \beta^{(1)}(s_2)$ we get then $\beta = \beta^{(2)}(s)$, $s_2 \leq s \leq s_f$ and $t = t^{(2)}(s), s_2 \leq s \leq s_f$.

The results are shown in Fig. 3 and Fig. 4. Obviously, the t-s-transformation and the optimal velocity profile are changed, after the new information is obtained.

4.3 Numerical results II: NN-approximation

As mentioned above, due to its too high demand for computing time, the standard method can not be used on-line. To overcome this barrier, we make use of neural networks approximation.

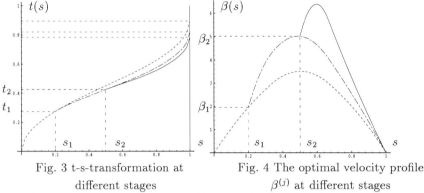

Fig. 3 t-s-transformation at different stages

Fig. 4 The optimal velocity profile $\beta^{(j)}$ at different stages

The stochastic trajectory planning problem is solved off-line by using different initial values s_j, β_j, $\bar{m}_l$ and σ_l. Using these data, we can train the neural network [7]. Consequently, the neural network yields the optimal coefficients and then the optimal velocity profile in real time, if the actual information about the initial values is available at time t_j. Fig. 5 and Fig. 6 show the results of this method. The solid curves represent the neural network approximation and the dashed line are from standard parameter optimization method. From these figures one can see that the neural network approximation yields a very heigh accuracy!

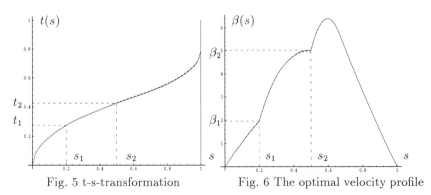

Fig. 5 t-s-transformation

Fig. 6 The optimal velocity profile

4.4 Consideration of the real-time capability of NN-approximation

In order to consider the real-time capability of NN-approximation, we perform 200 identical computations. The mean value of the computing time for finding the coefficients $\gamma^{(l)} = (\gamma_t^{(l)})_{t=1,\ldots T_l}$ is 0.00025 seconds and the mean value of the computing time for calculating the new initial values, the new velocity profile and the new controls (at

60 points) at one stage is 0.0163 seconds. Obviously, this shows that the above method can be applied in real-time.

If the point-to-point problem is considered, the computing time will not increase essentially, because in this case only some more computations for finding the coefficients of the configuration variable $q = q(s)$ are necessary. The computing time for calculating the new initial values, the new velocity profile and the new controls will not change.

References

[1] Marti, K., Qu, S.: Optimal Trajectory Planning for Robot Considering Stochastic Parameters and Disturbances - Computation of an Efficient Open-Loop Strategy, J. Intelligent and Robotic Systems, Vol. 15, pp.19-23, 1996

[2] Pfeiffer, F., Johanni, R.: A Concept for Manipulator Trajectory Planning, IEEE J. Robot. Automat. RA-3(3), pp. 115-123, 1987

[3] Qu, S.: Optimale Bahnplanung unter Berücksichtigung stochastischer Parameterschwankungen, VDI-Verlag, Düsseldorf, 1995

[4] Qu, S.: Stochastic Trajectory Planning for Robots and its Application on "Manutec R3". In: K.Marti, P. Kall (eds.): Stochastic Programming Methods and Technical Application, Springer-Verlag, 1998, LNEMS, Vol. 458, pp 382-393

[5] Türk, M.: Zur Modellierung der Dynamik von Robotern mit rotatorischen Gelenken, VDI-Verlag, Düsseldorf, 1990

[6] Zell,A.: Simulation neuronaler Netze, Addison-Wesley, Bonn, 1994

[7] Zell,A.: Stuttgart Neural Network Simulator: User Manual University of Stuttgart, 1995

VISWANATH RAMAKRISHNA AND HERSCHEL RABITZ
Control of molecular dynamics

1 Introduction

Quantum theory and control theory are two well-developed theories. However, until recently the interaction betweeen these two fields has been limited. This is partly due to the fact that even though any laboratory experiment (in the quantum regime or otherwise) is necessarily performed under carefully controlled environments, they are not perceived as being the result of a careful choice of inputs, [7].

On the other hand any "design" that has gone into the typical experiments has, hitherto, been more intuitive and not quite the result of the systematic design process that is the subject matter of control theory. The principal exception to this trend is the area of site-specific chemistry. However, it would be remiss of us if we were not to point out that there have been others who have contributed to this interaction. First, Mielnik and his collaborators (see [20, 9, 21] for instance) have been led to the study of controlled quantum dynamics in their studies of the foundations and paradoxes of quantum theory. Then there is the excellent text, [4] which considers several interesting control problems in the quantum regime. Secondly, there is the little known, but interesting paper of Lubkin, [19], which can certainly be considered as a paper about controlling quantum motion. Last, but not the least, there is the article of W. Lamb Jr, [17], whose content is very suggestive of quantum control.

Ever since lasers were first discovered chemists have always wanted to use them as part of their arsenal in enhancing product specificity of chemical reactions. Early attempts at using lasers consisted of intuiting the required incident frequencies of the laser pulses. However, this oversimplified thinking usually led to very poor product specificity (see Ch 3 of [1] for more details). The last decade, however, has seen a veritable resurrection of this goal of chemists, [29, 8]. There are two principal reasons behind this renewed interest in the field. First there have been spectacular advances in the area of pulse crafting, [29]. It is now well within the realms of current technology to create femtosecond pulses with a highly specified Fourier profile. Secondly, there has been a systematic employment of the tools of control theory (together with new tools suggested by researchers in this field) based on the realization that intuition alone will not do the job. These techniques from control theory will be explained in greater detail later in this paper. Together, these two innovations have at least added significantly to the practice of one of the most hallowed professions in chemistry, namely, inverting spectroscopic data for obtaining better fits for intermolecular potentials, [15].

Since, the bulk of this paper is about explaining various ideas from control theory, as they apply to a generic multilevel quantum system, we will provide here a brief list of theoretical studies of the applications of control theory to specific problems mostly (but not all) drawn from chemical physics. This list is by no means exhaustive and is meant to be more suggestive of the kinds of possible applications. The creation of $|JM>$ or superposition rotational states starting from the $|00>$ state via microwave fields is descibed in [14]; an application to semiconductor heterostructures is considered in [12]; a use of the so-called H_∞ control design to control a large polyatomic chain of molecules is studied in [2]: an application of optimal control to find regular orbits in the presence of irregular trajectories for a classical dynamical system is described in [3]: an application to situations of the "lasing without inversion" genre may be found in [28]: an application to the creation of specified coherent states on a higher electronic surface for purposes of spectroscopy in Li_2 in the weak field limit is studied in [26] etc., Finally, control theory is applicable to the creation of quantum logic gates and cryptosystems.

2 Controlled Quantum Mechanical Systems

Consider the following Schrödinger equation with an external time dependent interaction in it:

$$i\hbar\dot{\psi} = (H_0 + H_I)\psi \tag{2.1}$$

In the above equation H_0 is the internal hamiltonian and consists of a kinetic energy term and a potential energy term $V(x)$. The interaction term is assumed to an external electromagnetic term aligned along the dipole axis. If, as usual, we approximate the wavefunction as $\psi(x,t) = \sum_{k=1}^{N} \phi_k(x)c_k(t)$, where the $\phi_k(x), k = 1,\ldots,N$ are normalized eigenfunctions of H_0 (corresponding to eigenvalues or energy levels

$E_k, k = 1, \ldots, N$), then we get a corresponding evolution equation for the c_k's:

$$\dot{c} = Ac + Bc\epsilon(t) \tag{2.2}$$

In Equation (0.2.2)a c is the N complex vector whose components are the c_k's: A and B are $N \times N$ skew-Hermitian matrices obtained from the matrix representation of H_0 and H_I respectively and $\epsilon(t)$ is the external field which will serve as the control. Thus, A in particular is a diagonal matrix with purely imaginary diagonal elements (since we are working with the eigenbasis of H_0).

Associated to this is the corresponding (truncated) equation for the unitary generator:

$$\dot{U} = AU + BU\epsilon(t) \tag{2.3}$$

Of course, U is a $N \times N$ unitary matrix. The basic premise, underlying this paper, is that several concrete chemical objectives can be rephrased as the question of finding an appropriate $\epsilon(t)$ which will cause the state, c, of Equation (0.2.2) to move from some initial condition to a desired final condition c_f in some finite time T. In other words, chemical and spectrosocopic objectives admit a reformulation in terms of an appropriately chosen finite-dimensional model and the corresponding matrix representations. This is a reasonable assumption. However, these are not the only models of importance in molecular control. In the section on tracking for instance we will use the Heisenberg equations of motion (though this is not necessary) and thus, it is assumed that we do have at our disposal coordinate expressions for both H_0 and H_I (in particular, the dipole operator). Similarly, when dealing with large polytaomic molecules classical mechanical models or Ehrenfest's equations are used. Finally, in keeping with issues of feasibility and economy it is reasonable to study the weak field limit. This leads to a different model from a control theoretic point of view (see the subsection on controllability below for more on the difference), [26].

The basic control problem then in molecular control is then that of determining : i) if it is possible to drive the state of either Equation (0.2.2) or Equation (0.2.3) from a given initial condition to a desired final condition in finite time via an appropriate choic of the external control (this is called the problem of controllability); and ii) if the answer to i) is yes, the external control which will do the job. We will address i) in the next subsection. There are several techniques for ii) - path planning, tracking and its variations; and the most common technique optimal control. we will address path planning in the next subsection. The other two techniques will be discussed in separate subsections subsequent to the next one.

Before discussing controllability, it is useful write down the equation of the general controlled system of ordinary differential equations, since that will be helpful in explaining the differences between somne of the models considered above:

$$\dot{x} = f(x) + \sum_{k=1}^{m} g_k(x)\epsilon_i(t) \qquad (2.4)$$

In Equation (0.2.4) f plays the role of H_0 and is called the drift vector field. In Equation (0.2.4) there are more than one external control, and correspondingly there are more than one channels, $g_k(x), k = 1, \ldots, m$ for them to interact with the system.

2.1 Controllability

With the notation of the previous section, we are now ready to provide some answers to the question of controllability. The first issue that needs to be settled is the class of controls, $\epsilon_i(t)$, that will be allowed. The most general class is that of the so-called admissible controls - loosely speaking these are controls which will render the differential equation (0.2.4) to have unique solutions. An important result, see [10], states that for deciding controllability we may as well restrict our attention to the class of piecewise constant controls. With this fact at hand, the problem of deciding controllability becomes almost algebraic. Indeed, associated to the control system is an involutive distribution (i.e., a possibly infinite-dimensional Lie algebra of vector fields) Δ which is the Lie algebra generated by the drift vector field f and the control vector fields g_i. This is a distribution, usually of varying rank, which typically satisfies (certainly so in the real analytic case) the sufficient conditions for it to possess maximal integral manifolds through each point of the state space. The first result states that the set of all points which the state of (0.2.4) can be steered to, starting from the initial condition x_0, in finite time, is contained in the maximal integral manifold S_{x_0} which passes through x_0. However, it will usually only contain an open subset of S_{x_0} and not all of S_{x_0}, since we can only move forward in time. When $f = 0$ (i.e., the system is driftless) this is not a source of problems, and in this case the set of states reachable from x_0 is all of S_{x_0}. The case of driftless systems is now being investigated by various communities under various banners - nonholonomic systems, sub-Riemannian geometry, singular Riemannian geometry etc., However, most practical systems, including those of relevance to molecular control do have drift.

When there is drift there are not too many situations where one can assert that the set of states reachable from x_0 is all of S_{x_0}. Some important exceptions are: i) the case where $f(x)$ is a linear vector field and the g_i's are constant vector fields - such systems are called linear control systems, and arise in molecular control in the weak field limit, [26]; ii) the case where f and g_i are left (or right) invariant vector fields on Lie groups and the finite-dimensional Lie algebra generated by them is compact. In the compact case one also has the stronger assertion that the set of states reachable (say, from the identity) is the same when the controls are bounded; iii) when f can be written as a classical Hamiltonian vector field (with respect to the standard symplectic structure on R^{2N}) for a proper Hamiltonian. In all these sufficient cases if $dim \Delta(x) = n$, where n is the dimension of the ambient state space, then S_{x_0} equals the entire state space (and we have what is called complete controllability) and this criterion can be frequently checked purely algebraically. For an examination of the controllability of systems (0.2.2) and (0.2.3) see [23]. For system (0.2.2) we do not even need the Lie algebra generated by the A and B matrices to satisfy any extra features (such as semisimplicity) to assert the appropriate statement about reachable sets - this is due to the special structure of the A matrix. For an application to the universality of quantum gates we refer the reader to [24].

Once we have decided that a certain final state x_f is indeed reachable from x_0, we may ask for constructive procedures to generate controls which will drive the state from x_0 to x_f. In this section, we will briefly describe what is known about the so-called ab initio methods. These methods attempt to generate the $\epsilon_i(t)$'s directly from the Lie algebra structure of the system (0.2.4). The best case occurs when the control system is linear - in this case there is even a closed form formula for the desired control.

For a derivation and application of the appropriate formula in the molecular control context we refer the reader to [26]. For drift free systems there is an extensive literature on the usgage of various expansion formulae (BCH, Magnus expansions etc.,) along with techniques such as averaging to generate the driving controls. We recommend [16]. When the system has drift, as is the case with (0.2.2) there is precious little known. Some of the techniques for driftless systems may be combined with the property that the drift of (0.2.2) has an ergodic flow to obtain heuristics for generating controls, see [25].

2.2 Tracking and Inversion

In most instances we are not so much interested in the attainment of a given final state as we are in achieving a desired final value of an observable's average. For instance, if we wish to break a bond in a diatomic molecule we could attempt to do so by requiring that the average value of the internuclear distance behave like an anti-damped oscillator. In more complicated cases, prescribing the behaviour of the requisite observables requires a great degree of intuition. However, once this is done the required external field can be obtained without any costly computations.

Suppose we have $i\hbar\dot{\psi} = [H_0 - \epsilon(t)\mu]\psi(x,t)$ (here μ is the dipole operator). Let O_j be the operators associated to some physical observables - in *exact tracking* we demand that their average values $< O_j(t) >$ exactly equal, over the time interval $[0,T]$, certain *a priori* given functions of time $y_j(t)$. Following, [11, 22], we obtain from Heisenberg's equations of motion:

$$i\hbar \frac{d}{dt} < O_j > = < [O_j, H_0] > -\epsilon(t) < [O_j, \mu] >$$

If there is just one observable and $< [O_j, \mu] >$ is not identically zero then the above equation can be inverted, as the left hand side must equal $\dot{y}_j$, to obetain $\epsilon(t)$. If $< [O_j, \mu] >$ is identically zero, we keep differentiating the Heisenberg equations of motion till such time as the field appears explicitly in them and then we proceed as above. If there is more than one j then the above method will not work, unless there is a compatibility between the observables and the prescribed trajectories $y_j(t)$'s. However, we can still retain some of the simplicity of the method by combining tracking with other techniques. Various extensions of tracking have been studied, for molecular control purposes, in [6]. At any rate, the above equation (even when $j > 1$) can be written as a Volterra integral equation of the first kind $\int_0^t K(j,\hat{t})\epsilon(\hat{t})d\hat{t} = g_j(t)$, where the kernel K is $i < [O_j, \mu] >_t$ and $g_j(t) = \hbar(y_j(t) - y_j(0)) + \int_0^t i < [O_j, H_0] >_{\hat{t}} d\hat{t}$.

Tracking is essentially inversion. It can be used to find either $V(x)$ or $\mu(x)$. Indeed, suppose (as is the case experimentally) we probe the molecular system via a sequence of *known* fields $\epsilon_k(t)$ and observe the behaviour of a single observable 0 under the influence of these fields. We then het a sequence of functions of time $y_k(t)$ (the only difference from the control implementation of tracking is that these $y_k(t)$ are not prescirbed functions of time but data observed in the laboratory). We can once again employ the tracking formalism to obtain either $V(x)$ or $\mu(x)$ (assuming that the other is known), ie.e, the role of $\epsilon(t)$ is now assumed by either $V(x)$ or $\mu(x)$. For further details see [5, 18].

2.3 Other Techniques

For reasons of brevity we will content ourselves with just providing a reference to the very important techniques of optimal control and learning. The articles [29, 8] discuss their roles in the molecular context. Several of the applications cited in the introduction rely heavily on optimal control, in particular.

3 Acknowledgements

This work was supported in part by a grant from ARO and NSA - this support is gratefully acknowledged.

References

[1] P. Ball, *Designing The Molecular World*, Princeton University Press, 1994.

[2] J. Beumee and H. Rabitz, J. Math Chem, 14, 1993, 405.

[3] J. Botina and H. Rabitz, Phys. Rev. Lett.,, **75**, 16, 2948, 1995.

[4] A. Butkovsky and I. Samoilenko, *Control of Quantum Mechanical Processes and Systems*, Kluwer Acdemic, 1984.

[5] L. Caudill, H. Rabitz and A. Askar, Inverse Probl, **92**, 1099, 1994.

[6] Y. Chen, P. Gross, V. Ramakrishna, H. Rabitz, and K. Mease, J. Chem. Phys.,102,8001 (1995).

[7] J. Clark, in Condensed Matter Theories, Vol 11, E. V. Ludena editor, Nova Science Publishers, Commack, NY, 1996.

[8] M. Dahleh, A. Peirce, H. Rabitz, and V. Ramakrishna, Proceedings of the IEEE, 7, January 1996.

[9] D. Fernandez and B. Mielnik, J. Math Physics, **35** (5), 2083, 1994.

[10] K. Grasse and H. Sussmann, in *Nonlinear Controllability and Optimal Control*, H. Sussmann editor, Marcel Dekker, 1991.

[11] P. Gross, H. Singh, H. Rabitz, K. Mease, and G.M. Huang, Phys. Rev. A, 47, 4593 (1993).

[12] P. Gross, V. Ramakrishna, E. Vilallonga, H. Rabitz, M. Littman, S.A. Lyon, and M. Shayegan, Phys. Rev. B, 49, 11100 (1994).

[13] P. Gross, D. Neuhauser, and H. Rabitz, J. Chem. Phys., 96, 2834 (1992).

[14] R.S. Judson, K.K. Lehmann, Rabitz, and W. Warren, J. Molec. Structure, 223, 425 (1990).

[15] B. Kohler, V. Yakovlev, J. Che, J. Krause, M. Messina, K. Wilson, N. Schwentner, R. Whitnell and Y. Yan, *Phys. Rev. Lett.*, **74**, 3360, 1995, and Private Communication.

[16] G. Lafferriere and H. Sussmann, Proceedings of the 1991 IEEE Control and Decision Conference, Brighton, IEEE Press.

[17] W. E. Lamb, Jr, Phys Today, **22** (4), 23, 1969.

[18] Z.-M. Lu and H. Rabitz, J. Phys. Chem., 99, 13731 (1995).

[19] E. Lubkin, J. Math Physics, **15**, 663, 1974.

[20] B. Mielnik, J. Math Physics, **27** (9), 2290, 1986.

[21] B. Mielnik and D. Fernandez, "Exponential Formulae and Effective Operations", Reprint of talk given at 4th ICCSUR, June 1995.

[22] C, Ong, G. Huang, T. tarn and J. Clark, Math Systems Theory, **17**, 335, 1984.

[23] V. Ramakrishna, M.V. Salapaka, M. Dahleh, A. Peirce, and H. Rabitz, Phys. Rev. A, 51, 960 (1995).

[24] V. Ramakrishna and H. Rabitz, Phys Rev A, October 1996.

[25] V. Ramakrishna and H. Rabitz, Article in Preparation.

[26] L. Shen, S. Shi, and H. Rabitz, J. Phys. Chem., 97, 12114 (1993).

[27] N. Wang and H. Rabitz, in press, Phys. Rev. A.

[28] N. Wang and H. Rabitz, in press, Phys. Rev. A.

[29] W. Warren, H. Rabitz, and M. Dahleh, Science, 259, 1581 (1993).

Program in Math Sciences, Univ of TX at Dallas; Department of Chemistry, Princeton University

Index

A

Abnormal bang-bang extremals, 126–133
Absolute continuity, 50–51
Absorbing state, 268
Abstract set constraints, 209–212
Abundant subsets, of generalized control systems, 108–116
Adaptive fuzzy c-regression model (AFCR), 378, 379
Adaptive stochastic path planning, for robots, 486–494
Admissible controls, 171
Admissible processes, 163, 164
Aeronautics, 3, 459–467
AIMMS, 481
AI-OR hybrid, 366–375
Air charge density, 436
Air flow model, cylinder, 411
Air quality, ozone model, 477–484
Air-to-fuel ratio, 436, 440
Ammonia, 478, 479, 480
AMPL, 481
AOT40, 477
Applications. *See also* Automotive systems
    manufacturing systems, 271–279, 280–288, 468–476
    molecular dynamics, 495–501
    ozone model, 477–484
    production systems, 289–296
    robotic systems, 391–398, 486–494
    VTOL plane, 459–467
Arbitrage, absence of, 34
Arbitrary system, 238, 240, 243
Aspects, as a problem element, 362
Asplund spaces, 190, 192, 194, 195
Assets, 33
Asymptotic comparative performance, 329–331
Asymptotic stability, 20–26
ATM switches, 304
Aubin's Viability Theorem, 181
Automotive systems
    dynamic clustering technique with onboard traffic monitoring, 418–426
    hot wire mass air flow sensor modeling, 409–416
    identification of powertrain noise in low SNR environment, 401–408
    intake flow in variable geometry turbocharged engines, 436–444
    Smith predictor for electric EGR valve control system, 446–454
    zirconium oxygen exhaust sensors, 427–434

B

Backlog inventory, 289, 468
Back-Propagation-Algorithm, 490–491
Banach lattice, 135, 137
Banach space, 20, 21, 22, 23, 26–27, 38, 126, 137
    in characerization of weak sharp minima, 207–208
    generalized differentiation and, 189, 190, 192, 193, 194, 196
    open-loop Nash equilibrium and, 150
Bang-bang extremals, 126–133
Bang-bang trajectories, 171
Bases, of systems, 238
Basic condition, as a model element type, 363
Bayesian method, 345, 351
Bending moment, 81, 82
Bijection, 200, 202, 204
Binomial case, in process sampling, 343
Bishop-Phelps density theorem, 192
Black & Scholes model, 33, 34
Bonds, 33
Boolean states, 384, 385, 386, 387–388
Borel sets, 102, 106, 145, 203
Boundaries
    free, 62
    Lipschitzian, 53, 55, 95
    in yttria-stabilized zirconia, 430–433
Boundary layer, 3, 428, 434
Boundary smoothness, hidden, 53–57
Boundary value problem, in partial differential equations, 17
Bounded circuit, 280
Broken extremals, 171–179
Brownian motion, 271, 273, 277, 294
Bump functions, 192–193
Burst size, 290

C

Caccioppoli sets, 62, 66
Calculus of Variations, 126
Calling sequence, 360
Call option, 34
Carathéodory integrand, 145
Carbon monoxide, 436
Cartesian robot, 391–392
Cauchy problem, 31
Céa-Malanowski problem, 62
Cellular networks, tracking protocols, 261–269
CFSQP, 481, 482
Chain rule, 194
Chance constraint, 487
Change point, sequential, 345–353
Channel flow, 4–6
Chattering phenomena, 384
Circuits
    electronic, in MAFS, 412
    in manufacturing systems, 280, 281, 283, 284–287
Clarke directional derivative, 217
Clarke generalized gradient, 110, 114
Clarke subdifferential, 104, 216, 217
Classification, in fuzzy clustering, 378–380
Climate forecasting, 3
Closed-loop system
    in emissions control, 440
    in robotics, 393
Closed path, 281

Clustering
  Dynamic Clustering Algorithm, 420, 422–423
  dynamic technique, 418–426
  fuzzy, 376–382
Clusters, moving, 418–419
Coastal erosion, 62
Coderivatives, 190–191
Coloring, rate, 300–302
Column-sufficient, 245
Comb filters, 402, 403
Communication networks, 237–243, 253
Compact sets, convex, 135–143
Compensators, for discrete-time delay systems, 77–79
Complementarity problems, linear, 245–251
Computer systems
  in automobile traffic monitoring, 418–419
  interactive fuzzy modeling, 376–382
  memory management in, 254
  transient distributions of cumulative rate and impulsed based reward, 298–305
Concepts, development of hierarchy, 369–370
Cones
  Clarke tangent, 201
  contingent, 110, 207–208
  normal, 92, 204–206, 209
  polar, of decomposable sets, 204–206
  recession, 217
  Ursescu tangent, 207–208
Conjugate points, for optimal control problems, 162–169
Conopt, 481
Continuity. See also Lipschitz continuity
  absolute, 50–51
  Hölder, 225–233
Continuous trading, 34
Controllability
  in molecular dynamics, 498–500
  of robotic systems, 395–398
Controller design
  minimax linear-quadraic, 117–124
  robust, for a VTOL, 459–467
Control theory, 4
  abundant subsets of generalized control systems, 108–116
  feedback algorithms, 9, 10, 11
  linear control systems, 20–27
  molecular dynamics and, 495–501
  for monitoring the mean of a multivariate normal distribution, 327–334
  nontrivial maximum principle and, 100–106
  Ritz type discretizations and, 91–98
  shape control of elastic plates, 80–87
  state-dependent constraints, conjugate points and, 162–169
  transformation to uncontrolled diffusion process, 273
  of turbulent flows, 3–11
Control volume, 67
Convergence
  of discrete viability kernel, 182
  of the finite horizon relaxed problems, 157–158
  in L^p-closed sets, 199
  of the Proximal Point Algorithms, 231–233
  in Ritz type discretizations, 94–96, 98
Converging resource, 281, 285–286

Convex compact sets, directed sets and, 135–143
Convexity
  infinite horizon case and, 159, 160
  L-convex, 202
Convolution kernels, 9
Corector-step, 250
Cost
  linear, absolute value, 293
  paging, 265
  quadratic, 292–293
  in trading, 35
Cost function
  broken extremals and, 178
  in cellular networks, 261
  in computer systems, 303
  of dams, 63
  in economic process control, 336–343
  in failure-prone manufacturing systems, 470
  in failure-prone production systems, 289–296
  in minimax problems, 121, 124
  quadratic, 118, 292–293
  reduction of ground-level ozone, 478, 480
  in robotics, 487
  in shape control of elastic plates, 83
  turbulence of fluids, 5, 8, 10
Coulomb equation, 65
Counter-flow, in manufacturing systems, 283–284
Courant-Friedrichs-Lewry stability conditions, 8
C programming languange, 481, 482
Crisp structural models, 377
Critical directions, 128
Cumulative reward, transient distributions of, 298–305
Curvatures
  bounds on, 241–243
  dependent/independent, 238
CUSUM procedure, 328, 329, 330, 331–332, 345–348, 349–350, 352–353
Cylindrical domains, wave equation in, 53–57

D

Dams
  physical quantities and their constraints, 63–64
  safety factor and shape optimization, 62–70
Data gaps, missing values in decision-making, 323–324
Datko theorem, 26
Deadlock detection algorithm, in manufacturing systems, 280–288
Dead-Time Compensator (DTC), 446
Decision-making
  evaluation of options, 318–326
  in KONWERK, 374
Decomposable sets, L^1-closed, 198–206
Decomposition, 73–75, 77, 108, 145, 198
Degeneracy of linear complementarity problems, 246
Degenerate hyperbolic equations, 31
Delays, discrete-time delay systems, 71–79
De Leeuw-Glicksberg Decomposition Theorem, 23, 25
Demand, uncertain, 289–296
Demyanov's difference, 136
Density theorem, 192

Index

Dependence curvature, 238
Diesel engines, 436–444
Diffeomorphism, 176
Diffusion approximation, 293–294
Dimensions
 of clusters, 380–381
 infinite, 189–197, 498, 499
 as a problem element, 362
Directed intervals, 136–137
Directed sets, differences of convex compact sets and, 135–143
Directed supporting face, 135
Directional derivative, 218, 471–472
Dirichlet boundary conditions, 431
Disappearing solution, 17–18
Discrete-time delay systems, 71–79
Discretizations, Ritz type, 91–98
Discriminating domain, 186
Displacement, 18
Distance-based protocol, in cellular networks, 262, 267–269
Disturbances
 in gyroscope feedback control, 462
 in process control, 336, 337–338
Diverging resource, 281
Domains
 cylindrical, 53–57
 discriminating, 186
 effective, 216
 nonsmooth, 53–61
 optimization problem of, 80–82
 transformation of, 43–44
 Viability, 181, 182
 Victory, 187
Dore-Venni theorem, 39
Drag
 in skin friction, 3
 in turbulence of fluids, 7, 8
Duality, in minimax problems, 117–119, 124
Dual systems, 238, 242, 243
Dunford-Pettis Theorem, 155
Dynamical system, with state constraints, 180–188
Dynamic Clustering Algorithm (DCA), 420, 422–423
Dynamic Clustering Method (DCM), 419–420, 424, 425, 426
Dynamic clustering technique, 418–426
Dynamic programming equation, in failure-prone manufacturing systems, 471–472
Dynamic Storage Allocation problem, 253–258

E

Ecological risk assessment, ozone, 477–478
Economic models, 33–40
Economic process control, 336–343
Effective domain, 216
Effective stress, on dams, 64
Effective stress tensor, 64
EGR. *See* Exhaust gas recirculation
Ehrenfest equations, 497
Eigenfunctions
 on controlled quantum mechanics, 496
 in discrete-time delay systems, 78–79
 in fuzzy clustering, 379–380

 of infinitesimal generators, 12
 of linear absolute value cost, 293
 subspace and, 18, 23
Ekeland's variational principle, 228
Elastic plates, shape control of, 80–87
Electric potential, 429, 431
Electroneutrality, 427, 428, 429, 431–433
Elemental detail tables, 360
Elliptic fuzzy models, 381–382
Emissions, automobile, 427–434, 436–444
Emitter, 478
Endowment, initial, 34
Engines, turbocharged, 436–444
Environmental factors, 272, 273, 278
Environments, modeling of, 357–365, 366–375
Epigraph, 216
Equality constraints, 212–214
Erosion, 62
Essential limit sets, 97
Euclidean space, 53, 93
Euler models, 18, 95, 96, 97
Euro cycle, 440
Europe
 option pricing, 33–40
 ozone model, 477–484
Exact tracking, 500
Exclusions, 280
Exhaust gas recirculation (EGR)
 in turbocharged engines, 436–437, 440–442
 in valve control systems, 446–454
Exhaust sensors, zirconium oxygen, 427–434
Exponential stability, 21, 26–27
Exponential synchronous time average (ESTA), 403
Extended Global Maximum Principle, 127–129
Extractor approach, 53
Extremality-controllability theorem, 108
Extremal principles, 191, 194
Extremals, 164, 165–166, 171–179

F

Failure-prone manufacturing systems, 468–476
Failure-prone production systems, 289–296
Feasible state, 282
Feasible vector, 245, 246
Feedback
 for minimax problems, 117–124, 177
 optimal, 123–124
 pure (idealized) rate, 12
Feedback control
 in emissions monitoring and engine operation, 440, 441
 in robotic systems, 391
 turbulence of fluids, 9, 10, 11
 use of self-straining material for, 18
 in VTOL gyroscopes, 460, 461–462, 465–467
 in wear processes, 271
Feedforward controller design, 463–465
Feller-Myadera-Phillips theorem, 21, 22
Fermi level, 431
Filter, for dam stability and sliding, 62, 68, 69
Finance, economic models, 33–40
Fire protection, passive, 309–317
Firewall design, 312–313
Fixed Point theorem, 144

507

Flexible manufacturing systems (FMS), 468
Flow
 air, in MAFS, 409–416
 channel, 4–6
 counter-flow, 283–284
 intake, in variable geometry turbocharged engines, 436–444
 laminar, 10, 11
 in production systems, probability arguments, 290–291
 turbulent, 3–11
Flowshop configuration, 468
Fluid models
 in communication systems, 304–305
 of failure-prone machines, 289–296
Fluids, turbulence control of, 3–11
Fluxes, approximation of, 433
Fold and cusp-singularities, 171
Fourier profile, 495
Fréchet functions, 192–193, 194
Free boundary problems, 62, 63
FTP cycle, 440
Fuel
 air-to-fuel ratio, 436
 burnt gas fractions, 437–438, 440
Fuzzy c-regression model (FCRM), 378
Fuzzy modeling, interactive, 376–382
"Fuzzy" regions, 68

G

Games, 144, 152, 186–187
GAMS, 481
Gas dynamics, transonic, 29–31
Gaussian white noise, 419, 421
Gelfand theorem, 22
Generalised conjugate points, 162, 163, 166–167
Generalized Clark subdifferential, 216, 217
Generalized control systems, abundant subsets of, 108–116
Generalized likelihood ratio, 328, 329, 331, 332–334
Generic similarity, 360–361
Geoffrion model, 360
Global System for Mobile Communication (GSM), 262
Grades of determination, as a problem element, 362
Greedy Algorithms, 239–241, 243
Greiner-Nagel theorem, 13
Growth rate, of a function, 225, 226, 227
Guide vanes, 437
Gyroscopes, 460

H

H∞ algorithm, 459, 461, 463, 467, 496
Hadamard differentiability, 83, 84
Hahn-Banach Theorem, 23
Hamiltonian function, 91, 103, 114, 129, 164, 172
Hamilton-Jacobi-Bellman equation, 153, 171, 176, 295–296, 468, 472
Hamilton-Jacobi equations, 92, 144, 185
Handoff problem, 264
Harmonics (orders), 401–403
Hausdorff-distance, 136
Hausdorff metric, 114

Heat transfer rate, 438–439, 444
Hedging strategy, 34, 294
Heisenberg equations, 497, 500
Helicopters (VTOL), 459–467
Hermitian matrices, 497
Hidden boundary smoothness, in hyperbolic tangential problems, 53–61
Hidden Markov Chain model, 345, 351
Hidden shape derivative, in the wave equation, 42–52
Hierarchical Data Format, 483
High-order approximations, for abnormal bang-bang extremals, 126–133
Hilbert space, 12, 13, 20, 25, 26–27
 linear discrete-time systems, 71, 75, 77, 78
 Nash equilibria and, 147
 wave equation in non-cylindrical domains, 57
Hille-Phillips operational calculus, 16–17
Hille-Yosida theorem, 21, 22
Hold-all, 53
Hölder continuity, of inverse subdifferentials, 225–233
Hölder space, 36
Horizon control
 infinite case, 153–160
 receding, 120–123
Horizontal linear complementarity problems, 245–251
Hörmander's energy method, 31
Hot wire mass air flow sensor modeling (MAFS), 409–416
Hydrocarbons, 436, 440
Hyperbolic equations, degenerate, 31
Hyperbolic tangential problems, hidden boundary smoothness in, 53–61
Hypothesis, multiple alternative, 345–353

I

Impending deadlock, 283
Impulse based reward, 298–305
Inclusions
 differential, 114–116
 non-convex, 108
Independence curvature, 238
Independence system, 237–238, 240
Indicators
 of air quality, 478
 in decision-making, 318, 319, 320–323, 326
Inequality constraints, 212–214, 216
Infeasible-interior-point method, 246–247, 248
Infinite dimension
 Lie algebra of vector fields, 498, 499
 variational analysis in, 189–197
Infinite horizon case, 153–160
Infinitesimal generators, 12, 13
Information matrix, 318
Information model, in decision-making, 318–320
Insulation, in firewall design, 313
Intake flow, in variable geometry turbocharged engines, 436–444
Integral indicator, in decision-making, 319, 320–323
Interactive fuzzy modeling, 376–382
Interior-point methods, 246
Interior-point paths, analysis of, 247
Interpolation, set-valued, 142
Interpolation spaces, 38–39

Index

Interpretation pattern, as a model element type, 363
Inventory, 289, 468
Inverse subdifferentials, 225–233
Inversion, in molecular dynamics, 500–501
Invertibility conditions, 93
IS-41, 262
Isomorphisms, 78
Item partitioning, in storage allocation, 256–257
Ito equation, 271

J

Jensen's method, 300

K

Kakutani's Fixed Point theorem, 144, 148
Kalman filtering, 419, 420, 421
Karush-Kuhn-Tucker system (KKT), 92, 94
Kato formula, 15
K-connectedness, 237, 239
Kernel, 9, 126, 181–183, 187, 501
Kirchhoff plate model, 80–87
Kolmogorov equation, 275, 277
KONWERK tool-box, 366–375
Kuhn-Tucker conditions, 207, 208, 213
Kuhn uncontrolled process, 272, 273, 277
Kullback-Leibler distances, 345
Kuratowski-Painlevé limits, 189, 225
Kushner approach, 472

L

Lag
  in MAFS, 409–416
  in turbocharged engines, 440
Lagrange function, 92, 94, 121, 145, 328
Lagrange multipliers, 212
Lagrange polynomials, 142
Laminar flow of fluids, stabilization, 10, 11
Landslides, 62, 64–65
Laplace-Beltrami operator, wave equation for the, 58–61
Laplace-Stieltjes transforms, 22, 24
Laplace transforms, 291–292, 298
Large signal data, 409, 413–416
Lasar radar, 418, 423–424
Lasers, use in molecular dynamics, 495
L^1-closed decomposable sets
L-convex, 201–202
Lebesgue measurable, 146, 149, 179
Lebesgue point, 105
Lebesgue space, 35, 36
Leland model, 35
Levy "stable" density, 17
Lidar beam tracking, 418, 423–424
Lie algebra, 498, 499
Limits, properties of, 44
Linear complementarity problems, 245–251
Linear control systems, stabilization of, 20–27
Linear cost, 293
Linear mappings, 78–79
Linear Matrix Inequalities (LMI), 119
Linear time approximation algorithm, for storage allocation, 253–258
Lipschitz continuity, 103, 108, 136, 192, 226–230. *See also* Pseudo-Lipschitzian property
Lipschitz continuous curves, 97–98
Lipschitz functions, 154–155, 195, 207–208, 211, 470
Lipschitzian boundary, 53, 55, 95
Locally g-controllable, 109
Local minimum, strong, 171–179
Location
  of particle, in tracking, 421
  of sensor, 388–390
  update, 262, 265, 267–269
Loose uniformly round renorm (LUR), 193
Low-order models, 9
LS method, 461–462
Lunardi result, 39
Lyapunov theorem, 20
Lyusternik Theorem, 126

M

Machinery
failure-prone, 289–296, 468–476
wear processes in, 271–279
MAFS (hot wire mass air flow sensor modeling), 409–416
Magnetohydrodynamics, 3
Malanowski's convergence theorem, 96
Manufacturing systems. *See also* Production systems
  deadlock detection algorithm, 280–288
  failure-prone, 468–476
  flexible, 468
  wear processes in machinery, 271–279
Manutrec r3, 492
Mapping. *See also* Set-valued maps
  best reply, 144
  linear, 78–79
  Marchaud, 181, 183, 184, 185, 187
  open, 394
  projection, 61
  proximal, of Moreau, 232
  surjective, 93
  turbocharged engine data, 439, 444
Marchaud map, 181, 183, 184, 185, 187
Markov process, 263, 267, 289, 299, 300, 345, 351, 468
Material derivative method
  definition, 48
  or shape sensitivity analysis, elastic plates, 83
  for smooth data, 45–48
Mathematica, 482
Mathematical Programming, 359
Matroid, 238, 241, 242
Maturity, of assets, 34
Maximum Principle
  nontrivial, 100–106
  Pontryagin, 171
  second-order extended, 127–129
  Weak, 164
Mean, monitoring of, in multivariate normal distributions, 327–334
Mean rate of return, 33
Memory management in computer systems, 254
Metlab's Optimization Toolbox, 118
Minima, weak sharp, 207–214, 216

Minimax linear-quadratic controller design, 117–124
Minimax optimal control problems, relaxed, 153–160
Minimisers, weak local, 163
Minkowski addition, 135
Minos, 481
Missing values, in decision-making, 323–324
Mittag-Leffler function, 15
Mixed integer problem, 368
Mobile terminals, tracking, 261–269
Mobility matrix, 263
Model element types, 363
Model generator, 481–482
Modeling
  environments, 357–365, 366–375
  hybrid AI-OR approach, 366–375
  interactive fuzzy, 376–382
  mathematical, for robot arm control, 391–398
  rule-based design and 2D-analysis, 384–390
Modeling by Construction, 357, 362–364
Modeling by Example, 357
Molecular dynamics, 495–501
Monitoring policy, 337–338, 339
Monotone linear complementarity problems, 245, 247
Monte Carlo simulations, 327, 332, 334
Morawetz existence theorem, 30
Multi-decimation method (MD), 461
Multi-dimensional Shiryayev-Roberts procedure, 345, 348, 350, 352
Multiple alternative hypothesis, 345–353
Multiplexor queue, 304
Multivariate control charts, 327–334
Mutual exclusions, 280

N

Nash equilibrium, open-loop, 144–152
Navier-Stokes equations, turbulence of fluids, 3, 4, 6, 8, 9
Nearest neighborhood method, 422, 425–426
Nernst formula, 427
Nernst-Planck theory, 427, 429
Networks
  neural, 486–494
  optimization models and algorithms of, 237–243
Nitra River, KONWERK analysis, 366–375
Nitrogen, oxides of (NO_x), 436, 437, 440, 447, 478–480
Noise
  powertrain, 401–408
  white, 8, 337, 401, 419, 421
Nonanticipative strategies, 186
Non-convex calculus, 194–195
Non-convex inclusions, 108
Nondegeneracy, of the state constraint maximum principle, 101
Nonlinear control systems, open-loop Nash equilibrium for, 144–152
Nonlinear optimization, 477–484
Nonlinear programming, weak sharp minima of, 207–214
Nonsmooth data, shape derivative for, 49–52
Nonsmooth domains, 53–61
Nonsmooth functions, variational analysis and, 189
Nonsmooth programming, second-order necessary conditions in, 216–223
Nontrivial maximum principle, 100–106
NP-hardness, 239, 254–256
Numerical simulations
  Hölder continuity and, 225
  turbulence control of fluids, 6–8
Nyquist plot, 446, 447

O

Observability, of robotic systems, 395–398
Observers, for discrete-time delay systems, 75–77
Offshore structures, fire protection on, 309–317
Open-loop control, 4, 446
Open-loop Nash equilibrium, 144–152
Operations Research approach, 357
  AI-OR hybrid, 366–375
  to model analysis, 483, 484
Operator topology, 21
OPTIBEAM, 309, 310, 314–316, 317
Optical fiber networks, 253
Optimal control problems
  ground-level ozone reduction, 481
  nontrivial maximum principle for, 100–106
  relaxed minimax, 153–160
  reliability-based, 309–317
  risk sensitivity in, 271–279
Optimal feedback, 123–124
Optimal hedging point, 294
Optimality conditions
  in failure-prone manufacturing systems, 470–472
  Hölder continuity and, 225
  second-order necessary, 219
  in shape control, 82–83
Optimal monitoring policies, 339
Optimization
  models and algorithms for constructing reliable networks, 237–243
  non-linear, 477–484
  real-time, by neural networks, 486–494
Optimization Toolbox, 118
Option pricing, 33–40
Options, evaluation of, in decision-making, 318–326
OPTIWALL, 309, 310, 312–314, 317
OR. See Operations Research
Orders (harmonics), 401–403
Oxides of nitrogen. See Nitrogen, oxides of
Oxygen sensors, 427–434
Oxygen vacancies, 428, 429–430, 433, 434
Ozone model, 477–484

P

Paging process, in cellular networks, 261–262, 265, 266
Paley-Wiener theorem, 14
Parallel mutual exclusions, 280
Parametric optimization problems, 87
Part counter-flow, 283
Partial differential equations, boundary value problem, 17
Particle tracking algorithm, 420, 421–422
Particulate matter (PM), 436
Passive fire protection (PFP), 309–317
Path, 281

Index

Path planning, stochastic, 486–494
Payoff, 34
Penalization techniques, 101
Performance, comparison in, 331–334
Periodicity, of semigroups, 23, 24, 25
Petri nets, 280
Phase distribution, 267
PID controller, 446, 449–451, 452, 453
Plane strain hypothesis, 68
POD-based models, 9
Point source emissions to surface waters, modeling of, 366–375
Poisson coefficient, 68, 431
Poisson process, 264, 266, 289, 342–343
Pollutants
  forecasting of, 3
  ground ozone model, 477–484
  see page of, 62
Polyhedrals, 135
Pontryagin Mamimum Principle, 108, 114, 171
Post-evaluation of fire control system, 311, 314, 316
Powertrain noise, 401–408
Prediction algorithm, 420, 423
Preprocessing, of non-linear optimization problems, 477, 483
Pressure
  on dams, 63, 69
  in fluid turbulence control, 4
Pricing problems, 33–40
Primary deadlock, 283
Prime Ideal Theorem, 22
Primitive element, 360
Probability density equations, 290–292
Problem elements, six types of, 362
Problem name, as a model element type, 363
Process control, economic, 336–343
Production route, 280
Production systems, failure-prone, with uncertain demand, 289–296
Profitability, process, 336, 338–339
Projection mapping, 61
Proximal mapping, 232
Proximal Point Algorithm (PPA), 226–227, 231–233
Pseudo-Lipschitzian property, 195, 196, 197, 227
Pulse crafting, 495
Pure rate feedback, 12

Q

Q-superlinearly, 251
Quadratic cost, 292–293
Quantum theory, 495–501
Queue, in communication systems, 304

R

Radar, 418
Radio networks, storage algorithm for, 253–254
Rado-Edmonds Theorem, 241
Radon measure, 103
RAINS model, 478, 484
Randomization, 300
Rank functions, 238
Rate coloring, 300–302
Rate plus impulse model, 298–305

Reachable set, definition, 109
Real-time optimization, 486–494
Receding horizon control, 120–123
Recentering technique, 250
Receptor, 478
Recession cone, 217
Regression, fuzzy clustering and, 378–380
Regularity
  in model analysis, 483–484
  in transonic gas dynamics, 29–31
  uniform strong, 94–95
  of velocity potential, 29–31
Regularization technique, 80
Relations, as a problem element, 362
Relations between concepts, development of compositional hierarchy, 370–371
Relaxation
  in minimax optimal control problems, 153–160
  of the shape problem for dams, 66–67
  of the shape problem for elastic plates, 85–87
Reliability
  economic process control, 336–343
  fire protection on offshore topsides, 309–317
  monitoring the mean of a multivariate normal distribution, 327–334
  options in decision-making, 318–326
  sequential change point detection, 345–353
Repair
  in failure-prone manufacturing systems, 468–476
  in failure-prone production systems, 289–296
Resources, in manufacturing systems, 281–287
Reward, cumulative rate plus impulse based, 298–305
Reynold number, turbulence control of fluids, 4, 6
Riccati equation, 26, 421
Riemannian geometry, 499
Riesz basis property, 12, 18
Rishel formulation, 468
Rishel model of wear, 271
Risk assessment, ozone, 477–478
Risk profile, 34
Risk sensitivity, in wear processes, 271–279
Ritz type discretizations, 91–98
Rivers, modeling of, 366–375
Robotic systems
  adaptive stochastic path planning, 486–494
  mathematical model for control of, 391–398
Robust control techniques
  based on Smith Predictor, 446–454
  stochastic optimization and, 487
  turbulence of fluids, 9–10, 11
  for a VTOL plane, 459–467
Rosen's "trick," 146–147
Row-sufficient, 245
Rule-based design, in tank systems, 386–390

S

Safe counter-flow, 284
Safe move, in manufacturing systems, 283, 286, 288
Safety factor, resistance to dam ruptures, 63, 64–65
Sauce-type VTOL plane, 459–467
Scalarization, in minimax problems, 117–119, 124
Scaling, in minimax problems, 119–120
Schrödinger equation, 496

511

Second-order conditions, in nonsmooth programming, 216–223
Second-order extended Maximum Principle, 127–129
Semiconductors, 428, 431
Semigroups
  construction of, 14–16
  definition, 12
  tability of, and stabilization of linear control systems, 20–27
  superstability of, 12–18
Sensitivity analyses
  Hölder continuity and, 225
  in pre-evaluation of fire control system, 311
  shape, 80, 83–85
  in VTOL plane control, 465
Sensors
  emissions, 440
  location of, 388–390
  in onboard traffic monitoring, 418, 419
  zirconium oxygen, 427–434
Sequential change point detection and estimation, 345–353
Sequential mutual exclusions (SME), 280
Service robots, 489
Set-valued maps, 142, 181, 182, 203–204
Set-Valued Numerical Analysis, 181, 182
Shadow prices, 175
Shape derivative
  characterization, 48–49
  definition, 48
  for nonsmooth data, 49–52
Shape differentiability, 42
Shape optimization
  of dams, 62–70
  of elastic plates, 80–87
Shares, 34
Shiryayev-Roberts procedure, 345, 348, 350, 352
Sign, in decision-making, 318, 319, 320
Signal-to-noise ratios (SNR), powertrain noise, 404–405, 406, 407, 408
Simple circuit, 281, 283
Simple path, 281
Simple Sequential Processes with Resources, 280
Sinclair Theorem, 13–14
Siphons, 180
SISO system, 120
Site-specific chemistry, 495
Skin friction drag, 3
Sliding line, 64, 65, 69
Small signal (Bode) data, 409, 410, 412, 413
Smart string, 18
Smart structure theory, 12, 18
Smith Predictor, 446–454
Smooth data, material derivative for, 45–48
Smoothness
  assumptions, 127
  in hyperbolic tangential problems, 53–61
  polynomial approximations for, 127
Sobolev space, 35, 36
Soil mechanics, of dams, 62–70
Sonic line, 29
Spanning tree problems, 237
Spectroscopic data, 496
Spectrum decomposition, 73–75, 77
Sprinkler systems, 310
Stability
  asymptotic, 20–26
  of a dam, 65
  for discrete-time delay systems, 72–75, 77–79
  in dynamical systems with state constraints, 187–188
  exponential, 21, 26–27
  of semigroups, and stabilization of linear control systems, 20–27
  superstability, 12–18
Stabilizability, for linear discrete-time systems, 73–75
Staircase functions, 230
State constraints
  dynamical system with, 180–188
  nontrivial maximum principle for optimal control problems with, 100–106
State-dependent control constraints, 162–169
State equation, turbulence control of fluids, 5
State of the system, definition, 5
Step response data, 409, 413–416
Stochastic path planning, adaptive, 486–494
Stochastic processes. *See also* Hamilton-Jacobi-Bellman equation; White noise
  n particle tracking, 421
  probability of disturbance in, 336
  standard deviation of, 277
  wear, 271
Stocks, 33
Stopping rule, 327
Storage allocation, linear time approximation algorithm for, 253–258
Strike price, 34
Strong local minimum, of broken extremals, 171–179
Structural engineering, dams, 62–70
Structural factors, in fire protection, 309, 310
Structured Modeling, 357, 358–362
Structured Modeling Language, 359
Subdifferentials, inverse, 225–233
Subgraph, in manufacturing systems, 281
Submanifolds, 53
Subset Sum problem (SSP), 254–255
Subspace, 18, 35
  abnormal bang-bang extremals and, 132
  in discrete-time delay systems, 74, 78
Sufficient linear complementarity problems, 245–246
Sulphur oxides, 478, 479
Superstability of semigroups, 12–18
Surjective mapping, 93
Surjectivity condition, 126
Surplus inventory, 289, 468
Symmetry, in minimax problems, 119–120
Synchronous time averaging, in identification of powertrain noise, 401–408

T

Tank systems, 384–390
Tauberian theorems, 22, 23
Taylor-polynomial, 249
Telecommunication systems, 253
Terminal manifold, 171
Thermohydraulics, 3
Time
  discrete-time delay systems, 71–79

Index

linear approximation algorithm, for storage allocation, 253–258
minimum, harmonic oscillator time-optimally in, 126
as a problem element, 362
protocols, in cellular networks, 262, 264–266
real-time optimization, 486–494
Time of Crisis problem, 184–186
Timoshenko models of structures, 12, 18
Topology
 in discrete-time delay systems, 73–75
 operator, 21
 in relaxation of the shape problem for dams, 66–67
Torque, 438, 439, 440
Torsion, 18
Tracking
 of mobile terminals, 261–269
 in molecular dynamics, 500–501
Traffic control, on communication networks, 253–258
Traffic monitoring, dynamic clustering technique for, 418–426
Trajectory planning, for robots, 486–487
Transformation
 of domains, 43–44
 properties of, 44, 58
Transient distributions, of cumulative rate, 298–305
Transonic gas dynamics, 29–31
Transportation modeling, 358–359
Transportation systems, onboard traffic monitoring, 418–426
Transported problems, 43–44
Triangle-inequality, 250
Tricomi equation, 30, 31
Turbocharged engines, 436–444
Turbulent flows, control of, 3–11
Turbulent kinetic energy, 5, 7
2D analysis of a hybrid tank system, 384–390
Two-norm-discrepancy phenomenon, 92, 98

U

Uncertain demand, 289–296
Uniform Boundedness Principle, 27
Uniform exponential stability, 21
Uniformization technique, 300
Universe, 53
Update event, 264
Ursescu tangent cones, 207–208

V

Vacancy concentration, 428, 429–430, 433, 434
Value-function, 177, 178
Valve control systems, in engines, 446–454
Variable geometry turbocharger (VGT), 437, 440–442

Variables, as a problem element, 362
Variational analysis, in infinite dimensions, 189–197
Variational principles, 191
Vector case, in wear processes, 273–277
Vehicles. *See* Automotive systems
Velocity, in dynmaic clustering technique, 420, 426
Velocity potential, in transonic gas dynamics, 29–31
Verification functions, 92
Vertex-disjoint-chains, 237
Vertex graph, 241
Vertical take-off and landing plane (VTOL), 459–467
Viability Domain, 181, 182
Viability Kernel, 181, 182–183, 187
Viability Theory, 180–181
Vibration, in robotic systems, 391
Victory domain, 187
Volatile organic compounds (VOC), 478, 480, 481
Volatility of assets, 33
Voltage regulation, by rate gyroscope, 460
Voltage response data, 410, 415
Volterra integral equation, 501
VTOL (vertical take-off and landing plane), 459–467

W

Walls, near-wall turbulence of fluids, 6, 9
Wave equation
 in cylindrical domains, 53–57
 hidden shape derivative in the, 42–52
 for the Laplace-Beltrami operator, 58–61
 in non-cylindrical domains, 57–58
Weak local minimisers, 163, 165
Weak Maximum Principle, 164
Weak sharp minima, 207–214, 216
Wear-out phenomena, 336
Wear processes, risk sensitive optimal control of, 271–279
Weight
 in decision-making, 318, 319, 320
 in feedback controller design, 461–462
Wellposedness, of robotic systems, 395–396
White noise, 8, 337, 401, 419, 421
Whitney's criterion, 237–243
Wiener processes, 34, 273
Wind noise, "pure," 404
WWS analysis, 357

Y

Young module, 68
Yttria-stabilized zirconia, 427, 428

Z

Zirconium oxygen exhaust sensors, 427–434